Studies in Computational Intelligence

Volume 624

Series editor

Janusz Kacprzyk, Polish Academy of Sciences, Warsaw, Poland
e-mail: kacprzyk@ibspan.waw.pl

About this Series

The series "Studies in Computational Intelligence" (SCI) publishes new developments and advances in the various areas of computational intelligence—quickly and with a high quality. The intent is to cover the theory, applications, and design methods of computational intelligence, as embedded in the fields of engineering, computer science, physics and life sciences, as well as the methodologies behind them. The series contains monographs, lecture notes and edited volumes in computational intelligence spanning the areas of neural networks, connectionist systems, genetic algorithms, evolutionary computation, artificial intelligence, cellular automata, self-organizing systems, soft computing, fuzzy systems, and hybrid intelligent systems. Of particular value to both the contributors and the readership are the short publication timeframe and the worldwide distribution, which enable both wide and rapid dissemination of research output.

More information about this series at http://www.springer.com/series/7092

George A. Anastassiou · Ioannis K. Argyros

Intelligent Numerical Methods: Applications to Fractional Calculus

George A. Anastassiou
Department of Mathematical Sciences
The University of Memphis
Memphis, TN
USA

Ioannis K. Argyros
Department of Mathematical Sciences
Cameron University
Lawton, OK
USA

ISSN 1860-949X   ISSN 1860-9503   (electronic)
Studies in Computational Intelligence
ISBN 978-3-319-26720-3   ISBN 978-3-319-26721-0   (eBook)
DOI 10.1007/978-3-319-26721-0

Library of Congress Control Number: 2015955857

Springer Cham Heidelberg New York Dordrecht London

Springer International Publishing AG Switzerland is part of Springer Science+Business Media (www.springer.com)

This monograph is dedicated to: Angela, Christopher, Gus, Michael, and Peggy

Preface

It is a well-known fact that there exist functions that have nowhere first order derivative, but possess continuous Riemann-Liouville and Caputo fractional derivatives of all orders less than one, e.g., the famous Weierstrass function, see Chap. 1, [9, 18], p. 50.

This striking phenomenon motivated the authors to study Newton-like and other similar numerical methods, which involve fractional derivatives and fractional integral operators, for the first time studied in the literature. All for the purpose to solve numerically equations whose associated functions can be also non-differentiable in the ordinary sense.

That is among others extending the classical Newton method theory which requires usual differentiability of function.

In this monograph we present the complete recent work of the past three years of the authors on Numerical Analysis and Fractional Calculus. It is the natural outgrowth of their related publications. Chapters are self-contained and can be read independently and several advanced courses can be taught out of this book. An extensive list of references is given per chapter. The topics covered are from A to Z of this research area, all studied for the first time by the authors.

The list of presented topics of our related studies follows.

Newton-like methods on generalized Banach spaces and applications in fractional calculus.
Semilocal convergence of Newton-like methods under general conditions with applications in fractional calculus.
On the convergence of iterative methods with applications in generalized fractional calculus.
A fixed point technique for some iterative algorithm with applications to generalized right fractional calculus.
Approximating fixed points with applications in k-fractional calculus.
Generalized g-fractional calculus and iterative methods.
A unified convergence analysis for a certain family of iterative algorithms with applications to fractional calculus.

A convergence analysis for extended iterative algorithms with applications to fractional and vector calculus.

A convergence analysis for a certain family of extended iterative methods with applications to modified fractional calculus.

A convergence analysis for secant-like methods with applications to modified fractional calculus.

Semilocal convergence of secant-type methods with applications to modified g-fractional calculus.

On the convergence of secant-like algorithms with applications to generalized fractional calculus.

Generalized g-fractional calculus of Canavati-type and secant-like methods.

A convergence analysis for some iterative algorithms with applications to fractional calculus.

Convergence for iterative methods on Banach spaces of a convergence structure with applications to fractional calculus.

Local convergence analysis of inexact Gauss–Newton method for singular systems of equations under majorant and center-majorant condition.

The asymptotic mesh independence principle of Newton's method under weaker conditions.

Ball convergence of a sixth order iterative method with one parameter for solving equations under weak conditions.

Improved semilocal convergence of Broyden's method with regularly continuous divided differences.

Left general fractional monotone approximation theory.

Right general fractional monotone approximation.

Univariate left general high order fractional monotone approximation.

Univariate right general high order fractional monotone approximation theory.

Advanced fractional Taylor's formulae.

Generalized Canavati type fractional Taylor's formulae.

The last two topics were developed to be used in several chapters of this monograph.

The book's results are expected to find applications in many areas of applied mathematics, stochastics, computer science, and engineering. As such this monograph is suitable for researchers, graduate students, and seminars in the above subjects, also to be in all science and engineering libraries.

The preparation of the book took place during 2014–2015 in Memphis, Tennessee and Lawton, Oklahoma, USA.

We would like to thank Prof. Alina Lupas of University of Oradea, Romania, for checking and reading the manuscript.

October 2015   George A. Anastassiou
                                                                                Ioannis K. Argyros

Contents

About the Authors

George A. Anastassiou was born in Athens, Greece in 1952. He received his B.SC. degree in Mathematics from Athens University, Greece in 1975. He received his Diploma in Operations Research from Southampton University, UK in 1976. He also received his MA in Mathematics from University of Rochester, USA in 1981. He was awarded his Ph.D. in Mathematics from University of Rochester, USA in 1984. During 1984–1986 he served as a visiting assistant professor at the University of Rhode Island, USA. Since 1986 till now 2015, he is a faculty member at the University of Memphis, USA. He is currently a full Professor of Mathematics since 1994.

His research area is "Computational Analysis" in the very broad sense. He has published over 400 research articles in international mathematical journals and over 27 monographs, proceedings, and textbooks in well-known publishing houses. Several awards have been awarded to George Anastassiou. In 2007 he received the Honorary Doctoral Degree from University of Oradea, Romania. He is associate editor in over 60 international mathematical journals and editor-in-chief in 3 journals, most notably in the well-known "Journal of Computational Analysis and Applications."

Ioannis K. Argyros was born in Athens, Greece in 1956. He received his B.SC. degree in Mathematics from Athens University, Greece in 1979. He also received his MA in Mathematics from University of Georgia, USA in 1983. He was awarded his Ph.D. in Mathematics from University of Georgia, USA in 1984. During 1984–1986 he served as a visiting assistant professor at the University of Iowa, USA. During 1986–1990 he also served as an assistant professor at the New Mexico State University, USA. Since 1990 till now 2015, he is a faculty member at Cameron University, USA. He is currently a full Professor of Mathematics since 1994. His research area is "Computational Mathematics" in the very broad sense. He has published over 850 research articles in national and international mathematical journals and over 25 monographs, proceedings, and textbooks in well-known reputable publishing houses. Several recognitions have been awarded to Ioannis K. Argyros. In 2001 he received the "Distinguished Research Award" from the Southwest Oklahoma Advanced Technology Association. He is associate editor in over 50 international mathematical journals, most notably in the well-known "Applied Mathematics and Computation" and editor-in-chief in 1 journal.

Chapter 1
Newton-Like Methods on Generalized Banach Spaces and Fractional Calculus

We present a semilocal convergence study of Newton-like methods on a generalized Banach space setting to approximate a locally unique zero of an operator. Earlier studies such as [6–8, 15] require that the operator involved is Fréchet-differentiable. In the present study we assume that the operator is only continuous. This way we extend the applicability of Newton-like methods to include fractional calculus and problems from other areas. Moreover, we obtain under the same or weaker conditions: weaker sufficient convergence criteria; tighter error bounds on the distances involved and an at least as precise in formations on the location of the solution. Special cases are provided where the old convergence criteria cannot apply but the new criteria can apply to locate zeros of operators. Some applications include fractional calculus involving the Riemann-Liouville fractional integral and the Caputo fractional derivative. Fractional calculus is very important for its applications in many applied sciences. It follows [5].

1.1 Introduction

We present a semilocal convergence analysis for Newton-like methods on a generalized Banach space setting to approximate a zero of an operator. A generalized norm is defined to be an operator from a linear space into a partially order Banach space (to be precised in Sect. 1.2). Earlier studies such as [6–8, 15] for Newton's method have shown that a more precise convergence analysis is obtained when compared to the real norm theory. However, the main assumption is that the operator involved is Fréchet-differentiable. This hypothesis limits the applicability of Newton's method. In the present study we only assume the continuity of the operator. This may be expand the applicability of these methods. Our approach allows the extension of Newton-like methods in fractional calculus and other areas (see Sect. 1.4) not possible before (since the operator must be Fréchet-differentiable). Moreover, we

© Springer International Publishing Switzerland 2016
G.A. Anastassiou and I.K. Argyros, *Intelligent Numerical Methods:*
Applications to Fractional Calculus, Studies in Computational Intelligence 624,
DOI 10.1007/978-3-319-26721-0_1

obtain the following advantages over the earlier mentioned studies using Newton's method:

  (i) Weaker sufficient semilocal convergence criteria.
 (ii) Tighter error bounds on the distances involved.
(iii) An at least as precise information on the location of the zero.

Moreover, we show that the advantages (ii) are possible even if our Newton-like methods are reduced to Newton's method.

Furthermore, the advantages (i–iii) are obtained under the same or less computational cost.

The rest of the chapter is organized as follows: Sect. 1.2 contains the basic concepts on generalized Banach spaces and auxiliary results on inequalities and fixed points. In Sect. 1.3 we present the semilocal convergence analysis of Newton-like methods. Finally, in the concluding Sects. 1.4 and 1.5, we present special cases, favorable comparisons to earlier results and applications in some areas including fractional calculus.

1.2 Generalized Banach Spaces

We present some standard concepts that are needed in what follows to make the paper as self contained as possible. More details on generalized Banach spaces can be found in [6–8, 15], and the references there in.

Definition 1.1 A generalized Banach space is a triplet $(x, E, /\cdot/)$ such that

  (i) X is a linear space over $\mathbb{R}\,(\mathbb{C})$. (ii) $E = (E, K, \|\cdot\|)$ is a partially ordered Banach space, i.e.

(ii_1) $(E, \|\cdot\|)$ is a real Banach space,

(ii_2) E is partially ordered by a closed convex cone K,

(iii_3) The norm $\|\cdot\|$ is monotone on K. (iii) The operator $/\cdot/ : X \to K$ satisfies
$/x/ = 0 \Leftrightarrow x = 0,\ /\theta x/ = |\theta|\,/x/,$
$/x + y/ \le /x/ + /y/$ for each $x, y \in X, \theta \in \mathbb{R}(\mathbb{C})$.

  (iv) X is a Banach space with respect to the induced norm $\|\cdot\|_i := \|\cdot\| \cdot /\cdot/.$

Remark 1.2 The operator $/\cdot/$ is called a generalized norm. In view of (iii) and (ii_3) $\|\cdot\|_i$, is a real norm. In the rest of this paper all topological concepts will be understood with respect to this norm.

Let $L\left(X^j, Y\right)$ stand for the space of j-linear symmetric and bounded operators from X^j to Y, where X and Y are Banach spaces. For X, Y partially ordered $L_+\left(X^j, Y\right)$ stands for the subset of monotone operators P such that

$$0 \le a_i \le b_i \Rightarrow P\left(a_1, \ldots, a_j\right) \le P\left(b_1, \ldots, b_j\right). \qquad (1.2.1)$$

Definition 1.3 The set of bounds for an operator $Q \in L(X, X)$ on a generalized Banach space $(X, E, /\cdot/)$ the set of bounds is defined to be:

$$B(Q) := \{P \in L_+(E, E), /Qx/ \le P/x/ \quad \text{for each } x \in X\}. \tag{1.2.2}$$

Let $D \subset X$ and $T : D \to D$ be an operator. If $x_0 \in D$ the sequence $\{x_n\}$ given by

$$x_{n+1} := T(x_n) = T^{n+1}(x_0) \tag{1.2.3}$$

is well defined. We write in case of convergence

$$T^\infty(x_0) := \lim \left(T^n(x_0) \right) = \lim_{n \to \infty} x_n. \tag{1.2.4}$$

We need some auxiliary results on inequations.

Lemma 1.4 *Let $(E, K, \|\cdot\|)$ be a partially ordered Banach space, $\xi \in K$ and $M, N \in L_+(E, E)$.*
(i) Suppose there exists $r \in K$ such that

$$R(r) := (M + N)r + \xi \le r \tag{1.2.5}$$

and

$$(M + N)^k r \to 0 \quad \text{as } k \to \infty. \tag{1.2.6}$$

Then, $b := R^\infty(0)$ is well defined satisfies the equation $t = R(t)$ and is the smaller than any solution of the inequality $R(s) \le s$.
   (ii) Suppose there exists $q \in K$ and $\theta \in (0, 1)$ such that $R(q) \le \theta q$, then there exists $r \le q$ satisfying (i).

Proof (i) Define sequence $\{b_n\}$ by $b_n = R^n(0)$. Then, we have by (1.2.5) that $b_1 = R(0) = \xi \le r \Rightarrow b_1 \le r$. Suppose that $b_k \le r$ for each $k = 1, 2, \ldots, n$. Then, we have by (1.2.5) and the inductive hypothesis that $b_{n+1} = R^{n+1}(0) = R(R^n(0)) = R(b_n) = (M + N)b_n + \xi \le (M + N)r + \xi \le r \Rightarrow b_{n+1} \le r$. Hence, sequence $\{b_n\}$ is bounded above by r. Set $P_n = b_{n+1} - b_n$. We shall show that

$$P_n \le (M + N)^n r \quad \text{for each } n = 1, 2, \ldots \tag{1.2.7}$$

We have by the definition of P_n and (1.2.6) that

$$P_1 = R^2(0) - R(0) = R(R(0)) - R(0)$$

$$= R(\xi) - R(0) = \int_0^1 R'(t\xi)\xi dt \le \int_0^1 R'(\xi)\xi dt$$

$$\le \int_0^1 R'(r)r dt \le (M + N)r,$$

which shows (1.2.7) for $n = 1$. Suppose that (1.2.7) is true for $k = 1, 2, \ldots, n$. Then, we have in turn by (1.2.6) and the inductive hypothesis that

$$P_{k+1} = R^{k+2}(0) - R^{k+1}(0) = R^{k+1}(R(0)) - R^{k+1}(0) =$$

$$R^{k+1}(\xi) - R^{k+1}(0) = R\left(R^k(\xi)\right) - R\left(R^k(0)\right) =$$

$$\int_0^1 R'\left(R^k(0) + t\left(R^k(\xi) - R^k(0)\right)\right)\left(R^k(\xi) - R^k(0)\right)dt \le$$

$$R'\left(R^k(\xi)\right)\left(R^k(\xi) - R^k(0)\right) = R'\left(R^k(\xi)\right)\left(R^{k+1}(0) - R^k(0)\right) \le$$

$$R'(r)\left(R^{k+1}(0) - R^k(0)\right) \le (M+N)(M+N)^k r = (M+N)^{k+1} r,$$

which completes the induction for (1.2.7). It follows that $\{b_n\}$ is a complete sequence in a Banach space and as such it converges to some b. Notice that $R(b) = R\left(\lim_{n\to\infty} R^n(0)\right) = \lim_{n\to\infty} R^{n+1}(0) = b \Rightarrow b$ solves the equation $R(t) = t$. We have that $b_n \le r \Rightarrow b \le r$, where r a solution of $R(r) \le r$. Hence, b is smaller than any solution of $R(s) \le s$.

(ii) Define sequences $\{v_n\}, \{w_n\}$ by $v_0 = 0$, $v_{n+1} = R(v_n)$, $w_0 = q$, $w_{n+1} = R(w_n)$. Then, we have that

$$0 \le v_n \le v_{n+1} \le w_{n+1} \le w_n \le q, \tag{1.2.8}$$
$$w_n - v_n \le \theta^n (q - v_n)$$

and sequence $\{v_n\}$ is bounded above by q. Hence, it converges to some r with $r \le q$. We also get by (1.2.8) that $w_n - v_n \to 0$ as $n \to \infty \Rightarrow w_n \to r$ as $n \to \infty$. □

We also need the auxiliary result for computing solutions of fixed point problems.

Lemma 1.5 *Let* $(X, (E, K, \|\cdot\|), /\cdot/)$ *be a generalized Banach space, and* $P \in B(Q)$ *be a bound for* $Q \in L(X, X)$. *Suppose there exist* $y \in X$ *and* $q \in K$ *such that*

$$Pq + /y/ \le q \text{ and } P^k q \to 0 \quad as \ k \to \infty. \tag{1.2.9}$$

Then, $z = T^\infty(0)$, $T(x) := Qx + y$ *is well defined and satisfies:* $z = Qz + y$ *and* $/z/ \le P/z/ + /y/ \le q$. *Moreover,* z *is the unique solution in the subspace* $\{x \in X | \exists \theta \in \mathbb{R} : \{x\} \le \theta q\}$.

The proof can be found in [15, Lemma 3.2].

1.3 Semilocal Convergence

Let $(X, (E, K, \|\cdot\|), /\cdot/)$ and Y be generalized Banach spaces, $D \subset X$ an open subset, $G : D \to Y$ a continuous operator and $A(\cdot) : D \to L(X, Y)$. A zero of operator G is to be determined by a Newton-like method starting at a point $x_0 \in D$. The results are presented for an operator $F = JG$, where $J \in L(Y, X)$. The iterates are determined through a fixed point problem:

$$x_{n+1} = x_n + y_n, \; A(x_n) y_n + F(x_n) = 0 \qquad (1.3.1)$$
$$\Leftrightarrow y_n = T(y_n) := (I - A(x_n)) y_n - F(x_n).$$

Let $U(x_0, r)$ stand for the ball defined by

$$U(x_0, r) := \{x \in X : /x - x_0/ \le r\}$$

for some $r \in K$.

Next, we present the semilocal convergence analysis of Newton-like method (1.3.1) using the preceding notation.

Theorem 1.6 *Let $F : D \subset X$, $A(\cdot) : D \to L(X, Y)$ and $x_0 \in D$ be as defined previously. Suppose:*

(H_1) There exists an operator $M \in B(I - A(x))$ for each $x \in D$.
(H_2) There exists an operator $N \in L_+(E, E)$ satisfying for each $x, y \in D$

$$/F(y) - F(x) - A(x)(y - x)/ \le N/y - x/.$$

(H_3) There exists a solution $r \in K$ of

$$R_0(t) := (M + N)t + /F(x_0)/ \le t.$$

(H_4) $U(x_0, r) \subseteq D$.
(H_5) $(M + N)^k r \to 0$ as $k \to \infty$.
Then, the following hold:
(C_1) The sequence $\{x_n\}$ defined by

$$x_{n+1} = x_n + T_n^\infty(0), \; T_n(y) := (I - A(x_n))y - F(x_n) \qquad (1.3.2)$$

is well defined, remains in $U(x_0, r)$ for each $n = 0, 1, 2, \ldots$ and converges to the unique zero of operator F in $U(x_0, r)$.

(C_2) An apriori bound is given by the null-sequence $\{r_n\}$ defined by $r_0 := r$ and for each $n = 1, 2, \ldots$

$$r_n = P_n^\infty(0), \; P_n(t) = Mt + Nr_{n-1}.$$

(C_3) An aposteriori bound is given by the sequence $\{s_n\}$ defined by

$$s_n := R_n^\infty (0), \quad R_n (t) = (M + N) t + N a_{n-1},$$

$$b_n := /x_n - x_0/ \le r - r_n \le r,$$

where

$$a_{n-1} := /x_n - x_{n-1}/ \quad \text{for each } n = 1, 2, \ldots$$

Proof Let us define for each $n \in \mathbb{N}$ the statement:

(I_n) $x_n \in X$ and $r_n \in K$ are well defined and satisfy

$$r_n + a_{n-1} \le r_{n-1}.$$

We use induction to show (I_n). The statement (I_1) is true: By Lemma 1.4 and (H_3), (H_5) there exists $q \le r$ such that:

$$Mq + /F (x_0) / = q \text{ and } M^k q \le M^k r \to 0 \quad \text{as } k \to \infty.$$

Hence, by Lemma 1.5 x_1 is well defined and we have $a_0 \le q$. Then, we get the estimate

$$P_1 (r - q) = M (r - q) + N r_0$$

$$\le Mr - Mq + Nr = R_0 (r) - q$$

$$\le R_0 (r) - q = r - q.$$

It follows with Lemma 1.4 that r_1 is well defined and

$$r_1 + a_0 \le r - q + q = r = r_0.$$

Suppose that (I_j) is true for each $j = 1, 2, \ldots, n$. We need to show the existence of x_{n+1} and to obtain a bound q for a_n. To achieve this notice that:

$$Mr_n + N (r_{n-1} - r_n) = Mr_n + Nr_{n-1} - Nr_n = P_n (r_n) - Nr_n \le r_n.$$

Then, it follows from Lemma 1.4 that there exists $q \le r_n$ such that

$$q = Mq + N (r_{n-1} - r_n) \text{ and } (M + N)^k q \to 0, \quad \text{as } k \to \infty. \qquad (1.3.3)$$

By (I_j) it follows that

$$b_n - /x_n - x_0/ \leq \sum_{j=0}^{n-1} a_j \leq \sum_{j=0}^{n-1} (r_j - r_{j+1}) = r - r_n \leq r.$$

Hence, $x_n \in U(x_0, r) \subset D$ and by (H_1) M is a bound for $I - A(x_n)$.
We can write by (H_2) that

$$/F(x_n)/ = /F(x_n) - F(x_{n-1}) - A(x_{n-1})(x_n - x_{n-1})/$$

$$\leq Na_{n-1} \leq N(r_{n-1} - r_n). \tag{1.3.4}$$

It follows from (1.3.3) and (1.3.4) that

$$Mq + /F(x_n)/ \leq q.$$

By Lemma 1.5, x_{n+1} is well defined and $a_n \leq q \leq r_n$. In view of the definition of r_{n+1} we have that

$$P_{n+1}(r_n - q) = P_n(r_n) - q = r_n - q,$$

so that by Lemma 1.4, r_{n+1} is well defined and

$$r_{n+1} + a_n \leq r_n - q + q = r_n,$$

which proves (I_{n+1}). The induction for (I_n) is complete. Let $m \geq n$, then we obtain in turn that

$$/x_{m+1} - x_n/ \leq \sum_{j=n}^{m} a_j \leq \sum_{j=n}^{m} (r_j - r_{j+1}) = r_n - r_{m+1} \leq r_n. \tag{1.3.5}$$

Moreover, we get inductively the estimate

$$r_{n+1} = P_{n+1}(r_{n+1}) \leq P_{n+1}(r_n) \leq (M+N)r_n \leq \cdots \leq (M+N)^{n+1} r.$$

It follows from (H_5) that $\{r_n\}$ is a null-sequence. Hence, $\{x_n\}$ is a complete sequence in a Banach space X by (1.3.5) and as such it converges to some $x^* \in X$. By letting $m \to \infty$ in (1.3.5), we deduce that $x^* \in U(x_n, r_n)$. Furthermore, (1.3.4) shows that x^* is a zero of F. Hence, (C_1) and (C_2) are proved.

In view of the estimate

$$R_n(r_n) \leq P_n(r_n) \leq r_n$$

the apriori, bound of (C_3) is well defined by Lemma 1.4. That is s_n is smaller in general than r_n. The conditions of Theorem 1.6 are satisfied for x_n replacing x_0. A solution of the inequality of (C_2) is given by s_n (see (1.3.4)). It follows from (1.3.5) that the conditions of Theorem 1.6 are easily verified. Then, it follows from (C_1) that $x^* \in U(x_n, s_n)$ which proves (C_3).   □

In general the aposterior, estimate is of interest. Then, condition (H_5) can be avoided as follows:

Proposition 1.7 *Suppose: condition (H_1) of Theorem 1.6 is true.*
  (H_3') There exists $s \in K$, $\theta \in (0, 1)$ such that

$$R_0(s) = (M + N)s + /F(x_0)/ \le \theta s.$$

(H_4') $U(x_0, s) \subset D$.
  Then, there exists $r \le s$ satisfying the conditions of Theorem 1.6. Moreover, the zero x^* of F is unique in $U(x_0, s)$.

Remark 1.8 (i) Notice that by Lemma 1.4 $R_n^\infty(0)$ is the smallest solution of $R_n(s) \le s$. Hence any solution of this inequality yields on upper estimate for $R_n^\infty(0)$. Similar inequalities appear in (H_2) and (H_2').

(ii) The weak assumptions of Theorem 1.6 do not imply the existence of $A(x_n)^{-1}$. In practice the computation of $T_n^\infty(0)$ as a solution of a linear equation is no problem and the computation of the expensive or impossible to compute in general $A(x_n)^{-1}$ is not needed.

(iii) We can used the following result for the computation of the aposteriori estimates. The proof can be found in [15, Lemma 4.2] by simply exchanging the definitions of R.

Lemma 1.9 *Suppose that the conditions of Theorem 1.6 are satisfied. If $s \in K$ is a solution of $R_n(s) \le s$, then $q := s - a_n \in K$ and solves $R_{n+1}(q) \le q$. This solution might be improved by $R_{n+1}^k(q) \le q$ for each $k = 1, 2, \ldots$.*

1.4   Special Cases and Applications

Application 1.10 *The results obtained in earlier studies such as [6–8, 15] require that operator F (i.e. G) is Fréchet-differentiable. This assumption limits the applicability of the earlier results. In the present study we only require that F is a continuous operator. Hence, we have extended the applicability of Newton-like methods to classes of operators that are only continuous. Moreover, as we will show next by specializing F to be a Fréchet-differentiable operator (i.e. $F'(x_n) = A(x_n)$) our Theorem 1.6 improves earlier results. Indeed, first of all notice that Newton-like method defined by (1.3.1) reduces to Newton's method:*

$$x_{n+1} = x_n + y_n, \ F'(x_n) y_n + F(x_n) = 0 \tag{1.4.1}$$
$$\Leftrightarrow y_n = T_n(y_n) := (I - F'(x_n)) y_n - F(x_n).$$

Next, we present Theorem 2.1 from [15] and the specialization of our Theorem 1.6, so we can compare them.

Theorem 1.11 *Let* $F : D \to X$ *be a Fréchet-differentiable operator and* $x_0 \in D$. *Suppose that the following conditions hold:*
$(\overline{H}_1)$ *There exists an operator* $M_0 \in B(I - F'(x_0))$.
$(\overline{H}_2)$ *There exists an operator* $N_1 \in L_+(E^2, E)$ *satisfying for*

$$x, y \in D, z \in X : / (F'(x) - F'(y)) z / \leq 2N_1 (/x - y/, /z/).$$

$(\overline{H}_3)$ *There exists a solution* $c \in K$ *of the inequality*

$$\overline{R}_0(c) := M_0 c + N_1 c^2 + / F(x_0) / \leq c.$$

$(\overline{H}_4)$ $U(x_0, c) \subseteq D$.
$(\overline{H}_5)$ $(M_0 + 2N_1 c)^k c \to 0$ *as* $k \to \infty$.
Then, the following hold
$(\overline{C}_1)$ *The sequence* $\{x_n\}$ *generated by (1.4.1) is well defined and converges to a unique zero of* F *in* $U(x_0, c)$.
$(\overline{C}_2)$ *An a priori bound is given by the null-sequence* $\{c_n\}$ *defined by*

$$c_0 = c, \ c_n := \overline{P}_n^{\infty}(0),$$
$$\overline{P}_n(t) : = M_0 t + 2N_1 (c - c_{n-1}) t + N_1 c_{n-1}^2.$$

$(\overline{C}_3)$ *An a posteriori bound is given by sequence* $\{d_n\}$ *defined by*

$$d_n = \overline{R}_n^{\infty}(0), \overline{R}_n(t) := M_0 t + 2N_1 b_n t + N_1 t^2 + N_1 a_{n-1}^2,$$

where sequences $\{a_n\}$ *and* $\{b_n\}$ *we defined previously.*

Theorem 1.12 *Let* $F : D \to X$ *be a Fréchet-differentiable operator and* $x_0 \in D$. *Suppose that the following conditions hold:*
$(\widetilde{H}_1)$ *There exists an operator* $M_1 \in B(I - F'(x))$ *for each* $x \in D$.
$(\widetilde{H}_2)$ *There exists an operator* $N_2 \in L_+(E, E)$ *satisfying for each* $x, y \in D$

$$/ F(y) - F(x) - F'(x)(y - x) / \leq N_2 / y - x/.$$

$(\widetilde{H}_3)$ *There exists a solution* $\widetilde{r} \in K$ *of*

$$\widetilde{R}_0(t) := (M_1 + N_2) t + / F(x_0) / \leq t.$$

$(\widetilde{H}_4)$ $U\,(x_0, \widetilde{r}) \subseteq D$.
$(\widetilde{H}_5)$ $(M_1 + N_2)^k \, \widetilde{r} \to 0$ as $k \to \infty$.
Then, the following hold:

$(\widetilde{C}_1)$ *The sequence* $\{x_n\}$ *generated by (1.4.1) is well defined and converges to a unique zero of F in* $U\,(x_o, \widetilde{r})$.

$(\widetilde{C}_2)$ *An appriori bound is given by* $\widetilde{r}_0 = \widetilde{r}, \widetilde{r}_n := \widetilde{P}_n^\infty\,(0), \widetilde{P}_n\,(t) = M_1 t + N_2 \widetilde{r}_{n-1}$.

$(\widetilde{C}_3)$ *An a posteriori bound is given by the sequence* $\{\widetilde{s}_n\}$ *defined by* $\widetilde{s}_n := \widetilde{R}_n^\infty\,(0)$, $\widetilde{R}_n\,(t) = (M_1 + N_2)\,t + N_2 a_{n-1}$.

We can now compare the two preceding theorems. Notice that we can write

$$/\,F\,(y) - F\,(x) - F'\,(x)\,(y - x)\,/ =$$

$$\Big/\int_0^1 \big[F'\,(x + \theta\,(y - x)) - F'\,(x)\big]\,(y - x)\,dt\,\Big/.$$

Then, it follows from $(\overline{H}_2)$, $(\widetilde{H}_2)$ and preceding estimate that

$$N_2 \le N_1/p/, \quad \text{for each } p \in X,$$

holds in general. In particular, we have that

$$N_2 \le N_1 c. \tag{1.4.2}$$

Moreover, we get in turn by $(\overline{H}_1)$, $(\overline{H}_2)$ and $(\overline{H}_5)$ that

$$/1 - F'\,(x)\,/ \le /I - F'\,(x_0)\,/ + /F'\,(x_0) - F'\,(x)\,/ \tag{1.4.3}$$
$$\le M_0 + 2N_1/x - x_0/ \le M_0 + 2N_1 c.$$

Therefore, by $(\widetilde{H}_1)$ and (1.4.3), we obtain that

$$M_1 \le M_0 + 2N_1 c \tag{1.4.4}$$

holds in general.

Then, in view of (1.4.2), (1.4.4) and the $(\overline{H})$, $(\widetilde{H})$ hypotheses we deduce that

$$\overline{R}_0\,(c) \le c \Rightarrow \widetilde{R}_0\,(\widetilde{r}) \le \widetilde{r} \tag{1.4.5}$$

$$(M_0 + 2N_1 c)^k\,c \to 0 \ \Rightarrow \ (M_1 + N_2)^k\,\widetilde{r} \to 0 \tag{1.4.6}$$

but not necessarily vice versa unless if equality holds in (1.4.2) and (1.4.4);

$$\widetilde{r} \le c, \tag{1.4.7}$$

$$\widetilde{r}_n \le c_n, \tag{1.4.8}$$

and

$$\widetilde{s}_n \leq d_n. \qquad (1.4.9)$$

Notice also that strict inequality holds in (1.4.8) or (1.4.9) if strict inequality holds in (1.4.2) or (1.4.4).

Estimates (1.4.5)–(1.4.9) justify the advantages of our approach over the earlier studies as already stated in the introduction of this study.

Next, we show that the results of Theorem 2.1 in [15], i.e. of Theorem 1.11 can be improved under the same hypotheses by noticing that in view of $(\overline{H}_2)$.

$(\overline{H}_2^0)$ There exists an operator $N_0 \in L_+ (E^2, E)$ satisfying for $x \in D, z \in X$,

$$/ \left(F'(x) - F'(x_0) \right) z / \leq 2N_0 \left(/x - x_0/, /z/ \right).$$

Moreover,

$$N_0 \leq N_1 \qquad (1.4.10)$$

holds in general and $\frac{N_1}{N_0}$ can be arbitrarily large [4, 6–8].

It is worth noticing that $(\overline{H}_2^0)$ is not an additional to $(\overline{H}_2)$ hypothesis, since in practice the computation of N_1 requires the computation of N_0 as a special case. Using now $(\overline{H}_2^0)$ and $(\overline{H}_1)$ we get that

$$/I - F'(x) / \leq /I - F'(x_0) / + /F'(x_0) - F'(x) / \leq M_0 + 2N_0/x - x_0/.$$

Hence, $M_0 + 2N_0 b_n$, $M_0 + 2N_0 (c - c_n)$ can be used as a bounds for $I - F'(x_n)$ instead of $M_0 + 2N_1 b_n$, $M_0 + 2N_1 (c - c_n)$, respectively.

Notice also that

$$M_0 + 2N_0 b_n \leq M_0 + 2N_1 b_n \qquad (1.4.11)$$

and

$$M_0 + 2N_0 (c - c_n) \leq M_0 + 2N_1 (c - c_n). \qquad (1.4.12)$$

Then, with the above changes and following the proof of Theorem 2.1 in [15], we arrive at the following improvement:

Theorem 1.13 *Suppose that the conditions of Theorem 1.11 hold but with N_1 replaced by the at most as large N_0. Then, the conclusions $(\overline{C}_1)$–$(\overline{C}_3)$,*

$$\overline{c}_n \leq c_n \qquad (1.4.13)$$

and

$$\overline{d}_n \leq d_n, \qquad (1.4.14)$$

where the sequences $\{\bar{c}_n\}$, $\{\bar{d}_n\}$ are defined by

$$\bar{c}_0 = c, \ \bar{c}_n := \overline{\overline{P}}_n^{\infty}(0), \ \overline{\overline{P}}_n(t) := M_0 t + 2N_0(c - c_{n-1})t + N_1 c_{n-1}^2,$$

$$\bar{d}_n = \overline{\overline{R}}_n^{\infty}(0), \ \overline{\overline{R}}_n(t) := M_0 t + 2N_0 b_n t + N_1 t^2 + N_1 a_{n-1}^2.$$

Remark 1.14 Notice that estimates (1.4.13) and (1.4.14) follow by a simple inductive argument using (1.4.11) and (1.4.12). Moreover, strict inequality holds in (1.4.13) (for $n \geq 1$) and in (1.4.14) (for $n > 1$) if strict inequality holds in (1.4.11) or (1.4.12). Hence, again we obtain better apriori and aposteriori bounds under the same hypotheses $(\overline{H})$.

Condition $(\bar{H}_5)$ has been weakened since $N_0 \leq N_1$. It turns out that condition $(\bar{H}_3)$ can be weakened and sequences $\{c_n\}$ and $\{d_n\}$ can be replaced by more precise sequences as follows: Define operators Q_0, Q_1, Q_2, H_1, H_2 on D by

$$(\bar{\bar{H}}_3) Q_0(t) := M_0 t + /F(x_0)/$$

Suppose that there exists a solution $\mu_0 \in K$ of the inequality

$$Q_0(\mu_0) \leq \mu_0.$$

There exists a solution $\mu_1 \in K$ with $\mu_1 \leq \mu_0$ of the inequality

$$Q_1(t) \leq t,$$

where

$$Q_1(t) := M_0 t + 2N_0(\mu_0 - t)t + N_0 \mu_0^2.$$

There exists a solution $\mu_2 = \mu \in K$ with $\mu \leq \mu_1$ such that

$$Q_2(t) \leq t,$$

where

$$Q_2(t) := M_0 t + 2N_0(\mu - t)t + N_1 \mu_1^2.$$

Moreover, define operators on D by

$$H_1(t) := M_0 t, \quad H_2(t) := Q_1(t),$$

$$H_n(t) := M_0 t + 2N_0(\mu - \mu_{n-1})t + N_1 \mu_{n-1}^2, \quad n = 3, 4, \ldots$$

and

$$Q_n(t) := M_0 t + 2 N_0 b_n t + N_1 t^2 + N_1 a_{n-1}.$$

Furthermore, define sequences $\{\bar{\bar{c}}_n\}$ and $\{\bar{\bar{d}}_n\}$ by

$$\bar{\bar{c}}_n := H_n^\infty(0) \quad \text{and} \quad \bar{\bar{d}}_n := Q_n^\infty(0)$$

Then, the proof of Theorem 4.2 goes on through in this setting to arrive at:

Theorem 1.15 *Suppose that the conditions of Theorem 4.2 are satisfied but with* $c, (\bar{H}_3) - (\bar{H}_5)$ *replaced by* $\mu, (\bar{\bar{H}}_3),$

$(\bar{\bar{H}}_4)$ $U(x_0, \mu) \subseteq D$ $(\bar{\bar{H}}_5)$ $(M_0 + N_0 \mu)^k \mu \to 0$ *as* $k \to \infty,$ *respectively.*

Then, the conclusions of Theorem 4.2 hold with sequences $\{\bar{\bar{c}}_n\}$ *and* $\{\bar{\bar{d}}_n\}$ *replacing* $\{c_n\}$ *and* $\{d_n\}$ *respectively. Moreover, we have that*

$$\bar{\bar{c}}_n \leq \bar{c}_n \leq c_n,$$

$$\bar{\bar{d}}_n \leq \bar{d}_n \leq d_n,$$

and

$$\mu \leq c.$$

Clearly the new error bounds are more precise; the information on the location of the solution x^* *at least as precise and the sufficient convergence criteria* $(\bar{\bar{H}}_3)$ *and* $(\bar{\bar{H}}_5)$ *weaker than* $(\bar{H}_3)$ *and* $(\bar{H}_5),$ *respectively.*

Example 1.16 The j-dimensional space $\mathbb{R}^j$ is a classical example of a generalized Banach space. The generalized norm is defined by componentwise absolute values. Then, as ordered Banach space we set $E = \mathbb{R}^j$ with componentwise ordering with e.g. the maximum norm. A bound for a linear operator (a matrix) is given by the corresponding matrix with absolute values. Similarly, we can define the "N" operators. Let $E = \mathbb{R}$. That is we consider the case of a real normed space with norm denoted by $\|\cdot\|$. Let us see how the conditions of Theorems 1.6 and 4.4 look like.

Theorem 1.17 (H_1) $\|I - A(x)\| \leq M$ *for some* $M \geq 0.$
(H_2) $\|F(y) - F(x) - A(x)(y - x)\| \leq N \|y - x\|$ *for some* $N \geq 0.$
(H_3) $M + N < 1,$

$$r = \frac{\|F(x_0)\|}{1 - (M + N)}. \tag{1.4.15}$$

(H_4) $U(x_0, r) \subseteq D.$
(H_5) $(M + N)^k r \to 0$ *as* $k \to \infty,$ *where* r *is given by (1.4.15).*
Then, the conclusions of Theorem 1.6 hold.

Theorem 1.18 $(\overline{H}_1)$ $\left\| I - F'(x_0) \right\| \le M_0$ *for some* $M_0 \in [0, 1)$.
$(\overline{H}_2)$ $\left\| F'(x) - F'(x_0) \right\| \le 2N_0 \left\| x - x_0 \right\|$,
$\left\| F'(x) - F'(y) \right\| \le 2N_1 \left\| x - y \right\|$, *for some* $N_0 \ge 0$ *and* $N_1 > 0$.
$(\overline{H}_3)$

$$4N_1 \left\| F(x_0) \right\| \le (1 - M_0)^2, \tag{1.4.16}$$

$$c = \frac{1 - M_0 - \sqrt{(1 - M_0)^2 - 4N_1 \left\| F(x_0) \right\|}}{2N_1}. \tag{1.4.17}$$

$(\overline{H}_4)$ $U(x_0, c) \subseteq D$.
$(\overline{H}_5)$ $(M_0 + 2N_0 c)^k c \to 0$ *as* $k \to \infty$, *where* c *is defined by* (1.4.17).
Then, the conclusions of Theorem 4.4 hold.

Remark 1.19 Condition (1.4.16) is a Newton-Kantorovich type hypothesis appearing as a sufficient semilocal convergence hypothesis in connection to Newton-like methods. In particular, if $F'(x_0) = I$, then $M_0 = 0$ and (1.4.16) reduces to the famous for its simplicity and clarity Newton-Kantorovich hypothesis

$$4N_1 \left\| F(x_0) \right\| \le 1 \tag{1.4.18}$$

appearing in the study of Newton's method [1, 2, 6–8, 10–17].

1.5   Applications to Fractional Calculus

Based on [18], it makes sense to study Newton-like numerical methods.

Thus, our presented earlier semilocal convergence Newton-like general methods, see Theorem 4.8, apply in the next two fractional settings given that the following inequalities are fulfilled:

$$\left\| 1 - A(x) \right\|_\infty \le \gamma_0 \in (0, 1), \tag{1.5.1}$$

and

$$\left| F(y) - F(x) - A(x)(y - x) \right| \le \gamma_1 \left| y - x \right|, \tag{1.5.2}$$

where $\gamma_0, \gamma_1 \in (0, 1)$, furthermore

$$\gamma = \gamma_0 + \gamma_1 \in (0, 1), \tag{1.5.3}$$

for all $x, y \in [a^*, b]$.

Here, we consider $a < a^* < b$.

The specific functions $A(x)$, $F(x)$ will be described next.

I) Let $\alpha > 0$ and $f \in L_\infty([a, b])$. The Riemann-Liouville integral ([9], p. 13) is given by

$$\left(J_a^\alpha f\right)(x) = \frac{1}{\Gamma(\alpha)} \int_a^x (x - t)^{\alpha-1} f(t) \, dt, \quad x \in [a, b]. \tag{1.5.4}$$

Then

$$\left|\left(J_a^\alpha f\right)(x)\right| \le \frac{1}{\Gamma(\alpha)} \left(\int_a^x (x - t)^{\alpha-1} |f(t)| \, dt\right)$$

$$\le \frac{1}{\Gamma(\alpha)} \left(\int_a^x (x - t)^{\alpha-1} \, dt\right) \|f\|_\infty = \frac{1}{\Gamma(\alpha)} \frac{(x - a)^\alpha}{\alpha} \|f\|_\infty \tag{1.5.5}$$

$$= \frac{(x - a)^\alpha}{\Gamma(\alpha + 1)} \|f\|_\infty = (\xi_1).$$

Clearly

$$\left(J_a^\alpha f\right)(a) = 0. \tag{1.5.6}$$

$$(\xi_1) \le \frac{(b - a)^\alpha}{\Gamma(\alpha + 1)} \|f\|_\infty. \tag{1.5.7}$$

That is

$$\left\|J_a^\alpha f\right\|_{\infty, [a,b]} \le \frac{(b - a)^\alpha}{\Gamma(\alpha + 1)} \|f\|_\infty < \infty, \tag{1.5.8}$$

i.e. J_a^α is a bounded linear operator.

By [3], p. 388, we get that $\left(J_a^\alpha f\right)$ is a continuous function over $[a, b]$ and in particular over $[a^*, b]$. Thus there exist $x_1, x_2 \in [a^*, b]$ such that

$$\left(J_a^\alpha f\right)(x_1) = \min\left(J_a^\alpha f\right)(x), \tag{1.5.9}$$
$$\left(J_a^\alpha f\right)(x_2) = \max\left(J_a^\alpha f\right)(x), \quad x \in [a^*, b].$$

We assume that

$$\left(J_a^\alpha f\right)(x_1) > 0. \tag{1.5.10}$$

Hence

$$\left\|J_a^\alpha f\right\|_{\infty, [a^*, b]} = \left(J_a^\alpha f\right)(x_2) > 0. \tag{1.5.11}$$

Here it is

$$J(x) = mx, \quad m \ne 0. \tag{1.5.12}$$

Therefore the equation

$$Jf(x) = 0, \quad x \in [a^*, b],$$
(1.5.13)

has the same solutions as the equation

$$F(x) := \frac{Jf(x)}{2\left(J_a^\alpha f\right)(x_2)} = 0, \quad x \in [a^*, b].$$
(1.5.14)

Notice that

$$J_a^\alpha \left(\frac{f}{2\left(J_a^\alpha f\right)(x_2)} \right)(x) = \frac{\left(J_a^\alpha f\right)(x)}{2\left(J_a^\alpha f\right)(x_2)} \le \frac{1}{2} < 1, \quad x \in [a^*, b].$$
(1.5.15)

Call

$$A(x) := \frac{\left(J_a^\alpha f\right)(x)}{2\left(J_a^\alpha f\right)(x_2)}, \quad \forall x \in [a^*, b].$$
(1.5.16)

We notice that

$$0 < \frac{\left(J_a^\alpha f\right)(x_1)}{2\left(J_a^\alpha f\right)(x_2)} \le A(x) \le \frac{1}{2}, \quad \forall x \in [a^*, b].$$
(1.5.17)

Hence the first condition (1.5.1) is fulfilled

$$|1 - A(x)| = 1 - A(x) \le 1 - \frac{\left(J_a^\alpha f\right)(x_1)}{2\left(J_a^\alpha f\right)(x_2)} =: \gamma_0, \quad \forall x \in [a^*, b].$$
(1.5.18)

Clearly $\gamma_0 \in (0, 1)$.

Next we assume that $F(x)$ is a contraction, i.e.

$$|F(x) - F(y)| \le \lambda |x - y|; \quad \text{all } x, y \in [a^*, b],$$
(1.5.19)

and $0 < \lambda < \frac{1}{2}$.

Equivalently we have

$$|Jf(x) - Jf(y)| \le 2\lambda \left(J_a^\alpha f\right)(x_2) |x - y|, \quad \text{all } x, y \in [a^*, b].$$
(1.5.20)

We observe that

$$|F(y) - F(x) - A(x)(y - x)| \le |F(y) - F(x)| + |A(x)| |y - x| \le$$

$$\lambda |y - x| + |A(x)| |y - x| = (\lambda + |A(x)|) |y - x| =: (\psi_1), \quad \forall x, y \in [a^*, b].$$
(1.5.21)

We have that

$$\left| \left(J_a^\alpha f \right) (x) \right| \le \frac{(b-a)^\alpha}{\Gamma (\alpha + 1)} \| f \|_\infty < \infty, \quad \forall x \in \left[a^*, b \right]. \tag{1.5.22}$$

Hence

$$|A (x)| = \frac{\left| \left(J_a^\alpha f \right) (x) \right|}{2 \left(J_a^\alpha f \right) (x_2)} \le \frac{(b-a)^\alpha \| f \|_\infty}{2\Gamma (\alpha + 1) \left(\left(J_a^\alpha f \right) (x_2) \right)} < \infty, \quad \forall x \in \left[a^*, b \right]. \tag{1.5.23}$$

Therefore, we get

$$(\psi_1) \le \left(\lambda + \frac{(b-a)^a \| f \|_\infty}{2\Gamma (\alpha + 1) \left(\left(J_a^\alpha f \right) (x_2) \right)} \right) |y - x|, \quad \forall x, y \in \left[a^*, b \right]. \tag{1.5.24}$$

Call

$$0 < \gamma_1 := \lambda + \frac{(b-a)^a \| f \|_\infty}{2\Gamma (\alpha + 1) \left(\left(J_a^\alpha f \right) (x_2) \right)}, \tag{1.5.25}$$

choosing $(b - a)$ small enough we can make $\gamma_1 \in (0, 1)$, fulfilling (1.5.2).

Next we call and we need that

$$0 < \gamma := \gamma_0 + \gamma_1 = 1 - \frac{\left(J_a^\alpha f \right) (x_1)}{2 \left(J_a^\alpha f \right) (x_2)} + \lambda + \frac{(b-a)^a \| f \|_\infty}{2\Gamma (\alpha + 1) \left(\left(J_a^\alpha f \right) (x_2) \right)} < 1, \tag{1.5.26}$$

equivalently,

$$\lambda + \frac{(b-a)^a \| f \|_\infty}{2\Gamma (\alpha + 1) \left(\left(J_a^\alpha f \right) (x_2) \right)} < \frac{\left(J_a^\alpha f \right) (x_1)}{2 \left(J_a^\alpha f \right) (x_2)}, \tag{1.5.27}$$

equivalently,

$$2\lambda \left(J_a^\alpha f \right) (x_2) + \frac{(b-a)^a \| f \|_\infty}{\Gamma (\alpha + 1)} < \left(J_a^\alpha f \right) (x_1), \tag{1.5.28}$$

which is possible for small λ, $(b - a)$. That is $\gamma \in (0, 1)$, fulfilling (1.5.3). So our numerical method converges and solves (1.5.13).

II) Let again $a < a^* < b, \alpha > 0, m = \lceil \alpha \rceil$ ($\lceil \cdot \rceil$ ceiling function), $\alpha \notin \mathbb{N}$, $G \in C^{m-1} ([a, b]), 0 \ne G^{(m)} \in L_\infty ([a, b])$. Here we consider the Caputo fractional derivative (see [3], p. 270),

$$D_{*a}^\alpha G (x) = \frac{1}{\Gamma (m - \alpha)} \int_a^x (x - t)^{m-\alpha-1} G^{(m)} (t) \, dt. \tag{1.5.29}$$

By [3], p. 388, $D_{*a}^\alpha G$ is a continuous function over $[a, b]$ and in particular continuous over $[a^*, b]$. Notice that by [4], p. 358 we have that $D_{*a}^\alpha G (a) = 0$.

Therefore there exist $x_1, x_2 \in [a^*, b]$ such that $D_{*a}^\alpha G(x_1) = \min D_{*a}^\alpha G(x)$, and $D_{*a}^\alpha G(x_2) = \max D_{*a}^\alpha G(x)$, for $x \in [a^*, b]$.

We assume that

$$D_{*a}^\alpha G(x_1) > 0. \tag{1.5.30}$$

(i.e. $D_{*a}^\alpha G(x) > 0, \forall x \in [a^*, b]$).

Furthermore

$$\left\| D_{*a}^\alpha G \right\|_{\infty, [a^*, b]} = D_{*a}^\alpha G(x_2). \tag{1.5.31}$$

Here it is

$$J(x) = mx, \quad m \neq 0. \tag{1.5.32}$$

The equation

$$JG(x) = 0, \quad x \in [a^*, b], \tag{1.5.33}$$

has the same set of solutions as the equation

$$F(x) := \frac{JG(x)}{2D_{*a}^\alpha G(x_2)} = 0, \quad x \in [a^*, b]. \tag{1.5.34}$$

Notice that

$$D_{*a}^\alpha \left(\frac{G(x)}{2D_{*a}^\alpha G(x_2)} \right) = \frac{D_{*a}^\alpha G(x)}{2D_{*a}^\alpha G(x_2)} \leq \frac{1}{2} < 1, \quad \forall x \in [a^*, b]. \tag{1.5.35}$$

We call

$$A(x) := \frac{D_{*a}^\alpha G(x)}{2D_{*a}^\alpha G(x_2)}, \quad \forall x \in [a^*, b]. \tag{1.5.36}$$

We notice that

$$0 < \frac{D_{*a}^\alpha G(x_1)}{2D_{*a}^\alpha G(x_2)} \leq A(x) \leq \frac{1}{2}. \tag{1.5.37}$$

Hence, the first condition (1.5.1) is fulfilled

$$|1 - A(x)| = 1 - A(x) \leq 1 - \frac{D_{*a}^\alpha G(x_1)}{2D_{*a}^\alpha G(x_2)} =: \gamma_0, \quad \forall x \in [a^*, b]. \tag{1.5.38}$$

Clearly $\gamma_0 \in (0, 1)$.

Next, we assume that $F(x)$ is a contraction over $[a^*, b]$, i.e.

$$|F(x) - F(y)| \leq \lambda |x - y|; \quad \forall x, y \in [a^*, b], \qquad (1.5.39)$$

and $0 < \lambda < \frac{1}{2}$.

Equivalently we have

$$|JG(x) - JG(y)| \leq 2\lambda \left(D_{*a}^\alpha G(x_2) \right) |x - y|, \quad \forall x, y \in [a^*, b]. \qquad (1.5.40)$$

We observe that

$$|F(y) - F(x) - A(x)(y - x)| \leq |F(y) - F(x)| + |A(x)| |y - x| \leq$$

$$\lambda |y - x| + |A(x)| |y - x| = (\lambda + |A(x)|) |y - x| =: (\xi_2), \quad \forall x, y \in [a^*, b]. \qquad (1.5.41)$$

We observe that

$$\left| D_{*a}^\alpha G(x) \right| \leq \frac{1}{\Gamma(m - \alpha)} \int_a^x (x - t)^{m - \alpha - 1} \left\| G^{(m)}(t) \right\| dt$$

$$\leq \frac{1}{\Gamma(m - \alpha)} \left(\int_a^x (x - t)^{m - \alpha - 1} dt \right) \left\| G^{(m)} \right\|_\infty$$

$$= \frac{1}{\Gamma(m - \alpha)} \frac{(x - a)^{m - \alpha}}{(m - \alpha)} \left\| G^{(m)} \right\|_\infty$$

$$= \frac{1}{\Gamma(m - \alpha + 1)} (x - a)^{m - \alpha} \left\| G^{(m)} \right\|_\infty \leq \frac{(b - a)^{m - \alpha}}{\Gamma(m - \alpha + 1)} \left\| G^{(m)} \right\|_\infty. \qquad (1.5.42)$$

That is

$$\left| D_{*a}^\alpha G(x) \right| \leq \frac{(b - a)^{m - \alpha}}{\Gamma(m - \alpha + 1)} \left\| G^{(m)} \right\|_\infty < \infty, \quad \forall x \in [a, b]. \qquad (1.5.43)$$

Hence, $\forall x \in [a^*, b]$ we get that

$$|A(x)| = \frac{\left| D_{*a}^\alpha G(x) \right|}{2 D_{*a}^\alpha G(x_2)} \leq \frac{(b - a)^{m - \alpha}}{2\Gamma(m - \alpha + 1)} \frac{\left\| G^{(m)} \right\|_\infty}{D_{*a}^\alpha G(x_2)} < \infty. \qquad (1.5.44)$$

Consequently we observe

$$(\xi_2) \leq \left(\lambda + \frac{(b - a)^{m - \alpha}}{2\Gamma(m - \alpha + 1)} \frac{\left\| G^{(m)} \right\|_\infty}{D_{*a}^\alpha G(x_2)} \right) |y - x|, \quad \forall x, y \in [a^*, b]. \qquad (1.5.45)$$

Call

$$0 < \gamma_1 := \lambda + \frac{(b-a)^{m-\alpha}}{2\Gamma(m-\alpha+1)} \frac{\left\| G^{(m)} \right\|_\infty}{D_{*a}^\alpha G(x_2)}, \qquad (1.5.46)$$

choosing $(b-a)$ small enough we can make $\gamma_1 \in (0,1)$. So (1.5.2) is fulfilled.
   Next, we call and need

$$0 < \gamma := \gamma_0 + \gamma_1 = 1 - \frac{D_{*a}^\alpha G(x_1)}{2 D_{*a}^\alpha G(x_2)} + \lambda + \frac{(b-a)^{m-\alpha}}{2\Gamma(m-\alpha+1)} \frac{\left\| G^{(m)} \right\|_\infty}{D_{*a}^\alpha G(x_2)} < 1, \qquad (1.5.47)$$

equivalently we find,

$$\lambda + \frac{(b-a)^{m-\alpha}}{2\Gamma(m-\alpha+1)} \frac{\left\| G^{(m)} \right\|_\infty}{D_{*a}^\alpha G(x_2)} < \frac{D_{*a}^\alpha G(x_1)}{2 D_{*a}^\alpha G(x_2)}, \qquad (1.5.48)$$

or,

$$2\lambda D_{*a}^\alpha G(x_2) + \frac{(b-a)^{m-\alpha}}{\Gamma(m-\alpha+1)} \left\| G^{(m)} \right\|_\infty < D_{*a}^\alpha G(x_1), \qquad (1.5.49)$$

which is possible for small λ, $(b-a)$.
   That is $\gamma \in (0,1)$, fulfilling (1.5.3). Hence Eq. (1.5.33) can be solved with our presented numerical methods.

References

1. S. Amat, S. Busquier, Third-order iterative methods under Kantorovich conditions. J. Math. Anal. Appl. **336**, 243–261 (2007)
2. S. Amat, S. Busquier, S. Plaza, Chaotic dynamics of a third-order Newton-like method. J. Math. Anal. Appl. **366**(1), 164–174 (2010)
3. G. Anastassiou, *Fractional Differentiation Inequalities* (Springer, New York, 2009)
4. G. Anastassiou, *Intelligent Mathematics: Computational Analysis* (Springer, Heidelberg, 2011)
5. G. Anastassiou, I. Argyros, *Newton-Like Methods on Generalized Banach Spaces and Applications in Fractional Calculus* (2015)
6. I.K. Argyros, Newton-like methods in partially ordered linear spaces. J. Approx. Theory Appl. **9**(1), 1–10 (1993)
7. I.K. Argyros, Results on controlling the residuals of perturbed Newton-like methods on Banach spaces with a convergence structure. Southwest J. Pure Appl. Math. **1**, 32–38 (1995)
8. I.K. Argyros, *Convergence and Applications of Newton-Like Iterations* (Springer, New York, 2008)
9. K. Diethelm, *The Analysis of Fractional Differential Equations*. Lecture Notes in Mathematics, vol. 2004, 1st edn. (Springer, New York, 2010)
10. J.A. Ezquerro, J.M. Gutierrez, M.Á. Hernández, N. Romero, M.J. Rubio, The Newton method: from Newton to Kantorovich (Spanish). Gac. R. Soc. Mat. Esp. **13**, 53–76 (2010)
11. J.A. Ezquerro, M.Á. Hernández, Newton-like methods of high order and domains of semilocal and global convergence. Appl. Math. Comput. **214**(1), 142–154 (2009)

12. L.V. Kantorovich, G.P. Akilov, *Functional Analysis in Normed Spaces* (Pergamon Press, New York, 1964)
13. A.A. Magreñán, Different anomalies in a Jarratt family of iterative root finding methods. Appl. Math. Comput. **233**, 29 38 (2014)
14. A.A. Magreñán, A new tool to study real dynamics: the convergence plane. Appl. Math. Comput. **248**, 215–224 (2014)
15. P.W. Meyer, Newton's method in generalized Banach spaces. Numer. Func. Anal. Optimiz. **9**, **3, 4**, 244–259 (1987)
16. F.A. Potra, V. Ptak, *Nondiscrete Induction and Iterative Processes* (Pitman, London, 1984)
17. P.D. Proinov, New general convergence theory for iterative processes and its applications to Newton-Kantorovich type theorems. J. Complex. **26**, 3–42 (2010)
18. B. Ross, S. Samko, E. Love, Functions that have no first order derivatives might have fractional derivative might have fractional derivatives of all orders less than one. Real Anal. Exchange **20**(2), 140–157 (1994–1995)

Chapter 2
Semilocal Convegence of Newton-Like Methods and Fractional Calculus

We present a semilocal convergence study of Newton-like methods on a generalized Banach space setting to approximate a locally unique zero of an operator. Earlier studies such as [6–8, 15] require that the operator involved is Fré chet-differentiable. In the present study we assume that the operator is only continuous. This way we extend the applicability of Newton-like methods to include fractional calculus and problems from other areas. Some applications include fractional calculus involving the Riemann-Liouville fractional integral and the Caputo fractional derivative. Fractional calculus is very important for its applications in many applied sciences. It follows [5].

2.1 Introduction

We present a semilocal convergence analysis for Newton-like methods on a generalized Banach space setting to approximate a zero of an operator. The semilocal convergence is, based on the information around an initial point, to give conditions ensuring the convergence of the method. A generalized norm is defined to be an operator from a linear space into a partially order Banach space (to be precised in Sect. 2.2). Earlier studies such as [6–8, 15] for Newton's method have shown that a more precise convergence analysis is obtained when compared to the real norm theory. However, the main assumption is that the operator involved is Fréchet-differentiable. This hypothesis limits the applicability of Newton's method. In the present study we only assume the continuity of the operator. This may be expand the applicability of these methods.

The rest of the chapter is organized as follows: Sect. 2.2 contains the basic concepts on generalized Banach spaces and auxiliary results on inequalities and fixed points. In Sect. 2.3 we present the semilocal convergence analysis of Newton-like methods. Finally, in the concluding Sects. 2.4 and 2.5, we present special cases and applications in fractional calculus.

© Springer International Publishing Switzerland 2016

G.A. Anastassiou and I.K. Argyros, *Intelligent Numerical Methods:*
Applications to Fractional Calculus, Studies in Computational Intelligence 624,
DOI 10.1007/978-3-319-26721-0_2

2.2 Generalized Banach Spaces

We present some standard concepts that are needed in what follows to make the paper as self contained as possible. More details on generalized Banach spaces can be found in [6–8, 15], and the references there in.

Definition 2.1 A generalized Banach space is a triplet $(x, E, /\cdot/)$ such that
(i) X is a linear space over $\mathbb{R}\,(\mathbb{C})$.
(ii) $E = (E, K, \|\cdot\|)$ is a partially ordered Banach space, i.e.
(ii$_1$) $(E, \|\cdot\|)$ is a real Banach space,
(ii$_2$) E is partially ordered by a closed convex cone K,
(ii$_3$) The norm $\|\cdot\|$ is monotone on K.
(iii) The operator $/\cdot/ : X \to K$ satisfies
$/x/ = 0 \Leftrightarrow x = 0, /\theta x/ = |\theta|\,/x/,$
$/x + y/ \le /x/ + /y/$ for each $x, y \in X, \theta \in \mathbb{R}(\mathbb{C})$.
(iv) X is a Banach space with respect to the induced norm $\|\cdot\|_i := \|\cdot\| \cdot /\cdot/$.

Remark 2.2 The operator $/\cdot/$ is called a generalized norm. In view of (iii) and (ii$_3$) $\|\cdot\|_i$, is a real norm. In the rest of this paper all topological concepts will be understood with respect to this norm.

Let $L\left(X^j, Y\right)$ stand for the space of j-linear symmetric and bounded operators from X^j to Y, where X and Y are Banach spaces. For X, Y partially ordered
$L_+\left(X^j, Y\right)$ stands for the subset of monotone operators P such that

$$0 \le a_i \le b_i \Rightarrow P\left(a_1, \ldots, a_j\right) \le P\left(b_1, \ldots, b_j\right). \qquad (2.2.1)$$

Definition 2.3 The set of bounds for an operator $Q \in L\,(X, X)$ on a generalized
Banach space $(X, E, /\cdot/)$ the set of bounds is defined to be:

$$B\,(Q) := \left\{P \in L_+\,(E, E),\ \ /Qx/ \le P/x/ \text{ for each } x \in X\right\}. \qquad (2.2.2)$$

Let $D \subset X$ and $T : D \to D$ be an operator. If $x_0 \in D$ the sequence $\{x_n\}$ given by

$$x_{n+1} := T\,(x_n) = T^{n+1}\,(x_0) \qquad (2.2.3)$$

is well defined. We write in case of convergence

$$T^\infty\,(x_0) := \lim\,(T^n\,(x_0)) = \lim_{n\to\infty} x_n. \qquad (2.2.4)$$

We need some auxiliary results on inequations.

Lemma 2.4 *Let $(E, K, \|\cdot\|)$ be a partially ordered Banach space, $\xi \in K$ and
$M, N \in L_+\,(E, E)$.*
  (i) Suppose there exists $r \in K$ such that

$$R(r) := (M + N)r + \xi \le r \tag{2.2.5}$$

and

$$(M + N)^k r \to 0 \quad as \ k \to \infty. \tag{2.2.6}$$

Then, $b := R^\infty(0)$ is well defined satisfies the equation $t = R(t)$ and is the smaller than any solution of the inequality $R(s) \le s$.

(ii) Suppose there exists $q \in K$ and $\theta \in (0, 1)$ such that $R(q) \le \theta q$, then there exists $r \le q$ satisfying (i).

Proof (i) Define sequence $\{b_n\}$ by $b_n = R^n(0)$. Then, we have by (2.2.5) that $b_1 = R(0) = \xi \le r \Rightarrow b_1 \le r$. Suppose that $b_k \le r$ for each $k = 1, 2, \ldots, n$. Then, we have by (2.2.5) and the inductive hypothesis that $b_{n+1} = R^{n+1}(0) = R(R^n(0)) = R(b_n) = (M + N)b_n + \xi \le (M + N)r + \xi \le r \Rightarrow b_{n+1} \le r$. Hence, sequence $\{b_n\}$ is bounded above by r. Set $P_n = b_{n+1} - b_n$. We shall show that

$$P_n \le (M + N)^n r \quad for \ each \ n = 1, 2, \ldots \tag{2.2.7}$$

We have by the definition of P_n and (2.2.6) that

$$P_1 = R^2(0) - R(0) = R(R(0)) - R(0)$$

$$= R(\xi) - R(0) = \int_0^1 R'(t\xi)\xi dt \le \int_0^1 R'(\xi)\xi dt$$

$$\le \int_0^1 R'(r)r dt \le (M + N)r,$$

which shows (2.2.7) for $n = 1$. Suppose that (2.2.7) is true for $k = 1, 2, \ldots, n$. Then, we have in turn by (2.2.6) and the inductive hypothesis that

$$P_{k+1} = R^{k+2}(0) - R^{k+1}(0) = R^{k+1}(R(0)) - R^{k+1}(0) =$$

$$R^{k+1}(\xi) - R^{k+1}(0) = R(R^k(\xi)) - R(R^k(0)) =$$

$$\int_0^1 R'(R^k(0) + t(R^k(\xi) - R^k(0)))(R^k(\xi) - R^k(0)) dt \le$$

$$R'(R^k(\xi))(R^k(\xi) - R^k(0)) = R'(R^k(\xi))(R^{k+1}(0) - R^k(0)) \le$$

$$R'(r)(R^{k+1}(0) - R^k(0)) \le (M + N)(M + N)^k r = (M + N)^{k+1} r,$$

which completes the induction for (2.2.7). It follows that $\{b_n\}$ is a complete sequence in a Banach space and as such it converges to some b. Notice that $R(b) = R(\lim_{n \to \infty} R^n(0)) = \lim_{n \to \infty} R^{n+1}(0) = b \Rightarrow b$ solves the equation

$R(t) = t$. We have that $b_n \leq r \Rightarrow b \leq r$, where r a solution of $R(r) \leq r$. Hence, b is smaller than any solution of $R(s) \leq s$.

(ii) Define sequences $\{v_n\}$, $\{w_n\}$ by $v_0 = 0$, $v_{n+1} = R(v_n)$, $w_0 = q$, $w_{n+1} = R(w_n)$. Then, we have that

$$0 \leq v_n \leq v_{n+1} \leq w_{n+1} \leq w_n \leq q, \qquad (2.2.8)$$
$$w_n - v_n \leq \theta^n (q - v_n)$$

and sequence $\{v_n\}$ is bounded above by q. Hence, it converges to some r with $r \leq q$. We also get by (2.2.8) that $w_n - v_n \to 0$ as $n \to \infty \Rightarrow w_n \to r$ as $n \to \infty$. $\qquad \square$

We also need the auxiliary result for computing solutions of fixed point problems.

Lemma 2.5 *Let $(X, (E, K, \|\cdot\|), /\cdot/)$ be a generalized Banach space, and $P \in B(Q)$ be a bound for $Q \in L(X, X)$. Suppose there exists $y \in X$ and $q \in K$ such that*

$$Pq + /y/ \leq q \text{ and } P^k q \to 0 \quad \text{as } k \to \infty. \qquad (2.2.9)$$

Then, $z = T^\infty(0)$, $T(x) := Qx + y$ is well defined and satisfies: $z = Qz + y$ and $/z/ \leq P/z/ + /y/ \leq q$. Moreover, z is the unique solution in the subspace $\{x \in X | \exists \ \theta \in \mathbb{R} : \{x\} \leq \theta q\}$.

The proof can be found in [15, Lemma 3.2].

2.3 Semilocal Convergence

Let $(X, (E, K, \|\cdot\|), /\cdot/)$ and Y be generalized Banach spaces, $D \subset X$ an open subset, $G : D \to Y$ a continuous operator and $A(\cdot) : D \to L(X, Y)$. A zero of operator G is to be determined by a Newton-like method starting at a point $x_0 \in D$. The results are presented for an operator $F = JG$, where $J \in L(Y, X)$. The iterates are determined through a fixed point problem:

$$x_{n+1} = x_n + y_n, \quad A(x_n) y_n + F(x_n) = 0 \qquad (2.3.1)$$
$$\Leftrightarrow y_n = T(y_n) := (I - A(x_n)) y_n - F(x_n).$$

Let $U(x_0, r)$ stand for the ball defined by

$$U(x_0, r) := \left\{ x \in X : /x - x_0/ \leq r \right\}$$

for some $r \in K$.

Next, we present the semilocal convergence analysis of Newton-like method (2.3.1) using the preceding notation.

Theorem 2.6 *Let* $F : D \subset X$, $A(\cdot) : D \to L(X, Y)$ *and* $x_0 \in D$ *be as defined previously. Suppose:*

(H_1) There exists an operator $M \in B(I - A(x))$ *for each* $x \in D$

(H_2) There exists an operator $N \in L_+(E, E)$ *satisfying for each* $x, y \in D$

$$\big/ F(y) - F(x) - A(x)(y - x) \big/ \leq N \big/ y - x \big/.$$

(H_3) There exists a solution $r \in K$ *of*

$$R_0(t) := (M + N)t + \big/ F(x_0) \big/ \leq t.$$

(H_4) $U(x_0, r) \subseteq D$.

(H_5) $(M + N)^k r \to 0$ as $k \to \infty$.

Then, the following hold:

(C_1) The sequence $\{x_n\}$ defined by

$$x_{n+1} = x_n + T_n^\infty(0), \quad T_n(y) := (I - A(x_n))y - F(x_n) \qquad (2.3.2)$$

is well defined, remains in $U(x_0, r)$ *for each* $n = 0, 1, 2, \ldots$ *and converges to the unique zero of operator* F *in* $U(x_0, r)$.

(C_2) An apriori bound is given by the null-sequence $\{r_n\}$ defined by $r_0 := r$ and for each $n = 1, 2, \ldots$

$$r_n = P_n^\infty(0), \quad P_n(t) = Mt + Nr_{n-1}.$$

(C_3) An aposteriori bound is given by the sequence $\{s_n\}$ defined by

$$s_n := R_n^\infty(0), \quad R_n(t) = (M + N)t + Na_{n-1},$$

$$b_n := \big/ x_n - x_0 \big/ \leq r - r_n \leq r,$$

where

$$a_{n-1} := \big/ x_n - x_{n-1} \big/ \quad \text{for each } n = 1, 2, \ldots$$

Proof Let us define for each $n \in \mathbb{N}$ the statement:

(I_n) $x_n \in X$ and $r_n \in K$ are well defined and satisfy

$$r_n + a_{n-1} \leq r_{n-1}.$$

We use induction to show (I_n). The statement (I_1) is true: By Lemma 2.4 and (H_3), (H_5) there exists $q \leq r$ such that:

$$Mq + \big/ F(x_0) \big/ = q \quad \text{and} \quad M^k q \leq M^k r \to 0 \quad \text{as } k \to \infty.$$

Hence, by Lemma 2.5 x_1 is well defined and we have $a_0 \leq q$. Then, we get the estimate

$$P_1 (r - q) = M (r - q) + N r_0$$

$$\leq M r - M q + N r = R_0 (r) - q$$

$$\leq R_0 (r) - q = r - q.$$

It follows with Lemma 2.4 that r_1 is well defined and

$$r_1 + a_0 \leq r - q + q = r = r_0.$$

Suppose that (I_j) is true for each $j = 1, 2, \ldots, n$. We need to show the existence of x_{n+1} and to obtain a bound q for a_n. To achieve this notice that:

$$M r_n + N (r_{n-1} - r_n) = M r_n + N r_{n-1} - N r_n = P_n (r_n) - N r_n \leq r_n.$$

Then, it follows from Lemma 2.4 that there exists $q \leq r_n$ such that

$$q = M q + N (r_{n-1} - r_n) \quad \text{and} \quad (M + N)^k q \rightarrow 0, \quad \text{as } k \rightarrow \infty. \tag{2.3.3}$$

By (I_j) it follows that

$$b_n = \big/ x_n - x_0 \big/ \leq \sum_{j=0}^{n-1} a_j \leq \sum_{j=0}^{n-1} (r_j - r_{j+1}) = r - r_n \leq r.$$

Hence, $x_n \in U (x_0, r) \subset D$ and by (H_1) M is a bound for $I - A (x_n)$.
We can write by (H_2) that

$$\big/ F (x_n) \big/ = \big/ F (x_n) - F (x_{n-1}) - A (x_{n-1}) (x_n - x_{n-1}) \big/$$

$$\leq N a_{n-1} \leq N (r_{n-1} - r_n). \tag{2.3.4}$$

It follows from (2.3.3) and (2.3.4) that

$$M q + \big/ F (x_n) \big/ \leq q.$$

By Lemma 2.5, x_{n+1} is well defined and $a_n \leq q \leq r_n$. In view of the definition of r_{n+1} we have that

$$P_{n+1} (r_n - q) = P_n (r_n) - q = r_n - q,$$

so that by Lemma 2.4, r_{n+1} is well defined and

$$r_{n+1} + a_n < r_n - q + q = r_n,$$

which proves (I_{n+1}). The induction for (I_n) is complete. Let $m \geq n$, then we obtain in turn that

$$\left/ x_{m+1} - x_n \right/ \leq \sum_{j=n}^{m} a_j \leq \sum_{j=n}^{m} (r_j - r_{j+1}) = r_n - r_{m+1} \leq r_n. \qquad (2.3.5)$$

Moreover, we get inductively the estimate

$$r_{n+1} = P_{n+1}(r_{n+1}) \leq P_{n+1}(r_n) \leq (M+N) r_n \leq \cdots \leq (M+N)^{n+1} r.$$

It follows from (H_5) that $\{r_n\}$ is a null-sequence. Hence, $\{x_n\}$ is a complete sequence in a Banach space X by (2.3.5) and as such it converges to some $x^* \in X$. By letting $m \to \infty$ in (2.3.5) we deduce that $x^* \in U(x_n, r_n)$. Furthermore, (2.3.4) shows that x^* is a zero of F. Hence, (C_1) and (C_2) are proved.

In view of the estimate

$$R_n(r_n) \leq P_n(r_n) \leq r_n$$

the apriori, bound of (C_3) is well defined by Lemma 2.4. That is s_n is smaller in general than r_n. The conditions of Theorem 2.6 are satisfied for x_n replacing x_0. A solution of the inequality of (C_2) is given by s_n (see (2.3.4)). It follows from (2.3.5) that the conditions of Theorem 2.6 are easily verified. Then, it follows from (C_1) that $x^* \in U(x_n, s_n)$ which proves (C_3). $\qquad \square$

In general the aposterior, estimate is of interest. Then, condition (H_5) can be avoided as follows:

Proposition 2.7 *Suppose: condition (H_1) of Theorem 2.6 is true.*
(H_3') There exists $s \in K$, $\theta \in (0, 1)$ such that

$$R_0(s) = (M+N)s + \left/ F(x_0) \right/ \leq \theta s.$$

(H_4') $U(x_0, s) \subset D$.
Then, there exists $r \leq s$ satisfying the conditions of Theorem 2.6. Moreover, the zero x^* of F is unique in $U(x_0, s)$.

Remark 2.8 (i) Notice that by Lemma 2.4 $R_n^\infty(0)$ is the smallest solution of $R_n(s) \leq s$. Hence any solution of this inequality yields on upper estimate for $R_n^\infty(0)$. Similar inequalities appear in (H_2) and (H_2').

(ii) The weak assumptions of Theorem 2.6 do not imply the existence of $A(x_n)^{-1}$. In practice the computation of $T_n^\infty(0)$ as a solution of a linear equation is no problem and the computation of the expensive or impossible to compute in general $A(x_n)^{-1}$ is not needed.

(iii) We can used the following result for the computation of the aposteriori esti-mates. The proof can be found in [15, Lemma 4.2] by simply exchanging the defin-itions of R.

Lemma 2.9 *Suppose that the conditions of Theorem 2.6 are satisfied. If $s \in K$ is a solution of $R_n (s) \leq s$, then $q := s - a_n \in K$ and solves $R_{n+1} (q) \leq q$. This solution might be improved by $R_{n+1}^k (q) \leq q$ for each $k = 1, 2, \ldots$.*

2.4 Special Cases and Applications

Application 2.10 *The results obtained in earlier studies such as [6–8, 15] require that operator F (i.e. G) is Fré chet-differentiable. This assumption limits the applica-bility of the earlier results. In the present study we only require that F is a continu-ous operator. Hence, we have extended the applicability of Newton-like methods to classes of operators that are only continuous. If $A (x) = F' (x)$ Newton-like method (2.3.1) reduces to Newton's method considered in [15].*

Example 2.11 The j-dimensional space $\mathbb{R}^j$ is a classical example of a generalized Banach space. The generalized norm is defined by componentwise absolute values. Then, as ordered Banach space we set $E = \mathbb{R}^j$ with componentwise ordering with e.g. the maximum norm. A bound for a linear operator (a matrix) is given by the cor-responding matrix with absolute values. Similarly, we can define the "N" operators. Let $E = \mathbb{R}$. That is we consider the case of a real normed space with norm denoted by $\|\cdot\|$. Let us see how the conditions of Theorem 2.6 look like.

Theorem 2.12 $(H_1) \|I - A (x)\| \leq M$ for some $M \geq 0$.
  $(H_2) \|F (y) - F (x) - A (x) (y - x)\| \leq N \|y - x\|$ for some $N \geq 0$.
  $(H_3) M + N < 1$,

$$r = \frac{\|F (x_0)\|}{1 - (M + N)}. \tag{2.4.1}$$

  $(H_4) U (x_0, r) \subseteq D$.
  $(H_5) (M + N)^k r \to 0$ as $k \to \infty$, where r is given by (2.4.1).
  Then, the conclusions of Theorem 2.6 hold.

2.5 Applications to Fractional Calculus

Our presented earlier semilocal convergence Newton-like general methods, see Theo-rem 2.12, apply in the next two fractional settings given that the following inequalities are fulfilled:

$$\|1 - A (x)\|_\infty \leq \gamma_0 \in (0, 1), \tag{2.5.1}$$

and

$$|F(y) - F(x) - A(x)(y - x)| \leq \gamma_1 |y - x|, \tag{2.5.2}$$

where $\gamma_0, \gamma_1 \in (0, 1)$, furthermore

$$\gamma = \gamma_0 + \gamma_1 \in (0, 1), \tag{2.5.3}$$

for all $x, y \in [a, b^*]$.

Here we consider $a < b^* < b$.

The specific functions $A(x)$, $F(x)$ will be described next.

(I) Let $\alpha > 0$ and $f \in L_\infty([a, b])$. The right Riemann-Liouville integral ([4], pp. 333–354) is given by

$$\left(J_b^\alpha f\right)(x) := \frac{1}{\Gamma(\alpha)} \int_x^b (t - x)^{\alpha-1} f(t) \, dt, \quad x \in [a, b]. \tag{2.5.4}$$

Then

$$\left|\left(J_b^\alpha f\right)(x)\right| \leq \frac{1}{\Gamma(\alpha)} \left(\int_x^b (t - x)^{\alpha-1} |f(t)| \, dt\right)$$

$$\leq \frac{1}{\Gamma(\alpha)} \left(\int_x^b (t - x)^{\alpha-1} \, dt\right) \|f\|_\infty = \frac{1}{\Gamma(\alpha)} \frac{(b - x)^\alpha}{\alpha} \|f\|_\infty \tag{2.5.5}$$

$$= \frac{(b - x)^\alpha}{\Gamma(\alpha + 1)} \|f\|_\infty = (\xi_1).$$

Clearly

$$\left(J_b^\alpha f\right)(b) = 0. \tag{2.5.6}$$

$$(\xi_1) \leq \frac{(b - a)^\alpha}{\Gamma(\alpha + 1)} \|f\|_\infty. \tag{2.5.7}$$

That is

$$\left\|J_b^\alpha f\right\|_{\infty,[a,b]} \leq \frac{(b - a)^\alpha}{\Gamma(\alpha + 1)} \|f\|_\infty < \infty, \tag{2.5.8}$$

i.e. J_b^α is a bounded linear operator.

By [3] we get that $\left(J_b^\alpha f\right)$ is a continuous function over $[a, b]$ and in particular over $[a, b^*]$. Thus there exist $x_1, x_2 \in [a, b^*]$ such that

$$\left(J_b^\alpha f\right)(x_1) = \min \left(J_b^\alpha f\right)(x),$$
$$\left(J_b^\alpha f\right)(x_2) = \max \left(J_b^\alpha f\right)(x), \quad x \in [a, b^*]. \tag{2.5.9}$$

We assume that

$$\left(J_b^\alpha f\right)(x_1) > 0. \tag{2.5.10}$$

Hence

$$\left\| J_b^\alpha f \right\|_{\infty,[a,b^*]} = \left(J_b^\alpha f \right)(x_2) > 0. \tag{2.5.11}$$

Here it is

$$J(x) = mx, \quad m \neq 0. \tag{2.5.12}$$

Therefore the equation

$$Jf(x) = 0, \quad x \in [a, b^*], \tag{2.5.13}$$

has the same solutions as the equation

$$F(x) := \frac{Jf(x)}{2 \left(J_b^\alpha f \right)(x_2)} = 0, \quad x \in [a, b^*]. \tag{2.5.14}$$

Notice that

$$J_b^\alpha \left(\frac{f}{2 \left(J_b^\alpha f \right)(x_2)} \right)(x) = \frac{\left(J_b^\alpha f \right)(x)}{2 \left(J_b^\alpha f \right)(x_2)} \leq \frac{1}{2} < 1, \quad x \in [a, b^*]. \tag{2.5.15}$$

Call

$$A(x) := \frac{\left(J_b^\alpha f \right)(x)}{2 \left(J_b^\alpha f \right)(x_2)}, \quad \forall x \in [a, b^*]. \tag{2.5.16}$$

We notice that

$$0 < \frac{\left(J_b^\alpha f \right)(x_1)}{2 \left(J_b^\alpha f \right)(x_2)} \leq A(x) \leq \frac{1}{2}, \quad \forall x \in [a, b^*]. \tag{2.5.17}$$

Hence the first condition (2.5.1) is fulfilled

$$|1 - A(x)| = 1 - A(x) \leq 1 - \frac{\left(J_b^\alpha f \right)(x_1)}{2 \left(J_b^\alpha f \right)(x_2)} =: \gamma_0, \quad \forall x \in [a, b^*]. \tag{2.5.18}$$

Clearly $\gamma_0 \in (0, 1)$.

Next we assume that $F(x)$ is a contraction, i.e.

$$|F(x) - F(y)| \leq \lambda |x - y|; \quad \text{all } x, y \in [a, b^*], \tag{2.5.19}$$

and $0 < \lambda < \frac{1}{2}$.

Equivalently, we have

$$|Jf(x) - Jf(y)| \leq 2\lambda \left(J_b^\alpha f \right)(x_2) |x - y|, \quad \text{all } x, y \in [a, b^*]. \tag{2.5.20}$$

We observe that

$$|F(y) - F(x) - A(x)(y - x)| \le |F(y) - F(x)| + |A(x)| |y - x| \le$$

$$\lambda |y - x| + |A(x)| |y - x| = (\lambda + |A(x)|) |y - x| =: (\psi_1), \quad \forall x, y \in [a, b^*].$$
(2.5.21)

We have that

$$\left|\left(J_b^\alpha f\right)(x)\right| \le \frac{(b-a)^\alpha}{\Gamma(\alpha+1)} \|f\|_\infty < \infty, \quad \forall x \in [a, b^*]. \tag{2.5.22}$$

Hence

$$|A(x)| = \frac{\left|\left(J_b^\alpha f\right)(x)\right|}{2\left(J_b^\alpha f\right)(x_2)} \le \frac{(b-a)^\alpha \|f\|_\infty}{2\Gamma(\alpha+1)\left(\left(J_b^\alpha f\right)(x_2)\right)} < \infty, \quad \forall x \in [a, b^*]. \tag{2.5.23}$$

Therefore we get

$$(\psi_1) \le \left(\lambda + \frac{(b-a)^a \|f\|_\infty}{2\Gamma(\alpha+1)\left(\left(J_b^\alpha f\right)(x_2)\right)}\right) |y - x|, \quad \forall x, y \in [a, b^*]. \tag{2.5.24}$$

Call

$$0 < \gamma_1 := \lambda + \frac{(b-a)^a \|f\|_\infty}{2\Gamma(\alpha+1)\left(\left(J_b^\alpha f\right)(x_2)\right)}, \tag{2.5.25}$$

choosing $(b - a)$ small enough we can make $\gamma_1 \in (0, 1)$, fulfilling (2.5.2).

Next we call and we need that

$$0 < \gamma := \gamma_0 + \gamma_1 = 1 - \frac{\left(J_b^\alpha f\right)(x_1)}{2\left(J_b^\alpha f\right)(x_2)} + \lambda + \frac{(b-a)^a \|f\|_\infty}{2\Gamma(\alpha+1)\left(\left(J_b^\alpha f\right)(x_2)\right)} < 1, \tag{2.5.26}$$

equivalently,

$$\lambda + \frac{(b-a)^a \|f\|_\infty}{2\Gamma(\alpha+1)\left(\left(J_b^\alpha f\right)(x_2)\right)} < \frac{\left(J_b^\alpha f\right)(x_1)}{2\left(J_b^\alpha f\right)(x_2)}, \tag{2.5.27}$$

equivalently,

$$2\lambda \left(J_b^\alpha f\right)(x_2) + \frac{(b-a)^a \|f\|_\infty}{\Gamma(\alpha+1)} < \left(J_b^\alpha f\right)(x_1), \tag{2.5.28}$$

which is possible for small λ, $(b - a)$. That is $\gamma \in (0, 1)$, fulfilling (2.5.3). So our numerical method converges and solves (2.5.13).

(II) Let again $a < b^* < b$, $\alpha > 0$, $m = \lceil \alpha \rceil$ ($\lceil \cdot \rceil$ ceiling function), $\alpha \notin \mathbb{N}$, $G \in C^{m-1}([a, b])$, $0 \ne G^{(m)} \in L_\infty([a, b])$. Here we consider the right Caputo fractional derivative (see [4], p. 337),

$$D_{b-}^{\alpha} G\left(x\right) = \frac{(-1)^m}{\Gamma\left(m - \alpha\right)} \int_x^b \left(t - x\right)^{m-\alpha-1} G^{(m)}\left(t\right) dt. \tag{2.5.29}$$

By [3] $D_{b-}^{\alpha} G$ is a continuous function over $[a, b]$ and in particular continuous over $[a, b^*]$. Notice that by [4], p. 358 we have that $D_{b-}^{\alpha} G\left(b\right) = 0$.

Therefore there exist $x_1, x_2 \in [a, b^*]$ such that $D_{b-}^{\alpha} G\left(x_1\right) = \min D_{b-}^{\alpha} G\left(x\right)$, and $D_{b-}^{\alpha} G\left(x_2\right) = \max D_{b-}^{\alpha} G\left(x\right)$, for $x \in [a, b^*]$.

We assume that

$$D_{b-}^{\alpha} G\left(x_1\right) > 0. \tag{2.5.30}$$

(i.e. $D_{b-}^{\alpha} G\left(x\right) > 0, \forall\, x \in [a, b^*]$).

Furthermore

$$\left\| D_{b-}^{\alpha} G \right\|_{\infty, [a, b^*]} = D_{b-}^{\alpha} G\left(x_2\right). \tag{2.5.31}$$

Here it is

$$J\left(x\right) = mx, \quad m \neq 0. \tag{2.5.32}$$

The equation

$$JG\left(x\right) = 0, \quad x \in \left[a, b^*\right], \tag{2.5.33}$$

has the same set of solutions as the equation

$$F\left(x\right) := \frac{JG\left(x\right)}{2D_{b-}^{\alpha} G\left(x_2\right)} = 0, \quad x \in \left[a, b^*\right]. \tag{2.5.34}$$

Notice that

$$D_{b-}^{\alpha}\left(\frac{G\left(x\right)}{2D_{b-}^{\alpha} G\left(x_2\right)}\right) = \frac{D_{b-}^{\alpha} G\left(x\right)}{2D_{b-}^{\alpha} G\left(x_2\right)} \leq \frac{1}{2} < 1, \quad \forall\, x \in \left[a, b^*\right]. \tag{2.5.35}$$

We call

$$A\left(x\right) := \frac{D_{b-}^{\alpha} G\left(x\right)}{2D_{b-}^{\alpha} G\left(x_2\right)}, \quad \forall\, x \in \left[a, b^*\right]. \tag{2.5.36}$$

We notice that

$$0 < \frac{D_{b-}^{\alpha} G\left(x_1\right)}{2D_{b-}^{\alpha} G\left(x_2\right)} \leq A\left(x\right) \leq \frac{1}{2}. \tag{2.5.37}$$

Hence the first condition (2.5.1) is fulfilled

$$\left| 1 - A\left(x\right) \right| = 1 - A\left(x\right) \leq 1 - \frac{D_{b-}^{\alpha} G\left(x_1\right)}{2D_{b-}^{\alpha} G\left(x_2\right)} =: \gamma_0, \quad \forall\, x \in \left[a, b^*\right]. \tag{2.5.38}$$

Clearly $\gamma_0 \in (0, 1)$.

Next we assume that $F(x)$ is a contraction over $[a, b^*]$, i.e.

$$|F(x) - F(y)| \leq \lambda |x - y|; \quad \forall x, y \in [a, b^*], \tag{2.5.39}$$

and $0 < \lambda < \frac{1}{2}$.

Equivalently we have

$$|JG(x) - JG(y)| \leq 2\lambda \left(D_{b-}^{\alpha} G(x_2)\right)|x - y|, \quad \forall x, y \in [a, b^*]. \tag{2.5.40}$$

We observe that

$$|F(y) - F(x) - A(x)(y - x)| \leq |F(y) - F(x)| + |A(x)||y - x| \leq$$

$$\lambda |y - x| + |A(x)||y - x| = (\lambda + |A(x)|)|y - x| =: (\xi_2), \quad \forall x, y \in [a, b^*]. \tag{2.5.41}$$

Then, we get that

$$\left|D_{b-}^{\alpha} G(x)\right| \leq \frac{1}{\Gamma(m - \alpha)} \int_x^b (t - x)^{m-\alpha-1} \left|G^{(m)}(t)\right| dt$$

$$\leq \frac{1}{\Gamma(m - \alpha)} \left(\int_x^b (t - x)^{m-\alpha-1} dt\right) \left\|G^{(m)}\right\|_{\infty}$$

$$= \frac{1}{\Gamma(m - \alpha)} \frac{(b - x)^{m-\alpha}}{(m - \alpha)} \left\|G^{(m)}\right\|_{\infty}$$

$$= \frac{1}{\Gamma(m - \alpha + 1)} (b - x)^{m-\alpha} \left\|G^{(m)}\right\|_{\infty} \leq \frac{(b - a)^{m-\alpha}}{\Gamma(m - \alpha + 1)} \left\|G^{(m)}\right\|_{\infty}. \tag{2.5.42}$$

That is

$$\left|D_{b-}^{\alpha} G(x)\right| \leq \frac{(b - a)^{m-\alpha}}{\Gamma(m - \alpha + 1)} \left\|G^{(m)}\right\|_{\infty} < \infty, \quad \forall x \in [a, b]. \tag{2.5.43}$$

Hence, $\forall x \in [a, b^*]$ we get that

$$|A(x)| = \frac{\left|D_{b-}^{\alpha} G(x)\right|}{2D_{b-}^{\alpha} G(x_2)} \leq \frac{(b - a)^{m-\alpha}}{2\Gamma(m - \alpha + 1)} \frac{\left\|G^{(m)}\right\|_{\infty}}{D_{b-}^{\alpha} G(x_2)} < \infty. \tag{2.5.44}$$

Consequently we observe

$$(\xi_2) \leq \left(\lambda + \frac{(b - a)^{m-\alpha}}{2\Gamma(m - \alpha + 1)} \frac{\left\|G^{(m)}\right\|_{\infty}}{D_{b-}^{\alpha} G(x_2)}\right)|y - x|, \quad \forall x, y \in [a, b^*]. \tag{2.5.45}$$

Call

$$0 < \gamma_1 := \lambda + \frac{(b-a)^{m-\alpha}}{2\Gamma(m-\alpha+1)} \frac{\left\| G^{(m)} \right\|_\infty}{D_{b-}^\alpha G(x_2)}, \tag{2.5.46}$$

choosing $(b-a)$ small enough we can make $\gamma_1 \in (0, 1)$. So (2.5.2) is fulfilled.

Next we call and need

$$0 < \gamma := \gamma_0 + \gamma_1 = 1 - \frac{D_{b-}^\alpha G(x_1)}{2D_{b-}^\alpha G(x_2)} + \lambda + \frac{(b-a)^{m-\alpha}}{2\Gamma(m-\alpha+1)} \frac{\left\| G^{(m)} \right\|_\infty}{D_{b-}^\alpha G(x_2)} < 1, \tag{2.5.47}$$

equivalently we find,

$$\lambda + \frac{(b-a)^{m-\alpha}}{2\Gamma(m-\alpha+1)} \frac{\left\| G^{(m)} \right\|_\infty}{D_{b-}^\alpha G(x_2)} < \frac{D_{b-}^\alpha G(x_1)}{2D_{b-}^\alpha G(x_2)}, \tag{2.5.48}$$

so,

$$2\lambda D_{b-}^\alpha G(x_2) + \frac{(b-a)^{m-\alpha}}{\Gamma(m-\alpha+1)} \left\| G^{(m)} \right\|_\infty < D_{b-}^\alpha G(x_1), \tag{2.5.49}$$

which is possible for small λ, $(b-a)$.

That is $\gamma \in (0, 1)$, fulfilling (2.5.3). Hence Eq. (2.5.33) can be solved with our presented numerical methods.

References

1. S. Amat, S. Busquier, S. Plaza, Chaotic dynamics of a third-order Newton-like method. J. Math. Anal. Applic. **366**(1), 164–174 (2010)
2. G. Anastassiou, *Fractional Differentiation Inequalities* (Springer, New York, 2009)
3. G.A. Anastassiou, Fractional representation formulae and right fractional inequalities. Math. Comput. Model. **54**(11–12), 3098–3115 (2011)
4. G. Anastassiou, *Intelligent Mathematics: Computational Analysis* (Springer, Heidelberg, 2011)
5. G. Anastassiou, I. Argyros, *Semilocal convergence of Newton-like methods under general conditions with applications in fractional calculus* (submitted) (2015)
6. I.K. Argyros, Newton-like methods in partially ordered linear spaces. J. Approx. Th. Applic. **9**(1), 1–10 (1993)
7. I.K. Argyros, Results on controlling the residuals of perturbed Newton-like methods on banach spaces with a convergence structure. Southwest J. Pure Appl. Math. **1**, 32–38 (1995)
8. I.K. Argyros, *Convergence and Applications of Newton-like Iterations* (Springer, New York, 2008)
9. K. Diethelm, *The Analysis of Fractional DifferentialEquations*. Lecture Notes in Mathematics, vol. 2004, 1st edn. (Springer, New York, 2010)
10. J.A. Ezquerro, J.M. Gutierrez, M.Á. Hernández, N. Romero, M.J. Rubio, The Newton method: from Newton to Kantorovich (Spanish). Gac. R. Soc. Mat. Esp. **13**, 53–76 (2010)
11. J.A. Ezquerro, M.Á. Hernández, Newton-like methods of high order and domains of semilocal and global convergence. Appl. Math. Comput. **214**(1), 142–154 (2009)
12. L.V. Kantorovich, G.P. Akilov, *Functional Analysis in Normed Spaces* (Pergamon Press, New York, 1964)

13. A.A. Magreñán, Different anomalies in a Jarratt family of iterative root finding methods. Appl. Math. Comput. **233**, 29–38 (2014)
14. A.A. Magreñán, A new tool to study real dynamics: the convergence plane. Appl. Math. Comput. **248**, 215–224 (2014)
15. P.W. Meyer, Newton's method in generalized Banach spaces. Numer. Func. Anal. Optimiz. **9**(3–4), 244–259 (1987)
16. F.A. Potra, V. Ptak, *Nondiscrete Induction and Iterative Processes* (Pitman Publ, London, 1984)
17. P.D. Proinov, New general convergence theory for iterative processes and its applications to Newton-Kantorovich type theorems. J. Complex. **26**, 3–42 (2010)

Chapter 3
Convergence of Iterative Methods and Generalized Fractional Calculus

We present a semilocal convergence study of some iterative methods on a generalized Banach space setting to approximate a locally unique zero of an operator. Earlier studies such as [8–10, 15] require that the operator involved is Fréchet-differentiable. In the present study we assume that the operator is only continuous. This way we extend the applicability of these methods to include generalized fractional calculus and problems from other areas. Some applications include generalized fractional calculus involving the Riemann-Liouville fractional integral and the Caputo fractional derivative. Fractional calculus is very important for its applications in many applied sciences. It follows [7].

3.1 Introduction

Many problems in Computational sciences can be formulated as an operator equation using Mathematical Modelling [4, 10, 12, 16]. The zeros of these operators can rarely be found in closed form. That is why most solution methods are usually iterative.

The semilocal convergence is, based on the information around an initial point, to give conditions ensuring the convergence of the method.

We present a semilocal convergence analysis for some iterative methods on a generalized Banach space setting to approximate a zero of an operator. A generalized norm is defined to be an operator from a linear space into a partially order Banach space (to be precised in Sect. 3.2). Earlier studies such as [8–10, 15] for Newton's method have shown that a more precise convergence analysis is obtained when compared to the real norm theory. However, the main assumption is that the operator involved is Fré chet-differentiable. This hypothesis limits the applicability of Newton's method. In the present study we only assume the continuity of the operator. This may be expand the applicability of these methods.

The rest of the chapter is organized as follows: Sect. 3.2 contains the basic concepts on generalized Banach spaces and auxiliary results on inequalities and fixed points.

© Springer International Publishing Switzerland 2016

G.A. Anastassiou and I.K. Argyros, *Intelligent Numerical Methods: Applications to Fractional Calculus*, Studies in Computational Intelligence 624, DOI 10.1007/978-3-319-26721-0_3

In Sect. 3.3 we present the semilocal convergence analysis of Newton-like methods. Finally, in the concluding Sects. 3.4 and 3.5, we present special cases and applications in generalized fractional calculus.

3.2 Generalized Banach Spaces

We present some standard concepts that are needed in what follows to make the paper as self contained as possible. More details on generalized Banach spaces can be found in [8–10, 15], and the references there in.

Definition 3.1 A generalized Banach space is a triplet $(x, E, / \cdot /)$ such that
(i) X is a linear space over $\mathbb{R}(\mathbb{C})$.
(ii) $E = (E, K, \|\cdot\|)$ is a partially ordered Banach space, i.e.
(ii$_1$) $(E, \|\cdot\|)$ is a real Banach space, ·
(ii$_2$) E is partially ordered by a closed convex cone K,
(ii$_3$) The norm $\|\cdot\|$ is monotone on K.
(iii) The operator $/ \cdot / : X \to K$ satisfies
$/x/ = 0 \Leftrightarrow x = 0, /\theta x/ = |\theta| /x/,$
$/x + y/ \leq /x/ + /y/$ for each $x, y \in X, \theta \in \mathbb{R}(\mathbb{C})$.
(iv) X is a Banach space with respect to the induced norm $\|\cdot\|_i := \|\cdot\| \cdot / \cdot /$.

Remark 3.2 The operator $/ \cdot /$ is called a generalized norm. In view of (iii) and (ii$_3$) $\|\cdot\|_i$, is a real norm. In the rest of this paper all topological concepts will be understood with respect to this norm.

Let $L\left(X^j, Y\right)$ stand for the space of j-linear symmetric and bounded operators from X^j to Y, where X and Y are Banach spaces. For X, Y partially ordered $L_+\left(X^j, Y\right)$ stands for the subset of monotone operators P such that

$$0 \leq a_i \leq b_i \Rightarrow P\left(a_1, \ldots, a_j\right) \leq P\left(b_1, \ldots, b_j\right). \tag{3.2.1}$$

Definition 3.3 The set of bounds for an operator $Q \in L(X, X)$ on a generalized Banach space $(X, E, / \cdot /)$ the set of bounds is defined to be:

$$B(Q) := \{P \in L_+(E, E), /Qx/ \leq P/x/ \text{ for each } x \in X\}. \tag{3.2.2}$$

Let $D \subset X$ and $T : D \to D$ be an operator. If $x_0 \in D$ the sequence $\{x_n\}$ given by

$$x_{n+1} := T(x_n) = T^{n+1}(x_0) \tag{3.2.3}$$

is well defined. We write in case of convergence

$$T^\infty (x_0) := \lim \left(T^n (x_0) \right) = \lim_{n \to \infty} x_n. \tag{3.2.4}$$

We need some auxiliary results on inequations.

Lemma 3.4 *Let* $(E, K, \|\cdot\|)$ *be a partially ordered Banach space,* $\xi \in K$ *and* $M, N \in L_+ (E, E)$.
   (i) Suppose there exists $r \in K$ *such that*

$$R (r) := (M + N) r + \xi \leq r \tag{3.2.5}$$

and

$$(M + N)^k r \to 0 \quad as \ k \to \infty. \tag{3.2.6}$$

Then, $b := R^\infty (0)$ *is well defined satisfies the equation* $t = R (t)$ *and is the smaller than any solution of the inequality* $R (s) \leq s$.
   (ii) Suppose there exists $q \in K$ *and* $\theta \in (0, 1)$ *such that* $R (q) \leq \theta q$, *then there exists* $r \leq q$ *satisfying (i).*

Proof (i) Define sequence $\{b_n\}$ by $b_n = R^n (0)$. Then, we have by (3.2.5) that $b_1 = R (0) = \xi \leq r \Rightarrow b_1 \leq r$. Suppose that $b_k \leq r$ for each $k = 1, 2, \ldots, n$. Then, we have by (3.2.5) and the inductive hypothesis that $b_{n+1} = R^{n+1} (0) = R (R^n (0)) = R (b_n) = (M + N) b_n + \xi \leq (M + N) r + \xi \leq r \Rightarrow b_{n+1} \leq r$. Hence, sequence $\{b_n\}$ is bounded above by r. Set $P_n = b_{n+1} - b_n$. We shall show that

$$P_n \leq (M + N)^n r \quad \text{for each } n = 1, 2, \ldots \tag{3.2.7}$$

We have by the definition of P_n and (3.2.6) that

$$P_1 = R^2 (0) - R (0) = R (R (0)) - R (0)$$

$$= R (\xi) - R (0) = \int_0^1 R' (t\xi) \xi dt \leq \int_0^1 R' (\xi) \xi dt$$

$$\leq \int_0^1 R' (r) r dt \leq (M + N) r,$$

which shows (3.2.7) for $n = 1$. Suppose that (3.2.7) is true for $k = 1, 2, \ldots, n$. Then, we have in turn by (3.2.6) and the inductive hypothesis that

$$P_{k+1} = R^{k+2} (0) - R^{k+1} (0) = R^{k+1} (R (0)) - R^{k+1} (0) =$$

$$R^{k+1} (\xi) - R^{k+1} (0) = R \left(R^k (\xi) \right) - R \left(R^k (0) \right) =$$

$$\int_0^1 R' \left(R^k (0) + t \left(R^k (\xi) - R^k (0) \right) \right) \left(R^k (\xi) - R^k (0) \right) dt \le$$

$$R' \left(R^k (\xi) \right) \left(R^k (\xi) - R^k (0) \right) = R' \left(R^k (\xi) \right) \left(R^{k+1} (0) - R^k (0) \right) \le$$

$$R' (r) \left(R^{k+1} (0) - R^k (0) \right) \le (M + N) (M + N)^k r = (M + N)^{k+1} r,$$

which completes the induction for (3.2.7). It follows that $\{b_n\}$ is a complete sequence in a Banach space and as such it converges to some b. Notice that $R(b) = R(\lim_{n \to \infty} R^n (0)) = \lim_{n \to \infty} R^{n+1} (0) = b \Rightarrow b$ solves the equation $R(t) = t$. We have that $b_n \le r \Rightarrow b \le r$, where r a solution of $R(r) \le r$. Hence, b is smaller than any solution of $R(s) \le s$.

(ii) Define sequences $\{v_n\}$, $\{w_n\}$ by $v_0 = 0$, $v_{n+1} = R(v_n)$, $w_0 = q$, $w_{n+1} = R(w_n)$. Then, we have that

$$0 \le v_n \le v_{n+1} \le w_{n+1} \le w_n \le q, \tag{3.2.8}$$
$$w_n - v_n \le \theta^n (q - v_n)$$

and sequence $\{v_n\}$ is bounded above by q. Hence, it converges to some r with $r \le q$. We also get by (3.2.8) that $w_n - v_n \to 0$ as $n \to \infty \Rightarrow w_n \to r$ as $n \to \infty$. □

We also need the auxiliary result for computing solutions of fixed point problems.

Lemma 3.5 *Let* $(X, (E, K, \|\cdot\|), / \cdot /)$ *be a generalized Banach space, and* $P \in B(Q)$ *be a bound for* $Q \in L(X, X)$. *Suppose there exists* $y \in X$ *and* $q \in K$ *such that*

$$Pq + /y/ \le q \text{ and } P^k q \to 0 \quad as \ k \to \infty. \tag{3.2.9}$$

Then, $z = T^\infty (0)$, $T(x) := Qx + y$ *is well defined and satisfies:* $z = Qz + y$ *and* $/z/ \le P/z/ + /y/ \le q$. *Moreover,* z *is the unique solution in the subspace* $\{x \in X | \exists \theta \in \mathbb{R} : \{x\} \le \theta q\}$.

The proof can be found in [15, Lemma 3.2].

3.3 Semilocal Convergence

Let $(X, (E, K, \|\cdot\|), / \cdot /)$ and Y be generalized Banach spaces, $D \subset X$ an open subset, $G : D \to Y$ a continuous operator and $A(\cdot) : D \to L(X, Y)$. A zero of operator G is to be determined by a Newton-like method starting at a point $x_0 \in D$. The results are presented for an operator $F = JG$, where $J \in L(Y, X)$. The iterates are determined through a fixed point problem:

$$x_{n+1} = x_n + y_n, \ A(x_n) y_n + F(x_n) = 0 \tag{3.3.1}$$
$$\Leftrightarrow y_n = T(y_n) := (I - A(x_n)) y_n - F(x_n).$$

Let $U(x_0, r)$ stand for the ball defined by

$$U(x_0, r) := \{x \in X : /x - x_0/ \le r\}$$

for some $r \in K$.

Next, we present the semilocal convergence analysis of Newton-like method (3.3.1) using the preceding notation.

Theorem 3.6 *Let $F : D \subset X$, $A(\cdot) : D \to L(X, Y)$ and $x_0 \in D$ be as defined previously. Suppose:*

(H_1) There exists an operator $M \in B(I - A(x))$ for each $x \in D$.

(H_2) There exists an operator $N \in L_+(E, E)$ satisfying for each $x, y \in D$

$$/F(y) - F(x) - A(x)(y - x)/ \le N/y - x/.$$

(H_3) There exists a solution $r \in K$ of

$$R_0(t) := (M + N)t + /F(x_0)/ \le t.$$

(H_4) $U(x_0, r) \subseteq D$.

(H_5) $(M + N)^k r \to 0$ as $k \to \infty$.

Then, the following hold:

(C_1) The sequence $\{x_n\}$ defined by

$$x_{n+1} = x_n + T_n^\infty(0), \quad T_n(y) := (I - A(x_n))y - F(x_n) \tag{3.3.2}$$

is well defined, remains in $U(x_0, r)$ for each $n = 0, 1, 2, \ldots$ and converges to the unique zero of operator F in $U(x_0, r)$.

(C_2) An apriori bound is given by the null-sequence $\{r_n\}$ defined by $r_0 := r$ and for each $n = 1, 2, \ldots$

$$r_n = P_n^\infty(0), \quad P_n(t) = Mt + Nr_{n-1}.$$

(C_3) An aposteriori bound is given by the sequence $\{s_n\}$ defined by

$$s_n := R_n^\infty(0), \quad R_n(t) = (M + N)t + Na_{n-1},$$

$$b_n := /x_n - x_0/ \le r - r_n \le r,$$

where

$$a_{n-1} := /x_n - x_{n-1}/ \quad \text{for each } n = 1, 2, \ldots$$

Proof Let us define for each $n \in \mathbb{N}$ the statement:
  (I_n) $x_n \in X$ and $r_n \in K$ are well defined and satisfy

$$r_n + a_{n-1} \leq r_{n-1}.$$

We use induction to show (I_n). The statement (I_1) is true: By Lemma 3.4 and (H_3), (H_5) there exists $q \leq r$ such that:

$$Mq + /F(x_0)/ = q \text{ and } M^k q \leq M^k r \to 0 \quad \text{as } k \to \infty.$$

Hence, by Lemma 3.5 x_1 is well defined and we have $a_0 \leq q$. Then, we get the estimate

$$P_1(r - q) = M(r - q) + Nr_0$$

$$\leq Mr - Mq + Nr = R_0(r) - q$$

$$\leq R_0(r) - q = r - q.$$

It follows with Lemma 3.4 that r_1 is well defined and

$$r_1 + a_0 \leq r - q + q = r = r_0.$$

Suppose that (I_j) is true for each $j = 1, 2, \ldots, n$. We need to show the existence of x_{n+1} and to obtain a bound q for a_n. To achieve this notice that:

$$Mr_n + N(r_{n-1} - r_n) = Mr_n + Nr_{n-1} - Nr_n = P_n(r_n) - Nr_n \leq r_n.$$

Then, it follows from Lemma 3.4 that there exists $q \leq r_n$ such that

$$q = Mq + N(r_{n-1} - r_n) \text{ and } (M + N)^k q \to 0, \quad \text{as } k \to \infty. \tag{3.3.3}$$

By (I_j) it follows that

$$b_n = /x_n - x_0/ \leq \sum_{j=0}^{n-1} a_j \leq \sum_{j=0}^{n-1} (r_j - r_{j+1}) = r - r_n \leq r.$$

Hence, $x_n \in U(x_0, r) \subset D$ and by (H_1) M is a bound for $I - A(x_n)$.
  We can write by (H_2) that

$$/F(x_n)/ = /F(x_n) - F(x_{n-1}) - A(x_{n-1})(x_n - x_{n-1})/$$

$$\leq Na_{n-1} \leq N(r_{n-1} - r_n). \tag{3.3.4}$$

It follows from (3.3.3) and (3.3.4) that

$$Mq + /F(x_n)/ \leq q.$$

By Lemma 3.5, x_{n+1} is well defined and $a_n \leq q \leq r_n$. In view of the definition of r_{n+1} we have that

$$P_{n+1}(r_n - q) = P_n(r_n) - q = r_n - q,$$

so that by Lemma 3.4, r_{n+1} is well defined and

$$r_{n+1} + a_n \leq r_n - q + q = r_n,$$

which proves (I_{n+1}). The induction for (I_n) is complete. Let $m \geq n$, then we obtain in turn that

$$/x_{m+1} - x_n/ \leq \sum_{j=n}^{m} a_j \leq \sum_{j=n}^{m} (r_j - r_{j+1}) = r_n - r_{m+1} \leq r_n. \tag{3.3.5}$$

Moreover, we get inductively the estimate

$$r_{n+1} = P_{n+1}(r_{n+1}) \leq P_{n+1}(r_n) \leq (M + N) r_n \leq \cdots \leq (M + N)^{n+1} r.$$

It follows from (H_5) that $\{r_n\}$ is a null-sequence. Hence, $\{x_n\}$ is a complete sequence in a Banach space X by (3.3.5) and as such it converges to some $x^* \in X$. By letting $m \to \infty$ in (3.3.5) we deduce that $x^* \in U(x_n, r_n)$. Furthermore, (3.3.4) shows that x^* is a zero of F. Hence, (C_1) and (C_2) are proved.

In view of the estimate

$$R_n(r_n) \leq P_n(r_n) \leq r_n$$

the apriori, bound of (C_3) is well defined by Lemma 3.4. That is s_n is smaller in general than r_n. The conditions of Theorem 3.6 are satisfied for x_n replacing x_0. A solution of the inequality of (C_2) is given by s_n (see (3.3.4)). It follows from (3.3.5) that the conditions of Theorem 3.6 are easily verified. Then, it follows from (C_1) that $x^* \in U(x_n, s_n)$ which proves (C_3). $\qquad\square$

In general the aposterior, estimate is of interest. Then, condition (H_5) can be avoided as follows:

Proposition 3.7 *Suppose: condition (H_1) of Theorem 3.6 is true.*
(H'_3) There exists $s \in K$, $\theta \in (0, 1)$ such that

$$R_0(s) = (M + N)s + /F(x_0)/ \leq \theta s.$$

(H'_4) $U(x_0, s) \subset D$.

Then, there exists $r \leq s$ satisfying the conditions of Theorem 3.6. Moreover, the zero x^* of F is unique in $U(x_0, s)$.

Remark 3.8 (i) Notice that by Lemma 3.4 $R_n^\infty(0)$ is the smallest solution of $R_n(s) \leq s$. Hence any solution of this inequality yields on upper estimate for $R_n^\infty(0)$. Similar inequalities appear in (H$_2$) and (H$_2'$).

(ii) The weak assumptions of Theorem 3.6 do not imply the existence of $A(x_n)^{-1}$. In practice the computation of $T_n^\infty(0)$ as a solution of a linear equation is no problem and the computation of the expensive or impossible to compute in general $A(x_n)^{-1}$ is not needed.

(iii) We can used the following result for the computation of the aposteriori estimates. The proof can be found in [15, Lemma 4.2] by simply exchanging the definitions of R.

Lemma 3.9 *Suppose that the conditions of Theorem 3.6 are satisfied. If $s \in K$ is a solution of $R_n(s) \leq s$, then $q := s - a_n \in K$ and solves $R_{n+1}(q) \leq q$. This solution might be improved by $R_{n+1}^k(q) \leq q$ for each $k = 1, 2, \ldots$.*

3.4 Special Cases and Applications

Application 3.10 *The results obtained in earlier studies such as [8–10, 15] require that operator F (i.e. G) is Fré chet-differentiable. This assumption limits the applicability of the earlier results. In the present study we only require that F is a continuous operator. Hence, we have extended the applicability of these methods to classes of operators that are only continuous.*

Example 3.11 The j-dimensional space $\mathbb{R}^j$ is a classical example of a generalized Banach space. The generalized norm is defined by componentwise absolute values. Then, as ordered Banach space we set $E = \mathbb{R}^j$ with componentwise ordering with e.g. the maximum norm. A bound for a linear operator (a matrix) is given by the corresponding matrix with absolute values. Similarly, we can define the "N" operators. Let $E = \mathbb{R}$. That is we consider the case of a real normed space with norm denoted by $\|\cdot\|$. Let us see how the conditions of Theorem 3.6 look like.

Theorem 3.12 $(H_1) \ \|I - A(x)\| \leq M$ for some $M \geq 0$.
  $(H_2) \ \|F(y) - F(x) - A(x)(y - x)\| \leq N \|y - x\|$ for some $N \geq 0$.
  $(H_3) \ M + N < 1$,

$$r = \frac{\|F(x_0)\|}{1 - (M + N)}. \tag{3.4.1}$$

  $(H_4) \ U(x_0, r) \subseteq D$.
  $(H_5) \ (M + N)^k r \to 0$ as $k \to \infty$, where r is given by (3.4.1).
  Then, the conclusions of Theorem 3.6 hold.

3.5 Applications to Generalized Fractional Calculus

We present some applications of Theorem 3.12 in this section.
Background
We use a lot here the following generalized fractional integral.

Definition 3.13 (*see also* [12], *p. 99*) The left generalized fractional integral of a function f with respect to given function g is defined as follows:

Let $a, b \in \mathbb{R}$, $a < b$, $\alpha > 0$. Here $g \in AC([a, b])$ (absolutely continuous functions) and is striclty increasing, $f \in L_\infty([a, b])$. We set

$$\left(I^\alpha_{a+;g} f\right)(x) = \frac{1}{\Gamma(\alpha)} \int_a^x (g(x) - g(t))^{\alpha-1} g'(t) f(t) \, dt, \quad x \ge a, \qquad (3.5.1)$$

clearly $\left(I^\alpha_{a+;g} f\right)(a) = 0$.

When g is the identity function id, we get that $I^\alpha_{a+;id} = I^\alpha_{a+}$, the ordinary left Riemann-Liouville fractional integral, where

$$\left(I^\alpha_{a+} f\right)(x) = \frac{1}{\Gamma(\alpha)} \int_a^x (x - t)^{\alpha-1} f(t) \, dt, \quad x \ge a, \qquad (3.5.2)$$

$\left(I^\alpha_{a+} f\right)(a) = 0$.

When, $g(x) = \ln x$ on $[a, b]$, $0 < a < b < \infty$, we get:

Definition 3.14 ([12], *p. 110*) Let $0 < a < b < \infty$, $\alpha > 0$. The left Hadamard fractional integral of order α is given by

$$\left(J^\alpha_{a+} f\right)(x) = \frac{1}{\Gamma(\alpha)} \int_a^x \left(\ln \frac{x}{y}\right)^{\alpha-1} \frac{f(y)}{y} \, dy, \quad x \ge a, \qquad (3.5.3)$$

where $f \in L_\infty([a, b])$.

We mention:

Definition 3.15 ([5]) The left fractional exponential integral is defined as follows:
Let $a, b \in \mathbb{R}$, $a < b$, $\alpha > 0$, $f \in L_\infty([a, b])$. We set

$$\left(I^\alpha_{a+;e^x} f\right)(x) = \frac{1}{\Gamma(\alpha)} \int_a^x (e^x - e^t)^{\alpha-1} e^t f(t) \, dt, \quad x \ge a. \qquad (3.5.4)$$

Definition 3.16 ([5]) Let $a, b \in \mathbb{R}$, $a < b$, $\alpha > 0$, $f \in L_\infty([a, b])$, $A > 1$. We give the fractional integral

$$\left(I^\alpha_{a+;A^x} f\right)(x) = \frac{\ln A}{\Gamma(\alpha)} \int_a^x (A^x - A^t)^{\alpha-1} A^t f(t) \, dt, \quad x \ge a. \qquad (3.5.5)$$

We also give:

Definition 3.17 ([5]) Let $\alpha, \sigma > 0, 0 \leq a < b < \infty, f \in L_\infty ([a, b])$. We set

$$\left(K^\alpha_{a+;x^\sigma} f\right)(x) = \frac{1}{\Gamma(\alpha)} \int_z^x (x^\sigma - t^\sigma)^{\alpha-1} f(t) \sigma t^{\sigma-1} dt, \quad x \geq a. \qquad (3.5.6)$$

We mention the following generalized fractional derivatives:

Definition 3.18 ([5]) Let $\alpha > 0$ and $\lceil \alpha \rceil = m$. Consider $f \in AC^m ([a, b])$ (space of functions f with $f^{(m-1)} \in AC ([a, b])$). We define the left generalized fractional derivative of f of order α as follows

$$\left(D^\alpha_{*a;g} f\right)(x) = \frac{1}{\Gamma(m-\alpha)} \int_a^x (g(x) - g(t))^{m-\alpha-1} g'(t) f^{(m)}(t) dt, \qquad (3.5.7)$$

for any $x \in [a, b]$, where Γ is the gamma function.
We set

$$D^m_{*a;g} f(x) = f^{(m)}(x), \qquad (3.5.8)$$

$$D^0_{*a;g} f(x) = f(x), \quad \forall x \in [a, b]. \qquad (3.5.9)$$

When $g = id$, then $D^\alpha_{*a} f = D^\alpha_{*a;id} f$ is the left Caputo fractional derivative.

So, we have the specific generalized left fractional derivatives.

Definition 3.19 ([5])

$$D^\alpha_{*a;\ln x} f(x) = \frac{1}{\Gamma(m-\alpha)} \int_a^x \left(\ln \frac{x}{y}\right)^{m-\alpha-1} \frac{f^{(m)}(y)}{y} dy, \quad x \geq a > 0, \qquad (3.5.10)$$

$$D^\alpha_{*a;e^x} f(x) = \frac{1}{\Gamma(m-\alpha)} \int_a^x (e^x - e^t)^{m-\alpha-1} e^t f^{(m)}(t) dt, \quad x \geq a, \qquad (3.5.11)$$

and

$$D^\alpha_{*a;A^x} f(x) = \frac{\ln A}{\Gamma(m-\alpha)} \int_a^x (A^x - A^t)^{m-\alpha-1} A^t f^{(m)}(t) dt, \quad x \geq a, \qquad (3.5.12)$$

$$\left(D^\alpha_{*a;x^\sigma} f\right)(x) = \frac{1}{\Gamma(m-\alpha)} \int_a^x (x^\sigma - t^\sigma)^{m-\alpha-1} \sigma t^{\sigma-1} f^{(m)}(t) dt, \quad x \geq a \geq 0. \qquad (3.5.13)$$

We make:

Remark 3.20 ([5]) Here $g \in AC([a, b])$ (absolutely continuous functions), g is increasing over $[a, b]$, $\alpha > 0$. Then

$$\int_a^x (g(x) - g(t))^{\alpha-1} g'(t) dt = \frac{(g(x) - g(a))^\alpha}{\alpha}, \quad \forall x \in [a, b]. \quad (3.5.14)$$

We mention

Theorem 3.21 ([5]) *Let* $\alpha > 0$, $\mathbb{N} \ni m = \lceil \alpha \rceil$, *and* $f \in C^m([a, b])$. *Then* $\left(D_{*a;g}^\alpha f \right)(x)$ *is continuous in* $x \in [a, b]$.

Results
(I) We notice the following

$$\left| \left(I_{a+;g}^\alpha f \right)(x) \right| \leq \frac{1}{\Gamma(\alpha)} \int_a^x (g(x) - g(t))^{\alpha-1} g'(t) |f(t)| dt$$

$$\leq \frac{\|f\|_\infty}{\Gamma(\alpha)} \int_a^x (g(x) - g(t))^{\alpha-1} g'(t) dt = \frac{\|f\|_\infty}{\Gamma(\alpha)} \frac{(g(x) - g(a))^\alpha}{\alpha} \quad (3.5.15)$$

$$= \frac{\|f\|_\infty}{\Gamma(\alpha+1)} (g(x) - g(a))^\alpha.$$

That is

$$\left| \left(I_{a+;g}^\alpha f \right)(x) \right| \leq \frac{\|f\|_\infty}{\Gamma(\alpha+1)} (g(x) - g(a))^\alpha \leq \|f\|_\infty \frac{(g(b) - g(a))^\alpha}{\Gamma(\alpha+1)}, \quad (3.5.16)$$

$\forall x \in [a, b]$.

In particular $\left(I_{a+;g}^\alpha f \right)(a) = 0$.
Clearly $I_{a+;g}^\alpha$ is a bounded linear operator.
We use

Theorem 3.22 ([6]) *Let* $r > 0$, $a < b$, $F \in L_\infty([a, b])$, $g \in AC([a, b])$ *and* g *is strictly increasing.*
Consider

$$G(s) := \int_a^s (g(s) - g(t))^{r-1} g'(t) F(t) dt, \quad \text{for all } s \in [a, b]. \quad (3.5.17)$$

Then $G \in C([a, b])$.

By Theorem 3.22, the function $\left(I_{a+;g}^\alpha f \right)$ is a continuous function over $[a, b]$. Consider $a < a^* < b$. Therefore $\left(I_{a+;g}^\alpha f \right)$ is also continuous over $[a^*, b]$.

Thus, there exist $x_1, x_2 \in [a^*, b]$ such that

$$\left(I_{a+;g}^{\alpha} f\right)(x_1) = \min \left(I_{a+;g}^{\alpha} f\right)(x), \tag{3.5.18}$$

$$\left(I_{a+;g}^{\alpha} f\right)(x_2) = \max \left(I_{a+;g}^{\alpha} f\right)(x), \quad x \in [a^*, b]. \tag{3.5.19}$$

We assume that

$$\left(I_{a+;g}^{\alpha} f\right)(x_1) > 0. \tag{3.5.20}$$

Hence

$$\left\| I_{a+;g}^{\alpha} f \right\|_{\infty, [a^*, b]} = \left(I_{a+;g}^{\alpha} f\right)(x_2) > 0. \tag{3.5.21}$$

Here it is

$$J(x) = mx, \quad m \neq 0. \tag{3.5.22}$$

Therefore the equation

$$Jf(x) = 0, \quad x \in [a^*, b], \tag{3.5.23}$$

has the same solutions as the equation

$$F(x) := \frac{Jf(x)}{2\left(I_{a+;g}^{\alpha} f\right)(x_2)} = 0, \quad x \in [a^*, b]. \tag{3.5.24}$$

Notice that

$$I_{a+;g}^{\alpha} \left(\frac{f}{2\left(I_{a+;g}^{\alpha} f\right)(x_2)} \right)(x) = \frac{\left(I_{a+;g}^{\alpha} f\right)(x)}{2\left(I_{a+;g}^{\alpha} f\right)(x_2)} \leq \frac{1}{2} < 1, \quad x \in [a^*, b]. \tag{3.5.25}$$

Call

$$A(x) := \frac{\left(I_{a+;g}^{\alpha} f\right)(x)}{2\left(I_{a+;g}^{\alpha} f\right)(x_2)}, \quad \forall x \in [a^*, b]. \tag{3.5.26}$$

Then, we get that

$$0 < \frac{\left(I_{a+;g}^{\alpha} f\right)(x_1)}{2\left(I_{a+;g}^{\alpha} f\right)(x_2)} \leq A(x) \leq \frac{1}{2}, \quad \forall x \in [a^*, b]. \tag{3.5.27}$$

We observe

$$|1 - A(x)| = 1 - A(x) \leq 1 - \frac{\left(I_{a+;g}^{\alpha} f\right)(x_1)}{2\left(I_{a+;g}^{\alpha} f\right)(x_2)} =: \gamma_0, \quad \forall\, x \in [a^*, b]. \quad (3.5.28)$$

Clearly $\gamma_0 \in (0, 1)$.
  I.e.

$$|1 - A(x)| \leq \gamma_0, \quad \forall\, x \in [a^*, b], \gamma_0 \in (0, 1). \quad (3.5.29)$$

Next we assume that $F(x)$ is a contraction, i.e.

$$|F(x) - F(y)| \leq \lambda |x - y|; \quad \forall\, x, y \in [a^*, b], \quad (3.5.30)$$

and $0 < \lambda < \frac{1}{2}$.
  Equivalently we have

$$|Jf(x) - Jf(y)| \leq 2\lambda \left(I_{a+;g}^{\alpha} f\right)(x_2) |x - y|, \quad \text{all } x, y \in [a^*, b]. \quad (3.5.31)$$

We observe that

$$|F(y) - F(x) - A(x)(y - x)| \leq |F(y) - F(x)| + |A(x)| |y - x| \leq$$

$$\lambda |y - x| + |A(x)| |y - x| = (\lambda + |A(x)|) |y - x| =: (\psi_1), \quad \forall\, x, y \in [a^*, b]. \quad (3.5.32)$$

By (3.5.16) we get

$$\left|\left(I_{a+;g}^{\alpha} f\right)(x)\right| \leq \frac{\|f\|_{\infty}}{\Gamma(\alpha + 1)} (g(b) - g(a))^{\alpha}, \quad \forall\, x \in [a^*, b]. \quad (3.5.33)$$

Hence

$$|A(x)| = \frac{\left|\left(I_{a+;g}^{\alpha} f\right)(x)\right|}{2\left(I_{a+;g}^{\alpha} f\right)(x_2)} \leq \frac{\|f\|_{\infty} (g(b) - g(a))^{\alpha}}{2\Gamma(\alpha + 1)\left(I_{a+;g}^{\alpha} f\right)(x_2)} < \infty, \quad \forall\, x \in [a^*, b]. \quad (3.5.34)$$

Therefore we get

$$(\psi_1) \leq \left(\lambda + \frac{\|f\|_{\infty} (g(b) - g(a))^{a}}{2\Gamma(\alpha + 1)\left(I_{a+;g}^{\alpha} f\right)(x_2)}\right) |y - x|, \quad \forall\, x, y \in [a^*, b]. \quad (3.5.35)$$

Call

$$0 < \gamma_1 := \lambda + \frac{\|f\|_\infty \left(g\left(b\right) - g\left(a\right)\right)^a}{2\Gamma\left(\alpha + 1\right)\left(I_{a+;g}^\alpha f\right)\left(x_2\right)}, \tag{3.5.36}$$

choosing $\left(g\left(b\right) - g\left(a\right)\right)$ small enough we can make $\gamma_1 \in \left(0, 1\right)$.

We have proved that

$$|F\left(y\right) - F\left(x\right) - A\left(x\right)\left(y - x\right)| \le \gamma_1 \left|y - x\right|, \quad \forall\, x, y \in \left[a^*, b\right], \gamma_1 \in \left(0, 1\right). \tag{3.5.37}$$

Next we call and we need that

$$0 < \gamma := \gamma_0 + \gamma_1 = 1 - \frac{\left(I_{a+;g}^\alpha f\right)\left(x_1\right)}{2\left(I_{a+;g}^\alpha f\right)\left(x_2\right)} + \lambda + \frac{\|f\|_\infty \left(g\left(b\right) - g\left(a\right)\right)^a}{2\Gamma\left(\alpha + 1\right)\left(I_{a+;g}^\alpha f\right)\left(x_2\right)} < 1, \tag{3.5.38}$$

$$\lambda + \frac{\|f\|_\infty \left(g\left(b\right) - g\left(a\right)\right)^a}{2\Gamma\left(\alpha + 1\right)\left(I_{a+;g}^\alpha f\right)\left(x_2\right)} < \frac{\left(I_{a+;g}^\alpha f\right)\left(x_1\right)}{2\left(I_{a+;g}^\alpha f\right)\left(x_2\right)}, \tag{3.5.39}$$

equivalently,

$$2\lambda\left(I_{a+;g}^\alpha f\right)\left(x_2\right) + \frac{\|f\|_\infty \left(g\left(b\right) - g\left(a\right)\right)^a}{\Gamma\left(\alpha + 1\right)} < \left(I_{a+;g}^\alpha f\right)\left(x_1\right), \tag{3.5.40}$$

which is possible for small λ, $\left(g\left(b\right) - g\left(a\right)\right)$. That is $\gamma \in \left(0, 1\right)$. So our method solves (3.5.23).

(II) Let $\alpha \notin \mathbb{N}$, $\alpha > 0$ and $\lceil \alpha \rceil = m$, $a < a^* < b$, $G \in AC^m\left(\left[a, b\right]\right)$, with $0 \ne G^{(m)} \in L_\infty\left(\left[a, b\right]\right)$. Here we consider the left generalized (Caputo type) fractional derivative:

$$\left(D_{*a;g}^\alpha G\right)\left(x\right) = \frac{1}{\Gamma\left(m - \alpha\right)} \int_a^x \left(g\left(x\right) - g\left(t\right)\right)^{m-\alpha-1} g'\left(t\right) G^{(m)}\left(t\right) dt, \tag{3.5.41}$$

for any $x \in \left[a, b\right]$.

By Theorem 3.22 we get that $\left(D_{*a;g}^\alpha G\right) \in C\left(\left[a, b\right]\right)$, in particular $\left(D_{*a;g}^\alpha G\right) \in C\left(\left[a^*, b\right]\right)$. Here notice that $\left(D_{*a;g}^\alpha G\right)\left(a\right) = 0$.

Therefore there exist $x_1, x_2 \in \left[a^*, b\right]$ such that $D_{*a;g}^\alpha G\left(x_1\right) = \min D_{*a;g}^\alpha G\left(x\right)$, and $D_{*a;g}^\alpha G\left(x_2\right) = \max D_{*a;g}^\alpha G\left(x\right)$, for $x \in \left[a^*, b\right]$.

We assume that

$$D_{*a;g}^\alpha G\left(x_1\right) > 0. \tag{3.5.42}$$

(i.e. $D_{*a;g}^\alpha G\left(x\right) > 0$, $\forall\, x \in \left[a^*, b\right]$).

Furthermore

$$\left\| D_{*a;g}^{\alpha} G \right\|_{\infty,[a^*,b]} = D_{*a;g}^{\alpha} G (x_2). \tag{3.5.43}$$

Here it is

$$J (x) = mx, \quad m \neq 0. \tag{3.5.44}$$

The equation

$$JG (x) = 0, \quad x \in \left[a^*, b \right], \tag{3.5.45}$$

has the same set of solutions as the equation

$$F (x) := \frac{JG (x)}{2 D_{*a;g}^{\alpha} G (x_2)} = 0, \quad x \in \left[a^*, b \right]. \tag{3.5.46}$$

Notice that

$$D_{*a;g}^{\alpha} \left(\frac{G (x)}{2 D_{*a;g}^{\alpha} G (x_2)} \right) = \frac{D_{*a;g}^{\alpha} G (x)}{2 D_{*a;g}^{\alpha} G (x_2)} \leq \frac{1}{2} < 1, \quad \forall x \in \left[a^*, b \right]. \tag{3.5.47}$$

We call

$$A (x) := \frac{D_{*a;g}^{\alpha} G (x)}{2 D_{*a;g}^{\alpha} G (x_2)}, \quad \forall x \in \left[a^*, b \right]. \tag{3.5.48}$$

We notice that

$$0 < \frac{D_{*a;g}^{\alpha} G (x_1)}{2 D_{*a;g}^{\alpha} G (x_2)} \leq A (x) \leq \frac{1}{2}. \tag{3.5.49}$$

Hence it holds

$$|1 - A (x)| = 1 - A (x) \leq 1 - \frac{D_{*a;g}^{\alpha} G (x_1)}{2 D_{*a;g}^{\alpha} G (x_2)} =: \gamma_0, \quad \forall x \in \left[a^*, b \right]. \tag{3.5.50}$$

Clearly $\gamma_0 \in (0, 1)$.

We have proved that

$$|1 - A (x)| \leq \gamma_0 \in (0, 1), \quad \forall x \in \left[a^*, b \right]. \tag{3.5.51}$$

Next we assume that $F (x)$ is a contraction over $[a^*, b]$, i.e.

$$|F (x) - F (y)| \leq \lambda |x - y|; \quad \forall x, y \in \left[a^*, b \right], \tag{3.5.52}$$

and $0 < \lambda < \frac{1}{2}$.

Equivalently we have

$$|JG(x) - JG(y)| \leq 2\lambda \left(D_{*a;g}^{\alpha} G(x_2) \right) |x - y|, \quad \forall\, x, y \in [a^*, b]. \quad (3.5.53)$$

We observe that

$$|F(y) - F(x) - A(x)(y - x)| \leq |F(y) - F(x)| + |A(x)| |y - x| \leq$$

$$\lambda |y - x| + |A(x)| |y - x| = (\lambda + |A(x)|) |y - x| =: (\xi_2), \quad \forall\, x, y \in [a^*, b].$$
$$(3.5.54)$$

We observe that

$$\left| D_{*a;g}^{\alpha} G(x) \right| \leq \frac{1}{\Gamma(m - \alpha)} \int_a^x (g(x) - g(t))^{m-\alpha-1} g'(t) \left| G^{(m)}(t) \right| dt$$

$$\leq \frac{1}{\Gamma(m - \alpha)} \left(\int_a^x (g(x) - g(t))^{m-\alpha-1} g'(t)\, dt \right) \left\| G^{(m)} \right\|_{\infty}$$

$$= \frac{1}{\Gamma(m - \alpha)} \frac{(g(x) - g(a))^{m-\alpha}}{(m - \alpha)} \left\| G^{(m)} \right\|_{\infty}$$

$$= \frac{1}{\Gamma(m - \alpha + 1)} (g(x) - g(a))^{m-\alpha} \left\| G^{(m)} \right\|_{\infty} \leq \frac{(g(b) - g(a))^{m-\alpha}}{\Gamma(m - \alpha + 1)} \left\| G^{(m)} \right\|_{\infty}.$$
$$(3.5.55)$$

That is

$$\left| D_{*a;g}^{\alpha} G(x) \right| \leq \frac{(g(b) - g(a))^{m-\alpha}}{\Gamma(m - \alpha + 1)} \left\| G^{(m)} \right\|_{\infty} < \infty, \quad \forall\, x \in [a, b]. \quad (3.5.56)$$

Hence, $\forall\, x \in [a^*, b]$ we get that

$$|A(x)| = \frac{\left| D_{*a;g}^{\alpha} G(x) \right|}{2 D_{*a;g}^{\alpha} G(x_2)} \leq \frac{(g(b) - g(a))^{m-\alpha}}{2\Gamma(m - \alpha + 1)} \frac{\left\| G^{(m)} \right\|_{\infty}}{D_{*a;g}^{\alpha} G(x_2)} < \infty. \quad (3.5.57)$$

Consequently we observe

$$(\xi_2) \leq \left(\lambda + \frac{(g(b) - g(a))^{m-\alpha}}{2\Gamma(m - \alpha + 1)} \frac{\left\| G^{(m)} \right\|_{\infty}}{D_{*a;g}^{\alpha} G(x_2)} \right) |y - x|, \quad \forall\, x, y \in [a^*, b].$$
$$(3.5.58)$$

Call

$$0 < \gamma_1 := \lambda + \frac{(g(b) - g(a))^{m-\alpha}}{2\Gamma(m - \alpha + 1)} \frac{\left\| G^{(m)} \right\|_{\infty}}{D_{*a;g}^{\alpha} G(x_2)}, \quad (3.5.59)$$

choosing $(g(b) - g(a))$ small enough we can make $\gamma_1 \in (0, 1)$.

We proved that

$$|F(y) - F(x) - A(x)(y - x)| \leq \gamma_1 |y - x|, \text{ where } \gamma_1 \in (0, 1), \quad \forall x, y \in [a^*, b].$$
(3.5.60)

Next we call and need

$$0 < \gamma := \gamma_0 + \gamma_1 = 1 - \frac{D_{*a;g}^\alpha G(x_1)}{2D_{*a;g}^\alpha G(x_2)} + \lambda + \frac{(g(b) - g(a))^{m-\alpha}}{2\Gamma(m - \alpha + 1)} \frac{\|G^{(m)}\|_\infty}{D_{*a;g}^\alpha G(x_2)} < 1,$$
(3.5.61)

equivalently we find,

$$\lambda + \frac{(g(b) - g(a))^{m-\alpha}}{2\Gamma(m - \alpha + 1)} \frac{\|G^{(m)}\|_\infty}{D_{*a;g}^\alpha G(x_2)} < \frac{D_{*a;g}^\alpha G(x_1)}{2D_{*a;g}^\alpha G(x_2)},$$
(3.5.62)

equivalently,

$$2\lambda D_{*a;g}^\alpha G(x_2) + \frac{(g(b) - g(a))^{m-\alpha}}{\Gamma(m - \alpha + 1)} \|G^{(m)}\|_\infty < D_{*a;g}^\alpha G(x_1),$$
(3.5.63)

which is possible for small λ, $(g(b) - g(a))$.

That is $\gamma \in (0, 1)$. Hence Eq. (3.5.45) can be solved with our presented numerical methods.

Conclusion:

Our presented earlier semilocal convergence Newton-like general methods, see Theorem 3.12, can apply in the above two generalized fractional settings since the following inequalities have been fulfilled:

$$\|1 - A(x)\|_\infty \leq \gamma_0,$$
(3.5.64)

and

$$|F(y) - F(x) - A(x)(y - x)| \leq \gamma_1 |y - x|,$$
(3.5.65)

where $\gamma_0, \gamma_1 \in (0, 1)$, furthermore it holds

$$\gamma = \gamma_0 + \gamma_1 \in (0, 1),$$
(3.5.66)

for all $x, y \in [a^*, b]$, where $a < a^* < b$.

The specific functions $A(x)$, $F(x)$ have been described above.

References

1. S. Amat, S. Busquier, Third-order iterative methods under Kantorovich conditions. J. Math. Anal. Applic. **336**, 243–261 (2007)
2. S. Amat, S. Busquier, S. Plaza, Chaotic dynamics of a third-order Newton-like method. J. Math. Anal. Applic. **366**(1), 164–174 (2010)
3. G. Anastassiou, *Fractional Differentiation Inequalities* (Springer, New York, 2009)
4. G. Anastassiou, *Intelligent Mathematics Computational Analysis* (Springer, Heidelberg, 2011)
5. G.A. Anastassiou, *Left General Fractional Monotone Approximation Theory* (2015) (submitted)
6. G.A. Anastassiou, *Univariate Left General higher order Fractional Monotone Approximation* (2015) (submitted)
7. G. Anastassiou, I. Argyros, *On the convergence of iterative methods with applications in generalized fractional calculus* (2015) (submitted)
8. I.K. Argyros, Newton-like methods in partially ordered linear spaces. J. Approx. Th. Applic. **9**(1), 1–10 (1993)
9. I.K. Argyros, Results on controlling the residuals of perturbed Newton-like methods on Banach spaces with a convergence structure. Southwest J. Pure Appl. Math. **1**, 32–38 (1995)
10. I.K. Argyros, *Convergence and Applications of Newton-like iterations* (Springer-Verlag Publ, New York, 2008)
11. J.A. Ezquerro, J.M. Gutierrez, M.A. Hernandez, N. Romero, M.J. Rubio, The Newton method: from Newton to Kantorovich (spanish). Gac. R. Soc. Mat. Esp. **13**, 53–76 (2010)
12. A.A. Kilbas, H.M. Srivastava, J.J. Trujillo, *Theory and Applications of Fractional differential equations*. North-Holland Mathematics Studies, vol. 2004 (Elsevier, New York, NY, USA, 2006)
13. A.A. Magrenan, Different anomalies in a Surrutt family of iterative root finding methods. Appl. Math. Comput. **233**, 29–38 (2014)
14. A.A. Magrenan, A new tool to study real dynamics: the convergence plane. Appl. Math. Comput. **248**, 215–224 (2014)
15. P.W. Meyer, *Newton's method in generalized Banach spaces*, Numer. Func. Anal. Optimiz. 9, 3 and 4, 244-259 (1987)
16. F.A. Potra, V. Ptak, *Nondiscrete induction and iterative processes* (Pitman Publ, London, 1984)

Chapter 4
Fixed Point Techniques and Generalized Right Fractional Calculus

We present a fixed point technique for some iterative algorithms on a generalized Banach space setting to approximate a locally unique zero of an operator. Earlier studies such as [8–10, 15] require that the operator involved is Fréchet-differentiable. In the present study we assume that the operator is only continuous. This way we extend the applicability of these methods to include right fractional calculus as well as problems from other areas. Some applications include fractional calculus involving right generalized fractional integral and the right Hadamard fractional integral. Fractional calculus is very important for its applications in many applied sciences. It follows [7].

4.1 Introduction

We present a semilocal convergence analysis for some fixed point iterative algorithms on a generalized Banach space setting to approximate a zero of an operator. The semilocal convergence is, based on the information around an initial point, to give conditions ensuring the convergence of the iterative algorithm. A generalized norm is defined to be an operator from a linear space into a partially order Banach space (to be precised in Sect. 4.2). Earlier studies such as [8–10, 15] for Newton's method have shown that a more precise convergence analysis is obtained when compared to the real norm theory. However, the main assumption is that the operator involved is Fréchet-differentiable. This hypothesis limits the applicability of Newton's method. In the present study using a fixed point technique (see iterative algorithm (4.3.1)), we show convergence by only assuming the continuity of the operator. This way we expand the applicability of these iterative algorithms.

The rest of the chapter is organized as follows: Sect. 4.2 contains the basic concepts on generalized Banach spaces and auxiliary results on inequalities and fixed points. In

© Springer International Publishing Switzerland 2016

G.A. Anastassiou and I.K. Argyros, *Intelligent Numerical Methods:*
Applications to Fractional Calculus, Studies in Computational Intelligence 624,
DOI 10.1007/978-3-319-26721-0_4

Sect. 4.3 we present the semilocal convergence analysis. Finally, in the concluding Sects. 4.4 and 4.5, we present special cases and applications in generalized right fractional calculus.

4.2   Generalized Banach Spaces

We present some standard concepts that are needed in what follows to make the paper as self contained as possible. More details on generalized Banach spaces can be found in [8–10, 15], and the references there in.

Definition 4.1 A generalized Banach space is a triplet $(x, E, / \cdot /)$ such that
(i) X is a linear space over $\mathbb{R}\,(\mathbb{C})$.
(ii) $E = (E, K, \|\cdot\|)$ is a partially ordered Banach space, i.e.
(ii$_1$) $(E, \|\cdot\|)$ is a real Banach space,
(ii$_2$) E is partially ordered by a closed convex cone K,
(ii$_3$) The norm $\|\cdot\|$ is monotone on K.
(iii) The operator $/ \cdot / : X \rightarrow K$ satisfies
$/x/ = 0 \Leftrightarrow x = 0, /\theta x/ = |\theta| /x/,$
$/x + y/ \leq /x/ + /y/$ for each $x, y \in X, \theta \in \mathbb{R}(\mathbb{C})$.
(iv) X is a Banach space with respect to the induced norm $\|\cdot\|_i := \|\cdot\| \cdot / \cdot /.$

Remark 4.2 The operator $/ \cdot /$ is called a generalized norm. In view of (iii) and (ii$_3$) $\|\cdot\|_i$, is a real norm. In the rest of this paper all topological concepts will be understood with respect to this norm.

Let $L\left(X^j, Y\right)$ stand for the space of j-linear symmetric and bounded operators from X^j to Y, where X and Y are Banach spaces. For X, Y partially ordered $L_+\left(X^j, Y\right)$ stands for the subset of monotone operators P such that

$$0 \leq a_i \leq b_i \Rightarrow P\left(a_1, \ldots, a_j\right) \leq P\left(b_1, \ldots, b_j\right). \tag{4.2.1}$$

Definition 4.3 The set of bounds for an operator $Q \in L\,(X, X)$ on a generalized Banach space $(X, E, / \cdot /)$ the set of bounds is defined to be:

$$B\,(Q) := \{P \in L_+\,(E, E),\, /Qx/ \leq P/x/ \text{ for each } x \in X\}. \tag{4.2.2}$$

Let $D \subset X$ and $T : D \rightarrow D$ be an operator. If $x_0 \in D$ the sequence $\{x_n\}$ given by

$$x_{n+1} := T\,(x_n) = T^{n+1}\,(x_0) \tag{4.2.3}$$

is well defined. We write in case of convergence

$$T^\infty (x_0) := \lim \left(T^n (x_0) \right) = \lim_{n \to \infty} x_n. \qquad (4.2.4)$$

We need some auxiliary results on inequations.

Lemma 4.4 *Let $(E, K, \|\cdot\|)$ be a partially ordered Banach space, $\xi \in K$ and $M, N \in L_+ (E, E)$.*

(i) Suppose there exists $r \in K$ such that

$$R(r) := (M + N) r + \xi \le r \qquad (4.2.5)$$

and

$$(M + N)^k r \to 0 \quad as \ k \to \infty. \qquad (4.2.6)$$

Then, $b := R^\infty (0)$ is well defined satisfies the equation $t = R(t)$ and is the smaller than any solution of the inequality $R(s) \le s$.

(ii) Suppose there exist $q \in K$ and $\theta \in (0, 1)$ such that $R(q) \le \theta q$, then there exists $r \le q$ satisfying (i).

Proof (i) Define sequence $\{b_n\}$ by $b_n = R^n (0)$. Then, we have by (4.2.5) that $b_1 = R(0) = \xi \le r \Rightarrow b_1 \le r$. Suppose that $b_k \le r$ for each $k = 1, 2, \ldots, n$. Then, we have by (4.2.5) and the inductive hypothesis that $b_{n+1} = R^{n+1} (0) = R(R^n (0)) = R(b_n) = (M + N) b_n + \xi \le (M + N) r + \xi \le r \Rightarrow b_{n+1} \le r$. Hence, sequence $\{b_n\}$ is bounded above by r. Set $P_n = b_{n+1} - b_n$. We shall show that

$$P_n \le (M + N)^n r \quad for \ each \ n = 1, 2, \ldots \qquad (4.2.7)$$

We have by the definition of P_n and (4.2.6) that

$$P_1 = R^2 (0) - R(0) = R(R(0)) - R(0)$$

$$= R(\xi) - R(0) = \int_0^1 R'(t\xi) \xi dt \le \int_0^1 R'(\xi) \xi dt$$

$$\le \int_0^1 R'(r) r dt \le (M + N) r,$$

which shows (4.2.7) for $n = 1$. Suppose that (4.2.7) is true for $k = 1, 2, \ldots, n$. Then, we have in turn by (4.2.6) and the inductive hypothesis that

$$P_{k+1} = R^{k+2} (0) - R^{k+1} (0) = R^{k+1} (R(0)) - R^{k+1} (0) =$$

$$R^{k+1} (\xi) - R^{k+1} (0) = R \left(R^k (\xi) \right) - R \left(R^k (0) \right) =$$

$$\int_0^1 R' \left(R^k \left(0 \right) + t \left(R^k \left(\xi \right) - R^k \left(0 \right) \right) \right) \left(R^k \left(\xi \right) - R^k \left(0 \right) \right) dt \le$$

$$R' \left(R^k \left(\xi \right) \right) \left(R^k \left(\xi \right) - R^k \left(0 \right) \right) = R' \left(R^k \left(\xi \right) \right) \left(R^{k+1} \left(0 \right) - R^k \left(0 \right) \right) \le$$

$$R' \left(r \right) \left(R^{k+1} \left(0 \right) - R^k \left(0 \right) \right) \le \left(M + N \right) \left(M + N \right)^k r = \left(M + N \right)^{k+1} r,$$

which completes the induction for (4.2.7). It follows that $\{b_n\}$ is a complete sequence in a Banach space and as such it converges to some b. Notice that $R \left(b \right) = R \left(\lim_{n \to \infty} R^n \left(0 \right) \right) = \lim_{n \to \infty} R^{n+1} \left(0 \right) = b \Rightarrow b$ solves the equation $R \left(t \right) = t$. We have that $b_n \le r \Rightarrow b \le r$, where r a solution of $R \left(r \right) \le r$. Hence, b is smaller than any solution of $R \left(s \right) \le s$.

(ii) Define sequences $\{v_n\}$, $\{w_n\}$ by $v_0 = 0$, $v_{n+1} = R \left(v_n \right)$, $w_0 = q$, $w_{n+1} = R \left(w_n \right)$. Then, we have that

$$0 \le v_n \le v_{n+1} \le w_{n+1} \le w_n \le q, \tag{4.2.8}$$

$$w_n - v_n \le \theta^n \left(q - v_n \right)$$

and sequence $\{v_n\}$ is bounded above by q. Hence, it converges to some r with $r \le q$. We also get by (4.2.8) that $w_n - v_n \to 0$ as $n \to \infty \Rightarrow w_n \to r$ as $n \to \infty$. $\square$

We also need the auxiliary result for computing solutions of fixed point problems.

Lemma 4.5 *Let* $(X, (E, K, \|\cdot\|), / \cdot /)$ *be a generalized Banach space, and* $P \in B \left(Q \right)$ *be a bound for* $Q \in L \left(X, X \right)$. *Suppose there exists* $y \in X$ *and* $q \in K$ *such that*

$$Pq + /y/ \le q \text{ and } P^k q \to 0 \text{ as } k \to \infty. \tag{4.2.9}$$

Then, $z = T^\infty \left(0 \right)$, $T \left(x \right) := Qx + y$ *is well defined and satisfies:* $z = Qz + y$ *and* $/z/ \le P/z/ + /y/ \le q$. *Moreover,* z *is the unique solution in the subspace* $\{x \in X | \exists \, \theta \in \mathbb{R} : \{x\} \le \theta q\}$.

The proof can be found in [15, Lemma 3.2].

4.3 Semilocal Convergence

Let $(X, (E, K, \|\cdot\|), / \cdot /)$ and Y be generalized Banach spaces, $D \subset X$ an open subset, $G : D \to Y$ a continuous operator and $A \left(\cdot \right) : D \to L \left(X, Y \right)$. A zero of operator G is to be determined by an iterative algorithm starting at a point $x_0 \in D$. The results are presented for an operator $F = JG$, where $J \in L \left(Y, X \right)$. The iterates are determined through a fixed point problem:

$$x_{n+1} = x_n + y_n, \, A \left(x_n \right) y_n + F \left(x_n \right) = 0 \tag{4.3.1}$$

$$\Leftrightarrow y_n = T \left(y_n \right) := \left(I - A \left(x_n \right) \right) y_n - F \left(x_n \right).$$

Let $U(x_0, r)$ stand for the ball defined by

$$U(x_0, r) := \{x \in X : /x - x_0/ \leq r\}$$

for some $r \in K$.

Next, we present the semilocal convergence analysis of iterative algorithm (4.3.1) using the preceding notation.

Theorem 4.6 *Let* $F : D \subset X$, $A(\cdot) : D \to L(X, Y)$ *and* $x_0 \in D$ *be as defined previously. Suppose:*

(H_1) *There exists an operator* $M \in B(I - A(x))$ *for each* $x \in D$.

(H_2) *There exists an operator* $N \in L_+(E, E)$ *satisfying for each* $x, y \in D$

$$/F(y) - F(x) - A(x)(y - x)/ \leq N/y - x/.$$

(H_3) *There exists a solution* $r \in K$ *of*

$$R_0(t) := (M + N)t + /F(x_0)/ \leq t.$$

(H_4) $U(x_0, r) \subseteq D$.

(H_5) $(M + N)^k r \to 0$ *as* $k \to \infty$.

Then, the following hold:

(C_1) *The sequence* $\{x_n\}$ *defined by*

$$x_{n+1} = x_n + T_n^\infty(0), \; T_n(y) := (I - A(x_n)) y - F(x_n) \qquad (4.3.2)$$

is well defined, remains in $U(x_0, r)$ *for each* $n = 0, 1, 2, \ldots$ *and converges to the unique zero of operator* F *in* $U(x_0, r)$.

(C_2) *An apriori bound is given by the null-sequence* $\{r_n\}$ *defined by* $r_0 := r$ *and for each* $n = 1, 2, \ldots$

$$r_n = P_n^\infty(0), \; P_n(t) = Mt + Nr_{n-1}.$$

(C_3) *An aposteriori bound is given by the sequence* $\{s_n\}$ *defined by*

$$s_n := R_n^\infty(0), \; R_n(t) = (M + N)t + Na_{n-1},$$

$$b_n := /x_n - x_0/ \leq r - r_n \leq r,$$

where

$$a_{n-1} := /x_n - x_{n-1}/ \; \text{for each } n = 1, 2, \ldots$$

Proof Let us define for each $n \in \mathbb{N}$ the statement:
(I_n) $x_n \in X$ and $r_n \in K$ are well defined and satisfy

$$r_n + a_{n-1} \leq r_{n-1}.$$

We use induction to show (I_n). The statement (I_1) is true: By Lemma 4.4 and (H_3), (H_5) there exists $q \leq r$ such that:

$$Mq + /F(x_0)/ = q \text{ and } M^k q \leq M^k r \to 0 \quad \text{as } k \to \infty.$$

Hence, by Lemma 4.5 x_1 is well defined and we have $a_0 \leq q$. Then, we get the estimate

$$P_1(r - q) = M(r - q) + Nr_0$$

$$\leq Mr - Mq + Nr = R_0(r) - q$$

$$\leq R_0(r) - q = r - q.$$

It follows with Lemma 4.4 that r_1 is well defined and

$$r_1 + a_0 \leq r - q + q = r = r_0.$$

Suppose that (I_j) is true for each $j = 1, 2, \ldots, n$. We need to show the existence of x_{n+1} and to obtain a bound q for a_n. To achieve this notice that:

$$Mr_n + N(r_{n-1} - r_n) = Mr_n + Nr_{n-1} - Nr_n = P_n(r_n) - Nr_n \leq r_n.$$

Then, it follows from Lemma 4.4 that there exists $q \leq r_n$ such that

$$q = Mq + N(r_{n-1} - r_n) \text{ and } (M + N)^k q \to 0, \quad \text{as } k \to \infty. \tag{4.3.3}$$

By (I_j) it follows that

$$b_n = /x_n - x_0/ \leq \sum_{j=0}^{n-1} a_j \leq \sum_{j=0}^{n-1} (r_j - r_{j+1}) = r - r_n \leq r.$$

Hence, $x_n \in U(x_0, r) \subset D$ and by (H_1) M is a bound for $I - A(x_n)$.
We can write by (H_2) that

$$/F(x_n)/ = /F(x_n) - F(x_{n-1}) - A(x_{n-1})(x_n - x_{n-1})/$$

$$\leq Na_{n-1} \leq N(r_{n-1} - r_n). \tag{4.3.4}$$

It follows from (4.3.3) and (4.3.4) that

$$Mq + /F(x_n)/ \leq q.$$

By Lemma 4.5, x_{n+1} is well defined and $a_n \leq q \leq r_n$. In view of the definition of r_{n+1} we have that

$$P_{n+1}(r_n - q) = P_n(r_n) - q = r_n - q,$$

so that by Lemma 4.4, r_{n+1} is well defined and

$$r_{n+1} + a_n \leq r_n - q + q = r_n,$$

which proves (I_{n+1}). The induction for (I_n) is complete. Let $m \geq n$, then we obtain in turn that

$$/x_{m+1} - x_n/ \leq \sum_{j=n}^{m} a_j \leq \sum_{j=n}^{m}(r_j - r_{j+1}) = r_n - r_{m+1} \leq r_n. \qquad (4.3.5)$$

Moreover, we get inductively the estimate

$$r_{n+1} = P_{n+1}(r_{n+1}) \leq P_{n+1}(r_n) \leq (M + N)r_n \leq \cdots \leq (M + N)^{n+1} r.$$

It follows from (H_5) that $\{r_n\}$ is a null-sequence. Hence, $\{x_n\}$ is a complete sequence in a Banach space X by (4.3.5) and as such it converges to some $x^* \in X$. By letting $m \to \infty$ in (4.3.5) we deduce that $x^* \in U(x_n, r_n)$. Furthermore, (4.3.4) shows that x^* is a zero of F. Hence, (C_1) and (C_2) are proved.

In view of the estimate

$$R_n(r_n) \leq P_n(r_n) \leq r_n$$

the apriori, bound of (C_3) is well defined by Lemma 4.4. That is s_n is smaller in general than r_n. The conditions of Theorem 4.6 are satisfied for x_n replacing x_0. A solution of the inequality of (C_2) is given by s_n (see (4.3.4)). It follows from (4.3.5) that the conditions of Theorem 4.6 are easily verified. Then, it follows from (C_1) that $x^* \in U(x_n, s_n)$ which proves (C_3). $\qquad \square$

In general the aposterior, estimate is of interest. Then, condition (H_5) can be avoided as follows:

Proposition 4.7 *Suppose: condition (H_1) of Theorem 4.6 is true.*
(H_3') There exists $s \in K$, $\theta \in (0, 1)$ such that

$$R_0(s) = (M + N)s + /F(x_0)/ \leq \theta s.$$

(H_4') $U(x_0, s) \subset D$.

Then, there exists $r \leq s$ satisfying the conditions of Theorem 4.6. Moreover, the zero x^* of F is unique in $U(x_0, s)$.

Remark 4.8 (i) Notice that by Lemma 4.4 $R_n^\infty(0)$ is the smallest solution of $R_n(s) \leq s$. Hence any solution of this inequality yields on upper estimate for $R_n^\infty(0)$. Similar inequalities appear in (H_2) and (H_2').

(ii) The weak assumptions of Theorem 4.6 do not imply the existence of $A(x_n)^{-1}$. In practice the computation of $T_n^\infty(0)$ as a solution of a linear equation is no problem and the computation of the expensive or impossible to compute in general $A(x_n)^{-1}$ is not needed.

(iii) We can used the following result for the computation of the aposteriori estimates. The proof can be found in [15, Lemma 4.2] by simply exchanging the definitions of R.

Lemma 4.9 *Suppose that the conditions of Theorem 4.6 are satisfied. If $s \in K$ is a solution of $R_n(s) \leq s$, then $q := s - a_n \in K$ and solves $R_{n+1}(q) \leq q$. This solution might be improved by $R_{n+1}^k(q) \leq q$ for each $k = 1, 2, \ldots$..*

4.4 Special Cases and Applications

Application 4.10 *The results obtained in earlier studies such as [8–10, 15] require that operator F (i.e. G) is Fréchet-differentiable. This assumption limits the applicability of the earlier results. In the present study we only require that F is a continuous operator. Hence, we have extended the applicability of the iterative algorithms include to classes of operators that are only continuous. If $A(x) = F'(x)$ iterative algorithm (4.3.1) reduces to Newton's method considered in [15].*

Example 4.11 The j-dimensional space $\mathbb{R}^j$ is a classical example of a generalized Banach space. The generalized norm is defined by componentwise absolute values. Then, as ordered Banach space we set $E = \mathbb{R}^j$ with componentwise ordering with e.g. the maximum norm. A bound for a linear operator (a matrix) is given by the corresponding matrix with absolute values. Similarly, we can define the "N" operators. Let $E = \mathbb{R}$. That is we consider the case of a real normed space with norm denoted by $\|\cdot\|$. Let us see how the conditions of Theorem 4.6 look like.

Theorem 4.12 $(H_1) \|I - A(x)\| \leq M$ for some $M \geq 0$.

$(H_2) \|F(y) - F(x) - A(x)(y - x)\| \leq N \|y - x\|$ for some $N \geq 0$.

$(H_3) M + N < 1$,

$$r = \frac{\|F(x_0)\|}{1 - (M + N)}. \tag{4.4.1}$$

$(H_4) U(x_0, r) \subseteq D$.

$(H_5) (M + N)^k r \to 0$ as $k \to \infty$, where r is given by (4.4.1).

Then, the conclusions of Theorem 4.6 hold.

4.5 Applications to Generalized Right Fractional Calculus

Background

We use Theorem 4.12 in this section.

We use here the following right generalized fractional integral.

Definition 4.13 (*see also* [12, p. 99]) The right generalized fractional integral of a function f with respect to given function g is defined as follows:

Let $a, b \in \mathbb{R}, a < b, \alpha > 0$. Here $g \in AC([a, b])$ (absolutely continuous functions) and is striclty increasing, $f \in L_\infty([a, b])$. We set

$$\left(I^\alpha_{b-;g} f\right)(x) = \frac{1}{\Gamma(\alpha)} \int_x^b (g(t) - g(x))^{\alpha-1} g'(t) f(t) \, dt, \quad x \leq b, \quad (4.5.1)$$

clearly $\left(I^\alpha_{b-;g} f\right)(b) = 0$.

When g is the identity function id, we get that $I^\alpha_{b-;id} = I^\alpha_{b-}$, the ordinary right Riemann-Liouville fractional integral, where

$$\left(I^\alpha_{b-} f\right)(x) = \frac{1}{\Gamma(\alpha)} \int_x^b (t - x)^{\alpha-1} f(t) \, dt, \quad x \leq b, \quad (4.5.2)$$

$\left(I^\alpha_{b-} f\right)(b) = 0$.

When, $g(x) = \ln x$ on $[a, b], 0 < a < b < \infty$, we get:

Definition 4.14 ([12, p. 110]) Let $0 < a < b < \infty, \alpha > 0$. The right Hadamard fractional integral of order α is given by

$$\left(J^\alpha_{b-} f\right)(x) = \frac{1}{\Gamma(\alpha)} \int_x^b \left(\ln \frac{y}{x}\right)^{\alpha-1} \frac{f(y)}{y} dy, \quad x \leq b, \quad (4.5.3)$$

where $f \in L_\infty([a, b])$.

We mention:

Definition 4.15 ([5]) The right fractional exponential integral is defined as follows: Let $a, b \in \mathbb{R}, a < b, \alpha > 0, f \in L_\infty([a, b])$. We set

$$\left(I^\alpha_{b-;e^x} f\right)(x) = \frac{1}{\Gamma(\alpha)} \int_x^b \left(e^t - e^x\right)^{\alpha-1} e^t f(t) \, dt, \quad x \leq b. \quad (4.5.4)$$

Definition 4.16 ([5]) Let $a, b \in \mathbb{R}, a < b, \alpha > 0, f \in L_\infty([a, b]), A > 1$. We give the right fractional integral

$$\left(I^\alpha_{b-;A^x} f\right)(x) = \frac{\ln A}{\Gamma(\alpha)} \int_x^b \left(A^t - A^x\right)^{\alpha-1} A^t f(t) \, dt, \quad x \leq b. \quad (4.5.5)$$

We also give:

Definition 4.17 ([5]) Let $\alpha, \sigma > 0, 0 \leq a < b < \infty, f \in L_\infty ([a, b])$. We set

$$\left(K_{b-;x^\sigma}^\alpha f\right)(x) = \frac{1}{\Gamma(\alpha)} \int_x^b (t^\sigma - x^\sigma)^{\alpha-1} f(t) \sigma t^{\sigma-1} dt, \quad x \leq b. \qquad (4.5.6)$$

We mention the following generalized right fractional derivatives.

Definition 4.18 ([5]) Let $\alpha > 0$ and $\lceil \alpha \rceil = m$ ($\lceil \cdot \rceil$ ceiling of the number). Consider $f \in AC^m ([a, b])$ (space of functions f with $f^{(m-1)} \in AC ([a, b])$). We define the right generalized fractional derivative of f of order α as follows

$$\left(D_{b-;g}^\alpha f\right)(x) = \frac{(-1)^m}{\Gamma(m-\alpha)} \int_x^b (g(t) - g(x))^{m-\alpha-1} g'(t) f^{(m)}(t) dt, \quad (4.5.7)$$

for any $x \in [a, b]$, where Γ is the gamma function.
We set

$$D_{b-;g}^m f(x) = (-1)^m f^{(m)}(x), \qquad (4.5.8)$$

$$D_{b-;g}^0 f(x) = f(x), \quad \forall x \in [a, b]. \qquad (4.5.9)$$

When $g = id$, then $D_{b-}^\alpha f = D_{b-;id}^\alpha f$ is the right Caputo fractional derivative.

So we have the specific generalized right fractional derivatives.

Definition 4.19 ([5])

$$D_{b-;\ln x}^\alpha f(x) = \frac{(-1)^m}{\Gamma(m-\alpha)} \int_x^b \left(\ln \frac{y}{x}\right)^{m-\alpha-1} \frac{f^{(m)}(y)}{y} dy, \quad 0 < a \leq x \leq b,$$
$$(4.5.10)$$

$$D_{b-;e^x}^\alpha f(x) = \frac{(-1)^m}{\Gamma(m-\alpha)} \int_x^b \left(e^t - e^x\right)^{m-\alpha-1} e^t f^{(m)}(t) dt, \quad a \leq x \leq b,$$
$$(4.5.11)$$

and

$$D_{b-;A^x}^\alpha f(x) = \frac{(-1)^m \ln A}{\Gamma(m-\alpha)} \int_x^b \left(A^t - A^x\right)^{m-\alpha-1} A^t f^{(m)}(t) dt, \quad a \leq x \leq b,$$
$$(4.5.12)$$

$$\left(D_{b-;x^\sigma}^\alpha f\right)(x) = \frac{(-1)^m}{\Gamma(m-\alpha)} \int_x^b (t^\sigma - x^\sigma)^{m-\alpha-1} \sigma t^{\sigma-1} f^{(m)}(t) dt, \quad 0 \leq a \leq x \leq b.$$
$$(4.5.13)$$

We make:

Remark 4.20 ([5]) Here $g \in AC([a, b])$ (absolutely continuous functions), g is increasing over $[a, b]$, $\alpha > 0$. Then

$$\int_x^b (g(t) - g(x))^{\alpha-1} g'(t) \, dt = \frac{(g(b) - g(x))^\alpha}{\alpha}, \quad \forall x \in [a, b]. \quad (4.5.14)$$

Finally we will use

Theorem 4.21 ([5]) *Let* $\alpha > 0$, $\mathbb{N} \ni m = \lceil \alpha \rceil$, *and* $f \in C^m([a, b])$. *Then* $\left(D_{b-;g}^\alpha f\right)(x)$ *is continuous in* $x \in [a, b]$, $-\infty < a < b < \infty$.

Results
(I) We notice the following ($a \leq x \leq b$):

$$\left|(I_{b-;g}^\alpha f)(x)\right| \leq \frac{1}{\Gamma(\alpha)} \int_x^b (g(t) - g(x))^{\alpha-1} g'(t) \, |f(t)| \, dt \quad (4.5.15)$$

$$\leq \frac{\|f\|_\infty}{\Gamma(\alpha)} \int_x^b (g(t) - g(x))^{\alpha-1} g'(t) \, dt = \frac{\|f\|_\infty}{\Gamma(\alpha)} \frac{(g(b) - g(x))^\alpha}{\alpha}$$

$$= \frac{\|f\|_\infty}{\Gamma(\alpha+1)} (g(b) - g(x))^\alpha \leq \frac{\|f\|_\infty}{\Gamma(\alpha+1)} (g(b) - g(a))^\alpha. \quad (4.5.16)$$

In particular it holds

$$\left(I_{b-;g}^\alpha f\right)(b) = 0, \quad (4.5.17)$$

and

$$\left\|I_{b-;g}^\alpha f\right\|_{\infty,[a,b]} \leq \frac{(g(b) - g(a))^\alpha}{\Gamma(\alpha+1)} \|f\|_\infty, \quad (4.5.18)$$

proving that $I_{b-;g}^\alpha$ is a bounded linear operator.
We use:

Theorem 4.22 ([6]) *Let* $r > 0$, $a < b$, $F \in L_\infty([a, b])$, $g \in AC([a, b])$ *and* g *is strictly increasing.*
Consider

$$B(s) := \int_s^b (g(t) - g(s))^{r-1} g'(t) F(t) \, dt, \quad \text{for all } s \in [a, b]. \quad (4.5.19)$$

Then $B \in C([a, b])$.

By Theorem 4.22, the function $\left(I_{b-;g}^\alpha f\right)$ is a continuous function over $[a, b]$. Consider $a < b^* < b$. Therefore $\left(I_{b-;g}^\alpha f\right)$ is also continuous over $[a, b^*]$.

Thus, there exist $x_1, x_2 \in [a, b^*]$ such that

$$\left(I_{b-;g}^{\alpha} f\right)(x_1) = \min \left(I_{b-;g}^{\alpha} f\right)(x), \tag{4.5.20}$$

$$\left(I_{b-;g}^{\alpha} f\right)(x_2) = \max \left(I_{b-;g}^{\alpha} f\right)(x), \quad \text{where } x \in [a, b^*]. \tag{4.5.21}$$

We assume that

$$\left(I_{b-;g}^{\alpha} f\right)(x_1) > 0. \tag{4.5.22}$$

Hence

$$\left\| I_{b-;g}^{\alpha} f \right\|_{\infty,[a,b^*]} = \left(I_{b-;g}^{\alpha} f\right)(x_2) > 0. \tag{4.5.23}$$

Here it is

$$J(x) = mx, \quad m \neq 0. \tag{4.5.24}$$

Therefore the equation

$$Jf(x) = 0, \quad x \in [a, b^*], \tag{4.5.25}$$

has the same solutions as the equation

$$F(x) := \frac{Jf(x)}{2\left(I_{b-;g}^{\alpha} f\right)(x_2)} = 0, \quad x \in [a, b^*]. \tag{4.5.26}$$

Notice that

$$I_{b-;g}^{\alpha} \left(\frac{f}{2\left(I_{b-;g}^{\alpha} f\right)(x_2)} \right)(x) = \frac{\left(I_{b-;g}^{\alpha} f\right)(x)}{2\left(I_{b-;g}^{\alpha} f\right)(x_2)} \leq \frac{1}{2} < 1, \quad x \in [a, b^*]. \tag{4.5.27}$$

Call

$$A(x) := \frac{\left(I_{b-;g}^{\alpha} f\right)(x)}{2\left(I_{b-;g}^{\alpha} f\right)(x_2)}, \quad \forall\, x \in [a, b^*]. \tag{4.5.28}$$

We notice that

$$0 < \frac{\left(I_{b-;g}^{\alpha} f\right)(x_1)}{2\left(I_{b-;g}^{\alpha} f\right)(x_2)} \leq A(x) \leq \frac{1}{2}, \quad \forall\, x \in [a, b^*]. \tag{4.5.29}$$

We observe

$$|1 - A(x)| = 1 - A(x) \le 1 - \frac{\left(I_{b-;g}^{\alpha} f\right)(x_1)}{2\left(I_{b-;g}^{\alpha} f\right)(x_2)} =: \gamma_0, \quad \forall x \in [a, b^*]. \quad (4.5.30)$$

Clearly $\gamma_0 \in (0, 1)$.
  I.e.

$$|1 - A(x)| \le \gamma_0, \quad \forall x \in [a, b^*], \gamma_0 \in (0, 1). \quad (4.5.31)$$

Next we assume that $F(x)$ is a contraction, i.e.

$$|F(x) - F(y)| \le \lambda |x - y|; \quad \forall x, y \in [a, b^*], \quad (4.5.32)$$

and $0 < \lambda < \frac{1}{2}$.
  Equivalently we have

$$|Jf(x) - Jf(y)| \le 2\lambda \left(I_{b-;g}^{\alpha} f\right)(x_2)|x - y|, \quad \text{all } x, y \in [a, b^*]. \quad (4.5.33)$$

We observe that

$$|F(y) - F(x) - A(x)(y - x)| \le |F(y) - F(x)| + |A(x)||y - x| \le$$

$$\lambda |y - x| + |A(x)||y - x| = (\lambda + |A(x)|)|y - x| =: (\psi_1), \quad \forall x, y \in [a, b^*]. \quad (4.5.34)$$

By (4.5.18) we get

$$\left|\left(I_{b-;g}^{\alpha} f\right)(x)\right| \le \frac{\|f\|_{\infty}}{\Gamma(\alpha+1)} (g(b) - g(a))^{\alpha}, \quad \forall x \in [a, b^*]. \quad (4.5.35)$$

Hence

$$|A(x)| = \frac{\left|\left(I_{b-;g}^{\dot{\alpha}} f\right)(x)\right|}{2\left(I_{b-;g}^{\alpha} f\right)(x_2)} \le \frac{\|f\|_{\infty} (g(b) - g(a))^{\alpha}}{2\Gamma(\alpha+1)\left(I_{b-;g}^{\alpha} f\right)(x_2)} < \infty, \quad \forall x \in [a, b^*]. \quad (4.5.36)$$

Therefore we get

$$(\psi_1) \le \left(\lambda + \frac{\|f\|_{\infty} (g(b) - g(a))^a}{2\Gamma(\alpha+1)\left(I_{b-;g}^{\alpha} f\right)(x_2)}\right)|y - x|, \quad \forall x, y \in [a, b^*]. \quad (4.5.37)$$

Call

$$0 < \gamma_1 := \lambda + \frac{\|f\|_\infty (g(b) - g(a))^a}{2\Gamma(\alpha+1)\left(I^\alpha_{b-;g}f\right)(x_2)}, \tag{4.5.38}$$

choosing $(g(b) - g(a))$ small enough we can make $\gamma_1 \in (0, 1)$.

We have proved that

$$|F(y) - F(x) - A(x)(y - x)| \le \gamma_1 |y - x|, \quad \forall\, x, y \in [a, b^*], \gamma_1 \in (0, 1). \tag{4.5.39}$$

Next we call and we need that

$$0 < \gamma := \gamma_0 + \gamma_1 = 1 - \frac{\left(I^\alpha_{b-;g}f\right)(x_1)}{2\left(I^\alpha_{b-;g}f\right)(x_2)} + \lambda + \frac{\|f\|_\infty (g(b) - g(a))^a}{2\Gamma(\alpha+1)\left(I^\alpha_{b-;g}f\right)(x_2)} < 1, \tag{4.5.40}$$

$$\lambda + \frac{\|f\|_\infty (g(b) - g(a))^a}{2\Gamma(\alpha+1)\left(I^\alpha_{b-;g}f\right)(x_2)} < \frac{\left(I^\alpha_{b-;g}f\right)(x_1)}{2\left(I^\alpha_{b-;g}f\right)(x_2)}, \tag{4.5.41}$$

equivalently,

$$2\lambda\left(I^\alpha_{b-;g}f\right)(x_2) + \frac{\|f\|_\infty (g(b) - g(a))^a}{\Gamma(\alpha+1)} < \left(I^\alpha_{b-;g}f\right)(x_1), \tag{4.5.42}$$

which is possible for small λ, and small $(g(b) - g(a))$. That is $\gamma \in (0, 1)$. So our method solves (4.5.25).

(II) Let $\alpha \notin \mathbb{N}$, $\alpha > 0$ and $\lceil \alpha \rceil = m$, $a < b^* < b$, $G \in AC^m([a, b])$, with $0 \ne G^{(m)} \in L_\infty([a, b])$. Here we consider the right generalized (Caputo type) fractional derivative:

$$\left(D^\alpha_{b-;g}G\right)(x) = \frac{(-1)^m}{\Gamma(m - \alpha)} \int_x^b (g(t) - g(x))^{m-\alpha-1} g'(t) G^{(m)}(t)\, dt, \tag{4.5.43}$$

for any $x \in [a, b]$.

By Theorem 4.22 we get that $\left(D^\alpha_{b-;g}G\right) \in C([a, b])$, in particular $\left(D^\alpha_{b-;g}G\right)$ $\in C([a, b^*])$. Here notice that $\left(D^\alpha_{b-;g}G\right)(b) = 0$.

Therefore there exist $x_1, x_2 \in [a^*, b]$ such that $D^\alpha_{b-;g}G(x_1) = \min D^\alpha_{b-;g}G(x)$, and $D^\alpha_{b-;g}G(x_2) = \max D^\alpha_{b-;g}G(x)$, for $x \in [a, b^*]$.

We assume that

$$D^\alpha_{b-;g}G(x_1) > 0. \tag{4.5.44}$$

(i.e. $D^\alpha_{b-;g}G(x) > 0, \forall\, x \in [a, b^*]$).

Furthermore

$$\left\| D_{b-;g}^{\alpha} G \right\|_{\infty,[a,b^*]} = D_{b-;g}^{\alpha} G(x_2). \tag{4.5.45}$$

Here it is

$$J(x) = mx, \quad m \neq 0. \tag{4.5.46}$$

The equation

$$JG(x) = 0, \quad x \in [a, b^*], \tag{4.5.47}$$

has the same set of solutions as the equation

$$F(x) := \frac{JG(x)}{2 D_{b-;g}^{\alpha} G(x_2)} = 0, \quad x \in [a, b^*]. \tag{4.5.48}$$

Notice that

$$D_{b-;g}^{\alpha} \left(\frac{G(x)}{2 D_{b-;g}^{\alpha} G(x_2)} \right) = \frac{D_{b-;g}^{\alpha} G(x)}{2 D_{b-;g}^{\alpha} G(x_2)} \leq \frac{1}{2} < 1, \quad \forall x \in [a, b^*]. \tag{4.5.49}$$

We call

$$A(x) := \frac{D_{b-;g}^{\alpha} G(x)}{2 D_{b-;g}^{\alpha} G(x_2)}, \quad \forall x \in [a, b^*]. \tag{4.5.50}$$

We notice that

$$0 < \frac{D_{b-;g}^{\alpha} G(x_1)}{2 D_{b-;g}^{\alpha} G(x_2)} \leq A(x) \leq \frac{1}{2}. \tag{4.5.51}$$

Hence it holds

$$|1 - A(x)| = 1 - A(x) \leq 1 - \frac{D_{b-;g}^{\alpha} G(x_1)}{2 D_{b-;g}^{\alpha} G(x_2)} =: \gamma_0, \quad \forall x \in [a, b^*]. \tag{4.5.52}$$

Clearly $\gamma_0 \in (0, 1)$.

We have proved that

$$|1 - A(x)| \leq \gamma_0 \in (0, 1), \quad \forall x \in [a, b^*]. \tag{4.5.53}$$

Next we assume that $F(x)$ is a contraction over $[a, b^*]$, i.e.

$$|F(x) - F(y)| \leq \lambda |x - y|; \quad \forall x, y \in [a, b^*], \tag{4.5.54}$$

and $0 < \lambda < \frac{1}{2}$.

Equivalently we have

$$|JG(x) - JG(y)| \leq 2\lambda \left(D^\alpha_{b-;g} G(x_2) \right) |x - y|, \quad \forall x, y \in [a, b^*]. \quad (4.5.55)$$

We observe that

$$|F(y) - F(x) - A(x)(y - x)| \leq |F(y) - F(x)| + |A(x)||y - x| \leq$$

$$\lambda |y - x| + |A(x)||y - x| = (\lambda + |A(x)|)|y - x| =: (\xi_2), \quad \forall x, y \in [a, b^*]. \quad (4.5.56)$$

We observe that

$$\left| D^\alpha_{b-;g} G(x) \right| \leq \frac{1}{\Gamma(m-\alpha)} \int_x^b (g(t) - g(x))^{m-\alpha-1} g'(t) \left| G^{(m)}(t) \right| dt$$

$$\leq \frac{1}{\Gamma(m-\alpha)} \left(\int_x^b (g(t) - g(x))^{m-\alpha-1} g'(t) dt \right) \left\| G^{(m)} \right\|_\infty$$

$$= \frac{1}{\Gamma(m-\alpha)} \frac{(g(b) - g(x))^{m-\alpha}}{(m-\alpha)} \left\| G^{(m)} \right\|_\infty$$

$$= \frac{1}{\Gamma(m-\alpha+1)} (g(b) - g(x))^{m-\alpha} \left\| G^{(m)} \right\|_\infty \leq \frac{(g(b) - g(a))^{m-\alpha}}{\Gamma(m-\alpha+1)} \left\| G^{(m)} \right\|_\infty. \quad (4.5.57)$$

That is

$$\left| D^\alpha_{b-;g} G(x) \right| \leq \frac{(g(b) - g(a))^{m-\alpha}}{\Gamma(m-\alpha+1)} \left\| G^{(m)} \right\|_\infty < \infty, \quad \forall x \in [a, b]. \quad (4.5.58)$$

Hence, $\forall x \in [a, b^*]$ we get that

$$|A(x)| = \frac{\left| D^\alpha_{b-;g} G(x) \right|}{2 D^\alpha_{b-;g} G(x_2)} \leq \frac{(g(b) - g(a))^{m-\alpha}}{2\Gamma(m-\alpha+1)} \frac{\left\| G^{(m)} \right\|_\infty}{D^\alpha_{b-;g} G(x_2)} < \infty. \quad (4.5.59)$$

Consequently we observe

$$(\xi_2) \leq \left(\lambda + \frac{(g(b) - g(a))^{m-\alpha}}{2\Gamma(m-\alpha+1)} \frac{\left\| G^{(m)} \right\|_\infty}{D^\alpha_{b-;g} G(x_2)} \right) |y - x|, \quad \forall x, y \in [a, b^*]. \quad (4.5.60)$$

Call

$$0 < \gamma_1 := \lambda + \frac{(g(b) - g(a))^{m-\alpha}}{2\Gamma(m-\alpha+1)} \frac{\left\| G^{(m)} \right\|_\infty}{D^\alpha_{b-;g} G(x_2)}, \quad (4.5.61)$$

choosing $(g(b) - g(a))$ small enough we can make $\gamma_1 \in (0, 1)$.

We proved that

$$|F(y) - F(x) - A(x)(y - x)| \leq \gamma_1 |y - x|, \text{ where } \gamma_1 \in (0, 1), \quad \forall x, y \in [a, b^*].$$
$$(4.5.62)$$

Next we call and need

$$0 < \gamma := \gamma_0 + \gamma_1 = 1 - \frac{D_{b-;g}^\alpha G(x_1)}{2 D_{b-;g}^\alpha G(x_2)} + \lambda + \frac{(g(b) - g(a))^{m-\alpha}}{2\Gamma(m - \alpha + 1)} \frac{\left\| G^{(m)} \right\|_\infty}{D_{b-;g}^\alpha G(x_2)} < 1,$$
$$(4.5.63)$$

equivalently we find,

$$\lambda + \frac{(g(b) - g(a))^{m-\alpha}}{2\Gamma(m - \alpha + 1)} \frac{\left\| G^{(m)} \right\|_\infty}{D_{b-;g}^\alpha G(x_2)} < \frac{D_{b-;g}^\alpha G(x_1)}{2 D_{b-;g}^\alpha G(x_2)}, \qquad (4.5.64)$$

equivalently,

$$2\lambda D_{b-;g}^\alpha G(x_2) + \frac{(g(b) - g(a))^{m-\alpha}}{\Gamma(m - \alpha + 1)} \left\| G^{(m)} \right\|_\infty < D_{b-;g}^\alpha G(x_1), \qquad (4.5.65)$$

which is possible for small λ, $(g(b) - g(a))$.

That is $\gamma \in (0, 1)$. Hence Eq. (4.5.47) can be solved with our presented iterative algorithms.

Conclusion:

Our presented earlier semilocal fixed point iterative algorithms, see Theorem 4.12, can apply in the above two generalized fractional settings since the following inequalities have been fulfilled:

$$\|1 - A(x)\|_\infty \leq \gamma_0, \qquad (4.5.66)$$

and

$$|F(y) - F(x) - A(x)(y - x)| \leq \gamma_1 |y - x|, \qquad (4.5.67)$$

where $\gamma_0, \gamma_1 \in (0, 1)$, furthermore it holds

$$\gamma = \gamma_0 + \gamma_1 \in (0, 1), \qquad (4.5.68)$$

for all $x, y \in [a, b^*]$, where $a < b^* < b$.

The specific functions $A(x)$, $F(x)$ have been described above.

References

1. S. Amat, S. Busquier, Third-order iterative methods under Kantorovich conditions. J. Math. Anal. Appl. **336**, 243–261 (2007)
2. S. Amat, S. Busquier, S. Plaza, Chaotic dynamics of a third-order Newton-like method. J. Math. Anal. Appl. **366**(1), 164–174 (2010)
3. G. Anastassiou, *Fractional Differentiation Inequalities* (Springer, New York, 2009)
4. G. Anastassiou, *Intelligent Mathematics Computational Analysis* (Springer, Heidelberg, 2011)
5. G.A. Anastassiou, *Right General Fractional Monotone Approximation Theory* (2015) (submitted)
6. G.A. Anastassiou, *Univariate Right General higher order Fractional Monotone Approximation* (2015) (submitted)
7. G. Anastassiou, I. Argyros, *A fixed point technique for some iterative algorithm with applications to generalized right fractional calculus* (2015) (submitted)
8. I.K. Argyros, Newton-like methods in partially ordered linear spaces. J. Approx. Th. Applic. **9**(1), 1–10 (1993)
9. I.K. Argyros, Results on controlling the residuals of perturbed Newton-like methods on Banach spaces with a convergence structure. Southwest J. Pure Appl. Math. **1**, 32–38 (1995)
10. I.K. Argyros, *Convergence and Applications of Newton-like iterations* (Springer-Verlag Publ, New York, 2008)
11. J.A. Ezquerro, J.M. Gutierrez, M.A. Hernandez, N. Romero, M.J. Rubio, The Newton method: from Newton to Kantorovich (spanish). Gac. R. Soc. Mat. Esp. **13**, 53–76 (2010)
12. A.A. Kilbas, H.M. Srivastava, J.J. Trujillo, *Theory and Applications of Fractional differential equations*. North-Holland Mathematics Studies, vol. 2004 (Elsevier, New York, NY, USA, 2006)
13. A.A. Magrenan, Different anomalies in a Surrutt family of iterative root finding methods. Appl. Math. Comput. **233**, 29–38 (2014)
14. A.A. Magrenan, A new tool to study real dynamics: the convergence plane. Appl. Math. Comput. **248**, 215–224 (2014)
15. P.W. Meyer, *Newton's method in generalized Banach spaces*, Numer. Func. Anal. Optimiz. **9**, **3, 4**, 244-259 (1987)
16. F.A. Potra, V. Ptak, *Nondiscrete induction and iterative processes* (Pitman Publ, London, 1984)

Chapter 5
Approximating Fixed Points and k-Fractional Calculus

We approximate fixed points of some iterative methods on a generalized Banach space setting. Earlier studies such as [6–8, 13] require that the operator involved is Fréchet-differentiable. In the present study we assume that the operator is only continuous. This way we extend the applicability of these methods to include generalized fractional calculus and problems from other areas. Some applications include generalized fractional calculus involving the Riemann-Liouville fractional integral and the Caputo fractional derivative. Fractional calculus is very important for its applications in many applied sciences. It follows [5].

5.1 Introduction

Many problems in Computational sciences can be formulated as an operator equation using Mathematical Modelling [8, 11, 14–16]. The fixed points of these operators can rarely be found in closed form. That is why most solution methods are usually iterative. The semilocal convergence is, based on the information around an initial point, to give conditions ensuring the convergence of the method.

We present a semilocal convergence analysis for some iterative methods on a generalized Banach space setting to approximate fixed point or a zero of an operator. A generalized norm is defined to be an operator from a linear space into a partially order Banach space (to be precised in Sect. 5.2). Earlier studies such as [6–8, 13] for Newton's method have shown that a more precise convergence analysis is obtained when compared to the real norm theory. However, the main assumption is that the operator involved is Fréchet-differentiable. This hypothesis limits the applicability of Newton's method. In the present study we only assume the continuity of the operator. This may be expand the applicability of these methods.

© Springer International Publishing Switzerland 2016   75
G.A. Anastassiou and I.K. Argyros, *Intelligent Numerical Methods:*
Applications to Fractional Calculus, Studies in Computational Intelligence 624,
DOI 10.1007/978-3-319-26721-0_5

The rest of the chapter is organized as follows: Sect. 5.2 contains the basic concepts on generalized Banach spaces and auxiliary results on inequalities and fixed points. In Sect. 5.3 we present the semilocal convergence analysis of these methods. Finally, in the concluding Sects. 5.4 and 5.5, we present special cases and applications in generalized fractional calculus.

5.2 Generalized Banach Spaces

We present some standard concepts that are needed in what follows to make the paper as self contained as possible. More details on generalized Banach spaces can be found in [6–8, 13], and the references there in.

Definition 5.1 A generalized Banach space is a triplet $(x, E, / \cdot /)$ such that
  (i) X is a linear space over $\mathbb{R}(\mathbb{C})$.
  (ii) $E = (E, K, \|\cdot\|)$ is a partially ordered Banach space, i.e.
  (ii$_1$) $(E, \|\cdot\|)$ is a real Banach space,
  (ii$_2$) E is partially ordered by a closed convex cone K,
  (ii$_3$) The norm $\|\cdot\|$ is monotone on K.
  (iii) The operator $/ \cdot / : X \to K$ satisfies
  $/x/ = 0 \Leftrightarrow x = 0, /\theta x/ = |\theta| /x/,$
  $/x + y/ \leq /x/ + /y/$ for each $x, y \in X, \theta \in \mathbb{R}(\mathbb{C})$.
  (iv) X is a Banach space with respect to the induced norm $\|\cdot\|_i := \|\cdot\| \cdot / \cdot /$.

Remark 5.2 The operator $/ \cdot /$ is called a generalized norm. In view of (iii) and (ii$_3$) $\|\cdot\|_i$, is a real norm. In the rest of this paper all topological concepts will be understood with respect to this norm.

Let $L\left(X^j, Y\right)$ stand for the space of j-linear symmetric and bounded operators from X^j to Y, where X and Y are Banach spaces. For X, Y partially ordered $L_+\left(X^j, Y\right)$ stands for the subset of monotone operators P such that

$$0 \leq a_i \leq b_i \Rightarrow P\left(a_1, \ldots, a_j\right) \leq P\left(b_1, \ldots, b_j\right). \tag{5.2.1}$$

Definition 5.3 The set of bounds for an operator $Q \in L(X, X)$ on a generalized Banach space $(X, E, / \cdot /)$ the set of bounds is defined to be:

$$B(Q) := \{P \in L_+(E, E), /Qx/ \leq P/x/ \text{ for each } x \in X\}. \tag{5.2.2}$$

Let $D \subset X$ and $T : D \to D$ be an operator. If $x_0 \in D$ the sequence $\{x_n\}$ given by

$$x_{n+1} := T(x_n) = T^{n+1}(x_0) \tag{5.2.3}$$

is well defined. We write in case of convergence

$$T^\infty (x_0) := \lim \left(T^n (x_0) \right) = \lim_{n \to \infty} x_n. \qquad (5.2.4)$$

We need some auxiliary results on inequations.

Lemma 5.4 *Let $(E, K, \|\cdot\|)$ be a partially ordered Banach space, $\xi \in K$ and $M, N \in L_+ (E, E)$.*
  (i) Suppose there exist $r \in K$ such that

$$R (r) := (M + N) r + \xi \leq r \qquad (5.2.5)$$

and

$$(M + N)^k r \to 0 \quad as \; k \to \infty. \qquad (5.2.6)$$

Then, $b := R^\infty (0)$ is well defined satisfies the equation $t = R (t)$ and is the smaller than any solution of the inequality $R (s) \leq s$.
  (ii) Suppose there exist $q \in K$ and $\theta \in (0, 1)$ such that $R (q) \leq \theta q$, then there exists $r \leq q$ satisfying (i).

Proof (i) Define sequence $\{b_n\}$ by $b_n = R^n (0)$. Then, we have by (5.2.5) that $b_1 = R (0) = \xi \leq r \Rightarrow b_1 \leq r$. Suppose that $b_k \leq r$ for each $k = 1, 2, \ldots, n$. Then, we have by (5.2.5) and the inductive hypothesis that $b_{n+1} = R^{n+1} (0) = R (R^n (0)) = R (b_n) = (M + N) b_n + \xi \leq (M + N) r + \xi \leq r \Rightarrow b_{n+1} \leq r$. Hence, sequence $\{b_n\}$ is bounded above by r. Set $P_n = b_{n+1} - b_n$. We shall show that

$$P_n \leq (M + N)^n r \quad \text{for each } n = 1, 2, \ldots \qquad (5.2.7)$$

We have by the definition of P_n and (5.2.6) that

$$P_1 = R^2 (0) - R (0) = R (R (0)) - R (0)$$

$$= R (\xi) - R (0) = \int_0^1 R' (t\xi) \xi dt \leq \int_0^1 R' (\xi) \xi dt$$

$$\leq \int_0^1 R' (r) r dt \leq (M + N) r,$$

which shows (5.2.7) for $n = 1$. Suppose that (5.2.7) is true for $k = 1, 2, \ldots, n$. Then, we have in turn by (5.2.6) and the inductive hypothesis that

$$P_{k+1} = R^{k+2} (0) - R^{k+1} (0) = R^{k+1} (R (0)) - R^{k+1} (0) =$$

$$R^{k+1} (\xi) - R^{k+1} (0) = R (R^k (\xi)) - R (R^k (0)) =$$

$$\int_0^1 R'\left(R^k\left(0\right) + t\left(R^k\left(\xi\right) - R^k\left(0\right)\right)\right)\left(R^k\left(\xi\right) - R^k\left(0\right)\right)dt \le$$

$$R'\left(R^k\left(\xi\right)\right)\left(R^k\left(\xi\right) - R^k\left(0\right)\right) = R'\left(R^k\left(\xi\right)\right)\left(R^{k+1}\left(0\right) - R^k\left(0\right)\right) \le$$

$$R'\left(r\right)\left(R^{k+1}\left(0\right) - R^k\left(0\right)\right) \le \left(M + N\right)\left(M + N\right)^k r = \left(M + N\right)^{k+1} r,$$

which completes the induction for (5.2.7). It follows that $\{b_n\}$ is a complete sequence in a Banach space and as such it converges to some b. Notice that $R\left(b\right) = R\left(\lim_{n \to \infty} R^n\left(0\right)\right) = \lim_{n \to \infty} R^{n+1}\left(0\right) = b \Rightarrow b$ solves the equation $R\left(t\right) = t$. We have that $b_n \le r \Rightarrow b \le r$, where r a solution of $R\left(r\right) \le r$. Hence, b is smaller than any solution of $R\left(s\right) \le s$.

(ii) Define sequences $\{v_n\}$, $\{w_n\}$ by $v_0 = 0$, $v_{n+1} = R\left(v_n\right)$, $w_0 = q$, $w_{n+1} = R\left(w_n\right)$. Then, we have that

$$0 \le v_n \le v_{n+1} \le w_{n+1} \le w_n \le q, \qquad (5.2.8)$$
$$w_n - v_n \le \theta^n\left(q - v_n\right)$$

and sequence $\{v_n\}$ is bounded above by q. Hence, it converges to some r with $r \le q$. We also get by (5.2.8) that $w_n - v_n \to 0$ as $n \to \infty \Rightarrow w_n \to r$ as $n \to \infty$. $\square$

We also need the auxiliary result for computing solutions of fixed point problems.

Lemma 5.5 *Let* $(X, (E, K, \|\cdot\|), / \cdot /)$ *be a generalized Banach space, and* $P \in B\left(Q\right)$ *be a bound for* $Q \in L\left(X, X\right)$. *Suppose there exists* $y \in X$ *and* $q \in K$ *such that*

$$Pq + /y/ \le q \text{ and } P^k q \to 0 \quad \text{as } k \to \infty. \qquad (5.2.9)$$

Then, $z = T^\infty\left(0\right)$, $T\left(x\right) := Qx + y$ *is well defined and satisfies:* $z = Qz + y$ *and* $/z/ \le P/z/ + /y/ \le q$. *Moreover,* z *is the unique solution in the subspace* $\{x \in X | \exists\, \theta \in \mathbb{R} : \{x\} \le \theta q\}$.

The proof can be found in [13, Lemma 3.2].

5.3 Semilocal Convergence

Let $(X, (E, K, \|\cdot\|), / \cdot /)$ and Y be generalized Banach spaces, $D \subset X$ an open subset, $G : D \to Y$ a continuous operator and $A\left(\cdot\right) : D \to L\left(X, Y\right)$. A zero of operator G is to be determined by a method starting at a point $x_0 \in D$. The results are presented for an operator $F = JG$, where $J \in L\left(Y, X\right)$. The iterates are determined through a fixed point problem:

$$x_{n+1} = x_n + y_n, \; A\left(x_n\right)y_n + F\left(x_n\right) = 0 \qquad (5.3.1)$$
$$\Leftrightarrow y_n = T\left(y_n\right) := \left(I - A\left(x_n\right)\right)y_n - F\left(x_n\right).$$

Let $U(x_0, r)$ stand for the ball defined by

$$U(x_0, r) := \{x \in X : /x - x_0/ \le r\}$$

for some $r \in K$.

Next, we present the semilocal convergence analysis of method (5.3.1) using the preceding notation.

Theorem 5.6 *Let* $F : D \subset X$, $A(\cdot) : D \to L(X, Y)$ *and* $x_0 \in D$ *be as defined previously. Suppose:*

(H_1) There exists an operator $M \in B(I - A(x))$ *for each* $x \in D$.

(H_2) There exists an operator $N \in L_+(E, E)$ *satisfying for each* $x, y \in D$

$$/F(y) - F(x) - A(x)(y - x)/ \le N/y - x/.$$

(H_3) There exists a solution $r \in K$ *of*

$$R_0(t) := (M + N)t + /F(x_0)/ \le t.$$

(H_4) $U(x_0, r) \subseteq D$.

(H_5) $(M + N)^k r \to 0$ as $k \to \infty$.

Then, the following hold:

(C_1) The sequence $\{x_n\}$ defined by

$$x_{n+1} = x_n + T_n^\infty(0), \quad T_n(y) := (I - A(x_n))y - F(x_n) \qquad (5.3.2)$$

is well defined, remains in $U(x_0, r)$ *for each* $n = 0, 1, 2, \ldots$ *and converges to the unique zero of operator F in $U(x_0, r)$.*

(C_2) An apriori bound is given by the null-sequence $\{r_n\}$ defined by $r_0 := r$ and for each $n = 1, 2, \ldots$

$$r_n = P_n^\infty(0), \quad P_n(t) = Mt + Nr_{n-1}.$$

(C_3) An aposteriori bound is given by the sequence $\{s_n\}$ defined by

$$s_n := R_n^\infty(0), \quad R_n(t) = (M + N)t + Na_{n-1},$$

$$b_n := /x_n - x_0/ \le r - r_n \le r,$$

where

$$a_{n-1} := /x_n - x_{n-1}/ \quad \text{for each } n = 1, 2, \ldots$$

Proof Let us define for each $n \in \mathbb{N}$ the statement:

(I_n) $x_n \in X$ and $r_n \in K$ are well defined and satisfy

$$r_n + a_{n-1} \leq r_{n-1}.$$

We use induction to show (I_n). The statement (I_1) is true: By Lemma 5.4 and (H_3), (H_5) there exists $q \leq r$ such that:

$$Mq + /F(x_0)/ = q \text{ and } M^k q \leq M^k r \to 0 \quad \text{as } k \to \infty.$$

Hence, by Lemma 5.5 x_1 is well defined and we have $a_0 \leq q$. Then, we get the estimate

$$P_1(r - q) = M(r - q) + Nr_0$$

$$\leq Mr - Mq + Nr = R_0(r) - q$$

$$\leq R_0(r) - q = r - q.$$

It follows with Lemma 5.4 that r_1 is well defined and

$$r_1 + a_0 \leq r - q + q = r = r_0.$$

Suppose that (I_j) is true for each $j = 1, 2, \ldots, n$. We need to show the existence of x_{n+1} and to obtain a bound q for a_n. To achieve this notice that:

$$Mr_n + N(r_{n-1} - r_n) = Mr_n + Nr_{n-1} - Nr_n = P_n(r_n) - Nr_n \leq r_n.$$

Then, it follows from Lemma 5.4 that there exists $q \leq r_n$ such that

$$q = Mq + N(r_{n-1} - r_n) \text{ and } (M + N)^k q \to 0, \text{ as } k \to \infty. \tag{5.3.3}$$

By (I_j) it follows that

$$b_n = /x_n - x_0/ \leq \sum_{j=0}^{n-1} a_j \leq \sum_{j=0}^{n-1} (r_j - r_{j+1}) = r - r_n \leq r.$$

Hence, $x_n \in U(x_0, r) \subset D$ and by (H_1) M is a bound for $I - A(x_n)$.

We can write by (H_2) that

$$/F(x_n)/ = /F(x_n) - F(x_{n-1}) - A(x_{n-1})(x_n - x_{n-1})/$$

$$\leq Na_{n-1} \leq N(r_{n-1} - r_n). \tag{5.3.4}$$

It follows from (5.3.3) and (5.3.4) that

$$Mq + /F(x_n)/ \leq q.$$

By Lemma 5.5, x_{n+1} is well defined and $a_n \leq q \leq r_n$. In view of the definition of r_{n+1} we have that

$$P_{n+1}(r_n - q) = P_n(r_n) - q = r_n - q,$$

so that by Lemma 5.4, r_{n+1} is well defined and

$$r_{n+1} + a_n \leq r_n - q + q = r_n,$$

which proves (I_{n+1}). The induction for (I_n) is complete. Let $m \geq n$, then we obtain in turn that

$$/x_{m+1} - x_n/ \leq \sum_{j=n}^{m} a_j \leq \sum_{j=n}^{m} (r_j - r_{j+1}) = r_n - r_{m+1} \leq r_n. \qquad (5.3.5)$$

Moreover, we get inductively the estimate

$$r_{n+1} = P_{n+1}(r_{n+1}) \leq P_{n+1}(r_n) \leq (M+N)r_n \leq \cdots \leq (M+N)^{n+1} r.$$

It follows from (H_5) that $\{r_n\}$ is a null-sequence. Hence, $\{x_n\}$ is a complete sequence in a Banach space X by (5.3.5) and as such it converges to some $x^* \in X$. By letting $m \to \infty$ in (5.3.5) we deduce that $x^* \in U(x_n, r_n)$. Furthermore, (5.3.4) shows that x^* is a zero of F. Hence, (C_1) and (C_2) are proved.

In view of the estimate

$$R_n(r_n) \leq P_n(r_n) \leq r_n$$

the apriori, bound of (C_3) is well defined by Lemma 5.4. That is s_n is smaller in general than r_n. The conditions of Theorem 5.6 are satisfied for x_n replacing x_0. A solution of the inequality of (C_2) is given by s_n (see (5.3.4)). It follows from (5.3.5) that the conditions of Theorem 5.6 are easily verified. Then, it follows from (C_1) that $x^* \in U(x_n, s_n)$ which proves (C_3). $\qquad \square$

In general the aposterior, estimate is of interest. Then, condition (H_5) can be avoided as follows:

Proposition 5.7 *Suppose: condition (H_1) of Theorem 5.6 is true.*
(H'_3) There exist $s \in K$, $\theta \in (0, 1)$ such that

$$R_0(s) = (M+N)s + /F(x_0)/ \leq \theta s.$$

(H'_4) $U(x_0, s) \subset D$.

Then, there exists $r \leq s$ satisfying the conditions of Theorem 5.6. Moreover, the zero x^* of F is unique in $U(x_0, s)$.

Remark 5.8 (i) Notice that by Lemma 5.4 $R_n^\infty(0)$ is the smallest solution of $R_n(s) \leq s$. Hence any solution of this inequality yields on upper estimate for $R_n^\infty(0)$. Similar inequalities appear in (H_2) and (H_2').

(ii) The weak assumptions of Theorem 5.6 do not imply the existence of $A(x_n)^{-1}$. In practice the computation of $T_n^\infty(0)$ as a solution of a linear equation is no problem and the computation of the expensive or impossible to compute in general $A(x_n)^{-1}$ is not needed.

(iii) We can used the following result for the computation of the aposteriori estimates. The proof can be found in [13, Lemma 4.2] by simply exchanging the definitions of R.

Lemma 5.9 *Suppose that the conditions of Theorem 5.6 are satisfied. If $s \in K$ is a solution of $R_n(s) \leq s$, then $q := s - a_n \in K$ and solves $R_{n+1}(q) \leq q$. This solution might be improved by $R_{n+1}^k(q) \leq q$ for each $k = 1, 2, \ldots$*

5.4   Special Cases and Applications

Application 5.10 *The results obtained in earlier studies such as [6–8, 13] require that operator F (i.e. G) is Fréchet-differentiable. This assumption limits the applicability of the earlier results. In the present study we only require that F is a continuous operator. Hence, we have extended the applicability of these methods to include classes of operators that are only continuous.*

Example 5.11 The j-dimensional space $\mathbb{R}^j$ is a classical example of a generalized Banach space. The generalized norm is defined by componentwise absolute values. Then, as ordered Banach space we set $E = \mathbb{R}^j$ with componentwise ordering with e.g. the maximum norm. A bound for a linear operator (a matrix) is given by the corresponding matrix with absolute values. Similarly, we can define the "N" operators. Let $E = \mathbb{R}$. That is we consider the case of a real normed space with norm denoted by $\|\cdot\|$. Let us see how the conditions of Theorem 5.6 look like.

Theorem 5.12 $(H_1) \|I - A(x)\| \leq M$ *for some $M \geq 0$.*
   $(H_2) \|F(y) - F(x) - A(x)(y - x)\| \leq N\|y - x\|$ *for some $N \geq 0$.*
   $(H_3) M + N < 1$,

$$r = \frac{\|F(x_0)\|}{1 - (M + N)}. \tag{5.4.1}$$

   $(H_4) U(x_0, r) \subseteq D$.
   $(H_5) (M + N)^k r \to 0$ *as $k \to \infty$, where r is given by (5.4.1).*
   Then, the conclusions of Theorem 5.6 hold.

5.5 Applications to *k*-Fractional Calculus

Background

We apply Theorem 5.12 in this section.

Let $f \in L_\infty([a, b])$, the k-left Riemann-Liouville fractional integral ([16]) of order $\alpha > 0$ is defined as follows:

$$_k J_{a+}^\alpha f(x) = \frac{1}{k\Gamma_k(\alpha)} \int_a^x (x - t)^{\frac{\alpha}{k} - 1} f(t)\, dt, \tag{5.5.1}$$

all $x \in [a, b]$, where $k > 0$, and $\Gamma_k(\alpha)$ is the k-gamma function given by $\Gamma_k(\alpha) = \int_0^\infty t^{\alpha-1} e^{-\frac{t^k}{k}}\, dt$.

It holds ([4]) $\Gamma_k(\alpha + k) = \alpha\Gamma_k(\alpha)$, $\Gamma(\alpha) = \lim_{k \to 1} \Gamma_k(\alpha)$, and we set $_k J_{a+}^0 f(x) = f(x)$.

Similarly, we define the k-right Riemann-Liouville fractional integral as

$$_k J_{b-}^\alpha f(x) = \frac{1}{k\Gamma_k(\alpha)} \int_x^b (t - x)^{\frac{\alpha}{k} - 1} f(t)\, dt, \tag{5.5.2}$$

for all $x \in [a, b]$, and we set $_k J_{b-}^0 f(x) = f(x)$.

Results

(I) Here we work with $_k J_{a+}^\alpha f(x)$. We observe that

$$\left| _k J_{a+}^\alpha f(x) \right| \le \frac{1}{k\Gamma_k(\alpha)} \int_a^x (x - t)^{\frac{\alpha}{k} - 1} |f(t)|\, dt$$

$$\le \frac{\|f\|_\infty}{k\Gamma_k(\alpha)} \int_a^x (x - t)^{\frac{\alpha}{k} - 1} dt = \frac{\|f\|_\infty}{k\Gamma_k(\alpha)} \frac{(x - a)^{\frac{\alpha}{k}}}{\frac{\alpha}{k}} \tag{5.5.3}$$

$$= \frac{\|f\|_\infty}{\Gamma_k(\alpha + k)} (x - a)^{\frac{\alpha}{k}} \le \frac{\|f\|_\infty}{\Gamma_k(\alpha + k)} (b - a)^{\frac{\alpha}{k}}.$$

We have proved that

$$_k J_{a+}^\alpha f(a) = 0, \tag{5.5.4}$$

and

$$\left\| _k J_{a+}^\alpha f \right\|_\infty \le \frac{(b - a)^{\frac{\alpha}{k}}}{\Gamma_k(\alpha + k)} \|f\|_\infty, \tag{5.5.5}$$

proving that $_k J_{a+}^\alpha$ is a bounded linear operator.

By [3], p. 388, we get that $\left(_k J_{a+}^\alpha f \right)$ is a continuous function over $[a, b]$ and in particular continuous over $[a^*, b]$, where $a < a^* < b$.

Thus, there exist x_1, $x_2 \in [a^*, b]$ such that

$$\left({_kJ_{a+}^\alpha f} \right)(x_1) = \min \left({_kJ_{a+}^\alpha f} \right)(x), \tag{5.5.6}$$

$$\left({_kJ_{a+}^\alpha f} \right)(x_2) = \max \left({_kJ_{a+}^\alpha f} \right)(x), \quad x \in \left[a^*, b \right]. \tag{5.5.7}$$

We assume that

$$\left({_kJ_{a+}^\alpha f} \right)(x_1) > 0. \tag{5.5.8}$$

Hence

$$\left\| {_kJ_{a+}^\alpha f} \right\|_{\infty, [a^*, b]} = \left({_kJ_{a+}^\alpha f} \right)(x_2) > 0. \tag{5.5.9}$$

Here it is

$$J(x) = mx, \quad m \neq 0. \tag{5.5.10}$$

Therefore the equation

$$Jf(x) = 0, \quad x \in \left[a^*, b \right], \tag{5.5.11}$$

has the same solutions as the equation

$$F(x) := \frac{Jf(x)}{2 \left({_kJ_{a+}^\alpha f} \right)(x_2)} = 0, \quad x \in \left[a^*, b \right]. \tag{5.5.12}$$

Notice that

$${_kJ_{a+}^\alpha} \left(\frac{f}{2 \left({_kJ_{a+}^\alpha f} \right)(x_2)} \right)(x) = \frac{\left({_kJ_{a+}^\alpha f} \right)(x)}{2 \left({_kJ_{a+}^\alpha f} \right)(x_2)} \leq \frac{1}{2} < 1, \quad x \in \left[a^*, b \right]. \tag{5.5.13}$$

Call

$$A(x) := \frac{\left({_kJ_{a+}^\alpha f} \right)'(x)}{2 \left({_kJ_{a+}^\alpha f} \right)(x_2)}, \quad \forall\, x \in \left[a^*, b \right]. \tag{5.5.14}$$

We notice that

$$0 < \frac{\left({_kJ_{a+}^\alpha f} \right)(x_1)}{2 \left({_kJ_{a+}^\alpha f} \right)(x_2)} \leq A(x) \leq \frac{1}{2}, \quad \forall\, x \in \left[a^*, b \right]. \tag{5.5.15}$$

Hence it holds

$$|1 - A(x)| = 1 - A(x) \leq 1 - \frac{\left({_kJ_{a+}^\alpha f} \right)(x_1)}{2 \left({_kJ_{a+}^\alpha f} \right)(x_2)} =: \gamma_0, \quad \forall\, x \in \left[a^*, b \right]. \tag{5.5.16}$$

Clearly $\gamma_0 \in (0, 1)$.

We have proved that

$$|1 - A(x)| < \gamma_0, \quad \forall x \in [a^*, b].$$ (5.5.17)

Next we assume that $F(x)$ is a contraction, i.e.

$$|F(x) - F(y)| \leq \lambda |x - y|; \quad \forall x, y \in [a^*, b],$$ (5.5.18)

and $0 < \lambda < \frac{1}{2}$.

Equivalently we have

$$|Jf(x) - Jf(y)| \leq 2\lambda \left({}_k J^\alpha_{a+} f \right)(x_2) |x - y|, \quad \text{all } x, y \in [a^*, b].$$ (5.5.19)

We observe that

$$|F(y) - F(x) - A(x)(y - x)| \leq |F(y) - F(x)| + |A(x)| |y - x| \leq$$

$$\lambda |y - x| + |A(x)| |y - x| = (\lambda + |A(x)|) |y - x| =: (\psi_1), \quad \forall x, y \in [a^*, b].$$ (5.5.20)

We have that

$$\left| \left({}_k J^\alpha_{a+} f \right)(x) \right| \leq \frac{(b - a)^{\frac{\alpha}{k}}}{\Gamma_k(\alpha + k)} \|f\|_\infty < \infty, \quad \forall x \in [a^*, b].$$ (5.5.21)

Hence

$$|A(x)| = \frac{\left| \left({}_k J^\alpha_{a+} f \right)(x) \right|}{2 \left({}_k J^\alpha_{a+} f \right)(x_2)} \leq \frac{(b - a)^{\frac{\alpha}{k}} \|f\|_\infty}{2\Gamma_k(\alpha + k) \left({}_k J^\alpha_{a+} f \right)(x_2)} < \infty, \quad \forall x \in [a^*, b].$$ (5.5.22)

Therefore we get

$$(\psi_1) \leq \left(\lambda + \frac{(b - a)^{\frac{\alpha}{k}} \|f\|_\infty}{2\Gamma_k(\alpha + k) \left({}_k J^\alpha_{a+} f \right)(x_2)} \right) |y - x|, \quad \forall x, y \in [a^*, b].$$ (5.5.23)

Call

$$0 < \gamma_1 := \lambda + \frac{(b - a)^{\frac{\alpha}{k}} \|f\|_\infty}{2\Gamma_k(\alpha + k) \left({}_k J^\alpha_{a+} f \right)(x_2)},$$ (5.5.24)

choosing $(b - a)$ small enough we can make $\gamma_1 \in (0, 1)$.

We have proved that

$$|F(y) - F(x) - A(x)(y - x)| \leq \gamma_1 |y - x|, \quad \forall x, y \in [a^*, b], \gamma_1 \in (0, 1).$$ (5.5.25)

Next we call and we need that

$$0 < \gamma := \gamma_0 + \gamma_1 = 1 - \frac{\left(_k J_{a+}^\alpha f\right)(x_1)}{2\left(_k J_{a+}^\alpha f\right)(x_2)} + \lambda + \frac{(b-a)^{\frac{\alpha}{k}} \|f\|_\infty}{2\Gamma_k(\alpha+k)\left(_k J_{a+}^\alpha f\right)(x_2)} < 1,$$

$$(5.5.26)$$

equivalently,

$$\lambda + \frac{(b-a)^{\frac{\alpha}{k}} \|f\|_\infty}{2\Gamma_k(\alpha+k)\left(_k J_{a+}^\alpha f\right)(x_2)} < \frac{\left(_k J_{a+}^\alpha f\right)(x_1)}{2\left(_k J_{a+}^\alpha f\right)(x_2)}, \qquad (5.5.27)$$

equivalently,

$$2\lambda \left(_k J_{a+}^\alpha f\right)(x_2) + \frac{(b-a)^{\frac{\alpha}{k}} \|f\|_\infty}{\Gamma_k(\alpha+k)} < \left(_k J_{a+}^\alpha f\right)(x_1), \qquad (5.5.28)$$

which is possible for small λ, $(b-a)$. That is $\gamma \in (0, 1)$. So our numerical method converges and solves (5.5.11).

II) Here we act on $_k J_{b-}^\alpha f(x)$, see (5.5.2).

Let $f \in L_\infty([a, b])$. We have that

$$\left|_k J_{b-}^\alpha f(x)\right| \le \frac{1}{k\Gamma_k(\alpha)} \int_x^b (t-x)^{\frac{\alpha}{k}-1} |f(t)| \, dt$$

$$\le \frac{\|f\|_\infty}{k\Gamma_k(\alpha)} \int_x^b (t-x)^{\frac{\alpha}{k}-1} \, dt = \frac{\|f\|_\infty}{k\Gamma_k(\alpha)} \frac{(b-x)^{\frac{\alpha}{k}}}{\frac{\alpha}{k}}$$

$$= \frac{\|f\|_\infty}{\Gamma_k(\alpha+k)} (b-x)^{\frac{\alpha}{k}} \le \frac{\|f\|_\infty}{\Gamma_k(\alpha+k)} (b-a)^{\frac{\alpha}{k}}. \qquad (5.5.29)$$

We observe that

$$_k J_{b-}^\alpha f(b) = 0, \qquad (5.5.30)$$

and

$$\left\|_k J_{b-}^\alpha f\right\|_\infty \le \frac{(b-a)^{\frac{\alpha}{k}}}{\Gamma_k(\alpha+k)} \|f\|_\infty. \qquad (5.5.31)$$

That is $_k J_{b-}^\alpha$ is a bounded linear operator.

Let here $a < b^* < b$.

By [4] we get that $_k J_{b-}^\alpha f$ is continuous over $[a, b]$, and in particular it is continuous over $[a, b^*]$.

Thus, there exist $x_1, x_2 \in [a, b^*]$ such that

$$\left(_k J_{b-}^\alpha f\right)(x_1) = \min\left(_k J_{b-}^\alpha f\right)(x), \qquad (5.5.32)$$

$$\left(_k J_{b-}^\alpha f\right)(x_2) = \max\left(_k J_{b-}^\alpha f\right)(x), x \in [a, b^*].$$

We assume that
$$\left({}_kJ^\alpha_{b-}f\right)(x_1) > 0. \tag{5.5.33}$$

Hence
$$\left\|{}_kJ^\alpha_{b-}f\right\|_{\infty,[a^*,b]} = \left({}_kJ^\alpha_{b-}f\right)(x_2) > 0. \tag{5.5.34}$$

Here it is
$$J(x) = mx, \quad m \neq 0. \tag{5.5.35}$$

Therefore the equation
$$Jf(x) = 0, \ x \in [a, b^*], \tag{5.5.36}$$

has the same solutions as the equation
$$F(x) := \frac{Jf(x)}{2\left({}_kJ^\alpha_{b-}f\right)(x_2)} = 0, \quad x \in [a, b^*]. \tag{5.5.37}$$

Notice that
$${}_kJ^\alpha_{b-}\left(\frac{f}{2\left({}_kJ^\alpha_{b-}f\right)(x_2)}\right)(x) = \frac{\left({}_kJ^\alpha_{b-}f\right)(x)}{2\left({}_kJ^\alpha_{b-}f\right)(x_2)} \leq \frac{1}{2} < 1, \quad x \in [a, b^*]. \tag{5.5.38}$$

Call
$$A(x) := \frac{\left({}_kJ^\alpha_{b-}f\right)(x)}{2\left({}_kJ^\alpha_{b-}f\right)(x_2)}, \quad \forall\, x \in [a, b^*]. \tag{5.5.39}$$

We notice that
$$0 < \frac{\left({}_kJ^\alpha_{b-}f\right)(x_1)}{2\left({}_kJ^\alpha_{b-}f\right)(x_2)} \leq A(x) \leq \frac{1}{2}, \quad \forall\, x \in [a, b^*]. \tag{5.5.40}$$

Hence we have
$$|1 - A(x)| = 1 - A(x) \leq 1 - \frac{\left({}_kJ^\alpha_{b-}f\right)(x_1)}{2\left({}_kJ^\alpha_{b-}f\right)(x_2)} =: \gamma_0, \quad \forall\, x \in [a, b^*]. \tag{5.5.41}$$

Clearly $\gamma_0 \in (0, 1)$.

We have proved that
$$|1 - A(x)| \leq \gamma_0, \quad \forall\, x \in [a, b^*], \gamma_0 \in (0, 1). \tag{5.5.42}$$

Next we assume that $F(x)$ is a contraction, i.e.

$$|F(x) - F(y)| \leq \lambda |x - y|; \quad \forall\, x, y \in [a, b^*], \qquad (5.5.43)$$

and $0 < \lambda < \frac{1}{2}$.

Equivalently we have

$$|Jf(x) - Jf(y)| \leq 2\lambda \left({}_k J_{b-}^\alpha f\right)(x_2) |x - y|, \quad \text{all } x, y \in [a, b^*]. \qquad (5.5.44)$$

We observe that

$$|F(y) - F(x) - A(x)(y - x)| \leq |F(y) - F(x)| + |A(x)| |y - x| \leq$$

$$\lambda |y - x| + |A(x)| |y - x| = (\lambda + |A(x)|) |y - x| =: (\psi_1), \quad \forall\, x, y \in [a, b^*].$$
$$(5.5.45)$$

We have that

$$\left|\left({}_k J_{b-}^\alpha f\right)(x)\right| \leq \frac{(b-a)^{\frac{\alpha}{k}}}{\Gamma_k(\alpha + k)} \|f\|_\infty < \infty, \quad \forall\, x \in [a, b^*]. \qquad (5.5.46)$$

Hence

$$|A(x)| = \frac{\left|\left({}_k J_{b-}^\alpha f\right)(x)\right|}{2\left({}_k J_{b-}^\alpha f\right)(x_2)} \leq \frac{(b-a)^{\frac{\alpha}{k}} \|f\|_\infty}{2\Gamma_k(\alpha + k)\left({}_k J_{b-}^\alpha f\right)(x_2)} < \infty, \quad \forall\, x \in [a, b^*].$$
$$(5.5.47)$$

Therefore we get

$$(\psi_1) \leq \left(\lambda + \frac{(b-a)^{\frac{\alpha}{k}} \|f\|_\infty}{2\Gamma_k(\alpha + k)\left({}_k J_{b-}^\alpha f\right)(x_2)}\right) |y - x|, \quad \forall\, x, y \in [a, b^*]. \quad (5.5.48)$$

Call

$$0 < \gamma_1 := \lambda + \frac{(b-a)^{\frac{\alpha}{k}} \|f\|_\infty}{2\Gamma_k(\alpha + k)\left({}_k J_{b-}^\alpha f\right)(x_2)}, \qquad (5.5.49)$$

choosing $(b - a)$ small enough we can make $\gamma_1 \in (0, 1)$.

We have proved that

$$|F(y) - F(x) - A(x)(y - x)| \leq \gamma_1 |y - x|, \quad \forall\, x, y \in [a, b^*], \gamma_1 \in (0, 1).$$
$$(5.5.50)$$

Next we call and we need that

$$0 < \gamma := \gamma_0 + \gamma_1 = 1 - \frac{\left({}_k J_{b-}^\alpha f\right)(x_1)}{2\left({}_k J_{b-}^\alpha f\right)(x_2)} + \lambda + \frac{(b-a)^{\frac{\alpha}{k}} \|f\|_\infty}{2\Gamma_k(\alpha + k)\left({}_k J_{b-}^\alpha f\right)(x_2)} < 1,$$
$$(5.5.51)$$

equivalently,

$$\lambda + \frac{(b-a)^{\frac{a}{k}} \|f\|_\infty}{2\Gamma_k(\alpha+k)(_kJ_{b-}^\alpha f)(x_2)} < \frac{(_kJ_b^\alpha f)(x_1)}{2(_kJ_{b-}^\alpha f)(x_2)}, \tag{5.5.52}$$

equivalently,

$$2\lambda(_kJ_{b-}^\alpha f)(x_2) + \frac{(b-a)^{\frac{\alpha}{k}} \|f\|_\infty}{\Gamma_k(\alpha+k)} < (_kJ_{b-}^\alpha f)(x_1), \tag{5.5.53}$$

which is possible for small λ, $(b-a)$. That is $\gamma \in (0, 1)$. So our numerical method converges and solves (5.5.36).

(III) Here we deal with the fractional M. Caputo-Fabrizio derivative defined as follows (see [10]):

let $0 < \alpha < 1$, $f \in C^1([0, b])$,

$$^{CF}D_*^\alpha f(t) = \frac{1}{1-\alpha} \int_0^t \exp\left(-\frac{\alpha}{1-\alpha}(t-s)\right) f'(s)\, ds, \tag{5.5.54}$$

for all $0 \le t \le b$.

Call

$$\gamma := \frac{\alpha}{1-\alpha} > 0. \tag{5.5.55}$$

I.e.

$$^{CF}D_*^\alpha f(t) = \frac{1}{1-\alpha} \int_0^t e^{-\gamma(t-s)} f'(s)\, ds, \quad 0 \le t \le b. \tag{5.5.56}$$

We notice that

$$\left|^{CF}D_*^\alpha f(t)\right| \le \frac{1}{1-\alpha} \left(\int_0^t e^{-\gamma(t-s)} ds\right) \|f'\|_\infty$$

$$= \frac{e^{-\gamma t}}{\alpha}(e^{\gamma t} - 1)\|f'\|_\infty = \frac{1}{\alpha}(1 - e^{-\gamma t})\|f'\|_\infty \le \left(\frac{1-e^{-\gamma b}}{\alpha}\right)\|f'\|_\infty. \tag{5.5.57}$$

That is

$$(^{CF}D_*^\alpha f)(0) = 0, \tag{5.5.58}$$

and

$$\left|^{CF}D_*^\alpha f(t)\right| \le \left(\frac{1-e^{-\gamma b}}{\alpha}\right)\|f'\|_\infty, \quad \forall t \in [0, b]. \tag{5.5.59}$$

Notice here that $1 - e^{-\gamma t}$, $t \ge 0$ is an increasing function.

Thus the smaller the t, the smaller it is $1 - e^{-\gamma t}$. We rewrite

$$^{CF}D_*^\alpha f(t) = \frac{e^{-\gamma t}}{1 - \alpha} \int_0^t e^{\gamma s} f'(s)\, ds, \tag{5.5.60}$$

proving that $\left(^{CF}D_*^\alpha f\right)$ is a continuous function over $[0, b]$, in particular it is continuous over $[a, b]$, where $0 < a < b$.

Therefore there exist $x_1, x_2 \in [a, b]$ such that

$$^{CF}D_*^\alpha f(x_1) = \min\ ^{CF}D_*^\alpha f(x), \tag{5.5.61}$$

and

$$^{CF}D_*^\alpha f(x_2) = \max\ ^{CF}D_*^\alpha f(x), \quad \text{for } x \in [a, b]. $$

We assume that

$$^{CF}D_*^\alpha f(x_1) > 0. \tag{5.5.62}$$

(i.e. $^{CF}D_*^\alpha f(x) > 0,\ \forall\, x \in [a, b]$).

Furthermore

$$\left\| ^{CF}D_*^\alpha f G \right\|_{\infty, [a,b]} =\ ^{CF}D_*^\alpha f(x_2). \tag{5.5.63}$$

Here it is

$$J(x) = mx, \quad m \neq 0. \tag{5.5.64}$$

The equation

$$Jf(x) = 0, \quad x \in [a, b], \tag{5.5.65}$$

has the same set of solutions as the equation

$$F(x) := \frac{Jf(x)}{^{CF}D_*^\alpha f(x_2)} = 0, \quad x \in [a, b]. \tag{5.5.66}$$

Notice that

$$^{CF}D_*^\alpha \left(\frac{f(x)}{2^{CF}D_*^\alpha f(x_2)} \right) = \frac{^{CF}D_*^\alpha f(x)}{2^{CF}D_*^\alpha f(x_2)} \leq \frac{1}{2} < 1, \quad \forall\, x \in [a, b]. \tag{5.5.67}$$

We call

$$A(x) := \frac{^{CF}D_*^\alpha f(x)}{2^{CF}D_*^\alpha f(x_2)}, \quad \forall\, x \in [a, b]. \tag{5.5.68}$$

We notice that

$$0 < \frac{^{CF}D_*^\alpha f(x_1)}{2^{CF}D_*^\alpha f(x_2)} \leq A(x) \leq \frac{1}{2}. \tag{5.5.69}$$

Furthermore it holds

$$|1 - A(x)| - 1 - A(x) \le 1 - \frac{{}^{CF}D_*^\alpha f(x_1)}{2^{CF}D_*^\alpha f(x_2)} =: \gamma_0, \quad \forall x \in [a, b]. \quad (5.5.70)$$

Clearly $\gamma_0 \in (0, 1)$.

We have proved that

$$|1 - A(x)| \le \gamma_0 \in (0, 1), \quad \forall x \in [a, b]. \quad (5.5.71)$$

Next we assume that $F(x)$ is a contraction over $[a, b]$, i.e.

$$|F(x) - F(y)| \le \lambda |x - y|; \quad \forall x, y \in [a, b], \quad (5.5.72)$$

and $0 < \lambda < \frac{1}{2}$.

Equivalently we have

$$|Jf(x) - Jf(y)| \le 2\lambda \left({}^{CF}D_*^\alpha f(x_2)\right) |x - y|, \quad \forall x, y \in [a, b]. \quad (5.5.73)$$

We observe that

$$|F(y) - F(x) - A(x)(y - x)| \le |F(y) - F(x)| + |A(x)| |y - x| \le$$

$$\lambda |y - x| + |A(x)| |y - x| = (\lambda + |A(x)|) |y - x| =: (\xi_2), \quad \forall x, y \in [a, b]. \quad (5.5.74)$$

Here we have

$$\left|\left({}^{CF}D_*^\alpha f\right)(x)\right| \le \left(\frac{1 - e^{-\gamma b}}{\alpha}\right) \|f'\|_\infty, \quad \forall t \in [a, b]. \quad (5.5.75)$$

Hence, $\forall x \in [a, b]$ we get that

$$|A(x)| = \frac{\left|{}^{CF}D_*^\alpha f(x)\right|}{2\left({}^{CF}D_*^\alpha f\right)(x_2)} \le \frac{(1 - e^{-\gamma b}) \|f'\|_\infty}{2\alpha \left({}^{CF}D_*^\alpha f\right)(x_2)} < \infty. \quad (5.5.76)$$

Consequently we observe

$$(\xi_2) \le \left(\lambda + \frac{(1 - e^{-\gamma b}) \|f'\|_\infty}{2\alpha \left({}^{CF}D_*^\alpha f\right)(x_2)}\right) |y - x|, \quad \forall x, y \in [a, b]. \quad (5.5.77)$$

Call

$$0 < \gamma_1 := \lambda + \frac{(1 - e^{-\gamma b}) \|f'\|_\infty}{2\alpha \left({}^{CF}D_*^\alpha f\right)(x_2)}, \quad (5.5.78)$$

choosing b small enough, we can make $\gamma_1 \in (0, 1)$.

We have proved

$$|F(y) - F(x) - A(x)(y-x)| \leq \gamma_1 |y-x|, \quad \gamma_1 \in (0, 1), \quad \forall\, x, y \in [a, b].$$
(5.5.79)

Next we call and need

$$0 < \gamma := \gamma_0 + \gamma_1 = 1 - \frac{{}^{CF}D_*^\alpha f(x_1)}{2 {}^{CF}D_*^\alpha f(x_2)} + \lambda + \frac{\left(1 - e^{-\gamma b}\right) \|f'\|_\infty}{2\alpha \left({}^{CF}D_*^\alpha f\right)(x_2)} < 1, \quad (5.5.80)$$

equivalently,

$$\lambda + \frac{\left(1 - e^{-\gamma b}\right) \|f'\|_\infty}{2\alpha \left({}^{CF}D_*^\alpha f\right)(x_2)} < \frac{{}^{CF}D_*^\alpha f(x_1)}{2 {}^{CF}D_*^\alpha f(x_2)},$$
(5.5.81)

equivalently,

$$2\lambda {}^{CF}D_*^\alpha f(x_2) + \frac{\left(1 - e^{-\gamma b}\right)}{\alpha} \|f'\|_\infty < {}^{CF}D_*^\alpha f(x_1),$$
(5.5.82)

which is possible for small λ, b.

We have proved that

$$\gamma = \gamma_0 + \gamma_1 \in (0, 1).$$
(5.5.83)

Hence Eq. (5.5.65) can be solved with our presented numerical methods.

Conclusion:

In all three applications we have proved that

$$|1 - A(x)| \leq \gamma_0 \in (0, 1),$$
(5.5.84)

and

$$|F(y) - F(x) - A(x)(y-x)| \leq \gamma_1 |y-x|,$$
(5.5.85)

where $\gamma_1 \in (0, 1)$, and

$$\gamma = \gamma_0 + \gamma_1 \in (0, 1),$$
(5.5.86)

for all $x, y \in [a^*, b], [a, b^*], [a, b]$, respectively.

Consequently, our presented Numerical methods here, Theorem 5.12, apply to solve

$$f(x) = 0.$$
(5.5.87)

References

1. S. Amat, S. Busquier, Third-order iterative methods under Kantorovich conditions. J. Math. Anal. Appl. **336**, 243–261 (2007)
2. S. Amat, S. Busquier, S. Plaza, Chaotic dynamics of a third-order Newton-like method. J. Math. Anal. Appl. **366**(1), 164–174 (2010)
3. G. Anastassiou, *Fractional Differentiation Inequalities* (Springer, New York, 2009)
4. G. Anastassiou, Fractional representation formulae and right fractional inequalities. Math. Comput. Model. **54**(11–12), 3098–3115 (2011)
5. G. Anastassiou, I. Argyros, *Approximating fixed points with applications in fractional calculus* (2015) (submitted)
6. I.K. Argyros, Newton-like methods in partially ordered linear spaces. J. Approx. Th. Appl. **9**(1), 1–10 (1993)
7. I.K. Argyros, Results on controlling the residuals of perturbed Newton-like methods on Banach spaces with a convergence structure. Southwest J. Pure Appl. Math. **1**, 32–38 (1995)
8. I.K. Argyros, *Convergence and Applications of Newton-like iterations* (Springer-Verlag Publ, New York, 2008)
9. J.A. Ezquerro, J.M. Gutierrez, M.A. Hernandez, N. Romero, M.J. Rubio, The Newton method: from Newton to Kantorovich (spanish). Gac. R. Soc. Mat. Esp. **13**, 53–76 (2010)
10. J. Losada, J.J. Nieto, Properties of a new fractional derivative without singular kernel. Progr. Fract. Differ. Appl. **1**(2), 87–92 (2015)
11. A.A. Magrenan, Different anomalies in a Surrutt family of iterative root finding methods. Appl. Math. Comput. **233**, 29–38 (2014)
12. A.A. Magrenan, A new tool to study real dynamics: the convergence plane. Appl. Math. Comput. **248**, 215–224 (2014)
13. P.W. Meyer, *Newton's method in generalized Banach spaces*, Numer. Func. Anal. Optimiz. **9**, **3**, **4**, 244-259 (1987)
14. S. Mukeen, G.M. Habibullah, k-Fractional integrals and Application. Int. J. Contemp. Math. Sci. **7**(2), 89–94 (2012)
15. F.A. Potra, V. Ptak, *Nondiscrete Induction and Iterative Processes* (Pitman Publ, London, 1984)
16. M. Zekisarikaya, A. Karaca, On the Riemann-Liouville fractional integral and applications. Intern. J. Stat. and Math. **1**(3), 33–43 (2014)

Chapter 6
Iterative Methods and Generalized
g-Fractional Calculus

We approximated solutions of some iterative methods on a generalized Banach space setting in [5]. Earlier studies such as [8–13] the operator involved is Fréchet-differentiable. In [5] we assumed that the operator is only continuous. This way we extended the applicability of these methods to include generalized fractional calculus and problems from other areas. In the present study applications include generalized g-fractional calculus. Fractional calculus is very important for its applications in many applied sciences. It follows [6].

6.1 Introduction

Many problems in Computational sciences can be formulated as an operator equation using Mathematical Modelling [9, 11, 14–16]. The fixed points of these operators can rarely be found in closed form. That is why most solution methods are usually iterative. The semilocal convergence is, based on the information around an initial point, to give conditions ensuring the convergence of the method.

We presented a semilocal convergence analysis for some iterative methods on a generalized Banach space setting in [5] to approximate fixed point or a zero of an operator. A generalized norm is defined to be an operator from a linear space into a partially order Banach space (to be precised in Sect. 6.2). Earlier studies such as [8–13] for Newton's method have shown that a more precise convergence analysis is obtained when compared to the real norm theory. However, the main assumption is that the operator involved is Fréchet-differentiable. This hypothesis limits the applicability of Newton's method. In [5] study we only assumed the continuity of the operator. This may be expanded the applicability of these methods.

© Springer International Publishing Switzerland 2016

G.A. Anastassiou and I.K. Argyros, *Intelligent Numerical Methods:*
Applications to Fractional Calculus, Studies in Computational Intelligence 624,
DOI 10.1007/978-3-319-26721-0_6

The rest of the chapter is organized as follows: Sect. 6.2 contains the basic concepts on generalized Banach spaces and the semilocal convergence analysis of these methods. Finally, in the concluding Sect. 6.3, we present special cases and applications in generalized g-fractional calculus.

6.2 Generalized Banach Spaces

We present some standard concepts that are needed in what follows to make the paper as self contained as possible. More details on generalized Banach spaces can be found in [5, 7–13], and the references there in.

Definition 6.1 A generalized Banach space is a triplet $(x, E, / \cdot /)$ such that
(i) X is a linear space over $\mathbb{R}\,(\mathbb{C})$.
(ii) $E = (E, K, \|\cdot\|)$ is a partially ordered Banach space, i.e.
(ii$_1$) $(E, \|\cdot\|)$ is a real Banach space,
(ii$_2$) E is partially ordered by a closed convex cone K,
(iii$_3$) The norm $\|\cdot\|$ is monotone on K.
(iii) The operator $/ \cdot / : X \to K$ satisfies
$/x/ = 0 \Leftrightarrow x = 0, /\theta x/ = |\theta| \, /x/,$
$/x + y/ \le /x/ + /y/$ for each $x, y \in X, \theta \in \mathbb{R}(\mathbb{C})$.
(iv) X is a Banach space with respect to the induced norm $\|\cdot\|_i := \|\cdot\| \cdot / \cdot /$.

Remark 6.2 The operator $/ \cdot /$ is called a generalized norm. In view of (iii) and (ii$_3$) $\|\cdot\|_i$, is a real norm. In the rest of this paper all topological concepts will be understood with respect to this norm.

Let $L\left(X^j, Y\right)$ stand for the space of j-linear symmetric and bounded operators from X^j to Y, where X and Y are Banach spaces. For X, Y partially ordered $L_+\left(X^j, Y\right)$ stands for the subset of monotone operators P such that

$$0 \le a_i \le b_i \Rightarrow P\left(a_1, \ldots, a_j\right) \le P\left(b_1, \ldots, b_j\right).$$

Definition 6.3 The set of bounds for an operator $Q \in L\,(X, X)$ on a generalized Banach space $(X, E, / \cdot /)$ the set of bounds is defined to be:

$$B\,(Q) := \{P \in L_+\,(E, E), /Qx/ \le P/x/ \quad \text{for each } x \in X\}.$$

Let $D \subset X$ and $T : D \to D$ be an operator. If $x_0 \in D$ the sequence $\{x_n\}$ given by

$$x_{n+1} := T\,(x_n) = T^{n+1}\,(x_0)$$

is well defined. We write in case of convergence

$$T^\infty\,(x_0) := \lim \left(T^n\,(x_0)\right) = \lim_{n \to \infty} x_n.$$

Let $(X, (E, K, \|\cdot\|), /\cdot/)$ and Y be generalized Banach spaces, $D \subset X$ an open subset, $G : D \to Y$ a continuous operator and $A(\cdot) : D \to L(X, Y)$. A zero of operator G is to be determined by a method starting at a point $x_0 \in D$. The results are presented for an operator $F = JG$, where $J \in L(Y, X)$. The iterates are determined through a fixed point problem:

$$x_{n+1} = x_n + y_n, \; A(x_n) y_n + F(x_n) = 0 \tag{6.2.1}$$
$$\Leftrightarrow y_n = T(y_n) := (I - A(x_n)) y_n - F(x_n).$$

Let $U(x_0, r)$ stand for the ball defined by

$$U(x_0, r) := \{x \in X : /x - x_0/ \leq r\}$$

for some $r \in K$.

Next, we state the semilocal convergence analysis of method (6.2.1) using the preceding notation.

Theorem 6.4 *[5] Let $F : D \subset X$, $A(\cdot) : D \to L(X, Y)$ and $x_0 \in D$ be as defined previously. Suppose:*
(H_1) There exists an operator $M \in B(I - A(x))$ for each $x \in D$.
(H_2) There exists an operator $N \in L_+(E, E)$ satisfying for each $x, y \in D$

$$/F(y) - F(x) - A(x)(y - x)/ \leq N/y - x/.$$

(H_3) There exists a solution $r \in K$ of

$$R_0(t) := (M + N)t + /F(x_0)/ \leq t.$$

(H_4) $U(x_0, r) \subseteq D$.
(H_5) $(M + N)^k r \to 0$ as $k \to \infty$.
Then, the following hold:
(C_1) The sequence $\{x_n\}$ defined by

$$x_{n+1} = x_n + T_n^\infty(0), \; T_n(y) := (I - A(x_n)) y - F(x_n) \tag{6.2.2}$$

is well defined, remains in $U(x_0, r)$ for each $n = 0, 1, 2, \ldots$ and converges to the unique zero of operator F in $U(x_0, r)$.

(C_2) An apriori bound is given by the null-sequence $\{r_n\}$ defined by $r_0 := r$ and for each $n = 1, 2, \ldots$

$$r_n = P_n^\infty(0), \; P_n(t) = Mt + Nr_{n-1}.$$

(C_3) An aposteriori bound is given by the sequence $\{s_n\}$ defined by

$$s_n := R_n^\infty(0), \; R_n(t) = (M + N)t + Na_{n-1},$$

$$b_n := /x_n - x_0/ \leq r - r_n \leq r,$$

where

$$a_{n-1} := /x_n - x_{n-1}/ \quad \text{for each } n = 1, 2, \ldots$$

Remark 6.5 The results obtained in earlier studies such as [8–13] require that operator F (i.e. G) is Fréchet-differentiable. This assumption limits the applicability of the earlier results. In the present study we only require that F is a continuous operator. Hence, we have extended the applicability of these methods to include classes of operators that are only continuous.

Example 6.6 The j-dimensional space $\mathbb{R}^j$ is a classical example of a generalized Banach space. The generalized norm is defined by componentwise absolute values. Then, as ordered Banach space we set $E = \mathbb{R}^j$ with componentwise ordering with e.g. the maximum norm. A bound for a linear operator (a matrix) is given by the corresponding matrix with absolute values. Similarly, we can define the "N" operators. Let $E = \mathbb{R}$. That is we consider the case of a real normed space with norm denoted by $\|\cdot\|$. Let us see how the conditions of Theorem 6.4 look like.

Theorem 6.7 (H_1) $\|I - A(x)\| \leq M$ *for some* $M \geq 0$.
  (H_2) $\|F(y) - F(x) - A(x)(y - x)\| \leq N \|y - x\|$ *for some* $N \geq 0$.
  (H_3) $M + N < 1$,

$$r = \frac{\|F(x_0)\|}{1 - (M + N)}. \tag{6.2.3}$$

  (H_4) $U(x_0, r) \subseteq D$.
  (H_5) $(M + N)^k r \to 0$ *as* $k \to \infty$, *where* r *is given by* (6.2.3).
  Then, the conclusions of Theorem 6.4 hold.

6.3 Applications to g-Fractional Calculus

We apply Theorem 6.7 in this section. Here basic concepts and facts come from [4] and Chap. 24. We need:

Definition 6.8 Let $\alpha > 0$, $\alpha \notin \mathbb{N}$, $\lceil \alpha \rceil = m$, $\lceil \cdot \rceil$ the ceiling of the number. Here $g \in AC([a, b])$ (absolutely continuous functions) and g is strictly increasing. Let $G : [a, b] \to \mathbb{R}$ such that $(G \circ g^{-1})^{(m)} \circ g \in L_\infty([a, b])$.
  We define the left generalized g-fractional derivative of G of order α as follows:

$$\left(D^\alpha_{a+;g} G\right)(x) :=$$

$$\frac{1}{\Gamma(m - \alpha)} \int_a^x (g(x) - g(t))^{m - \alpha - 1} g'(t) \left(G \circ g^{-1}\right)^{(m)} (g(t)) \, dt, \tag{6.3.1}$$

$a \leq x \leq b$, where Γ is the gamma function.

We also define the right generalized g-fractional derivative of G of order α as follows:

$$\left(D_{b-;g}^{\alpha}G\right)(x) :=$$

$$\frac{(-1)^{m}}{\Gamma(m-\alpha)}\int_{x}^{b}(g(t)-g(x))^{m-\alpha-1}g'(t)\left(G\circ g^{-1}\right)^{(m)}(g(t))\,dt, \qquad (6.3.2)$$

$a \le x \le b$.

Both $\left(D_{a+;g}^{\alpha}G\right)$, $\left(D_{b-;g}^{\alpha}G\right) \in C([a,b])$.

(I) Let $a < a^{*} < b$. In particular we have that $\left(D_{a+;g}^{\alpha}G\right) \in C([a^{*},b])$. We notice that

$$\left|\left(D_{a+;g}^{\alpha}G\right)(x)\right| \le$$

$$\frac{\left\|\left(G\circ g^{-1}\right)^{(m)}\circ g\right\|_{\infty,[a,b]}}{\Gamma(m-\alpha)}\left(\int_{a}^{x}(g(x)-g(t))^{m-\alpha-1}g'(t)\,dt\right) = \qquad (6.3.3)$$

$$\frac{\left\|\left(G\circ g^{-1}\right)^{(m)}\circ g\right\|_{\infty,[a,b]}}{\Gamma(m-\alpha)}\frac{(g(x)-g(a))^{m-\alpha}}{(m-\alpha)} =$$

$$\frac{\left\|\left(G\circ g^{-1}\right)^{(m)}\circ g\right\|_{\infty,[a,b]}}{\Gamma(m-\alpha+1)}(g(x)-g(a))^{m-\alpha}, \quad \forall\, x \in [a,b].$$

We have proved that

$$\left|\left(D_{a+;g}^{\alpha}G\right)(x)\right| \le \frac{\left\|\left(G\circ g^{-1}\right)^{(m)}\circ g\right\|_{\infty,[a,b]}}{\Gamma(m-\alpha+1)}(g(x)-g(a))^{m-\alpha}$$

$$\le \frac{\left\|\left(G\circ g^{-1}\right)^{(m)}\circ g\right\|_{\infty,[a,b]}}{\Gamma(m-\alpha+1)}(g(b)-g(a))^{m-\alpha} < \infty, \quad \forall\, x \in [a,b], \qquad (6.3.4)$$

in particular true $\forall\, x \in [a^{*},b]$.

We obtain that

$$\left(D_{a+;g}^{\alpha}G\right)(a) = 0. \qquad (6.3.5)$$

Therefore there exist $x_1, x_2 \in [a^{*},b]$ such that $D_{a+;g}^{\alpha}G(x_1) = \min D_{a+;g}^{\alpha}G(x)$, and $D_{a+;g}^{\alpha}G(x_2) = \max D_{a+;g}^{\alpha}G(x)$, for $x \in [a^{*},b]$.

We assume that

$$D_{a+;g}^{\alpha}G(x_1) > 0. \qquad (6.3.6)$$

(i.e. $D_{a+;g}^{\alpha} G (x) > 0, \forall x \in [a^*, b]$).

Furthermore

$$\left\| D_{a+;g}^{\alpha} G \right\|_{\infty, [a^*, b]} = D_{a+;g}^{\alpha} G (x_2) . \tag{6.3.7}$$

Here it is

$$J (x) = mx, \quad m \neq 0. \tag{6.3.8}$$

The equation

$$JG (x) = 0, \quad x \in [a^*, b], \tag{6.3.9}$$

has the same set of solutions as the equation

$$F (x) := \frac{JG (x)}{2D_{a+;g}^{\alpha} G (x_2)} = 0, \quad x \in [a^*, b]. \tag{6.3.10}$$

Notice that

$$D_{a+;g}^{\alpha} \left(\frac{G (x)}{2D_{a+;g}^{\alpha} G (x_2)} \right) = \frac{D_{a+;g}^{\alpha} G (x)}{2D_{a+;g}^{\alpha} G (x_2)} \leq \frac{1}{2} < 1, \quad \forall x \in [a^*, b]. \tag{6.3.11}$$

We call

$$A (x) := \frac{D_{a+;g}^{\alpha} G (x)}{2D_{a+;g}^{\alpha} G (x_2)}, \quad \forall x \in [a^*, b]. \tag{6.3.12}$$

We notice that

$$0 < \frac{D_{a+;g}^{\alpha} G (x_1)}{2D_{a+;g}^{\alpha} G (x_2)} \leq A (x) \leq \frac{1}{2}. \tag{6.3.13}$$

Hence it holds

$$|1 - A (x)| = 1 - A (x) \leq 1 - \frac{D_{a+;g}^{\alpha} G (x_1)}{2D_{a+;g}^{\alpha} G (x_2)} =: \gamma_0, \quad \forall x \in [a^*, b]. \tag{6.3.14}$$

Clearly $\gamma_0 \in (0, 1)$.

We have proved that

$$|1 - A (x)| \leq \gamma_0 \in (0, 1), \quad \forall \ x \in [a^*, b]. \tag{6.3.15}$$

Next we assume that $F (x)$ is a contraction over $[a^*, b]$, i.e.

$$|F (x) - F (y)| \leq \lambda |x - y|; \quad \forall x, y \in [a^*, b], \tag{6.3.16}$$

and $0 < \lambda < \frac{1}{2}$.

Equivalently we have

$$|JG(x) - JG(y)| \le 2\lambda \left(D_{a+;g}^{\alpha} G(x_2) \right) |x - y|, \quad \forall \, x, y \in \left[a^*, b \right]. \quad (6.3.17)$$

We observe that

$$|F(y) - F(x) - A(x)(y - x)| \le |F(y) - F(x)| + |A(x)| \, |y - x| \le$$

$$\lambda |y - x| + |A(x)| \, |y - x| = (\lambda + |A(x)|) \, |y - x| =: (\xi_1), \quad \forall \, x, y \in \left[a^*, b \right]. \quad (6.3.18)$$

Hence by (6.3.4), $\forall \, x \in [a^*, b]$ we get that

$$|A(x)| = \frac{\left| D_{a+;g}^{\alpha} G(x) \right|}{2 D_{a+;g}^{\alpha} G(x_2)} \le \frac{(g(b) - g(a))^{m-\alpha}}{2\Gamma(m - \alpha + 1)} \frac{\left\| \left(G \circ g^{-1} \right)^{(m)} \circ g \right\|_{\infty, [a,b]}}{D_{a+;g}^{\alpha} G(x_2)} < \infty. \quad (6.3.19)$$

Consequently we observe

$$(\xi_1) \le \left(\lambda + \frac{(g(b) - g(a))^{m-\alpha}}{2\Gamma(m - \alpha + 1)} \frac{\left\| \left(G \circ g^{-1} \right)^{(m)} \circ g \right\|_{\infty, [a,b]}}{D_{a+;g}^{\alpha} G(x_2)} \right) |y - x|, \quad (6.3.20)$$

$\forall \, x, y \in [a^*, b]$.

Call

$$0 < \gamma_1 := \lambda + \frac{(g(b) - g(a))^{m-\alpha}}{2\Gamma(m - \alpha + 1)} \frac{\left\| \left(G \circ g^{-1} \right)^{(m)} \circ g \right\|_{\infty, [a,b]}}{D_{a+;g}^{\alpha} G(x_2)}, \quad (6.3.21)$$

choosing $(g(b) - g(a))$ small enough we can make $\gamma_1 \in (0, 1)$.

We proved that

$$|F(y) - F(x) - A(x)(y - x)| \le \gamma_1 |y - x|, \quad \text{where } \gamma_1 \in (0, 1), \, \forall \, x, y \in \left[a^*, b \right]. \quad (6.3.22)$$

Next we call and need

$$0 < \gamma := \gamma_0 + \gamma_1 = 1 - \frac{D_{a+;g}^{\alpha} G(x_1)}{2 D_{a+;g}^{\alpha} G(x_2)} + \lambda +$$

$$\frac{(g(b) - g(a))^{m-\alpha}}{2\Gamma(m - \alpha + 1)} \frac{\left\| \left(G \circ g^{-1} \right)^{(m)} \circ g \right\|_{\infty, [a,b]}}{D_{a+;g}^{\alpha} G(x_2)} < 1, \quad (6.3.23)$$

equivalently we find,

$$\lambda + \frac{(g(b) - g(a))^{m-\alpha}}{2\Gamma(m-\alpha+1)} \frac{\left\| (G \circ g^{-1})^{(m)} \circ g \right\|_{\infty,[a,b]}}{D_{a+;g}^{\alpha} G(x_2)} < \frac{D_{a+;g}^{\alpha} G(x_1)}{2D_{a+;g}^{\alpha} G(x_2)}, \qquad (6.3.24)$$

equivalently,

$$2\lambda D_{a+;g}^{\alpha} G(x_2) + \frac{(g(b) - g(a))^{m-\alpha}}{\Gamma(m-\alpha+1)} \left\| (G \circ g^{-1})^{(m)} \circ g \right\|_{\infty,[a,b]} < D_{a+;g}^{\alpha} G(x_1),$$
$$(6.3.25)$$

which is possible for small λ, $(g(b) - g(a))$.

That is $\gamma \in (0, 1)$. Hence equation (6.3.9) can be solved with our presented iterative algorithms.

Conclusion 6.9 *(for (I))*

Our presented earlier semilocal results, see Theorem 6.7, can apply in the above generalized fractional setting for $g(x) = x$ for each $x \in [a, b]$ since the following inequalities have been fulfilled:

$$\|1 - A(x)\|_{\infty} \le \gamma_0, \qquad (6.3.26)$$

and

$$|F(y) - F(x) - A(x)(y-x)| \le \gamma_1 |y - x|, \qquad (6.3.27)$$

where $\gamma_0, \gamma_1 \in (0, 1)$, furthermore it holds

$$\gamma = \gamma_0 + \gamma_1 \in (0, 1), \qquad (6.3.28)$$

for all $x, y \in [a^*, b]$, where $a < a^* < b$.

The specific functions $A(x)$, $F(x)$ have been described above, see (6.3.12) and (6.3.10), respectively.

(II) Let $a < b^* < b$. In particular we have that $\left(D_{b-;g}^{\alpha} G \right) \in C([a, b^*])$. We notice that

$$\left| \left(D_{b-;g}^{\alpha} G \right)(x) \right| \le$$

$$\frac{\left\| (G \circ g^{-1})^{(m)} \circ g \right\|_{\infty,[a,b]}}{\Gamma(m-\alpha)} \left(\int_x^b (g(t) - g(x))^{m-\alpha-1} g'(t) \, dt \right) = \qquad (6.3.29)$$

$$\frac{\left\|\left(G \circ g^{-1}\right)^{(m)} \circ g\right\|_{\infty,[a,b]}}{\Gamma(m - \alpha + 1)} (g(b) - g(x))^{m-\alpha} \leq$$

$$\frac{\left\|\left(G \circ g^{-1}\right)^{(m)} \circ g\right\|_{\infty,[a,b]}}{\Gamma(m - \alpha + 1)} (g(b) - g(a))^{m-\alpha} < \infty, \quad \forall\, x \in [a, b], \quad (6.3.30)$$

in particular true $\forall\, x \in [a, b^*]$.

We obtain that

$$\left(D_{b-;g}^{\alpha} G\right)(b) = 0. \tag{6.3.31}$$

Therefore there exist $x_1, x_2 \in [a, b^*]$ such that $D_{b-;g}^{\alpha} G(x_1) = \min D_{b-;g}^{\alpha} G(x)$, and $D_{b-;g}^{\alpha} G(x_2) = \max D_{b-;g}^{\alpha} G(x)$, for $x \in [a, b^*]$.

We assume that

$$D_{b-;g}^{\alpha} G(x_1) > 0. \tag{6.3.32}$$

(i.e. $D_{b-;g}^{\alpha} G(x) > 0, \forall\, x \in [a, b^*]$).

Furthermore

$$\left\| D_{b-;g}^{\alpha} G \right\|_{\infty,[a,b^*]} = D_{b-;g}^{\alpha} G(x_2). \tag{6.3.33}$$

Here it is

$$J(x) = mx, \quad m \neq 0. \tag{6.3.34}$$

The equation

$$JG(x) = 0, \quad x \in [a, b^*], \tag{6.3.35}$$

has the same set of solutions as the equation

$$F(x) := \frac{JG(x)}{2D_{b-;g}^{\alpha} G(x_2)} = 0, \quad x \in [a, b^*]. \tag{6.3.36}$$

Notice that

$$D_{b-;g}^{\alpha} \left(\frac{G(x)}{2D_{b-;g}^{\alpha} G(x_2)} \right) = \frac{D_{b-;g}^{\alpha} G(x)}{2D_{b-;g}^{\alpha} G(x_2)} \leq \frac{1}{2} < 1, \quad \forall\, x \in [a, b^*]. \tag{6.3.37}$$

We call

$$A(x) := \frac{D_{b-;g}^{\alpha} G(x)}{2D_{b-;g}^{\alpha} G(x_2)}, \quad \forall\, x \in [a, b^*]. \tag{6.3.38}$$

We notice that

$$0 < \frac{D_{b-;g}^{\alpha} G(x_1)}{2D_{b-;g}^{\alpha} G(x_2)} \leq A(x) \leq \frac{1}{2}. \tag{6.3.39}$$

Hence it holds

$$|1 - A(x)| = 1 - A(x) \le 1 - \frac{D_{b-;g}^{\alpha} G(x_1)}{2 D_{b-;g}^{\alpha} G(x_2)} =: \gamma_0, \quad \forall x \in [a, b^*]. \quad (6.3.40)$$

Clearly $\gamma_0 \in (0, 1)$.

We have proved that

$$|1 - A(x)| \le \gamma_0 \in (0, 1), \quad \forall x \in [a, b^*]. \quad (6.3.41)$$

Next we assume that $F(x)$ is a contraction over $[a, b^*]$, i.e.

$$|F(x) - F(y)| \le \lambda |x - y|; \quad \forall x, y \in [a, b^*], \quad (6.3.42)$$

and $0 < \lambda < \frac{1}{2}$.

Equivalently we have

$$|JG(x) - JG(y)| \le 2\lambda \left(D_{b-;g}^{\alpha} G(x_2) \right) |x - y|, \quad \forall x, y \in [a, b^*]. \quad (6.3.43)$$

We observe that

$$|F(y) - F(x) - A(x)(y - x)| \le |F(y) - F(x)| + |A(x)| |y - x| \le$$

$$\lambda |y - x| + |A(x)| |y - x| = (\lambda + |A(x)|) |y - x| =: (\xi_2), \quad \forall x, y \in [a, b^*]. \quad (6.3.44)$$

Hence by (6.3.30), $\forall x \in [a, b^*]$ we get that

$$|A(x)| = \frac{\left| D_{b-;g}^{\alpha} G(x) \right|}{2 D_{b-;g}^{\alpha} G(x_2)} \le \frac{(g(b) - g(a))^{m-\alpha}}{2\Gamma(m - \alpha + 1)} \frac{\left\| (G \circ g^{-1})^{(m)} \circ g \right\|_{\infty,[a,b]}}{D_{b-;g}^{\alpha} G(x_2)} < \infty. \quad (6.3.45)$$

Consequently we observe

$$(\xi_2) \le \left(\lambda + \frac{(g(b) - g(a))^{m-\alpha}}{2\Gamma(m - \alpha + 1)} \frac{\left\| (G \circ g^{-1})^{(m)} \circ g \right\|_{\infty,[a,b]}}{D_{b-;g}^{\alpha} G(x_2)} \right) |y - x|, \quad (6.3.46)$$

$\forall x, y \in [a, b^*]$.

Call

$$0 < \gamma_1 := \lambda + \frac{(g(b) - g(a))^{m-\alpha}}{2\Gamma(m - \alpha + 1)} \frac{\left\| (G \circ g^{-1})^{(m)} \circ g \right\|_{\infty,[a,b]}}{D_{b-;g}^{\alpha} G(x_2)}, \quad (6.3.47)$$

choosing $(g(b) - g(a))$ small enough we can make $\gamma_1 \in (0, 1)$.

We proved that

$$|F(y) - F(x) - A(x)(y - x)| \le \gamma_1 |y - x|, \text{ where } \gamma_1 \in (0, 1), \quad \forall\, x, y \in [a, b^*].$$
(6.3.48)

Next we call and need

$$0 < \gamma := \gamma_0 + \gamma_1 = 1 - \frac{D_{b-;g}^{\alpha} G(x_1)}{2 D_{b-;g}^{\alpha} G(x_2)} + \lambda +$$

$$\frac{(g(b) - g(a))^{m-\alpha}}{2\Gamma(m - \alpha + 1)} \frac{\left\| (G \circ g^{-1})^{(m)} \circ g \right\|_{\infty, [a,b]}}{D_{b-;g}^{\alpha} G(x_2)} < 1,$$
(6.3.49)

equivalently we find,

$$\lambda + \frac{(g(b) - g(a))^{m-\alpha}}{2\Gamma(m - \alpha + 1)} \frac{\left\| (G \circ g^{-1})^{(m)} \circ g \right\|_{\infty, [a,b]}}{D_{b-;g}^{\alpha} G(x_2)} < \frac{D_{b-;g}^{\alpha} G(x_1)}{2 D_{b-;g}^{\alpha} G(x_2)},$$
(6.3.50)

equivalently,

$$2\lambda D_{b-;g}^{\alpha} G(x_2) + \frac{(g(b) - g(a))^{m-\alpha}}{\Gamma(m - \alpha + 1)} \left\| (G \circ g^{-1})^{(m)} \circ g \right\|_{\infty, [a,b]} < D_{b-;g}^{\alpha} G(x_1),$$
(6.3.51)

which is possible for small λ, $(g(b) - g(a))$.

That is $\gamma \in (0, 1)$. Hence equation (6.3.35) can be solved with our presented iterative algorithms.

Conclusion 6.10 *(for (II))*

Our presented earlier semilocal iterative methods, see Theorem 6.7, can apply in the above generalized fractional setting for $g(x) = x$ for each $x \in [a, b]$ since the following inequalities have been fulfilled:

$$\|1 - A(x)\|_{\infty} \le \gamma_0,$$
(6.3.52)

and

$$|F(y) - F(x) - A(x)(y - x)| \le \gamma_1 |y - x|,$$
(6.3.53)

where $\gamma_0, \gamma_1 \in (0, 1)$, furthermore it holds

$$\gamma = \gamma_0 + \gamma_1 \in (0, 1),$$
(6.3.54)

for all $x, y \in [a, b^*]$, where $a < b^* < b$.

The specific functions $A(x)$, $F(x)$ have been described above, see (6.3.38) and (6.3.36), respectively.

References

1. S. Amat, S. Busquier, S. Plaza, Chaotic dynamics of a third-order Newton-like method. J. Math. Anal. Appl. **366**(1), 164–174 (2010)
2. G. Anastassiou, *Fractional Differentiation Inequalities* (Springer, New York, 2009)
3. G. Anastassiou, Fractional representation formulae and right fractional inequalities. Math. Comput. Model. **54**(11–12), 3098–3115 (2011)
4. G. Anastassiou, Advanced fractional Taylor's formula. J. Comput. Anal. Appl (2016) (to appear)
5. G. Anastassiou, I. Argyros, *Convergence for Iterative Methods on Banach Spaces of Convergence Structure with Applications in Fractional Calculus*, SeMA doi:10.1007/s40324-015-0044-y
6. G. Anastassiou, I. Argyros, *Generalized g-fractional calculus and iterative methods* (2015) (submitted)
7. I.K. Argyros, Newton-like methods in partially ordered linear spaces. J. Approx. Th. Applic. **9**(1), 1–10 (1993)
8. I.K. Argyros, Results on controlling the residuals of perturbed Newton-like methods on banach spaces with a convergence structure. Southwest J. Pure Appl. Math. **1**, 32–38 (1995)
9. I.K. Argyros, *Convergence and Applications of Newton-Like Iterations* (Springer, New York, 2008)
10. J.A. Ezquerro, J.M. Gutierrez, M.A. Hernandez, N. Romero, M.J. Rubio, The Newton method: from Newton to Kantorovich (Spanish). Gac. R. Soc. Mat. Esp. **13**, 53–76 (2010)
11. A.A. Magrenan, Different anomalies in a Surrutt family of iterative root finding methods. Appl. Math. Comput. **233**, 29–38 (2014)
12. A.A. Magrenan, A new tool to study real dynamics: the convergence plane. Appl. Math. Comput. **248**, 215–224 (2014)
13. P.W. Meyer, Newton's method in generalized Banach spaces. Numer. Func. Anal. Optimiz. **9**(3–4), 244–259 (1987)
14. S. Mukeen, G.M. Habibullah, k-Fractional integrals and application. Int. J. Contemp. Math. Sci. **7**(2), 89–94 (2012)
15. F.A. Potra, V. Ptak, *Nondiscrete Induction and Iterative Processes* (Pitman Publ, London, 1984)
16. M. Zekisarikaya, A. Karaca, On the Riemann-Liouville fractional integral and applications. Intern. J. Stat. and Math. **1**(3), 33–43 (2014)

Chapter 7
Unified Convergence Analysis for Iterative Algorithms and Fractional Calculus

We present local and semilocal convergence results for some iterative algorithms in order to approximate a locally unique solution of a nonlinear equation in a Banach space setting. In earlier studies to operator involved is assumed to be at least once Fréchet-differentiable. In the present study, we assume that the operator is only continuous. This way we expand the applicability of these iterative algorithms. In the third part of the study we present some choices of the operators involved in fractional calculus where the operators satisfy the convergence conditions. It follows [5].

7.1 Introduction

In this study we are concerned with the problem of approximating a locally unique solution x^* of the nonlinear equation

$$F(x) = 0, \tag{7.1.1}$$

where F is a continuous operator defined on a subset D of a Banach space X with values in a Banach space Y.

A lot of problems in Computational Sciences and other disciplines can be brought in a form like (7.1.1) using Mathematical Modelling [8, 12, 16]. The solutions of such equations can be found in closed form only in special cases. That is why most solution methods for these equations are iterative. Iterative algorithms are usually studied based on semilocal and local convergence. The semilocal convergence matter is, based on the information around the initial point to give hypotheses ensuring the convergence of the iterative algorithm; while the local one is, based on the information

© Springer International Publishing Switzerland 2016   107
G.A. Anastassiou and I.K. Argyros, *Intelligent Numerical Methods:*
Applications to Fractional Calculus, Studies in Computational Intelligence 624,
DOI 10.1007/978-3-319-26721-0_7

around a solution, to find estimates of the radii of convergence balls as well as error bounds on the distances involved.

We introduce the iterative algorithm defined for each $n = 0, 1, 2, \ldots$ by

$$x_{n+1} = x_n - A (x_n)^{-1} F (x_n), \tag{7.1.2}$$

where $x_0 \in D$ is an initial point and $A (x) \in L (X, Y)$ the space of bounded linear operators from X into Y. There is a plethora on local as well as semilocal convergence theorems for iterative algorithm (7.1.2) provided that the operator A is an approximation to the Fréchet-derivative F' [1, 2, 6–16]. In the present study we do not assume that operator A is related to F'. This way we expand the applicability of iterative algorithm (7.1.2). Notice that many well known methods are special case of interative algorithm (7.1.2).

Newton's method: Choose $A (x) = F' (x)$ for each $x \in D$.

Steffensen's method: Choose $A (x) = [x, G (x); F]$, where $G : X \to X$ is a known operator and $[x, y; F]$ denotes a divided difference of order one [8, 12, 15].

The so called Newton-like methods and many other methods are special cases of iterative algorithm (7.1.2).

The rest of the chapter is organized as follows. The semilocal as well as the local convergence analysis of iterative algorithm (7.1.2) is given in Sect. 7.2. Some applications from fractional calculus are given in Sect. 7.3.

7.2   Convergence Analysis

We present the main semilocal convergence result for iterative algorithm (7.1.2).

Theorem 7.1 Let $F : D \subset X \to Y$ be a continuous operator and let $A (x) \in L (X, Y)$. Suppose that there exist $x_0 \in D$, $\eta \geq 0$, $p \geq 1$, a function $g : [0, \eta] \to [0, \infty)$ continuous and nondecreasing such that for each $x, y \in D$

$$A (x)^{-1} \in L (Y, X), \tag{7.2.1}$$

$$\left\| A (x_0)^{-1} F (x_0) \right\| \leq \eta, \tag{7.2.2}$$

$$\left\| A (y)^{-1} (F (y) - F (x) - A (x) (y - x)) \right\| \leq g (\|x - y\|) \|x - y\|^{p+1}, \tag{7.2.3}$$

$$q := g (\eta) \eta^p < 1 \tag{7.2.4}$$

and

$$\overline{U} (x_0, r) \subseteq D, \tag{7.2.5}$$

where,

$$r = \frac{\eta}{1 - q}. \tag{7.2.6}$$

Then, the sequence $\{x_n\}$ generated by iterative algorithm (7.1.2) is well defined, remains in $\overline{U}(x_0, r)$ for each $n = 0, 1, 2, \ldots$ and converges to some $x^* \in \overline{U}(x_0, r)$ such that

$$\|x_{n+1} - x_n\| \leq g(\|x_n - x_{n-1}\|) \|x_n - x_{n-1}\|^{p+1} \leq q \|x_n - x_{n-1}\| \qquad (7.2.7)$$

and

$$\|x_n - x^*\| \leq \frac{q^n \eta}{1 - q}. \qquad (7.2.8)$$

Proof The iterate x_1 is well defined by iterative algorithm (7.1.2) for $n = 0$ and (7.2.1) for $x = x_0$. We also have by (7.2.2) and (7.2.6) that $\|x_1 - x_0\| = \|A(x_0)^{-1} F(x_0)\| \leq \eta < r$, so we get that $x_1 \in \overline{U}(x_0, r)$ and x_2 is well defined (by (7.2.5)). Using (7.2.3) for $y = x_1$, $x = x_0$ and (7.2.4) we get that

$$\|x_2 - x_1\| = \|A(x_1)^{-1} [F(x_1) - F(x_0) - A(x_0)(x_1 - x_0)]\|$$

$$\leq g(\|x_1 - x_0\|) \|x_1 - x_0\|^{p+1} \leq q \|x_1 - x_0\|,$$

which shows (7.2.7) for $n = 1$. Then, we can have that

$$\|x_2 - x_0\| \leq \|x_2 - x_1\| + \|x_1 - x_0\| \leq q \|x_1 - x_0\| + \|x_1 - x_0\|$$

$$= (1 + q) \|x_1 - x_0\| \leq \frac{1 - q^2}{1 - q} \eta < r,$$

so $x_2 \in \overline{U}(x_0, r)$ and x_3 is well defined.

Assuming $\|x_{k+1} - x_k\| \leq q \|x_k - x_{k-1}\|$ and $x_{k+1} \in \overline{U}(x_0, r)$ for each $k = 1, 2, \ldots, n$ we get

$$\|x_{k+2} - x_{k+1}\| = \|A(x_{k+1})^{-1} [F(x_{k+1}) - F(x_k) - A(x_k)(x_{k+1} - x_k)]\|$$

$$\leq g(\|x_{k+1} - x_k\|) \|x_{k+1} - x_k\|^{p+1}$$

$$\leq g(\|x_1 - x_0\|) \|x_1 - x_0\|^p \|x_{k+1} - x_k\| \leq q \|x_{k+1} - x_k\|$$

and

$$\|x_{k+2} - x_0\| \leq \|x_{k+2} - x_{k+1}\| + \|x_{k+1} - x_k\| + \cdots + \|x_1 - x_0\|$$

$$\leq (q^{k+1} + q^k + \cdots + 1) \|x_1 - x_0\| \leq \frac{1 - q^{k+2}}{1 - q} \|x_1 - x_0\|$$

$$< \frac{\eta}{1 - q} = r,$$

which completes the induction for (7.2.7) and $x_{k+2} \in \overline{U}(x_0, r)$. We also have that
for $m \geq 0$

$$\|x_{n+m} - x_n\| \leq \|x_{n+m} - x_{n+m-1}\| + \cdots + \|x_{n+1} - x_n\|$$

$$\leq \left(q^{m-1} + q^{m-2} + \cdots + 1\right) \|x_{n+1} - x_n\|$$

$$\leq \frac{1 - q^m}{1 - q} q^n \|x_1 - x_0\|.$$

It follows that $\{x_n\}$ is a complete sequence in a Banach space X and as such it
converges to some $x^* \in \overline{U}(x_0, r)$ (since $\overline{U}(x_0, r)$ is a closed set). By letting $m \to \infty$,
we obtain (7.2.8).   □

Stronger hypotheses are needed to show that x^* is a solution of equation $F(x) = 0$.

Proposition 7.2 *Let* $F : D \subset X \to Y$ *be a continuous operator and let* $A(x) \in$
$L(X, Y)$. *Suppose that there exist* $x_0 \in D$, $\eta \geq 0$, $p \geq 1$, $\psi > 0$, *a function*
$g_1 : [0, \eta] \to [0, \infty)$ *continuous and nondecreasing such that for each* $x, y \in D$

$$A(x)^{-1} \in L(Y, X), \quad \left\|A(x)^{-1}\right\| \leq \psi, \quad \left\|A(x_0)^{-1} F(x_0)\right\| \leq \eta, \qquad (7.2.9)$$

$$\|F(y) - F(x) - A(x)(y - x)\| \leq \frac{g_1(\|x - y\|)}{\psi} \|x - y\|^{p+1}, \qquad (7.2.10)$$

$$q_1 := g_1(\eta) \eta^p < 1$$

and

$$\overline{U}(x_0, r_1) \subseteq D,$$

where,

$$r_1 = \frac{\eta}{1 - q_1}.$$

Then, the conclusions of Theorem 7.1 for sequence $\{x_n\}$ *hold with* g_1, q_1, r_1, *replacing*
g, q *and* r, *respectively. Moreover,* x^* *is a solution of the equation* $F(x) = 0$.

Proof Notice that

$$\left\|A(x_n)^{-1} \left[F(x_n) - F(x_{n-1}) - A(x_{n-1})(x_n - x_{n-1})\right]\right\|$$

$$\leq \left\|A(x_n)^{-1}\right\| \|F(x_n) - F(x_{n-1}) - A(x_{n-1})(x_n - x_{n-1})\|$$

$$\leq g_1(\|x_n - x_{n-1}\|) \|x_n - x_{n-1}\|^{p+1} \leq q_1 \|x_n - x_{n-1}\|.$$

Therefore, the proof of Theorem 7.1 can apply. Then, in view of the estimate

$$\|F(x_n)\| = \|F(x_n) - F(x_{n-1}) - A(x_{n-1})(x_n - x_{n-1})\| <$$

$$\frac{g_1(\|x_n - x_{n-1}\|)}{\psi}\|x_n - x_{n-1}\|^{p+1} \le q_1 \|x_n - x_{n-1}\|,$$

we deduce by letting $n \to \infty$ that $F(x^*) = 0$. $\qquad\qquad\qquad\square$

Concerning the uniqueness of the solution x^* we have the following result:

Proposition 7.3 *Under the hypotheses of Proposition 7.2, further suppose that*

$$q_1 r_1^p < 1. \qquad\qquad (7.2.11)$$

Then, x^* is the only solution of equation $F(x) = 0$ in $\overline{U}(x_0, r_1)$.

Proof The existence of the solution $x^* \in \overline{U}(x_0, r_1)$ has been established in Proposition 7.2. Let $y^* \in \overline{U}(x_0, r_1)$ with $F(y^*) = 0$. Then, we have in turn that

$$\|x_{n+1} - y^*\| = \|x_n - y^* - A(x_n)^{-1} F(x_n)\| =$$

$$\|A(x_n)^{-1}[A(x_n)(x_n - y^*) - F(x_n) + F(y^*)]\| \le$$

$$\|A(x_n)^{-1}\| \|F(y^*) - F(x_n) - A(x_n)(y^* - x_n)\| \le$$

$$\psi \frac{g_1(\|x_n - y^*\|)}{\psi}\|x_n - y^*\|^{p+1} \le q_1 r_1^p \|x_n - x^*\| < \|x_n - y^*\|,$$

so we deduce that $\lim_{n\to\infty} x_n = y^*$. But we have that $\lim_{n\to\infty} x_n = x^*$. Hence, we conclude that $x^* = y^*$. $\qquad\qquad\qquad\square$

Next, we present a local convergence analysis for the iterative algorithm (7.1.2).

Proposition 7.4 *Let $F : D \subset X \to Y$ be a continuous operator and let $A(x) \in L(X, Y)$. Suppose that there exist $x^* \in D$, $p \ge 1$, a function $g_2 : [0, \infty) \to [0, \infty)$ continuous and nondecreasing such that for each $x \in D$*

$$F(x^*) = 0, \quad A(x)^{-1} \in L(Y, X),$$

$$\|A(x)^{-1}[F(x) - F(x^*) - A(x)(x - x^*)]\| \le g_2(\|x - x^*\|)\|x - x^*\|^{p+1},$$
$$(7.2.12)$$

and

$$\overline{U}(x^*, r_2) \subseteq D,$$

where r_2 is the smallest positive solution of equation

$$h(t) := g_2(t) t^p - 1.$$

Then, sequence $\{x_n\}$ generated by algorithm (7.1.2) for $x_0 \in U(x^*, r_2) - \{x^*\}$ is well defined, remains in $U(x^*, r_2)$ for each $n = 0, 1, 2, \ldots$ and converges to x^*. Moreover, the following estimates hold

$$\|x_{n+1} - x^*\| \leq g_2(\|x_n - x^*\|) \|x_n - x^*\|^{p+1} < \|x_n - x^*\| < r_2.$$

Proof We have that $h(0) = -1 < 0$ and $h(t) \to +\infty$ as $t \to +\infty$. Then, it follows from the intermediate value theorem that function h has positive zeros. Denote by r_2 the smallest such zero. By hypothesis $x_0 \in U(x^*, r_2) - \{x^*\}$. Then, we get in turn that

$$\|x_1 - x^*\| = \|x_0 - x^* - A(x_0)^{-1} F(x_0)\| =$$

$$\|A(x_0)^{-1} [F(x^*) - F(x_0) - A(x_0)(x^* - x_0)]\| \leq$$

$$g_2(\|x_0 - x^*\|) \|x_0 - x^*\|^{p+1} < g_2(r_2) r_2^p \|x_0 - x^*\| =$$

$$\|x_0 - x^*\| < r_2,$$

which shows that $x_1 \in U(x^*, r_2)$ and x_2 is well defined. By a simple inductive argument as in the preceding estimate we get that

$$\|x_{k+1} - x^*\| = \|x_k - x^* - A(x_k)^{-1} F(x_k)\| \leq$$

$$\|A(x_k)^{-1} [F(x^*) - F(x_k) - A(x_k)(x^* - x_k)]\| \leq$$

$$g_2(\|x_k - x^*\|) \|x_k - x^*\|^{p+1} < g_2(r_2) r_2^p \|x_k - x^*\| = \|x_k - x^*\| < r_2,$$

which shows $\lim_{k\to\infty} x_k = x^*$ and $x_{k+1} \in U(x^*, r_2)$.   □

Remark 7.5 (a) Hypothesis (7.2.3) specializes to Newton-Mysowski-type, if $A(x) = F'(x)$ [8, 12, 15]. However, if F is not Fréchet-differentiable, then our results extend the applicability of iterative algorithm (7.1.2).

(b) Theorem 7.1 has practical value although we do not show that x^* is a solution of equation $F(x) = 0$, since this may be shown in another way.

(c) Hypothesis (7.2.12) can be replaced by the stronger

$$\|A(x)^{-1} [F(x) - F(y) - A(x)(x - y)]\| \leq g_2(\|x - y\|) \|x - y\|^{p+1}.$$

The preceding results can be extended to hold for two point iterative algorithms defined for each $n = 0, 1, 2, \ldots$ by

$$x_{n+1} = x_n - A(x_n, x_{n-1})^{-1} F(x_n), \tag{7.2.13}$$

where $x_{-1}, x_0 \in D$ are initial points and $A(w, v) \in L(X, Y)$ for each $v, w \in D$. If $A(w, v) = [w, v; F]$, then iterative algorithm (7.2.13) reduces to the popular secant method, where $[w, v; F]$ denotes a divided difference of order one for the operator F. Many other choices for A are also possible [8, 12, 16].

If we simply replace $A(x)$ by $A(y, x)$ in the proof of Proposition 7.2 we arrive at the following semilocal convergence result for iterative algorithm (7.2.13).

Theorem 7.6 Let $F : D \subset X \to Y$ be a continuous operator and let $A(y, x) \in L(X, Y)$ for each $x, y \in D$. Suppose that there exist $x_{-1}, x_0 \in D$, $\eta \geq 0$, $p \geq 1$, $\psi > 0$, a function $g_1 : [0, \eta] \to [0, \infty)$ continuous and nondecreasing such that for each $x, y \in D$:

$$A(y, x)^{-1} \in L(Y, X), \quad \left\| A(y, x)^{-1} \right\| \leq \psi, \tag{7.2.14}$$

$$\min \left\{ \|x_0 - x_{-1}\|, \left\| A(x_0, x_{-1})^{-1} F(x_0) \right\| \right\} \leq \eta,$$

$$\|F(y) - F(x) - A(y, x)(y - x)\| \leq \frac{g_1(\|x - y\|)}{\psi} \|x - y\|^{p+1}, \tag{7.2.15}$$

$$q_1 < 1, \quad q_1 r_1^p < 1$$

and

$$\overline{U}(x_0, r_1) \subseteq D,$$

where,

$$r_1 = \frac{\eta}{1 - q_1}$$

and q_1 is defined in Proposition 7.2.

Then, sequence $\{x_n\}$ generated by iterative algorithm (7.2.13) is well defined, remains in $\overline{U}(x_0, r_1)$ for each $n = 0, 1, 2, \ldots$ and converges to the only solution of equation $F(x) = 0$ in $\overline{U}(x_0, r_1)$. Moreover, the estimates (7.2.7) and (7.2.8) hold with g_1, q_1 replacing g and q, respectively.

Concerning, the local convergence of the iterative algorithm (7.2.13) we obtain the analogous to Proposition 7.4 result.

Proposition 7.7 Let $F : D \subset X \to Y$ be a continuous operator and let $A(y, x) \in L(X, Y)$. Suppose that there exist $x^* \in D$, $p \geq 1$, a function $g_2 : [0, \infty)^2 \to [0, \infty)$ continuous and nondecreasing such that for each $x, y \in D$

$$F(x^*) = 0, \quad A(y, x)^{-1} \in L(Y, X),$$

$$\left\| A\left(y,x\right)^{-1}\left[F\left(y\right)-F\left(x^{*}\right)-A\left(y,x\right)\left(y-x^{*}\right)\right]\right\| \leq$$

$$g_2\left(\left\|y-x^{*}\right\|,\left\|x-x^{*}\right\|\right)\left\|y-x^{*}\right\|^{p+1}$$

and

$$\overline{U}\left(x^{*},r_2\right)\subseteq D,$$

where r_2 is the smallest positive solution of equation

$$h\left(t\right):=g_2\left(t,t\right)t^p-1.$$

Then, sequence $\{x_n\}$ generated by algorithm (7.2.13) for $x_{-1}, x_0 \in U\left(x^{*}, r_2\right)-\{x^{*}\}$ is well defined, remains in $U\left(x^{*}, r_2\right)$ for each $n = 0, 1, 2, \ldots$ and converges to x^{*}. Moreover, the following estimates hold

$$\left\|x_{n+1}-x^{*}\right\| \leq g_2\left(\left\|x_n-x^{*}\right\|,\left\|x_{n-1}-x^{*}\right\|\right)\left\|x_n-x^{*}\right\|^{p+1}$$

$$< \left\|x_n-x^{*}\right\| < r_2.$$

Remark 7.8 In the next section we present some choices and properties of operator $A\left(y,x\right)$ from fractional calculus satisfying the crucial estimate (7.2.15) in the special case when,

$$g_1\left(t\right)=\frac{c_1\psi}{p+1}\text{ for some }c_1 > 0.$$

Hence, Theorem 7.6 can apply to solve equation $F\left(x\right)=0$. Other choices for operator $A\left(x\right)$ or operator $A\left(y,x\right)$ can be found in [5, 7, 8, 10–16].

7.3   Applications to Fractional Calculus

In this section we apply the earlier numerical methods to fractional calculus for solving $f\left(x\right)=0$.

Here we would like to establish for $[a,b]\subseteq\mathbb{R}$, $a < b$, $f \in C^p\left([a,b]\right)$, $p \in \mathbb{N}$, that

$$\left|f\left(y\right)-f\left(x\right)-A\left(x,y\right)\left(y-x\right)\right|\leq c_1\frac{|x-y|^{p+1}}{p+1},\qquad(7.3.1)$$

$\forall\,x,y\in[a,b]$, where $c_1 > 0$, and

$$\left|A\left(x,x\right)-A\left(y,y\right)\right|\leq c_2\left|x-y\right|,\qquad(7.3.2)$$

with $c_2 > 0$, $\forall\,x,y\in[a,b]$.

Above A stands for a differential operator to be defined and presented per case in the next, it will be denoted as $A_+ (f)$, $A_- (f)$ in the fractional cases, and $A_0 (f)$ in the ordinary case.

We examine the following cases:

(I) Here see [3], pp. 7–10.

Let $x, y \in [a, b]$ such that $x \geq y, \nu > 0, \nu \notin \mathbb{N}$, such that $p = [\nu]$, $[\cdot]$ the integral part, $\alpha = \nu - p$ $(0 < \alpha < 1)$.

Let $f \in C^p ([a, b])$ and define

$$\left(J_\nu^y f\right)(x) := \frac{1}{\Gamma(\nu)} \int_y^x (x - t)^{\nu - 1} f(t) \, dt, \quad y \leq x \leq b, \tag{7.3.3}$$

the left generalized Riemann-Liouville fractional integral.

Here Γ stands for the gamma function.

Clearly here it holds $\left(J_\nu^y f\right)(y) = 0$. We define $\left(J_\nu^y f\right)(x) = 0$ for $x < y$. By [3], p. 388, $\left(J_\nu^y f\right)(x)$ is a continuous function in x, for a fixed y.

We define the subspace $C_{y+}^\nu ([a, b])$ of $C^p ([a, b])$:

$$C_{y+}^\nu ([a, b]) := \left\{ f \in C^p ([a, b]) : J_{1-\alpha}^y f^{(p)} \in C^1 ([y, b]) \right\}. \tag{7.3.4}$$

So let $f \in C_{y+}^\nu ([a, b])$, we define the generalized ν—fractional derivative of f over $[y, b]$ as

$$D_y^\nu f = \left(J_{1-\alpha}^y f^{(p)}\right)', \tag{7.3.5}$$

that is

$$\left(D_y^\nu f\right)(x) = \frac{1}{\Gamma(1-\alpha)} \frac{d}{dx} \int_y^x (x - t)^{-\alpha} f^{(p)}(t) \, dt, \tag{7.3.6}$$

which exists for $f \in C_{y+}^\nu ([a, b])$, for $a \leq y \leq x \leq b$.

Here we consider $f \in C^p ([a, b])$ such that $f \in C_{y+}^\nu ([a, b])$, for every $y \in [a, b]$, which means also that $f \in C_{x+}^\nu ([a, b])$, for every $x \in [a, b]$ (i.e. exchange roles of x and y), we write that as $f \in C_+^\nu ([a, b])$.

That is

$$\left(D_x^\nu f\right)(y) = \frac{1}{\Gamma(1-\alpha)} \frac{d}{dy} \int_x^y (y - t)^{-\alpha} f^{(p)}(t) \, dt \tag{7.3.7}$$

exists for $f \in C_{x+}^\nu ([a, b])$, for $a \leq x \leq y \leq b$.

We mention the following left generalized fractional Taylor formula ($f \in C_{y+}^\nu ([a, b]), \nu > 1$).

It holds

$$f(x) - f(y) = \sum_{k=1}^{p-1} \frac{f^{(k)}(y)}{k!} (x - y)^k + \frac{1}{\Gamma(\nu)} \int_y^x (x - t)^{\nu - 1} \left(D_y^\nu f\right)(t) \, dt, \tag{7.3.8}$$

all $x, y \in [a, b]$ with $x \geq y$.

Similarly for $f \in C_{x+}^{\nu}([a, b])$ we have

$$f(y) - f(x) = \sum_{k=1}^{p-1} \frac{f^{(k)}(x)}{k!} (y - x)^k + \frac{1}{\Gamma(\nu)} \int_x^y (y - t)^{\nu-1} (D_x^{\nu} f)(t) dt,$$

(7.3.9)

all $x, y \in [a, b]$ with $y \geq x$.

So here we work with $f \in C^p([a, b])$, such that $f \in C_+^{\nu}([a, b])$.

We define the left linear fractional operator

$$(A_+(f))(x, y) := \begin{cases} \sum_{k=1}^{p-1} \frac{f^{(k)}(y)}{k!} (x - y)^{k-1} + (D_y^{\nu} f)(x) \frac{(x-y)^{\nu-1}}{\Gamma(\nu+1)}, & x > y, \\ \sum_{k=1}^{p-1} \frac{f^{(k)}(x)}{k!} (y - x)^{k-1} + (D_x^{\nu} f)(y) \frac{(y-x)^{\nu-1}}{\Gamma(\nu+1)}, & y > x, \\ f^{(p-1)}(x), x = y. \end{cases}$$

(7.3.10)

Notice that

$$|(A_+(f))(x, x) - (A_+(f))(y, y)| = \left| f^{(p-1)}(x) - f^{(p-1)}(y) \right|$$

(7.3.11)

$$\leq \left\| f^{(p)} \right\|_{\infty} |x - y|, \quad \forall\, x, y \in [a, b],$$

so that condition (7.3.2) is fulfilled.

Next we will prove condition (7.3.1). It is trivially true if $x = y$. So, we examine the case of $x \neq y$.

We distinguish the subcases:

(1) $x > y$: We observe that

$$|f(y) - f(x) - A_+(f)(x, y)(y - x)| =$$

$$|f(x) - f(y) - A_+(f)(x, y)(x - y)| \overset{\text{(by (7.3.8), (7.3.10))}}{=}$$

$$\left| \sum_{k=1}^{p-1} \frac{f^{(k)}(y)}{k!} (x - y)^k + \frac{1}{\Gamma(\nu)} \int_y^x (x - t)^{\nu-1} (D_y^{\nu} f)(t) dt - \right.$$

(7.3.12)

$$\left. \sum_{k=1}^{p-1} \frac{f^{(k)}(y)}{k!} (x - y)^k - (D_y^{\nu} f)(x) \frac{(x - y)^{\nu}}{\Gamma(\nu + 1)} \right| =$$

$$\left| \frac{1}{\Gamma(\nu)} \int_y^x (x - t)^{\nu-1} (D_y^{\nu} f)(t) dt - \frac{1}{\Gamma(\nu)} \int_y^x (x - t)^{\nu-1} (D_y^{\nu} f)(x) dt \right| =$$

$$\frac{1}{\Gamma(\nu)} \left| \int_y^x (x - t)^{\nu-1} ((D_y^{\nu} f)(t) - (D_y^{\nu} f)(x)) dt \right| \leq$$

(7.3.13)

$$\frac{1}{\Gamma(\nu)} \int_y^x (x-t)^{\nu-1} \left| \left(D_y^\nu f\right)(t) - \left(D_y^\nu f\right)(x) \right| dt$$

(we assume that

$$\left| \left(D_y^\nu f\right)(t) - \left(D_y^\nu f\right)(x) \right| \le \lambda_1(y) |t-x|^{p+1-\nu}, \tag{7.3.14}$$

for all $x, y, t \in [a, b]$ with $x \ge t \ge y$, with $\lambda_1(y) > 0$ and $\lim_{y\in[a,b]}\sup\lambda_1(y) =: \lambda_1 < \infty$, also it is $0 < p + 1 - \nu < 1$)

$$\le \frac{\lambda_1}{\Gamma(\nu)} \int_y^x (x-t)^{\nu-1} (x-t)^{p+1-\nu} dt = \tag{7.3.15}$$

$$\frac{\lambda_1}{\Gamma(\nu)} \int_y^x (x-t)^p dt = \frac{\lambda_1}{\Gamma(\nu)} \frac{(x-y)^{p+1}}{(p+1)}.$$

We have proved condition (7.3.1)

$$|f(y) - f(x) - A_+(f)(x, y)(y-x)| \le \frac{\lambda_1}{\Gamma(\nu)} \frac{(x-y)^{p+1}}{(p+1)}, \text{ for } x > y. \tag{7.3.16}$$

(2) $x < y$: We observe that

$$|f(y) - f(x) - (A_+(f))(x, y)(y-x)| \overset{\text{(by (7.3.9), (7.3.10))}}{=}$$

$$\left| \sum_{k=1}^{p-1} \frac{f^{(k)}(x)}{k!} (y-x)^k + \frac{1}{\Gamma(\nu)} \int_x^y (y-t)^{\nu-1} \left(D_x^\nu f\right)(t) dt - \tag{7.3.17}$$

$$\sum_{k=1}^{p-1} \frac{f^{(k)}(x)}{k!} (y-x)^k - \left(D_x^\nu f\right)(y) \frac{(y-x)^\nu}{\Gamma(\nu+1)} \right| =$$

$$\left| \frac{1}{\Gamma(\nu)} \int_x^y (y-t)^{\nu-1} \left(D_x^\nu f\right)(t) dt - \left(D_x^\nu f\right)(y) \frac{(y-x)^\nu}{\Gamma(\nu+1)} \right| =$$

$$\left| \frac{1}{\Gamma(\nu)} \int_x^y (y-t)^{\nu-1} \left(D_x^\nu f\right)(t) dt - \frac{1}{\Gamma(\nu)} \int_x^y (y-t)^{\nu-1} \left(D_x^\nu f\right)(y) dt \right| = \tag{7.3.18}$$

$$\frac{1}{\Gamma(\nu)} \left| \int_x^y (y-t)^{\nu-1} \left(\left(D_x^\nu f\right)(t) - \left(D_x^\nu f\right)(y) \right) dt \right| \le$$

$$\frac{1}{\Gamma(\nu)} \int_x^y (y-t)^{\nu-1} \left| (D_x^\nu f)(t) - (D_x^\nu f)(y) \right| dt$$

(we assume here that

$$\left| (D_x^\nu f)(t) - (D_x^\nu f)(y) \right| \le \lambda_2(x) |t-y|^{p+1-\nu}, \tag{7.3.19}$$

for all $x, y, t \in [a, b]$ with $y \ge t \ge x$, with $\lambda_2(x) > 0$ and $\lim_{x \in [a,b]} \sup \lambda_2(x) =: \lambda_2 < \infty$)

$$\le \frac{\lambda_2}{\Gamma(\nu)} \int_x^y (y-t)^{\nu-1} (y-t)^{p+1-\nu} dt = \tag{7.3.20}$$

$$\frac{\lambda_2}{\Gamma(\nu)} \int_x^y (y-t)^p dt = \frac{\lambda_2}{\Gamma(\nu)} \frac{(y-x)^{p+1}}{(p+1)}.$$

We have proved that

$$|f(y) - f(x) - (A_+(f))(x, y)(y-x)| \le \frac{\lambda_2}{\Gamma(\nu)} \frac{(y-x)^{p+1}}{(p+1)}, \tag{7.3.21}$$

for all $x, y \in [a, b]$ such that $y > x$.

Call $\lambda := \max(\lambda_1, \lambda_2)$.

Conclusion We have proved condition (7.3.1), in detail that

$$|f(y) - f(x) - (A_+(f))(x, y)(y-x)| \le \frac{\lambda}{\Gamma(\nu)} \frac{|x-y|^{p+1}}{(p+1)}, \quad \forall x, y \in [a, b]. \tag{7.3.22}$$

(II) Here see [4], p. 333, and again [4], pp. 345–348.

Let $x, y \in [a, b]$ such that $x \le y$, $\nu > 0$, $\nu \notin \mathbb{N}$, such that $p = [\nu]$, $\alpha = \nu - p$ $(0 < \alpha < 1)$.

Let $f \in C^p([a, b])$ and define

$$(J_{y-}^\nu f)(x) := \frac{1}{\Gamma(\nu)} \int_x^y (z-x)^{\nu-1} f(z) dz, \quad a \le x \le y, \tag{7.3.23}$$

the right generalized Riemann-Liouville fractional integral.

Define the subspace of functions

$$C_{y-}^\nu([a, b]) := \left\{ f \in C^p([a, b]) : J_{y-}^{1-\alpha} f^{(p)} \in C^1([a, y]) \right\}. \tag{7.3.24}$$

Define the right generalized ν—fractional derivative of f over $[a, y]$ as

$$D_{y-}^\nu f := (-1)^{p-1} \left(J_{y-}^{1-\alpha} f^{(p)} \right)'. \tag{7.3.25}$$

Notice that

$$J_{y-}^{1-\alpha} f^{(p)}(x) = \frac{1}{\Gamma(1-\alpha)} \int_x^y (z-x)^{-\alpha} f^{(p)}(z)\, dz, \qquad (7.3.26)$$

exists for $f \in C_{y-}^{\nu}([a,b])$, and

$$\left(D_{y-}^{\nu} f\right)(x) = \frac{(-1)^{p-1}}{\Gamma(1-\alpha)} \frac{d}{dx} \int_x^y (z-x)^{-\alpha} f^{(p)}(z)\, dz. \qquad (7.3.27)$$

I.e.

$$\cdot \left(D_{y-}^{\nu} f\right)(x) = \frac{(-1)^{p-1}}{\Gamma(p-\nu+1)} \frac{d}{dx} \int_x^y (z-x)^{p-\nu} f^{(p)}(z)\, dz. \qquad (7.3.28)$$

Here we consider $f \in C^p([a,b])$ such that $f \in C_{y-}^{\nu}([a,b])$, for every $y \in [a,b]$, which means also that $f \in C_{x-}^{\nu}([a,b])$, for every $x \in [a,b]$ (i.e. exchange roles of x and y), we write that as $f \in C_{-}^{\nu}([a,b])$.

That is

$$\left(D_{x-}^{\nu} f\right)(y) = \frac{(-1)^{p-1}}{\Gamma(p-\nu+1)} \frac{d}{dy} \int_y^x (z-y)^{p-\nu} f^{(p)}(z)\, dz \qquad (7.3.29)$$

exists for $f \in C_{x-}^{\nu}([a,b])$, for $a \le y \le x \le b$.

We mention the following right generalized fractional Taylor formula ($f \in C_{y-}^{\nu}([a,b]), \nu > 1$).

It holds

$$f(x) - f(y) = \sum_{k=1}^{p-1} \frac{f^{(k)}(y)}{k!}(x-y)^k + \frac{1}{\Gamma(\nu)} \int_x^y (z-x)^{\nu-1} \left(D_{y-}^{\nu} f\right)(z)\, dz, \qquad (7.3.30)$$

all $x, y \in [a,b]$ with $x \le y$.

Similarly for $f \in C_{x-}^{\nu}([a,b])$ we have

$$f(y) - f(x) = \sum_{k=1}^{p-1} \frac{f^{(k)}(x)}{k!}(y-x)^k + \frac{1}{\Gamma(\nu)} \int_y^x (z-y)^{\nu-1} \left(D_{x-}^{\nu} f\right)(z)\, dz, \qquad (7.3.31)$$

all $x, y \in [a,b]$ with $x \ge y$.

So here we work with $f \in C^p([a, b])$, such that $f \in C^\nu_-([a, b])$.
We define the right linear fractional operator

$$A_-(f)(x, y) := \begin{cases} \sum_{k=1}^{p-1} \frac{f^{(k)}(x)}{k!}(y-x)^{k-1} - \left(D^\nu_{x-}f\right)(y)\frac{(x-y)^{\nu-1}}{\Gamma(\nu+1)}, & x > y, \\ \sum_{k=1}^{p-1} \frac{f^{(k)}(y)}{k!}(x-y)^{k-1} - \left(D^\nu_{y-}f\right)(x)\frac{(y-x)^{\nu-1}}{\Gamma(\nu+1)}, & y > x, \\ f^{(p-1)}(x), & x = y. \end{cases}$$

$$(7.3.32)$$

Condition (7.3.2) is fulfilled, the same as in (7.3.11), now for $A_-(f)(x, x)$.
We would like to prove that

$$|f(x) - f(y) - (A_-(f))(x, y)(x - y)| \le c \cdot \frac{|x-y|^{p+1}}{p+1}, \qquad (7.3.33)$$

for any $x, y \in [a, b]$, where $c > 0$.

When $x = y$ the last condition (7.3.33) is trivial. We assume $x \ne y$.
We distinguish the subcases:

(1) $x > y$: We observe that

$$|(f(x) - f(y)) - (A_-(f))(x, y)(x - y)| = \qquad (7.3.34)$$

$$|(f(y) - f(x)) - (A_-(f))(x, y)(y - x)| =$$

$$\left| \left(\sum_{k=1}^{p-1} \frac{f^{(k)}(x)}{k!}(y-x)^k + \frac{1}{\Gamma(\nu)}\int_y^x (z-y)^{\nu-1}\left(D^\nu_{x-}f\right)(z)\,dz \right) - \right.$$

$$\left(\sum_{k=1}^{p-1} \frac{f^{(k)}(x)}{k!}(y-x)^{k-1} - \left(D^\nu_{x-}f\right)(y)\frac{(x-y)^{\nu-1}}{\Gamma(\nu+1)} \right)(y-x) \Bigg| = \qquad (7.3.35)$$

$$\left| \frac{1}{\Gamma(\nu)}\int_y^x (z-y)^{\nu-1}\left(D^\nu_{x-}f\right)(z)\,dz + \left(D^\nu_{x-}f\right)(y)\frac{(x-y)^{\nu-1}}{\Gamma(\nu+1)}(y-x) \right| =$$

$$\left| \frac{1}{\Gamma(\nu)}\int_y^x (z-y)^{\nu-1}\left(D^\nu_{x-}f\right)(z)\,dz - \left(D^\nu_{x-}f\right)(y)\frac{(x-y)^{\nu}}{\Gamma(\nu+1)} \right| =$$

$$\frac{1}{\Gamma(\nu)}\left| \int_y^x (z-y)^{\nu-1}\left(D^\nu_{x-}f\right)(z)\,dz - \int_y^x (z-y)^{\nu-1}\left(D^\nu_{x-}f\right)(y)\,dz \right| =$$

$$\frac{1}{\Gamma(\nu)}\left| \int_y^x (z-y)^{\nu-1}\left(\left(D^\nu_{x-}f\right)(z) - \left(D^\nu_{x-}f\right)(y)\right)\,dz \right| \le \qquad (7.3.36)$$

$$\frac{1}{\Gamma(\nu)}\int_y^x (z-y)^{\nu-1}\left|\left(D^\nu_{x-}f\right)(z) - \left(D^\nu_{x-}f\right)(y)\right|\,dz$$

(we assume that

$$\left|\left(D_{x-}^{\nu}f\right)(z) - \left(D_{x-}^{\nu}f\right)(y)\right| \leq \lambda_1 |z - y|^{p+1-\nu}, \qquad (7.3.37)$$

$\lambda_1 > 0$, for all $x, z, y \in [a, b]$ with $x \geq z \geq y$)

$$\leq \frac{\lambda_1}{\Gamma(\nu)} \int_y^x (z - y)^{\nu-1} (z - y)^{p+1-\nu} dz = \qquad (7.3.38)$$

$$\frac{\lambda_1}{\Gamma(\nu)} \int_y^x (z - y)^p dz = \frac{\lambda_1}{\Gamma(\nu)} \frac{(x - y)^{p+1}}{p+1} = \rho_1 \frac{(x - y)^{p+1}}{p+1},$$

where $\rho_1 := \frac{\lambda_1}{\Gamma(\nu)} > 0$.

We have proved, when $x > y$, that

$$|f(x) - f(y) - (A_-(f))(x, y)(x - y)| \leq \rho_1 \frac{(x - y)^{p+1}}{p+1}. \qquad (7.3.39)$$

(2) $y > x$: We observe that

$$|f(x) - f(y) - (A_-(f))(x, y)(x - y)| =$$

$$\left| \left(\sum_{k=1}^{p-1} \frac{f^{(k)}(y)}{k!} (x - y)^k + \frac{1}{\Gamma(\nu)} \int_x^y (z - x)^{\nu-1} \left(D_{y-}^{\nu}f\right)(z) dz \right) - \right.$$

$$\left. \left(\sum_{k=1}^{p-1} \frac{f^{(k)}(y)}{k!} (x - y)^{k-1} - \left(D_{y-}^{\nu}f\right)(x) \frac{(y - x)^{\nu-1}}{\Gamma(\nu+1)} \right) (x - y) \right| = \qquad (7.3.40)$$

$$\left| \frac{1}{\Gamma(\nu)} \int_x^y (z - x)^{\nu-1} \left(D_{y-}^{\nu}f\right)(z) dz - \left(D_{y-}^{\nu}f\right)(x) \frac{(y - x)^{\nu}}{\Gamma(\nu+1)} \right| =$$

$$\left| \frac{1}{\Gamma(\nu)} \int_x^y (z - x)^{\nu-1} \left(D_{y-}^{\nu}f\right)(z) dz - \frac{1}{\Gamma(\nu)} \int_x^y (z - x)^{\nu-1} \left(D_{y-}^{\nu}f\right)(x) dz \right| = \qquad (7.3.41)$$

$$\frac{1}{\Gamma(\nu)} \left| \int_x^y (z - x)^{\nu-1} \left(\left(D_{y-}^{\nu}f\right)(z) - \left(D_{y-}^{\nu}f\right)(x) \right) dz \right| \leq \qquad (7.3.42)$$

$$\frac{1}{\Gamma(\nu)} \int_x^y (z - x)^{\nu-1} \left|\left(D_{y-}^{\nu}f\right)(z) - \left(D_{y-}^{\nu}f\right)(x)\right| dz$$

(we assume that

$$\left| \left(D_{y-}^{\nu} f \right)(z) - \left(D_{y-}^{\nu} f \right)(x) \right| \leq \lambda_2 \left| z - x \right|^{p+1-\nu}, \tag{7.3.43}$$

$\lambda_2 > 0$, for all $y, z, x \in [a, b]$ with $y \geq z \geq x$)

$$\leq \frac{\lambda_2}{\Gamma(\nu)} \int_x^y (z - x)^{\nu-1} (z - x)^{p+1-\nu} \, dz = \tag{7.3.44}$$

$$\frac{\lambda_2}{\Gamma(\nu)} \int_x^y (z - x)^p \, dz = \frac{\lambda_2}{\Gamma(\nu)} \frac{(y - x)^{p+1}}{p+1}.$$

We have proved, for $y > x$, that

$$|f(x) - f(y) - (A_-(f))(x, y)(x - y)| \leq \rho_2 \frac{(y - x)^{p+1}}{p+1}, \tag{7.3.45}$$

where $\rho_2 := \frac{\lambda_2}{\Gamma(\nu)} > 0$.

Set $\lambda := \max(\lambda_1, \lambda_2)$ and $\rho := \frac{\lambda}{\Gamma(\nu)} > 0$.

Conclusion We have proved (7.3.1) that

$$|f(x) - f(y) - (A_-(f))(x, y)(x - y)| \leq \rho \frac{|x - y|^{p+1}}{p+1}, \tag{7.3.46}$$

for any $x, y \in [a, b]$.

(III) Let again $f \in C^p([a, b])$, $p \in \mathbb{N}$, $x, y \in [a, b]$.
By Taylor's formula we have

$$f(x) - f(y) =$$

$$\sum_{k=1}^{p} \frac{f^{(k)}(y)}{k!} (x - y)^k + \frac{1}{(p-1)!} \int_y^x (x - t)^{p-1} \left(f^{(p)}(t) - f^{(p)}(y) \right) dt, \tag{7.3.47}$$

$\forall \, x, y \in [a, b]$.

We define the function

$$(A_0(f))(x, y) := \begin{cases} \sum_{k=1}^{p} \frac{f^{(k)}(y)}{k!} (x - y)^{k-1}, & x \neq y, \\ f^{(p-1)}(x), & x = y. \end{cases} \tag{7.3.48}$$

Then it holds

$$|(A_0(f))(x,x) - (A_0(f))(y,y)| = \left| f^{(p-1)}(x) - f^{(p-1)}(y) \right| \qquad (7.3.49)$$

$$\leq \left\| f^{(p)} \right\|_\infty |x - y|, \quad \forall\, x, y \in [a, b],$$

so that condition (7.3.2) is fulfilled.

Next we observe that

$$|f(x) - f(y) - (A_0(f))(x,y)(x-y)| =$$

$$\left| \sum_{k=1}^{p} \frac{f^{(k)}(y)}{k!}(x-y)^k + \frac{1}{(p-1)!}\int_y^x (x-t)^{p-1}\left(f^{(p)}(t) - f^{(p)}(y) \right) dt \right.$$

$$(7.3.50)$$

$$\left. - \sum_{k=1}^{p} \frac{f^{(k)}(y)}{k!}(x-y)^k \right| =$$

$$\frac{1}{(p-1)!}\left| \int_y^x (x-t)^{p-1}\left(f^{(p)}(t) - f^{(p)}(y) \right) dt \right| =: (\xi). \qquad (7.3.51)$$

Here we assume that

$$\left| f^{(p)}(t) - f^{(p)}(y) \right| \leq c\,|t - y|, \quad \forall\, t, y \in [a,b]\,, c > 0. \qquad (7.3.52)$$

(1) Subcase of $x > y$: We have that

$$(\xi) \leq \frac{1}{(p-1)!}\int_y^x (x-t)^{p-1}\left| f^{(p)}(t) - f^{(p)}(y) \right| dt \leq$$

$$\frac{c}{(p-1)!}\int_y^x (x-t)^{p-1}(t-y)^{2-1}\, dt = \qquad (7.3.53)$$

$$c\frac{\Gamma(p)\Gamma(2)}{(p-1)!\,\Gamma(p+2)}(x-y)^{p+1} = c\frac{(p-1)!}{(p-1)!\,(p+1)!}(x-y)^{p+1}$$

$$= \frac{c(x-y)^{p+1}}{(p+1)!}.$$

Hence

$$(\xi) \leq c\frac{(x-y)^{p+1}}{(p+1)!}, \quad x > y. \qquad (7.3.54)$$

(2) Subcase of $y > x$.

We have that

$$(\xi) = \frac{1}{(p-1)!} \left| \int_x^y (t-x)^{p-1} \left(f^{(p)}(y) - f^{(p)}(t) \right) dt \right| \le \tag{7.3.55}$$

$$\frac{1}{(p-1)!} \int_x^y (t-x)^{p-1} \left| f^{(p)}(y) - f^{(p)}(t) \right| dt \le$$

$$\frac{c}{(p-1)!} \int_x^y (t-x)^{p-1} (y-t) dt =$$

$$\frac{c}{(p-1)!} \int_x^y (y-t)^{2-1} (t-x)^{p-1} dt = \tag{7.3.56}$$

$$\frac{c}{(p-1)!} \frac{\Gamma(2)\Gamma(p)}{\Gamma(p+2)} (y-x)^{p+1} = \frac{c}{(p-1)!} \frac{(p-1)!}{(p+1)!} (y-x)^{p+1}$$

$$= c \frac{(y-x)^{p+1}}{(p+1)!}.$$

That is

$$(\xi) \le c \frac{(y-x)^{p+1}}{(p+1)!}, \quad y > x. \tag{7.3.57}$$

Therefore it holds

$$(\xi) \le c \frac{|x-y|^{p+1}}{(p+1)!}, \quad \text{all } x, y \in [a,b] \text{ such that } x \ne y. \tag{7.3.58}$$

We have found that

$$|f(x) - f(y) - (A_0(f))(x,y)(x-y)| \le c \frac{|x-y|^{p+1}}{(p+1)!}, \quad c > 0, \tag{7.3.59}$$

for all $x \ne y$.

When $x = y$ inequality (7.3.59) holds trivially, so (7.3.1) it is true for any $x, y \in [a,b]$.

References

1. S. Amat, S. Busquier, Third-order iterative methods under Kantorovich conditions. J. Math. Anal. Appl. **336**, 243–261 (2007)
2. S. Amat, S. Busquier, S. Plaza, Chaotic dynamics of a third-order Newton-like method. J. Math. Anal. Appl. **366**(1), 164–174 (2010)

3. G. Anastassiou, *Fractional Differentiation Inequalities* (Springer, New York, 2009)
4. G. Anastassiou, *Intelligent Mathematics: Computational Analysis* (Springer, Heidelberg, 2011)
5. G. Anastassiou, I. Argyros, *A Unified Convergence Analysis for a Certain Family of Iterative Algorithms with Applications to Fractional Calculus* (2015) (submitted)
6. I.K. Argyros, Newton-like methods in partially ordered linear spaces. J. Approx. Theory Appl. **9**(1), 1–10 (1993)
7. I.K. Argyros, Results on controlling the residuals of perturbed Newton-like methods on Banach spaces with a convergence structure. Southwest J. Pure Appl. Math. **1**, 32–38 (1995)
8. I.K. Argyros, *Convergence and Applications of Newton-like iterations* (Springer, New York, 2008)
9. K. Diethelm, *The Analysis of Fractional Differential Equations*. Lecture Notes in Mathematics, vol. 2004, 1st edn. (Springer, New York, 2010)
10. J.A. Ezquerro, J.M. Gutierrez, M.A. Hernandez, N. Romero, M.J. Rubio, The Newton method: from Newton to Kantorovich (Spanish). Gac. R. Soc. Mat. Esp. **13**, 53–76 (2010)
11. J.A. Ezquerro, M.A. Hernandez, Newton-like methods of high order and domains of semilocal and global convergence. Appl. Math. Comput. **214**(1), 142–154 (2009)
12. L.V. Kantorovich, G.P. Akilov, *Functional Analysis in Normed Spaces* (Pergamon Press, New York, 1964)
13. A.A. Magrenan, Different anomalies in a Jarratt family of iterative root finding methods. Appl. Math. Comput. **233**, 29–38 (2014)
14. A.A. Magrenan, A new tool to study real dynamics: the convergence plane. Appl. Math. Comput. **248**, 215–224 (2014)
15. F.A. Potra, V. Ptak, *Nondiscrete Induction and Iterative Processes* (Pitman Publishing, London, 1984)
16. P.D. Proinov, New general convergence theory for iterative processes and its applications to Newton-Kantorovich type theorems. J. Complexity **26**, 3–42 (2010)

Chapter 8
Convergence Analysis for Extended Iterative Algorithms and Fractional and Vector Calculus

We give local and semilocal convergence results for some iterative algorithms in order to approximate a locally unique solution of a nonlinear equation in a Banach space setting. In earlier studies the operator involved is assumed to be at least once Fréchet-differentiable. In the present study, we assume that the operator is only continuous. This way we expand the applicability of iterative algorithms. In the third part of the study we present some choices of the operators involved in fractional calculus and vector calculus where the operators satisfy the convergence conditions. It follows [5].

8.1 Introduction

In this study we are concerned with the problem of approximating a locally unique solution x^* of the nonlinear equation

$$F(x) = 0, \qquad (8.1.1)$$

where F is a continuous operator defined on a subset D of a Banach space X with values in a Banach space Y.

A lot of problems in Computational Sciences and other disciplines can be brought in a form like (8.1.1) using Mathematical Modelling [8, 13, 17, 19, 20]. The solutions of such equations can be found in closed form only in special cases. That is why most solution methods for these equations are iterative. Iterative algorithms are usually studied based on semilocal and local convergence. The semilocal convergence matter is, based on the information around the initial point to give hypotheses ensuring the convergence of the iterative algorithm; while the local one is, based on the information around a solution, to find estimates of the radii of convergence balls as well as error bounds on the distances involved.

© Springer International Publishing Switzerland 2016
G.A. Anastassiou and I.K. Argyros, *Intelligent Numerical Methods:*
Applications to Fractional Calculus, Studies in Computational Intelligence 624,
DOI 10.1007/978-3-319-26721-0_8

We introduce the iterative algorithm defined for each $n = 0, 1, 2, \ldots$ by

$$x_{n+1} = x_n - (A\,(F)\,(x_n))^{-1}\,F\,(x_n)\,, \tag{8.1.2}$$

where $x_0 \in D$ is an initial point and $A\,(F)\,(x) \in L\,(X, Y)$ the space of bounded linear operators from X into Y. There is a plethora on local as well as semilocal convergence theorems for iterative algorithm (8.1.2) provided that the operator A is an approximation to the Fréchet-derivative F' [1, 2, 6–16]. In the present study we do not assume that operator A is related to F'. This way we expand the applicability of iterative algorithm (8.1.2). Notice that many well known methods are special case of interative algorithm (8.1.2).

Newton's method: Choose $A\,(F)\,(x) = F'\,(x)$ for each $x \in D$.

Steffensen's method: Choose $A\,(F)\,(x) = [x, G\,(x)\,;\,F]$, where $G : X \to X$ is a known operator and $[x, y;\,F]$ denotes a divided difference of order one [8, 13, 16].

The so called Newton-like methods and many other methods are special cases of iterative algorithm (8.1.2).

The rest of the chapter is organized as follows. The semilocal as well as the local convergence analysis of iterative algorithm (8.1.2) is given in Sect. 8.2. Some applications from fractional calculus are given in Sect. 8.3.

8.2 Convergence Analysis

We present the main semilocal convergence result for iterative algorithm (8.1.2).

Theorem 8.1 *Let $F : D \subset X \to Y$ be a continuous operator and let $A\,(F)\,(x) \in L\,(X, Y)$. Suppose that there exist $x_0 \in D$, $\eta \geq 0$, $p \geq 1$, a function $h : [0, \eta] \to [0, \infty)$ continuous and nondecreasing such that for each $x, y \in D$*

$$(A\,(F)\,(x))^{-1} \in L\,(Y, X)\,, \tag{8.2.1}$$

$$\left\| (A\,(F)\,(x_0))^{-1}\,F\,(x_0) \right\| \leq \eta, \tag{8.2.2}$$

$$\left\| (A\,(F)\,(y))^{-1}\,(F\,(y) - F\,(x) - A\,(F)\,(x)\,(y - x)) \right\| \leq h\,(\|x - y\|)\,\|x - y\|^{p+1}\,, \tag{8.2.3}$$

$$q := h\,(\eta)\,\eta^p < 1 \tag{8.2.4}$$

and

$$\overline{U}\,(x_0, r) \subseteq D\,, \tag{8.2.5}$$

where,

$$r = \frac{\eta}{1 - q}\,. \tag{8.2.6}$$

Then, the sequence $\{x_n\}$ generated by iterative algorithm (8.1.2) is well defined, remains in $\overline{U}(x_0, r)$ for each $n = 0, 1, 2, \ldots$ and converges to some $x^* \in \overline{U}(x_0, r)$ such that

$$\|x_{n+1} - x_n\| \le h(\|x_n - x_{n-1}\|) \|x_n - x_{n-1}\|^{p+1} \le q \|x_n - x_{n-1}\| \quad (8.2.7)$$

and

$$\|x_n - x^*\| \le \frac{q^n \eta}{1-q}. \quad (8.2.8)$$

Proof The iterate x_1 is well defined by iterative algorithm (8.1.2) for $n = 0$ and (8.2.1) for $x = x_0$. We also have by (8.2.2) and (8.2.6) that $\|x_1 - x_0\| = \left\| (A(F)(x_0))^{-1} F(x_0) \right\| \le \eta < r$, so we get that $x_1 \in \overline{U}(x_0, r)$ and x_2 is well defined (by (8.2.5)). Using (8.2.3) for $y = x_1$, $x = x_0$ and (8.2.4) we get that

$$\|x_2 - x_1\| = \left\| (A(F)(x_1))^{-1} [F(x_1) - F(x_0) - A(F)(x_0)(x_1 - x_0)] \right\|$$

$$\le h(\|x_1 - x_0\|) \|x_1 - x_0\|^{p+1} \le q \|x_1 - x_0\|,$$

which shows (8.2.7) for $n = 1$. Then, we can have that

$$\|x_2 - x_0\| \le \|x_2 - x_1\| + \|x_1 - x_0\| \le q \|x_1 - x_0\| + \|x_1 - x_0\|$$

$$= (1+q) \|x_1 - x_0\| \le \frac{1-q^2}{1-q} \eta < r,$$

so $x_2 \in \overline{U}(x_0, r)$ and x_3 is well defined.

Assuming $\|x_{k+1} - x_k\| \le q \|x_k - x_{k-1}\|$ and $x_{k+1} \in \overline{U}(x_0, r)$ for each $k = 1, 2, \ldots, n$ we get

$$\|x_{k+2} - x_{k+1}\| =$$

$$\left\| (A(F)(x_{k+1}))^{-1} [F(x_{k+1}) - F(x_k) - A(F)(x_k)(x_{k+1} - x_k)] \right\|$$

$$\le h(\|x_{k+1} - x_k\|) \|x_{k+1} - x_k\|^{p+1}$$

$$\le h(\|x_1 - x_0\|) \|x_1 - x_0\|^p \|x_{k+1} - x_k\| \le q \|x_{k+1} - x_k\|$$

and

$$\|x_{k+2} - x_0\| \le \|x_{k+2} - x_{k+1}\| + \|x_{k+1} - x_k\| + \cdots + \|x_1 - x_0\|$$

$$\le (q^{k+1} + q^k + \cdots + 1) \|x_1 - x_0\| \le \frac{1-q^{k+2}}{1-q} \|x_1 - x_0\|$$

$$< \frac{\eta}{1-q} = r,$$

which completes the induction for (8.2.7) and $x_{k+2} \in \overline{U}(x_0, r)$. We also have that for $m \geq 0$

$$\|x_{n+m} - x_n\| \leq \|x_{n+m} - x_{n+m-1}\| + \cdots + \|x_{n+1} - x_n\|$$

$$\leq \left(q^{m-1} + q^{m-2} + \cdots + 1\right) \|x_{n+1} - x_n\|$$

$$\leq \frac{1 - q^m}{1 - q} q^n \|x_1 - x_0\|.$$

It follows that $\{x_n\}$ is a complete sequence in a Banach space X and as such it converges to some $x^* \in \overline{U}(x_0, r)$ (since $\overline{U}(x_0, r)$ is a closed set). By letting $m \to \infty$, we obtain (8.2.8). $\qquad\square$

Stronger hypotheses are needed to show that x^* is a solution of equation $F(x) = 0$.

Proposition 8.2 *Let $F : D \subset X \to Y$ be a continuous operator and let $A(F)(x) \in L(X, Y)$. Suppose that there exist $x_0 \in D$, $\eta \geq 0$, $p \geq 1$, $\psi > 0$, a function $h_1 : [0, \eta] \to [0, \infty)$ continuous and nondecreasing such that for each $x, y \in D$*

$$A(F)(x)^{-1} \in L(Y, X), \quad \left\|(A(F)(x))^{-1}\right\| \leq \psi, \quad \left\|(A(F)(x_0))^{-1} F(x_0)\right\| \leq \eta,$$
$$\tag{8.2.9}$$

$$\|F(y) - F(x) - A(F)(x)(y - x)\| \leq \frac{h_1(\|x - y\|)}{\psi} \|x - y\|^{p+1}, \quad (8.2.10)$$

$$q_1 := h_1(\eta) \eta^p < 1$$

and

$$\overline{U}(x_0, r_1) \subseteq D,$$

where,

$$r_1 = \frac{\eta}{1 - q_1}.$$

Then, the conclusions of Theorem 8.1 for sequence $\{x_n\}$ hold with h_1, q_1, r_1, replacing h, q and r, respectively. Moreover, x^* is a solution of the equation $F(x) = 0$.

Proof Notice that

$$\left\|(A(F)(x_n))^{-1} \left[F(x_n) - F(x_{n-1}) - A(F)(x_{n-1})(x_n - x_{n-1})\right]\right\|$$

$$\leq \left\|(A(F)(x_n))^{-1}\right\| \|F(x_n) - F(x_{n-1}) - A(F)(x_{n-1})(x_n - x_{n-1})\|$$

$$\leq h_1(\|x_n - x_{n-1}\|) \|x_n - x_{n-1}\|^{p+1} \leq q_1 \|x_n - x_{n-1}\|.$$

Therefore, the proof of Theorem 8.1 can apply. Then, in view of the estimate

$$\|F(x_n)\| = \|F(x_n) - F(x_{n-1}) - A(F)(x_{n-1})(x_n - x_{n-1})\| \le$$

$$\frac{h_1(\|x_n - x_{n-1}\|)}{\psi} \|x_n - x_{n-1}\|^{p+1} \le q_1 \|x_n - x_{n-1}\|,$$

we deduce by letting $n \to \infty$ that $F(x^*) = 0$. □

Concerning the uniqueness of the solution x^* we have the following result:

Proposition 8.3 *Under the hypotheses of Proposition 8.2, further suppose that*

$$q_1 r_1^p < 1. \tag{8.2.11}$$

Then, x^* is the only solution of equation $F(x) = 0$ in $\overline{U}(x_0, r_1)$.

Proof The existence of the solution $x^* \in \overline{U}(x_0, r_1)$ has been established in Proposition 8.2. Let $y^* \in \overline{U}(x_0, r_1)$ with $F(y^*) = 0$. Then, we have in turn that

$$\|x_{n+1} - y^*\| = \|x_n - y^* - (A(F)(x_n))^{-1} F(x_n)\| =$$

$$\|(A(F)(x_n))^{-1} [A(F)(x_n)(x_n - y^*) - F(x_n) + F(y^*)]\| \le$$

$$\|(A(F)(x_n))^{-1}\| \|F(y^*) - F(x_n) - A(F)(x_n)(y^* - x_n)\| \le$$

$$\psi \frac{h_1(\|x_n - y^*\|)}{\psi} \|x_n - y^*\|^{p+1} \le q_1 r_1^p \|x_n - x^*\| < \|x_n - y^*\|,$$

so we deduce that $\lim_{n \to \infty} \lim x_n = y^*$. But we have that $\lim_{n \to \infty} \lim x_n = x^*$. Hence, we conclude that $x^* = y^*$. □

Next, we present a local convergence analysis for the iterative algorithm (8.1.2).

Proposition 8.4 *Let $F : D \subset X \to Y$ be a continuous operator and let $A(F)(x) \in L(X, Y)$. Suppose that there exist $x^* \in D$, $p \ge 1$, a function $h_2 : [0, \infty) \to [0, \infty)$ continuous and nondecreasing such that for each $x \in D$*

$$F(x^*) = 0, \quad (A(F)(x))^{-1} \in L(Y, X),$$

$$\|(A(F)(x))^{-1} [F(x) - F(x^*) - A(F)(x)(x - x^*)]\| \le$$

$$h_2(\|x - x^*\|) \|x - x^*\|^{p+1}, \tag{8.2.12}$$

and

$$\overline{U}(x^*, r_2) \subseteq D,$$

where r_2 is the smallest positive solution of equation

$$h^* (t) := h_2 (t) t^p - 1.$$

Then, sequence $\{x_n\}$ generated by algorithm (8.1.2) for $x_0 \in U (x^*, r_2) - \{x^*\}$ is well defined, remains in $U (x^*, r_2)$ for each $n = 0, 1, 2, \ldots$ and converges to x^*. Moreover, the following estimates hold

$$\|x_{n+1} - x^*\| \le h_2 (\|x_n - x^*\|) \|x_n - x^*\|^{p+1} < \|x_n - x^*\| < r_2.$$

Proof We have that $h^* (0) = -1 < 0$ and $h^* (t) \to +\infty$ as $t \to +\infty$. Then, it follows from the intermediate value theorem that function h^* has positive zeros. Denote by r_2 the smallest such zero. By hypothesis $x_0 \in U (x^*, r_2) - \{x^*\}$. Then, we get in turn that

$$\|x_1 - x^*\| = \|x_0 - x^* - (A (F) (x_0))^{-1} F (x_0)\| =$$

$$\|(A (F) (x_0))^{-1} \left[F (x^*) - F (x_0) - A (F) (x_0) (x^* - x_0) \right]\| \le$$

$$h_2 (\|x_0 - x^*\|) \|x_0 - x^*\|^{p+1} < h_2 (r_2) r_2^p \|x_0 - x^*\| =$$

$$\|x_0 - x^*\| < r_2,$$

which shows that $x_1 \in U (x^*, r_2)$ and x_2 is well defined. By a simple inductive argument as in the preceding estimate we get that

$$\|x_{k+1} - x^*\| = \|x_k - x^* - (A (F) (x_k))^{-1} F (x_k)\| \le$$

$$\|(A (F) (x_k))^{-1} \left[F (x^*) - F (x_k) - A (F) (x_k) (x^* - x_k) \right]\| \le$$

$$h_2 (\|x_k - x^*\|) \|x_k - x^*\|^{p+1} < h_2 (r_2) r_2^p \|x_k - x^*\| = \|x_k - x^*\| < r_2,$$

which shows $\lim_{k \to \infty} \lim x_k = x^*$ and $x_{k+1} \in U (x^*, r_2)$. □

Remark 8.5 (a) Hypothesis (8.2.3) specializes to Newton-Mysowski-type, if $A (F) (x) = F' (x)$ [8, 13, 17]. However, if F is not Fréchet-differentiable, then our results extend the applicability of iterative algorithm (8.1.2).

(b) Theorem 8.1 has practical value although we do not show that x^* is a solution of equation $F (x) = 0$, since this may be shown in another way.

(c) Hypothesis (8.2.12) can be replaced by the stronger

$$\|(A (F) (x))^{-1} [F (x) - F (y) - A (F) (x) (x - y)]\| \le h_2 (\|x - y\|) \|x - y\|^{p+1}.$$

The preceding results can be extended to hold for two point iterative algorithms defined for each $n = 0, 1, 2, \ldots$ by

$$x_{n+1} = x_n - (A(F)(x_n, x_{n-1}))^{-1} F(x_n), \qquad (8.2.13)$$

where $x_{-1}, x_0 \in D$ are initial points and $A(F)(w, v) \in L(X, Y)$ for each $v, w \in D$. If $A(F)(w, v) = [w, v; F]$, then iterative algorithm (8.2.13) reduces to the popular secant method, where $[w, v; F]$ denotes a divided difference of order one for the operator F. Many other choices for A are also possible [8, 13, 17].

If we simply replace $A(F)(x)$ by $A(F)(y, x)$ in the proof of Proposition 8.2 we arrive at the following semilocal convergence result for iterative algorithm (8.2.13).

Theorem 8.6 *Let $F : D \subset X \to Y$ be a continuous operator and let $A(F)(y, x) \in L(X, Y)$ for each $x, y \in D$. Suppose that there exist $x_{-1}, x_0 \in D$, $\eta \geq 0$, $p \geq 1$, $\psi > 0$, a function $h_1 : [0, \eta] \to [0, \infty)$ continuous and nondecreasing such that for each $x, y \in D$:*

$$(A(F)(y, x))^{-1} \in L(Y, X), \quad \left\| (A(F)(y, x))^{-1} \right\| \leq \psi, \qquad (8.2.14)$$

$$\min \left\{ \|x_0 - x_{-1}\|, \left\| (A(F)(x_0, x_{-1}))^{-1} F(x_0) \right\| \right\} \leq \eta,$$

$$\|F(y) - F(x) - A(F)(y, x)(y - x)\| \leq \frac{h_1(\|x - y\|)}{\psi} \|x - y\|^{p+1}, \quad (8.2.15)$$

$$q_1 < 1, \quad q_1 r_1^p < 1$$

and

$$\overline{U}(x_0, r_1) \subseteq D,$$

where,

$$r_1 = \frac{\eta}{1 - q_1}$$

and q_1 is defined in Proposition 8.2. Then, sequence $\{x_n\}$ generated by iterative algorithm (8.2.13) is well defined, remains in $\overline{U}(x_0, r_1)$ for each $n = 0, 1, 2, \ldots$ and converges to the only solution of equation $F(x) = 0$ in $\overline{U}(x_0, r_1)$. Moreover, the estimates (8.2.7) and (8.2.8) hold with h_1, q_1 replacing h and q, respectively.

Concerning, the local convergence of the iterative algorithm (8.2.13) we obtain the analogous to Proposition 8.4 result.

Proposition 8.7 *Let $F : D \subset X \to Y$ be a continuous operator and let $A(F)(y, x) \in L(X, Y)$. Suppose that there exist $x^* \in D$, $p \geq 1$, a function $h_2 : [0, \infty)^2 \to [0, \infty)$ continuous and nondecreasing such that for each $x, y \in D$*

$$F(x^*) = 0, \quad (A(F)(y, x))^{-1} \in L(Y, X),$$

$$\left\| \left(A\left(F \right) \left(y, x \right) \right)^{-1} \left[F\left(y \right) - F\left(x^{*} \right) - A\left(F \right) \left(y, x \right) \left(y - x^{*} \right) \right] \right\| \le$$

$$h_2 \left(\left\| y - x^{*} \right\|, \left\| x - x^{*} \right\| \right) \left\| y - x^{*} \right\|^{p+1}$$

and

$$\overline{U} \left(x^{*}, r_2 \right) \subseteq D,$$

where r_2 is the smallest positive solution of equation

$$h^{**} \left(t \right) := h_2 \left(t, t \right) t^{p} - 1.$$

Then, sequence $\{x_n\}$ generated by algorithm (8.2.13) for $x_{-1}, x_0 \in U\left(x^{*}, r_2 \right) - \{x^{*}\}$ is well defined, remains in $U\left(x^{*}, r_2 \right)$ for each $n = 0, 1, 2, \ldots$ and converges to x^{*}. Moreover, the following estimates hold

$$\left\| x_{n+1} - x^{*} \right\| \le h_2 \left(\left\| x_n - x^{*} \right\|, \left\| x_{n-1} - x^{*} \right\| \right) \left\| x_n - x^{*} \right\|^{p+1}$$

$$< \left\| x_n - x^{*} \right\| < r_2.$$

Remark 8.8 In the next section, we present some choices and properties of operator $A\left(F \right) \left(y, x \right)$ satisfying the crucial estimate (8.2.15). In particular, we choose

$$h_1 \left(t \right) = \frac{\rho \psi t}{p}, \quad \text{for some } \rho > 0, \text{ (see (8.3.22))},$$

$$h_1 \left(t \right) = \frac{c\psi}{\left(p + 1 \right)!}, \quad \text{for some } c > 0 \text{ (see (8.3.36))},$$

and

$$h_1 \left(t \right) = \frac{K \psi \gamma^{p+1}}{\left(p + 1 \right)!}, \text{ (see (8.3.66))},$$

if $\left| g\left(x \right) - g\left(y \right) \right| \le \gamma \left| x - y \right|$ for some $K > 0$ and $\gamma > 0$.

Hence, Theorem 8.6 can apply to solve equation $F\left(x \right) = 0$. Other choices for operator $A\left(F \right) \left(x \right)$ or operator $A\left(F \right) \left(y, x \right)$ can be found in [6–8, 11–18].

8.3   Applications to Fractional and Vector Calculus

We want to solve numerically

$$f\left(x \right) = 0. \tag{8.3.1}$$

(I) Application to Fractional Calculus

Let $p \in \mathbb{N} - \{1\}$ such that $p - 1 < \nu < p$, where $\nu \notin \mathbb{N}$, $\nu > 0$, i.e. $\lceil \nu \rceil = p$ ($\lceil \cdot \rceil$ ceiling of the number), $a < b$, $f \in C^p([a, b])$.

We define the following left Caputo fractional derivatives (see [3], p. 270)

$$\left(D_{*y}^{\nu} f\right)(x) := \frac{1}{\Gamma(p - \nu)} \int_y^x (x - t)^{p - \nu - 1} f^{(p)}(t)\, dt, \qquad (8.3.2)$$

when $x \geq y$, and

$$\left(D_{*x}^{\nu} f\right)(y) := \frac{1}{\Gamma(p - \nu)} \int_x^y (y - t)^{p - \nu - 1} f^{(p)}(t)\, dt, \qquad (8.3.3)$$

when $y \geq x$, where Γ is the gamma function.

We define also the linear operator

$$(A_1(f))(x, y) := \begin{cases} \sum_{k=1}^{p-1} \frac{f^{(k)}(y)}{k!}(x - y)^{k-1} + \left(D_{*y}^{\nu} f\right)(x) \frac{(x-y)^{\nu-1}}{\Gamma(\nu+1)}, & x > y, \\ \sum_{k=1}^{p-1} \frac{f^{(k)}(x)}{k!}(y - x)^{k-1} + \left(D_{*x}^{\nu} f\right)(y) \frac{(y-x)^{\nu-1}}{\Gamma(\nu+1)}, & y > x, \\ f^{(p-1)}(x), & x = y. \end{cases}$$

$$(8.3.4)$$

By left fractional Caputo Taylor's formula (see [9], p. 54 and [3], p. 395), we get that

$$f(x) - f(y) =$$

$$\sum_{k=1}^{p-1} \frac{f^{(k)}(y)}{k!}(x - y)^k + \frac{1}{\Gamma(\nu)} \int_y^x (x - t)^{\nu-1} D_{*y}^{\nu} f(t)\, dt, \quad \text{for } x > y, \quad (8.3.5)$$

and

$$f(y) - f(x) =$$

$$\sum_{k=1}^{p-1} \frac{f^{(k)}(x)}{k!}(y - x)^k + \frac{1}{\Gamma(\nu)} \int_x^y (y - t)^{\nu-1} D_{*x}^{\nu} f(t)\, dt, \quad \text{for } x < y. \quad (8.3.6)$$

Immediately, we observe that

$$|(A_1(f))(x, x) - (A_1(f))(y, y)| = \left| f^{(p-1)}(x) - f^{(p-1)}(y) \right| \qquad (8.3.7)$$

$$\leq \left\| f^{(p)} \right\|_\infty |x - y|, \qquad \forall x, y \in [a, b],$$

We would like to prove that

$$|f(x) - f(y) - (A_1(f))(x, y)(x - y)| \le c\frac{|x - y|^p}{p}, \tag{8.3.8}$$

for any $x, y \in [a, b]$ and some constant $0 < c < 1$.

When $x = y$, the last condition (8.3.8) is trivial.

We assume $x \ne y$. We distinguish the cases:

(1) $x > y$: We observe that

$$|f(x) - f(y) - (A_1(f))(x, y)(x - y)| = \tag{8.3.9}$$

$$\left| \sum_{k=1}^{p-1} \frac{f^{(k)}(y)}{k!}(x - y)^k + \frac{1}{\Gamma(\nu)} \int_y^x (x - t)^{\nu-1} D_{*y}^\nu f(t)\, dt - \right.$$

$$\left. \sum_{k=1}^{p-1} \frac{f^{(k)}(y)}{k!}(x - y)^k - (D_{*y}^\nu f)(x) \frac{(x - y)^\nu}{\Gamma(\nu + 1)} \right| =$$

$$\left| \frac{1}{\Gamma(\nu)} \int_y^x (x - t)^{\nu-1} (D_{*y}^\nu f)(t)\, dt - \frac{1}{\Gamma(\nu)} \int_y^x (x - t)^{\nu-1} (D_{*y}^\nu f)(x)\, dt \right| =$$

$$\frac{1}{\Gamma(\nu)} \left| \int_y^x (x - t)^{\nu-1} ((D_{*y}^\nu f)(t) - (D_{*y}^\nu f)(x))\, dt \right| \le$$

$$\frac{1}{\Gamma(\nu)} \int_y^x (x - t)^{\nu-1} \left| (D_{*y}^\nu f)(t) - (D_{*y}^\nu f)(x) \right| dt \tag{8.3.10}$$

(assume that

$$\left| (D_{*y}^\nu f)(t) - (D_{*y}^\nu f)(x) \right| \le \lambda_1 |t - x|^{p-\nu}, \tag{8.3.11}$$

for any $t, x, y \in [a, b] : x \ge t \ge y$, where $\lambda_1 < \Gamma(\nu)$, i.e. $\rho_1 := \frac{\lambda_1}{\Gamma(\nu)} < 1$)

$$\le \frac{\lambda_1}{\Gamma(\nu)} \int_y^x (x - t)^{\nu-1} (x - t)^{p-\nu}\, dt = \tag{8.3.12}$$

$$\frac{\lambda_1}{\Gamma(\nu)} \int_y^x (x - t)^{p-1}\, dt = \frac{\lambda_1}{\Gamma(\nu)} \frac{(x - y)^p}{p} = \rho_1 \frac{(x - y)^p}{p}. \tag{8.3.13}$$

We have proved that

$$|f(x) - f(y) - (A_1(f))(x, y)(x - y)| \le \rho_1 \frac{(x - y)^p}{p}, \tag{8.3.14}$$

where $0 < \rho_1 < 1$, and $x > y$.

(2) $x < y$: We observe that

$$|f(x) - f(y) - (A_1(f))(x, y)(x - y)| = \tag{8.3.15}$$

$$|f(y) - f(x) - (A_1(f))(x, y)(y - x)| =$$

$$\left| \sum_{k=1}^{p-1} \frac{f^{(k)}(x)}{k!}(y - x)^k + \frac{1}{\Gamma(\nu)} \int_x^y (y - t)^{\nu-1} (D_{*x}^\nu f)(t)\, dt - \right.$$

$$\left. \sum_{k=1}^{p-1} \frac{f^{(k)}(x)}{k!}(y - x)^k - (D_{*x}^\nu f)(y)\frac{(y - x)^\nu}{\Gamma(\nu + 1)} \right| =$$

$$\left| \frac{1}{\Gamma(\nu)} \int_x^y (y - t)^{\nu-1}(D_{*x}^\nu f)(t)\, dt - \frac{1}{\Gamma(\nu)} \int_x^y (y - t)^{\nu-1}(D_{*x}^\nu f)(y)\, dt \right| = \tag{8.3.16}$$

$$\frac{1}{\Gamma(\nu)} \left| \int_x^y (y - t)^{\nu-1} \left((D_{*x}^\nu f)(t) - (D_{*x}^\nu f)(y) \right) dt \right| \le$$

$$\frac{1}{\Gamma(\nu)} \int_x^y (y - t)^{\nu-1} \left| (D_{*x}^\nu f)(t) - (D_{*x}^\nu f)(y) \right| dt \tag{8.3.17}$$

(we assume that

$$\left| (D_{*x}^\nu f)(t) - (D_{*x}^\nu f)(y) \right| \le \lambda_2 |t - y|^{p-\nu}, \tag{8.3.18}$$

for any $t, y, x \in [a, b] : y \ge t \ge x$)

$$\le \frac{\lambda_2}{\Gamma(\nu)} \int_x^y (y - t)^{\nu-1}(y - t)^{p-\nu}\, dt =$$

$$\frac{\lambda_2}{\Gamma(\nu)} \int_x^y (y - t)^{p-1}\, dt = \frac{\lambda_2}{\Gamma(\nu)} \frac{(y - x)^p}{p}. \tag{8.3.19}$$

Assuming also

$$\rho_2 := \frac{\lambda_2}{\Gamma(\nu)} < 1 \tag{8.3.20}$$

(i.e. $\lambda_2 < \Gamma(\nu)$), we have proved that

$$|f(x) - f(y) - (A_1(f))(x, y)(x - y)| \le \rho_2 \frac{(y - x)^p}{p}, \quad \text{for } x < y. \quad (8.3.21)$$

Conclusion: Choosing $\lambda := \max(\lambda_1, \lambda_2)$ and $\rho := \frac{\lambda}{\Gamma(\nu)} < 1$, we have proved that

$$|f(x) - f(y) - (A_1(f))(x, y)(x - y)| \le \rho \frac{|x - y|^p}{p}, \quad \text{for any } x, y \in [a, b]. $$
$$(8.3.22)$$

(II) Application to Vector Calculus
(II$_1$) Background
(see [20], pp. 83–94)

Let $f(t)$ be a function defined on $[a, b] \subseteq \mathbb{R}$ taking values in a real or complex normed linear space $(X, \|\cdot\|)$. Then $f(t)$ is said to be differentiable at a point $t_0 \in [a, b]$ if the limit

$$f'(t_0) := \lim_{h \to 0} \frac{f(t_0 + h) - f(t_0)}{h} \quad (8.3.23)$$

exists in X, the convergence is in $\|\cdot\|$. This is called the derivative of $f(t)$ at $t = t_0$.

We call $f(t)$ differentiable on $[a, b]$, iff there exists $f'(t) \in X$ for all $t \in [a, b]$.

Similarly and inductively are defined higher order derivatives of f, denoted f'', $f^{(3)}, \ldots, f^{(k)}$, $k \in \mathbb{N}$, just as for numerical functions.

For all the properties of derivatives see [20], pp. 83–86.

From now let $(X, \|\cdot\|)$ be a Banach space and $f : [a, b] \to X$.

We define the vector valued Riemann integral $\int_a^b f(t)\,dt \in X$ as the limit of the vector valued Riemann sums in X, convergence is in $\|\cdot\|$. The definition is as for the numerical valued functions.

If $\int_a^b f(t)\,dt \in X$ we call f integrable on $[a, b]$. If $f \in C([a, b], X)$, then f is integrable, [20], p. 87.

For all other properties of vector valued Riemann integrals see [20], pp. 86–91. We mention some of them here.

Let f, g vector valued Riemann integrable functions, we have that (see [20], p. 88)

$$\int_a^b \alpha f(t)\,dt = \alpha \int_a^b f(t)\,dt, \quad \alpha \in \mathbb{R} \text{ or } \alpha \in \mathbb{C}, \quad (8.3.24)$$

$$\int_a^b (f(t) + g(t))\,dt = \int_a^b f(t)\,dt + \int_a^b g(t)\,dt \quad (8.3.25)$$

$$\int_a^c f(t)\,dt + \int_c^b f(t)\,dt = \int_a^b f(t)\,dt, \quad a < c < b, \quad (8.3.26)$$

$$\left\| \int_a^b f(t)\,dt \right\| \le (b - a) \max_{a \le t \le b} \| f(t) \|, \tag{8.3.27}$$

$$\left\| \int_a^b f(t)\,dt \right\| \le \int_a^b \| f(t) \|\,dt. \tag{8.3.28}$$

By [10], also we get by convention that

$$\int_\alpha^\beta f(t)\,dt = -\int_\beta^\alpha f(t)\,dt, \quad \text{for } a \le \beta \le \alpha \le b. \tag{8.3.29}$$

Let $f : [a, b] \to \mathbb{R}$ Riemann integrable function, i.e. $\int_a^b f(t)\,dt$ exists as a real number, and $c \in X$. Then clearly it holds

$$c \int_a^b f(x)\,dx = \int_a^b c f(x)\,dx \in X. \tag{8.3.30}$$

We define the space $C^p([a, b], X)$, $p \in \mathbb{N}$, of p-times continuously differentiable functions from $[a, b]$ into X; here continuity is with respect to $\|\cdot\|$ and defined in the usual way as for numerical functions.

Let $(X, \|\cdot\|)$ be a Banach space and $f \in C^p([a, b], X)$, then we have the vector valued Taylor's formula, see [20], pp. 93–94, and also [19], (IV, 9; 47).

It holds

$$f(y) - f(x) - f'(x)(y - x) - \frac{1}{2}f''(x)(y - x)^2 - \cdots - \frac{1}{(p-1)!}f^{(p-1)}(x)(y-x)^{p-1}$$

$$= \frac{1}{(p-1)!}\int_x^y (y-t)^{p-1} f^{(p)}(t)\,dt, \quad \forall\, x, y \in [a, b]. \tag{8.3.31}$$

In particular (8.3.31) is true when $X = \mathbb{R}^m$, $\mathbb{C}^m$, $m \in \mathbb{N}$, etc.

Clearly it holds that

$$f(y) - f(x) =$$

$$\sum_{k=1}^p \frac{f^{(k)}(x)}{k!}(y-x)^k + \frac{1}{(p-1)!}\int_x^y (y-t)^{p-1}\left(f^{(p)}(t) - f^{(p)}(x)\right)dt,$$

$$\tag{8.3.32}$$

$\forall\, x, y \in [a, b]$.

We will use (8.3.32).

We need also the mean value theorem for Banach space valued functions.

Theorem 8.9 (see [14], p. 3) *Let $f \in C([a, b], X)$, where X is a Banach space. Assume f' exists on $[a, b]$ and $\| f'(t) \| \le K$, $a < t < b$, then*

$$\| f(b) - f(a) \| \le K(b - a). \tag{8.3.33}$$

(**II$_2$**) From now on we assume that $f \in C^{p+1}\left([a, b], X\right)$, $p \in \mathbb{N}$.
We define the function

$$\left(A_2\left(f\right)\right)\left(x, y\right) := \begin{cases} \sum_{k=1}^{p} \frac{f^{(k)}(x)}{k!} \left(y - x\right)^{k-1}, & x \neq y, \\ f^{(p)}\left(x\right), & x = y. \end{cases} \tag{8.3.34}$$

Then it holds

$$\left\|\left(A_2\left(f\right)\right)\left(x, x\right) - \left(A_2\left(f\right)\right)\left(y, y\right)\right\| = \left\|f^{(p)}\left(x\right) - f^{(p)}\left(y\right)\right\|$$

$$\leq \left\|\left|f^{(p+1)}\right|\right\|_\infty \left|x - y\right|, \quad \forall\, x, y \in \left[a, b\right], \tag{8.3.35}$$

where

$$\left\|\left|f^{(p+1)}\right|\right\|_\infty := \sup_{t \in [a,b]} \left\|f^{(p+1)}\right\| < \infty.$$

We would like to prove that

$$\left\|f\left(y\right) - f\left(x\right) - \left(A_2\left(f\right)\right)\left(x, y\right)\left(y - x\right)\right\| \leq c \frac{\left|x - y\right|^{p+1}}{\left(p + 1\right)!}, \tag{8.3.36}$$

where $c > 0$.

When $x = y$ inequality (8.3.36) is trivially true. We will prove it for $x \neq y$.
We observe that

$$\left\|f\left(y\right) - f\left(x\right) - \left(A_2\left(f\right)\right)\left(x, y\right)\left(y - x\right)\right\| = \tag{8.3.37}$$

$$\left\|\sum_{k=1}^{p} \frac{f^{(k)}\left(x\right)}{k!} \left(y - x\right)^k + \frac{1}{\left(p - 1\right)!} \int_x^y \left(y - t\right)^{p-1} \left(f^{(p)}\left(t\right) - f^{(p)}\left(x\right)\right) dt\right.$$

$$\left. - \sum_{k=1}^{p} \frac{f^{(k)}\left(x\right)}{k!} \left(y - x\right)^k\right\| =$$

$$\frac{1}{\left(p - 1\right)!} \left\|\int_x^y \left(y - t\right)^{p-1} \left(f^{(p)}\left(t\right) - f^{(p)}\left(x\right)\right) dt\right\| =: \left(\xi\right). \tag{8.3.38}$$

Let $y > x$: we observe that

$$\left(\xi\right) \leq \frac{1}{\left(p - 1\right)!} \int_x^y \left(y - t\right)^{p-1} \left\|f^{(p)}\left(t\right) - f^{(p)}\left(x\right)\right\| dt$$

$$\leq \frac{\left\|\left|f^{(p+1)}\right|\right\|_\infty}{\left(p - 1\right)!} \int_x^y \left(y - t\right)^{p-1} \left(t - x\right)^{2-1} dt \tag{8.3.39}$$

$$= \frac{\left\| \left| f^{(p+1)} \right| \right\|_\infty}{(p-1)!} \frac{\Gamma(p)\,\Gamma(2)}{\Gamma(p+2)} (y-x)^{p+1} =$$

$$\frac{\left\| \left| f^{(p+1)} \right| \right\|_\infty}{(p-1)!} \frac{(p-1)!}{(p+1)!} (y-x)^{p+1} = \frac{\left\| \left| f^{(p+1)} \right| \right\|_\infty}{(p+1)!} (y-x)^{p+1}. \tag{8.3.40}$$

Hence,

$$(\xi) \le \frac{\left\| \left| f^{(p+1)} \right| \right\|_\infty}{(p+1)!} (y-x)^{p+1}, \quad \text{for } y > x. \tag{8.3.41}$$

Let now $x > y$: we observe that

$$(\xi) = \frac{1}{(p-1)!} \left\| \int_y^x (y-t)^{p-1} \left(f^{(p)}(t) - f^{(p)}(x) \right) dt \right\|$$

$$\le \frac{1}{(p-1)!} \int_y^x (t-y)^{p-1} \left\| f^{(p)}(t) - f^{(p)}(x) \right\| dt \tag{8.3.42}$$

$$\le \frac{\left\| \left| f^{(p+1)} \right| \right\|_\infty}{(p-1)!} \int_y^x (x-t)^{2-1} (t-y)^{p-1} dt$$

$$= \frac{\left\| \left| f^{(p+1)} \right| \right\|_\infty}{(p-1)!} \frac{\Gamma(p)\,\Gamma(2)}{\Gamma(p+2)} (x-y)^{p+1} =$$

$$\frac{\left\| \left| f^{(p+1)} \right| \right\|_\infty}{(p-1)!} \frac{(p-1)!}{(p+1)!} (x-y)^{p+1} = \frac{\left\| \left| f^{(p+1)} \right| \right\|_\infty}{(p+1)!} (x-y)^{p+1}. \tag{8.3.43}$$

We have proved that

$$(\xi) \le \frac{\left\| \left| f^{(p+1)} \right| \right\|_\infty}{(p+1)!} (x-y)^{p+1}, \quad \text{for } x > y. \tag{8.3.44}$$

Conclusion: We have proved the following:
Let $p \in \mathbb{N}$ and $f \in C^{p+1}([a,b])$. Then

$$\| f(y) - f(x) - (A_2(f))(x,y) \cdot (y-x) \| \tag{8.3.45}$$

$$\le \frac{\left\| \left| f^{(p+1)} \right| \right\|_\infty}{(p+1)!} |x-y|^{p+1}, \quad \forall x, y \in [a,b].$$

(III) Applications from Mathematical Analysis

In [4], pp. 400–402, we have proved the following general diverse Taylor's formula:

Theorem 8.10 *Let* $f, f', \ldots, f^{(p)};$ g, g' *be continuous from* $[a, b]$ *(or* $[b, a]$*) into* $\mathbb{R}$, $p \in \mathbb{N}$. *Assume* $(g^{-1})^{(k)}$, $k = 0, 1, \ldots, p$ *are continuous. Then*

$$f(b) = f(a) + \sum_{k=1}^{p-1} \frac{(f \circ g^{-1})^{(k)}(g(a))}{k!}(g(b) - g(a))^k + R_p(a, b), \quad (8.3.46)$$

where

$$R_p(a, b) = \frac{1}{(p-1)!} \int_a^b (g(b) - g(s))^{p-1} (f \circ g^{-1})^{(p)}(g(s)) g'(s) \, ds \quad (8.3.47)$$

$$= \frac{1}{(p-1)!} \int_{g(a)}^{g(b)} (g(b) - t)^{p-1} (f \circ g^{-1})^{(p)}(t) \, dt.$$

Theorem 8.10 will be applied next for $g(x) = e^x$. One can give similar applications for $g = \sin, \cos, \tan$, etc., over suitable intervals, see [4], p. 402.

Proposition 8.11 *Let* $f^{(p)}$ *continuous, from* $[a, b]$ *(or* $[b, a]$*) into* $\mathbb{R}$, $p \in \mathbb{N}$. *Then*

$$f(b) = f(a) + \sum_{k=1}^{p-1} \frac{[(f \circ \ln)^{(k)}(e^a)]}{k!} \cdot (e^b - e^a)^k + R_p(a, b), \quad (8.3.48)$$

where

$$R_p(a, b) = \frac{1}{(p-1)!} \int_{e^a}^{e^b} (e^b - t)^{p-1} (f \circ \ln)^{(p)}(t) \, dt \quad (8.3.49)$$

$$= \frac{1}{(p-1)!} \int_a^b (e^b - e^a)^{p-1} (f \circ \ln)^{(p)}(e^s) \cdot e^s ds.$$

We will use the following variant.

Theorem 8.12 *Let* $f, f', \ldots, f^{(p)}$, $p \in \mathbb{N}$ *and* g, g' *be continuous from* $[a, b]$ *into* $\mathbb{R}$, $p \in \mathbb{N}$. *Assume that* $(g^{-1})^{(k)}$, $k = 0, 1, \ldots, p$, *are continuous. Then*

$$f(\beta) - f(\alpha) = \sum_{k=1}^{p} \frac{(f \circ g^{-1})^{(k)}(g(\alpha))}{k!}(g(\beta) - g(\alpha))^k + R_p^*(\alpha, \beta), \quad (8.3.50)$$

where

$$R_p^*(\alpha, \beta) = \frac{1}{(p-1)!} \cdot$$

$$\int_\alpha^\beta (g(\beta) - g(s))^{p-1} \left((f \circ g^{-1})^{(p)}(g(s)) - (f \circ g^{-1})^{(p)}(g(\alpha)) \right) g'(s) \, ds$$

$$(8.3.51)$$

$$= \frac{1}{(p-1)!} \int_{g(\alpha)}^{g(\beta)} (g(\beta) - t)^{p-1} \left(\left(f \circ g^{-1} \right)^{(p)} (t) - \left(f \circ g^{-1} \right)^{(p)} (g(\alpha)) \right) dt,$$

$$(8.3.52)$$

$\forall \alpha, \beta \in [a, b]$.

Proof Easy. □

Remark 8.13 Call $l = f \circ g^{-1}$. Then $l, l', \ldots, l^{(p)}$ are continuous from $g([a, b])$ into $f([a, b])$.

Next we estimate $R_p^* (\alpha, \beta)$: We assume that

$$\left| \left(f \circ g^{-1} \right)^{(p)} (t) - \left(f \circ g^{-1} \right)^{(p)} (g(\alpha)) \right| \leq K |t - g(\alpha)|, \qquad (8.3.53)$$

$\forall t, g(\alpha) \in [g(a), g(b)]$ or $\forall t, g(\alpha) \in [g(b), g(a)]$ where $K > 0$.

We distinguish the cases:

(i) if $g(\beta) > g(\alpha)$, then

$$\left| R_p^* (\alpha, \beta) \right| \leq$$

$$\frac{1}{(p-1)!} \int_{g(\alpha)}^{g(\beta)} (g(\beta) - t)^{p-1} \left| \left(f \circ g^{-1} \right)^{(p)} (t) - \left(f \circ g^{-1} \right)^{(p)} (g(\alpha)) \right| dt \leq$$

$$\frac{K}{(p-1)!} \int_{g(\alpha)}^{g(\beta)} (g(\beta) - t)^{p-1} (t - g(\alpha))^{2-1} dt = \qquad (8.3.54)$$

$$\frac{K}{(p-1)!} \frac{\Gamma(p) \Gamma(2)}{\Gamma(p+2)} (g(\beta) - g(\alpha))^{p+1} =$$

$$\frac{K}{(p-1)!} \frac{(p-1)!}{(p+1)!} (g(\beta) - g(\alpha))^{p+1} = K \frac{(g(\beta) - g(\alpha))^{p+1}}{(p+1)!}. \qquad (8.3.55)$$

We have proved that

$$\left| R_p^* (\alpha, \beta) \right| \leq K \frac{(g(\beta) - g(\alpha))^{p+1}}{(p+1)!}, \qquad (8.3.56)$$

when $g(\beta) > g(\alpha)$.

(ii) if $g(\alpha) > g(\beta)$, then

$$\left| R_p^* (\alpha, \beta) \right| =$$

$$\frac{1}{(p-1)!} \left| \int_{g(\beta)}^{g(\alpha)} (t - g(\beta))^{p-1} \left(\left(f \circ g^{-1} \right)^{(p)} (t) - \left(f \circ g^{-1} \right)^{(p)} (g(\alpha)) \right) dt \right| \leq$$

$$\frac{1}{(p-1)!} \int_{g(\beta)}^{g(\alpha)} (t - g(\beta))^{p-1} \left| \left(f \circ g^{-1}\right)^{(p)} (t) - \left(f \circ g^{-1}\right)^{(p)} (g(\alpha)) \right| dt \le$$

$$\tag{8.3.57}$$

$$\frac{K}{(p-1)!} \int_{g(\beta)}^{g(\alpha)} (g(\alpha) - t)^{2-1} (t - g(\beta))^{p-1} dt =$$

$$\frac{K}{(p-1)!} \frac{\Gamma(2) \Gamma(p)}{\Gamma(p+2)} (g(\alpha) - g(\beta))^{p+1} = \tag{8.3.58}$$

$$\frac{K}{(p-1)!} \frac{(p-1)!}{(p+1)!} (g(\alpha) - g(\beta))^{p+1} = K \frac{(g(\alpha) - g(\beta))^{p+1}}{(p+1)!}. \tag{8.3.59}$$

We have proved that

$$\left| R_p^* (\alpha, \beta) \right| \le K \frac{(g(\alpha) - g(\beta))^{p+1}}{(p+1)!}, \tag{8.3.60}$$

whenever $g(\alpha) > g(\beta)$.

Conclusion: It holds

$$\left| R_p^* (\alpha, \beta) \right| \le K \frac{|g(\alpha) - g(\beta)|^{p+1}}{(p+1)!}, \tag{8.3.61}$$

$\forall \, \alpha, \beta \in [a, b]$.

Both sides of (8.3.61) equal zero when $\alpha = \beta$.

We define the following linear operator:

$$(A_3 (f)) (x, y) :=$$

$$\begin{cases} \sum_{k=1}^{p} \frac{\left(f \circ g^{-1}\right)^{(k)} (g(y))}{k!} (g(x) - g(y))^{k-1}, & \text{when } g(x) \ne g(y), \\ f^{(p-1)} (x), \, x = y, \end{cases} \tag{8.3.62}$$

for any $x, y \in [a, b]$.

Easily, we see that

$$|(A_3 (f)) (x, x) - (A_3 (f)) (y, y)| = \left| f^{(p-1)} (x) - f^{(p-1)} (y) \right|$$

$$\le \left\| f^{(p)} \right\|_{\infty} |x - y|, \quad \forall \, x, y \in [a, b]. \tag{8.3.63}$$

Next we observe that (case of $g(x) \ne g(y)$)

$$|f(x) - f(y) - (A_3 (f)) (x, y) \cdot (g(x) - g(y))| =$$

$$\left|\sum_{k=1}^{p} \frac{\left(f \circ g^{-1}\right)^{(k)}(g(y))}{k!}(g(x)-g(y))^{k}+R_{p}^{*}(y, x)\right.$$ (8.3.64)

$$\left.-\left(\sum_{k=1}^{p} \frac{\left(f \circ g^{-1}\right)^{(k)}(g(y))}{k!}(g(x)-g(y))^{k-1}(g(x)-g(y))\right)\right|=$$

$$\left|R_{p}^{*}(y, x)\right| \overset{(8.3.61)}{\leq} K \frac{|g(x)-g(y)|^{p+1}}{(p+1)!},$$ (8.3.65)

$\forall\, x, y \in [a, b] : g(x) \neq g(y)$.

We have proved that

$$|f(x)-f(y)-(A_3(f))(x, y) \cdot (g(x)-g(y))| \leq K \frac{|g(x)-g(y)|^{p+1}}{(p+1)!},$$ (8.3.66)

$\forall\, x, y \in [a, b]$

(the case $x = y$ is trivial).

We apply the above theory as follows:

(III$_1$) We define

$$(A_{31}(f))(x, y) := \begin{cases} \sum_{k=1}^{p} \frac{(f \circ \ln)^{(k)}(e^y)}{k!}(e^x - e^y)^{k-1}, & x \neq y, \\ f^{(p-1)}(x), & x = y, \end{cases}$$ (8.3.67)

for any $x, y \in [a, b]$.

Furthermore it holds

$$\left|f(x)-f(y)-(A_{31}(f))(x, y) \cdot \left(e^x - e^y\right)\right| \leq K_1 \frac{|e^x - e^y|^{p+1}}{(p+1)!},$$ (8.3.68)

$\forall\, x, y \in [a, b]$, where we assume that

$$\left|(f \circ \ln)^{(p)}(t) - (f \circ \ln)^{(p)}\left(e^y\right)\right| \leq K_1 \left|t - e^y\right|,$$ (8.3.69)

$\forall\, t, e^y \in \left[e^a, e^b\right]$, $a < b$, with $K_1 > 0$.

(III$_2$) Next let $f \in C^p\left(\left[-\frac{\pi}{2}+\varepsilon, \frac{\pi}{2}-\varepsilon\right]\right)$, $p \in \mathbb{N}$, $\varepsilon > 0$ small.

Here we define that

$$(A_{32}(f))(x, y) := \begin{cases} \sum_{k=1}^{p} \frac{\left(f \circ \sin^{-1}\right)^{(k)}(\sin y)}{k!}(\sin x - \sin y)^{k-1}, & \text{when } x \neq y, \\ f^{(p-1)}(x), & x = y, \end{cases}$$ (8.3.70)

for any $x, y \in \left[-\frac{\pi}{2}+\varepsilon, \frac{\pi}{2}-\varepsilon\right]$.

We assume that

$$\left| \left(f \circ \sin^{-1} \right)^{(p)} (t) - \left(f \circ \sin^{-1} \right)^{(p)} (\sin y) \right| \le K_2 \left| t - \sin y \right|, \qquad (8.3.71)$$

$\forall\, t, \sin y \in \left[\sin \left(-\frac{\pi}{2} + \varepsilon \right), \sin \left(\frac{\pi}{2} - \varepsilon \right) \right]$, where $K_2 > 0$.
It holds

$$\left| f(x) - f(y) - (A_{32}(f))(x, y) \cdot (\sin x - \sin y) \right| \le K_2 \frac{\left| \sin x - \sin y \right|^{p+1}}{(p+1)!}, \qquad (8.3.72)$$

$\forall\, x, y \in \left[-\frac{\pi}{2} + \varepsilon, \frac{\pi}{2} - \varepsilon \right]$.
   (III$_3$) Next let $f \in C^p \left([\varepsilon, \pi - \varepsilon] \right)$, $p \in \mathbb{N}$, $\varepsilon > 0$ small.
Here we define

$$(A_{33}(f))(x, y) := \begin{cases} \sum_{k=1}^{p} \frac{\left(f \circ \cos^{-1} \right)^{(k)} (\cos y)}{k!} (\cos x - \cos y)^{k-1}, & \text{when } x \ne y, \\ f^{(p-1)}(x), & x = y, \end{cases}$$

$$(8.3.73)$$

for any $x, y \in [\varepsilon, \pi - \varepsilon]$.
   We assume that

$$\left| \left(f \circ \cos^{-1} \right)^{(p)} (t) - \left(f \circ \cos^{-1} \right)^{(p)} (\cos y) \right| \le K_3 \left| t - \cos y \right|, \qquad (8.3.74)$$

$\forall\, t, \cos y \in [\cos \varepsilon, \cos (\pi - \varepsilon)]$, where $K_3 > 0$.
   It holds

$$\left| f(x) - f(y) - (A_{33}(f))(x, y) \cdot (\cos x - \cos y) \right| \le K_3 \frac{\left| \cos x - \cos y \right|^{p+1}}{(p+1)!}, \qquad (8.3.75)$$

$\forall\, x, y \in [\varepsilon, \pi - \varepsilon]$.
   Finally we give:
   (III$_4$) Let $f \in C^p \left(\left[-\frac{\pi}{2} + \varepsilon, \frac{\pi}{2} - \varepsilon \right] \right)$, $p \in \mathbb{N}$, $\varepsilon > 0$ small.
We define

$$(A_{34}(f))(x, y) := \begin{cases} \sum_{k=1}^{p} \frac{\left(f \circ \tan^{-1} \right)^{(k)} (\tan y)}{k!} (\tan x - \tan y)^{k-1}, & \text{when } x \ne y, \\ f^{(p-1)}(x), & x = y, \end{cases}$$

$$(8.3.76)$$

for any $x, y \in \left[-\frac{\pi}{2} + \varepsilon, \frac{\pi}{2} - \varepsilon \right]$.
   We assume that

$$\left| \left(f \circ \tan^{-1} \right)^{(p)} (t) - \left(f \circ \tan^{-1} \right)^{(p)} (\tan y) \right| \le K_4 \left| t - \tan y \right|, \qquad (8.3.77)$$

$\forall\, t, \tan y \in \left[\tan \left(-\frac{\pi}{2} + \varepsilon \right), \tan \left(\frac{\pi}{2} - \varepsilon \right) \right]$, where $K_4 > 0$.

It holds that

$$|f(x) - f(y) - (A_{34}(f))(x, y) \cdot (\tan x - \tan y)| \le K_4 \frac{|\tan x - \tan y|^{p+1}}{(p+1)!},$$

(8.3.78)

$\forall\, x, y \in \left[-\frac{\pi}{2} + \varepsilon, \frac{\pi}{2} - \varepsilon\right].$

References

1. S. Amat, S. Busquier, Third-order iterative methods under Kantorovich conditions. J. Math. Anal. Appl. **336**, 243–261 (2007)
2. S. Amat, S. Busquier, S. Plaza, Chaotic dynamics of a third-order Newton-like method. J. Math. Anal. Appl. **366**(1), 164–174 (2010)
3. G. Anastassiou, *Fractional Differentiation Inequalities* (Springer, New York, 2009)
4. G. Anastassiou, *Intelligent Mathematics: Computational Analysis* (Springer, Heidelberg, 2011)
5. G. Anastassiou, I. Argyros, *A Convergence Analysis for Extended Iterative Algorithms with Applications to Fractional and Vector Calculus* (2015) (submitted)
6. I.K. Argyros, Newton-like methods in partially ordered linear spaces. J. Approx. Theory Appl. **9**(1), 1–10 (1993)
7. I.K. Argyros, Results on controlling the residuals of perturbed Newton-like methods on Banach spaces with a convergence structure. Southwest J. Pure Appl. Math. **1**, 32–38 (1995)
8. I.K. Argyros, *Convergence and Applications of Newton-like iterations* (Springer, New York, 2008)
9. K. Diethelm, *The Analysis of Fractional Differential Equations*. Lecture Notes in Mathematics, vol. 2004, 1st edn. (Springer, New York, 2010)
10. B. Driver, *The Riemann Integral*, www.math.ucsd.edu/~bdriver/231-02-03/Lecture_Notes/Chap.4.pdf
11. J.A. Ezquerro, J.M. Gutierrez, M.A. Hernandez, N. Romero, M.J. Rubio, The Newton method: from Newton to Kantorovich (Spanish). Gac. R. Soc. Mat. Esp. **13**, 53–76 (2010)
12. J.A. Ezquerro, M.A. Hernandez, Newton-like methods of high order and domains of semilocal and global convergence. Appl. Math. Comput. **214**(1), 142–154 (2009)
13. L.V. Kantorovich, G.P. Akilov, *Functional Analysis in Normed Spaces* (Pergamon Press, New York, 1964)
14. G. Ladas, V. Laksmikantham, *Differential Equations in Abstract Spaces* (Academic Press, New York, 1972)
15. A.A. Magrenan, Different anomalies in a Jarratt family of iterative root finding methods. Appl. Math. Comput. **233**, 29–38 (2014)
16. A.A. Magrenan, A new tool to study real dynamics: the convergence plane. Appl. Math. Comput. **248**, 215–224 (2014)
17. F.A. Potra, V. Ptak, *Nondiscrete Induction and Iterative Processes* (Pitman Publishing, London, 1984)
18. P.D. Proinov, New general convergence theory for iterative processes and its applications to Newton-Kantorovich type theorems. J. Complexity **26**, 3–42 (2010)
19. L. Schwartz, *Analyse Mathematique* (Herman, Paris, 1967)
20. G. Shilov, *Elementary Functional Analysis* (The MIT Press, Cambridge, 1974)

Chapter 9
Convergence Analysis for Extended Iterative Algorithms and Fractional Calculus

We present local and semilocal convergence results for some extended methods in order to approximate a locally unique solution of a nonlinear equation in a Banach space setting. In earlier studies the operator involved is assumed to be at least once Fréchet-differentiable. In the present study, we assume that the operator is only continuous. This way we expand the applicability of these methods. In the third part of the study we present some choices of the operators involved in fractional calculus where the operators satisfy the convergence conditions. Moreover, we present a corrected version of the generalized fractional Taylor's formula given in [16]. It follows [5, 6].

9.1 Introduction

In this study we are concerned with the problem of approximating a locally unique solution x^* of the nonlinear equation

$$F(x) = 0, \qquad (9.1.1)$$

where F is a continuous operator defined on a subset D of a Banach space X with values in a Banach space Y.

A lot of problems in Computational Sciences and other disciplines can be brought in a form like (9.1.1) using Mathematical Modelling [9, 13, 17]. The solutions of such equations can be found in closed form only in special cases. That is why most solution methods for these equations are iterative. Iterative algorithms are usually studied based on semilocal and local convergence. The semilocal convergence matter is, based on the information around the initial point to give hypotheses ensuring the convergence of the method; while the local one is, based on the information around a solution, to find estimates of the radii of convergence balls as well as error bounds on the distances involved.

© Springer International Publishing Switzerland 2016

G.A. Anastassiou and I.K. Argyros, *Intelligent Numerical Methods: Applications to Fractional Calculus*, Studies in Computational Intelligence 624, DOI 10.1007/978-3-319-26721-0_9

We introduce the method defined for each $n = 0, 1, 2, \ldots$ by

$$x_{n+1} = x_n - A(x_n)^{-1} F(x_n),\qquad (9.1.2)$$

where $x_0 \in D$ is an initial point and $A(x) \in L(X, Y)$ the space of bounded linear operators from X into Y. There is a plethora on local as well as semilocal convergence theorems for method (9.1.2) provided that the operator A is an approximation to the Fréchet-derivative F' [1, 2, 7–17]. In the present study we do not assume that operator A is not necessarily related to F'. This way we expand the applicability of method (9.1.2). Notice that many well known methods are special case of method (9.1.2).

Newton's method: Choose $A(x) = F'(x)$ for each $x \in D$.

Steffensen's method: Choose $A(x) = [x, G(x); F]$, where $G : X \to X$ is a known operator and $[x, y; F]$ denotes a divided difference of order one [9, 13, 17].

The so called Newton-like methods and many other methods are special cases of method (9.1.2).

The rest of the chapter is organized as follows. The semilocal as well as the local convergence analysis of method (9.1.2) is given in Sect. 9.2. Some applications from fractional calculus are given in Sect. 9.3. In particular, we first correct the generalized fractional Taylor's formula, the integral version extracted from [16]. Then, we use the corrected formula in our applications.

9.2   Convergence Analysis

We present the main semilocal convergence result for method (9.1.2).

Theorem 9.1 *Let* $F : D \subset X \to Y$ *be a continuous operator and let* $A(x) \in L(X, Y)$. *Suppose that there exist* $x_0 \in D$, $\eta \geq 0$, $p \geq 1$, *a function* $g : [0, \eta] \to [0, \infty)$ *continuous and nondecreasing such that for each* $x, y \in D$

$$A(x)^{-1} \in L(Y, X),\qquad (9.2.1)$$

$$\left\| A(x_0)^{-1} F(x_0) \right\| \leq \eta,\qquad (9.2.2)$$

$$\left\| A(y)^{-1} (F(y) - F(x) - A(x)(y - x)) \right\| \leq g(\|x - y\|) \|x - y\|^{p+1},\quad (9.2.3)$$

$$q := g(\eta) \eta^p < 1\qquad (9.2.4)$$

and

$$\overline{U}(x_0, r) \subseteq D,\qquad (9.2.5)$$

where,

$$r = \frac{\eta}{1 - q}.\qquad (9.2.6)$$

Then, the sequence $\{x_n\}$ generated by method (9.1.2) is well defined, remains in $\overline{U}(x_0, r)$ for each $n = 0, 1, 2, \ldots$ and converges to some $x^* \in \overline{U}(x_0, r)$ such that

$$\|x_{n+1} - x_n\| \leq g(\|x_n - x_{n-1}\|) \|x_n - x_{n-1}\|^{p+1} \leq q \|x_n - x_{n-1}\| \qquad (9.2.7)$$

and

$$\|x_n - x^*\| \leq \frac{q^n \eta}{1 - q}. \qquad (9.2.8)$$

Proof The iterate x_1 is well defined by method (9.1.2) for $n = 0$ and (9.2.1) for $x = x_0$. We also have by (9.2.2) and (9.2.6) that $\|x_1 - x_0\| = \|A(x_0)^{-1} F(x_0)\| \leq \eta < r$, so we get that $x_1 \in \overline{U}(x_0, r)$ and x_2 is well defined (by (9.2.5)). Using (9.2.3) for $y = x_1$, $x = x_0$ and (9.2.4) we get that

$$\|x_2 - x_1\| = \|A(x_1)^{-1} [F(x_1) - F(x_0) - A(x_0)(x_1 - x_0)]\|$$

$$\leq g(\|x_1 - x_0\|) \|x_1 - x_0\|^{p+1} \leq q \|x_1 - x_0\|,$$

which shows (9.2.7) for $n = 1$. Then, we can have that

$$\|x_2 - x_0\| \leq \|x_2 - x_1\| + \|x_1 - x_0\| \leq q \|x_1 - x_0\| + \|x_1 - x_0\|$$

$$= (1 + q) \|x_1 - x_0\| \leq \frac{1 - q^2}{1 - q} \eta < r,$$

so $x_2 \in \overline{U}(x_0, r)$ and x_3 is well defined.

Assuming $\|x_{k+1} - x_k\| \leq q \|x_k - x_{k-1}\|$ and $x_{k+1} \in \overline{U}(x_0, r)$ for each $k = 1, 2, \ldots, n$ we get

$$\|x_{k+2} - x_{k+1}\| = \|A(x_{k+1})^{-1} [F(x_{k+1}) - F(x_k) - A(x_k)(x_{k+1} - x_k)]\|$$

$$\leq g(\|x_{k+1} - x_k\|) \|x_{k+1} - x_k\|^{p+1}$$

$$\leq g(\|x_1 - x_0\|) \|x_1 - x_0\|^p \|x_{k+1} - x_k\| \leq q \|x_{k+1} - x_k\|$$

and

$$\|x_{k+2} - x_0\| \leq \|x_{k+2} - x_{k+1}\| + \|x_{k+1} - x_k\| + \cdots + \|x_1 - x_0\|$$

$$\leq (q^{k+1} + q^k + \cdots + 1) \|x_1 - x_0\| \leq \frac{1 - q^{k+2}}{1 - q} \|x_1 - x_0\|$$

$$< \frac{\eta}{1 - q} = r,$$

which completes the induction for (9.2.7) and $x_{k+2} \in \overline{U}(x_0, r)$. We also have that for $m \geq 0$

$$\|x_{n+m} - x_n\| \leq \|x_{n+m} - x_{n+m-1}\| + \cdots + \|x_{n+1} - x_n\|$$

$$\leq \left(q^{m-1} + q^{m-2} + \cdots + 1\right) \|x_{n+1} - x_n\|$$

$$\leq \frac{1 - q^m}{1 - q} q^n \|x_1 - x_0\|.$$

It follows that $\{x_n\}$ is a complete sequence in a Banach space X and as such it converges to some $x^* \in \overline{U}(x_0, r)$ (since $\overline{U}(x_0, r)$ is a closed set). By letting $m \to \infty$, we obtain (9.2.8).   □

Stronger hypotheses are needed to show that x^* is a solution of equation $F(x) = 0$.

Proposition 9.2 *Let* $F : D \subset X \to Y$ *be a continuous operator and let* $A(x) \in L(X, Y)$. *Suppose that there exist* $x_0 \in D$, $\eta \geq 0$, $p \geq 1$, $\psi > 0$, *a function* $g_1 : [0, \eta] \to [0, \infty)$ *continuous and nondecreasing such that for each* $x, y \in D$

$$A(x)^{-1} \in L(Y, X), \quad \left\|A(x)^{-1}\right\| \leq \psi, \quad \left\|A(x_0)^{-1} F(x_0)\right\| \leq \eta, \qquad (9.2.9)$$

$$\|F(y) - F(x) - A(x)(y - x)\| \leq \frac{g_1(\|x - y\|)}{\psi} \|x - y\|^{p+1}, \qquad (9.2.10)$$

$$q_1 := g_1(\eta) \eta^p < 1$$

and

$$\overline{U}(x_0, r_1) \subseteq D,$$

where,

$$r_1 = \frac{\eta}{1 - q_1}.$$

Then, the conclusions of Theorem 9.1 for sequence $\{x_n\}$ *hold with* g_1, q_1, r_1, *replacing* g, q *and* r, *respectively. Moreover,* x^* *is a solution of the equation* $F(x) = 0$.

Proof Notice that

$$\left\|A(x_n)^{-1} \left[F(x_n) - F(x_{n-1}) - A(x_{n-1})(x_n - x_{n-1})\right]\right\|$$

$$\leq \left\|A(x_n)^{-1}\right\| \|F(x_n) - F(x_{n-1}) - A(x_{n-1})(x_n - x_{n-1})\|$$

$$\leq g_1(\|x_n - x_{n-1}\|) \|x_n - x_{n-1}\|^{p+1} \leq q_1 \|x_n - x_{n-1}\|.$$

Therefore, the proof of Theorem 9.1 can apply. Then, in view of the estimate

$$\|F(x_n)\| = \|F(x_n) - F(x_{n-1}) - A(x_{n-1})(x_n - x_{n-1})\| \le$$

$$\frac{g_1(\|x_n - x_{n-1}\|)}{\psi}\|x_n - x_{n-1}\|^{p+1} \le q_1\|x_n - x_{n-1}\|,$$

we deduce by letting $n \to \infty$ that $F(x^*) = 0$. $\qquad\square$

Concerning the uniqueness of the solution x^* we have the following result:

Proposition 9.3 *Under the hypotheses of Proposition 9.2, further suppose that*

$$q_1 r_1^p < 1. \tag{9.2.11}$$

Then, x^* is the only solution of equation $F(x) = 0$ in $\overline{U}(x_0, r_1)$.

Proof The existence of the solution $x^* \in \overline{U}(x_0, r_1)$ has been established in Proposition 9.2. Let $y^* \in \overline{U}(x_0, r_1)$ with $F(y^*) = 0$. Then, we have in turn that

$$\|x_{n+1} - y^*\| = \|x_n - y^* - A(x_n)^{-1}F(x_n)\| =$$

$$\|A(x_n)^{-1}[A(x_n)(x_n - y^*) - F(x_n) + F(y^*)]\| \le$$

$$\|A(x_n)^{-1}\|\|F(y^*) - F(x_n) - A(x_n)(y^* - x_n)\| \le$$

$$\psi\frac{g_1(\|x_n - y^*\|)}{\psi}\|x_n - y^*\|^{p+1} \le q_1 r_1^p\|x_n - x^*\| < \|x_n - y^*\|,$$

so we deduce that $\lim_{n\to\infty} x_n = y^*$. But we have that $\lim_{n\to\infty} x_n = x^*$. Hence, we conclude that $x^* = y^*$. $\qquad\square$

Next, we present a local convergence analysis for the method (9.1.2).

Proposition 9.4 *Let $F : D \subset X \to Y$ be a continuous operator and let $A(x) \in L(X, Y)$. Suppose that there exist $x^* \in D$, $p \ge 1$, a function $g_2 : [0, \infty) \to [0, \infty)$ continuous and nondecreasing such that for each $x \in D$*

$$F(x^*) = 0, \quad A(x)^{-1} \in L(Y, X),$$

$$\|A(x)^{-1}[F(x) - F(x^*) - A(x)(x - x^*)]\| \le g_2(\|x - x^*\|)\|x - x^*\|^{p+1}, \tag{9.2.12}$$

and

$$\overline{U}(x^*, r_2) \subseteq D,$$

where r_2 is the smallest positive solution of equation

$$h(t) := g_2(t) t^p - 1.$$

Then, sequence $\{x_n\}$ generated by method (9.1.2) for $x_0 \in U(x^*, r_2) - \{x^*\}$ is well defined, remains in $U(x^*, r_2)$ for each $n = 0, 1, 2, \ldots$ and converges to x^*. Moreover, the following estimates hold

$$\|x_{n+1} - x^*\| \le g_2(\|x_n - x^*\|) \|x_n - x^*\|^{p+1} < \|x_n - x^*\| < r_2.$$

Proof We have that $h(0) = -1 < 0$ and $h(t) \to +\infty$ as $t \to +\infty$. Then, it follows from the intermediate value theorem that function h has positive zeros. Denote by r_2 the smallest such zero. By hypothesis $x_0 \in U(x^*, r_2) - \{x^*\}$. Then, we get in turn that

$$\|x_1 - x^*\| = \|x_0 - x^* - A(x_0)^{-1} F(x_0)\| =$$

$$\|A(x_0)^{-1} [F(x^*) - F(x_0) - A(x_0)(x^* - x_0)]\| \le$$

$$g_2(\|x_0 - x^*\|) \|x_0 - x^*\|^{p+1} < g_2(r_2) r_2^p \|x_0 - x^*\| =$$

$$\|x_0 - x^*\| < r_2,$$

which shows that $x_1 \in U(x^*, r_2)$ and x_2 is well defined. By a simple inductive argument as in the preceding estimate we get that

$$\|x_{k+1} - x^*\| = \|x_k - x^* - A(x_k)^{-1} F(x_k)\| \le$$

$$\|A(x_k)^{-1} [F(x^*) - F(x_k) - A(x_k)(x^* - x_k)]\| \le$$

$$g_2(\|x_k - x^*\|) \|x_k - x^*\|^{p+1} < g_2(r_2) r_2^p \|x_k - x^*\| = \|x_k - x^*\| < r_2,$$

which shows $\lim_{k \to \infty} x_k = x^*$ and $x_{k+1} \in U(x^*, r_2)$.   □

Remark 9.5 (a) Hypothesis (9.2.3) specializes to Newton-Mysowski-type, if $A(x) = F'(x)$ [9, 13, 17]. However, if F is not Fréchet-differentiable, then our results extend the applicability of iterative algorithm (9.1.2).

(b) Theorem 9.1 has practical value although we do not show that x^* is a solution of equation $F(x) = 0$, since this may be shown in another way.

(c) Hypothesis (9.2.12) can be replaced by the stronger

$$\|A(x)^{-1} [F(x) - F(y) - A(x)(x - y)]\| \le g_2(\|x - y\|) \|x - y\|^{p+1}.$$

The preceding results can be extended to hold for two point methods defined for each $n = 0, 1, 2, \ldots$ by

$$x_{n+1} = x_n - A\,(x_n, x_{n-1})^{-1}\,F\,(x_n),\qquad(9.2.13)$$

where $x_{-1}, x_0 \in D$ are initial points and $A\,(w, v) \in L\,(X, Y)$ for each $v, w \in D$. If $A\,(w, v) = [w, v; F]$, then method (9.2.13) reduces to the popular secant method, where $[w, v; F]$ denotes a divided difference of order one for the operator F. Many other choices for A are also possible [9, 13, 17].

If we simply replace $A\,(x)$ by $A\,(y, x)$ in the proof of Proposition 9.2 we arrive at the following semilocal convergence result for method (9.2.13).

Theorem 9.6 *Let $F : D \subset X \to Y$ be a continuous operator and let $A\,(y, x) \in L\,(X, Y)$ for each $x, y \in D$. Suppose that there exist $x_{-1}, x_0 \in D$, $\eta \geq 0$, $p \geq 1$, $\psi > 0$, a function $g_1 : [0, \eta] \to [0, \infty)$ continuous and nondecreasing such that for each $x, y \in D$:*

$$A\,(y, x)^{-1} \in L\,(Y, X),\quad \left\| A\,(y, x)^{-1} \right\| \leq \psi,\qquad(9.2.14)$$

$$\min \left\{ \|x_0 - x_{-1}\|, \left\| A\,(x_0, x_{-1})^{-1}\,F\,(x_0) \right\| \right\} \leq \eta,$$

$$\|F\,(y) - F\,(x) - A\,(y, x)\,(y - x)\| \leq \frac{g_1\,(\|x - y\|)}{\psi}\,\|x - y\|^{p+1},\qquad(9.2.15)$$

$$q_1 < 1,\quad q_1 r_1^p < 1$$

and

$$\overline{U}\,(x_0, r_1) \subseteq D,$$

where,

$$r_1 = \frac{\eta}{1 - q_1}$$

and q_1 is defined in Proposition 9.2.

Then, sequence $\{x_n\}$ generated by method (9.2.13) is well defined, remains in $\overline{U}\,(x_0, r_1)$ for each $n = 0, 1, 2, \ldots$ and converges to the only solution of equation $F\,(x) = 0$ in $\overline{U}\,(x_0, r_1)$.

Moreover, the estimates (9.2.7) and (9.2.8) hold with g_1, q_1 replacing g and q, respectively.

Concerning, the local convergence of the iterative algorithm (9.2.13) we obtain the analogous to Proposition 9.4 result.

Proposition 9.7 *Let $F : D \subset X \to Y$ be a continuous operator and let $A(y, x) \in$ $L(X, Y)$. Suppose that there exist $x^* \in D$, $p \geq 1$, a function $g_2 : [0, \infty)^2 \to [0, \infty)$ continuous and nondecreasing such that for each $x, y \in D$*

$$F(x^*) = 0, \quad A(y, x)^{-1} \in L(Y, X),$$

$$\left\| A(y, x)^{-1} \left[F(y) - F(x^*) - A(y, x)(y - x^*) \right] \right\| \leq$$

$$g_2 \left(\|y - x^*\|, \|x - x^*\| \right) \|y - x^*\|^{p+1}$$

and

$$\overline{U}(x^*, r_2) \subseteq D,$$

where r_2 is the smallest positive solution of equation

$$h(t) := g_2(t, t) t^p - 1.$$

Then, sequence $\{x_n\}$ generated by method (9.2.13) for $x_{-1}, x_0 \in U(x^*, r_2) - \{x^*\}$ is well defined, remains in $U(x^*, r_2)$ for each $n = 0, 1, 2, \ldots$ and converges to x^*. Moreover, the following estimates hold

$$\|x_{n+1} - x^*\| \leq g_2 \left(\|x_n - x^*\|, \|x_{n-1} - x^*\| \right) \|x_n - x^*\|^{p+1}$$

$$< \|x_n - x^*\| < r_2.$$

Remark 9.8 In the next section we present some choices and properties of operator $A(y, x)$ from fractional calculus satisfying the crucial estimate (9.2.15) in the special case when,

$$g_1(t) = c\psi \quad \text{for some } c > 0 \text{ and each } t \geq 0.$$

(see the end of Sect. 9.3 for a possible definition of the constant c).

Hence, Theorem 9.6 can apply to solve equation $F(x) = 0$. Other choices for operator $A(x)$ or operator $A(y, x)$ can be found in [7–9, 11–17].

9.3 Applications to Fractional Calculus

Let $f : [a, b] \to \mathbb{R}$ such that $f^{(m)} \in L_\infty([a, b])$, the left Caputo fractional derivative of order $\alpha \notin \mathbb{N}$, $\alpha > 0$, $m = \lceil \alpha \rceil$ ($\lceil \cdot \rceil$ ceiling) is defined as follows:

$$\left(D_a^\alpha f \right)(x) = \frac{1}{\Gamma(m - \alpha)} \int_a^x (x - t)^{m - \alpha - 1} f^{(m)}(t) \, dt, \tag{9.3.1}$$

where Γ is the gamma function, $\forall \, x \in [a, b]$.

We observe that

$$\left| \left(D_a^\alpha f \right) (x) \right| \le \frac{1}{\Gamma (m - \alpha)} \int_a^x (x - t)^{m - \alpha - 1} \left| f^{(m)} (t) \right| dt$$

$$\le \frac{\left\| f^{(m)} \right\|_\infty}{\Gamma (m - \alpha)} \left(\int_a^x (x - t)^{m - \alpha - 1} dt \right) = \frac{\left\| f^{(m)} \right\|_\infty}{\Gamma (m - \alpha)} \frac{(x - a)^{m - \alpha}}{(m - \alpha)}$$

$$= \frac{\left\| f^{(m)} \right\|_\infty}{\Gamma (m - \alpha + 1)} (x - a)^{m - \alpha} . \tag{9.3.2}$$

We have proved that

$$\left| \left(D_a^\alpha f \right) (x) \right| \le \frac{\left\| f^{(m)} \right\|_\infty}{\Gamma (m - \alpha + 1)} (x - a)^{m - \alpha} \le \frac{\left\| f^{(m)} \right\|_\infty}{\Gamma (m - \alpha + 1)} (b - a)^{m - \alpha} . \tag{9.3.3}$$

Clearly then $\left(D_a^\alpha f \right) (a) = 0$.

Let $n \in \mathbb{N}$ we denote $D_a^{n\alpha} = D_a^\alpha D_a^\alpha \ldots D_a^\alpha$ (n-times).

Let us assume now that

$$D_a^{k\alpha} f \in C \left([a, b] \right), \quad k = 0, 1, \ldots, n + 1; n \in \mathbb{N}, 0 < \alpha \le 1. \tag{9.3.4}$$

By [16], we are able to extract the following interesting generalized fractional Caputo type Taylor's formula: (there it is assumed that $D_a^{k\alpha} f (x) \in C ((a, b])$, $k = 0, 1, \ldots, n + 1; 0 < \alpha \le 1$)

$$f (x) = \sum_{i=0}^n \frac{(x - a)^{i\alpha}}{\Gamma (i\alpha + 1)} \left(D_a^{i\alpha} f \right) (a) + \tag{9.3.5}$$

$$\frac{1}{\Gamma ((n + 1) \alpha)} \int_a^x (x - t)^{(n+1)\alpha - 1} \left(D_a^{(n+1)\alpha} f \right) (t) dt, \quad \forall x \in (a, b].$$

Notice that [16] has lots of typos or minor errors, which we fixed.

Under our assumption and conclusion, see (9.3.4), Taylor's formula (9.3.5) becomes

$$f (x) - f (a) = \sum_{i=2}^n \frac{(x - a)^{i\alpha}}{\Gamma (i\alpha + 1)} \left(D_a^{i\alpha} f \right) (a) +$$

$$\frac{1}{\Gamma ((n + 1) \alpha)} \int_a^x (x - t)^{(n+1)\alpha - 1} \left(D_a^{(n+1)\alpha} f \right) (t) dt, \quad \forall x \in (a, b], 0 < \alpha < 1. \tag{9.3.6}$$

Here we are going to operate more generally. Again we assume $0 < \alpha \leq 1$, and $f : [a, b] \to \mathbb{R}$, such that $f' \in C([a, b])$. We define the following left Caputo fractional derivatives:

$$\left(D_y^\alpha f\right)(x) = \frac{1}{\Gamma(1-\alpha)} \int_y^x (x-t)^{-\alpha} f'(t)\, dt, \qquad (9.3.7)$$

for any $x \geq y$; $x, y \in [a, b]$, and

$$\left(D_x^\alpha f\right)(y) = \frac{1}{\Gamma(1-\alpha)} \int_x^y (y-t)^{-\alpha} f'(t)\, dt, \qquad (9.3.8)$$

for any $y \geq x$; $x, y \in [a, b]$.

Notice $D_y^1 f = f'$, $D_x^1 f = f'$ by convention.

Clearly here $\left(D_y^\alpha f\right)$, $\left(D_x^\alpha f\right)$ are continuous functions over $[a, b]$, see [3], p. 388. We also make the convention that $\left(D_y^\alpha f\right)(x) = 0$, for $x < y$, and $\left(D_x^\alpha f\right)(y) = 0$, for $y < x$.

Here we assume that $D_y^{k\alpha} f$, $D_x^{k\alpha} f \in C([a, b])$, $k = 0, 1, \ldots, n+1$, $n \in \mathbb{N}$; $\forall\, x, y \in [a, b]$.

Then by (9.3.6) we obtain

$$f(x) - f(y) = \sum_{i=2}^n \frac{(x-y)^{i\alpha}}{\Gamma(i\alpha+1)} \left(D_y^{i\alpha} f\right)(y) +$$

$$\frac{1}{\Gamma((n+1)\alpha)} \int_y^x (x-t)^{(n+1)\alpha-1} \left(D_y^{(n+1)\alpha} f\right)(t)\, dt, \qquad (9.3.9)$$

$\forall\, x > y$; $x, y \in [a, b]$, $0 < \alpha < 1$.
And also it holds

$$f(y) - f(x) = \sum_{i=2}^n \frac{(y-x)^{i\alpha}}{\Gamma(i\alpha+1)} \left(D_x^{i\alpha} f\right)(x) +$$

$$\frac{1}{\Gamma((n+1)\alpha)} \int_x^y (y-t)^{(n+1)\alpha-1} \left(D_x^{(n+1)\alpha} f\right)(t)\, dt, \qquad (9.3.10)$$

$\forall\, y > x$; $x, y \in [a, b]$, $0 < \alpha < 1$.

We define the following linear operator

$$(A(f))(x, y) =$$

$$\begin{cases} \sum_{i=2}^{n} \frac{(x-y)^{i\alpha-1}}{\Gamma(i\alpha+1)} \left(D_y^{i\alpha} f\right)(y) + \left(D_y^{(n+1)\alpha} f(x)\right) \frac{(x-y)^{(n+1)\alpha-1}}{\Gamma((n+1)\alpha+1)}, & x > y, \\[2ex] \sum_{i=2}^{n} \frac{(y-x)^{i\alpha-1}}{\Gamma(i\alpha+1)} \left(D_x^{i\alpha} f\right)(x) + \left(D_x^{(n+1)\alpha} f(y)\right) \frac{(y-x)^{(n+1)\alpha-1}}{\Gamma((n+1)\alpha+1)}, & y > x, \\[2ex] f'(x), & \text{when } x = y, \end{cases} \tag{9.3.11}$$

$\forall\, x, y \in [a, b], 0 < \alpha < 1.$

We may assume that

$$|(A(f))(x, x) - (A(f))(y, y)| = |f'(x) - f'(y)| \tag{9.3.12}$$

$$\leq \Phi |x - y|, \quad \forall x, y \in [a, b], \text{ with } \Phi > 0.$$

We estimate and have:

(i) case of x > y:

$$|f(x) - f(y) - (A(f))(x, y)(x - y)| =$$

$$\left| \frac{1}{\Gamma((n+1)\alpha)} \int_y^x (x - t)^{(n+1)\alpha-1} \left(D_y^{(n+1)\alpha} f\right)(t)\, dt \tag{9.3.13} \right.$$

$$\left. - \left(D_y^{(n+1)\alpha} f(x)\right) \frac{(x - y)^{(n+1)\alpha}}{\Gamma((n+1)\alpha+1)} \right| =$$

$$\frac{1}{\Gamma((n+1)\alpha)} \left| \int_y^x (x - t)^{(n+1)\alpha-1} \left(\left(D_y^{(n+1)\alpha} f\right)(t) - \left(D_y^{(n+1)\alpha} f\right)(x)\right) dt \right|$$

$$\leq \frac{1}{\Gamma((n+1)\alpha)} \int_y^x (x - t)^{(n+1)\alpha-1} \left| D_y^{(n+1)\alpha} f(t) - \left(D_y^{(n+1)\alpha} f\right)(x)\right| dt$$

(we assume here that

$$\left| D_y^{(n+1)\alpha} f(t) - D_y^{(n+1)\alpha} f(x) \right| \leq \lambda_1 |t - x|, \tag{9.3.14}$$

$\forall\, t, x, y \in [a, b] : x \geq t \geq y, \text{ where } \lambda_1 > 0)$

$$\leq \frac{\lambda_1}{\Gamma((n+1)\alpha)} \int_y^x (x - t)^{(n+1)\alpha-1} (x - t)\, dt =$$

$$\frac{\lambda_1}{\Gamma((n+1)\alpha)} \int_y^x (x - t)^{(n+1)\alpha}\, dt = \frac{\lambda_1}{\Gamma((n+1)\alpha)} \frac{(x - y)^{(n+1)\alpha+1}}{((n+1)\alpha+1)}. \tag{9.3.15}$$

We have proved that

$$|f(x) - f(y) - (A(f))(x, y)(x - y)| \leq \frac{\lambda_1}{\Gamma((n+1)\alpha)} \frac{(x-y)^{(n+1)\alpha+1}}{((n+1)\alpha+1)},$$

(9.3.16)

for any $x, y \in [a, b] : x > y, 0 < \alpha < 1$.

(ii) case of x < y:

$$|f(x) - f(y) - (A(f))(x, y)(x - y)| =$$

$$|f(y) - f(x) - (A(f))(x, y)(y - x)| =$$

$$\left| \frac{1}{\Gamma((n+1)\alpha)} \int_x^y (y - t)^{(n+1)\alpha-1} \left(D_x^{(n+1)\alpha} f \right)(t) \, dt \right.$$

(9.3.17)

$$\left. - \left(D_x^{(n+1)\alpha} f(y) \right) \frac{(y-x)^{(n+1)\alpha}}{\Gamma((n+1)\alpha+1)} \right| =$$

$$\frac{1}{\Gamma((n+1)\alpha)} \left| \int_x^y (y - t)^{(n+1)\alpha-1} \left(\left(D_x^{(n+1)\alpha} f \right)(t) - \left(D_x^{(n+1)\alpha} f \right)(y) \right) dt \right|$$

$$\leq \frac{1}{\Gamma((n+1)\alpha)} \int_x^y (y - t)^{(n+1)\alpha-1} \left| \left(D_x^{(n+1)\alpha} f \right)(t) - \left(D_x^{(n+1)\alpha} f \right)(y) \right| dt$$

(we assume that

$$\left| \left(D_x^{(n+1)\alpha} f \right)(t) - \left(D_x^{(n+1)\alpha} f \right)(y) \right| \leq \lambda_2 |t - y|,$$

(9.3.18)

$\forall t, y, x \in [a, b] : y \geq t \geq x$, where $\lambda_2 > 0$)

$$\leq \frac{\lambda_2}{\Gamma((n+1)\alpha)} \int_x^y (y - t)^{(n+1)\alpha-1} (y - t) \, dt =$$

$$\frac{\lambda_2}{\Gamma((n+1)\alpha)} \int_x^y (y - t)^{(n+1)\alpha} \, dt = \frac{\lambda_2}{\Gamma((n+1)\alpha)} \frac{(y-x)^{(n+1)\alpha+1}}{((n+1)\alpha+1)}.$$

(9.3.19)

We have proved that

$$|f(x) - f(y) - A(f)(x, y)(x - y)| \leq \frac{\lambda_2}{\Gamma((n+1)\alpha)} \frac{(y-x)^{(n+1)\alpha+1}}{((n+1)\alpha+1)},$$

(9.3.20)

$\forall x, y \in [a, b] : y > x, 0 < \alpha < 1$.

Conclusion Let $\lambda := \max(\lambda_1, \lambda_2)$. It holds

$$|f(x) - f(y) - (\Lambda(f))(x, y)(x - y)| \le \frac{\lambda}{\Gamma((n+1)\alpha)} \frac{|x - y|^{(n+1)\alpha+1}}{((n+1)\alpha+1)},$$

(9.3.21)

$\forall\, x, y \in [a, b]$, where $0 < \alpha < 1, n \in \mathbb{N}$.

One may assume that $\frac{\lambda}{\Gamma((n+1)\alpha)} < 1$.

(Above notice that (9.3.21) is trivial when $x = y$.)

Now based on (9.3.12) and (9.3.21), we can apply our numerical methods presented in this chapter, to solve $f(x) = 0$.

To have $(n+1)\alpha + 1 \ge 2$, we need to take $1 > \alpha \ge \frac{1}{n+1}$, where $n \in \mathbb{N}$.

Then, returning back to Remark 9.8, we see that the constant c can be defined by

$$c = \frac{\lambda}{\Gamma((n+1)\alpha)[(n+1)\alpha+1]}$$

provided that $n = p, (p+1)\alpha \le p$ and

$$|y - x| \le 1 \text{ for each } x, y \in [a, b].$$

(9.3.22)

Notice that condition (9.3.22) can always be satisfied by choosing x, y (i.e. a, b) sufficiently close to each other.

References

1. S. Amat, S. Busquier, Third-order iterative methods under Kantorovich conditions. J. Math. Anal. Appl. **336**, 243–261 (2007)
2. S. Amat, S. Busquier, S. Plaza, Chaotic dynamics of a third-order Newton-like method. J. Math. Anal. Appl. **366**(1), 164–174 (2010)
3. G. Anastassiou, *Fractional Differentiation Inequalities* (Springer, New York, 2009)
4. G. Anastassiou, *Intelligent Mathematics: Computational Analysis* (Springer, Heidelberg, 2011)
5. G. Anastassiou, I. Argyros, A convergence analysis for a certain family of extended iterative methods: part I, theory. Annales Univ. Sci. Budapest Sect. Comp. (2015) (accepted)
6. G. Anastassiou, I. Argyros, A convergence analysis for a certain family of extended iterative methods: part II. Applications to fractional calculus. Annales Univ. Sci. Budapest Sect. Comp. (2015) (accepted)
7. I.K. Argyros, Newton-like methods in partially ordered linear spaces. J. Approx. Th. Appl. **9**(1), 1–10 (1993)
8. I.K. Argyros, Results on controlling the residuals of perturbed Newton-like methods on Banach spaces with a convergence structure. Southwest J. Pure Appl. Math. **1**, 32–38 (1995)
9. I.K. Argyros, *Convergence and Applications of Newton-Like Iterations* (Springer, New York, 2008)
10. K. Diethelm, *The Analysis of Fractional Differential Equations*. Lecture Notes in Mathematics, vol 2004, 1st edn. (Springer, New York, 2010)
11. J.A. Ezquerro, J.M. Gutierrez, M.A. Hernandez, N. Romero, M.J. Rubio, The Newton method: from Newton to Kantorovich (Spanish). Gac. R. Soc. Mat. Esp. **13**, 53–76 (2010)

12. J.A. Ezquerro, M.A. Hernandez, Newton-like methods of high order and domains of semilocal and global convergence. Appl. Math. Comput. **214**(1), 142–154 (2009)
13. L.V. Kantorovich, G.P. Akilov, *Functional Analysis in Normed Spaces* (Pergamon Press, New York, 1964)
14. A.A. Magrenan, Different anomalies in a Jarratt family of iterative root finding methods. Appl. Math. Comput. **233**, 29–38 (2014)
15. A.A. Magrenan, A new tool to study real dynamics: the convergence plane. Appl. Math. Comput. **248**, 215–224 (2014)
16. Z.M. Odibat, N.J. Shawagleh, Generalized Taylor's formula. Appl. Math. Comput. **186**, 286–293 (2007)
17. F.A. Potra, V. Ptak, *Nondiscrete Induction and Iterative Processes* (Pitman Publ, London, 1984)

Chapter 10
Secant-Like Methods and Fractional Calculus

We present local and semilocal convergence results for secant-like methods in order to approximate a locally unique solution of a nonlinear equation in a Banach space setting. In the last part of the study we present some choices of the operators involved in fractional calculus where the operators satisfy the convergence conditions. It follows [5].

10.1 Introduction

In this study we are concerned with the problem of approximating a locally unique solution x^* of the nonlinear equation

$$F(x) = 0, \tag{10.1.1}$$

where F is a continuous operator defined on a subset D of a Banach space X with values in a Banach space Y.

A lot of problems in Computational Sciences and other disciplines can be brought in a form like (10.1.1) using Mathematical Modelling [8, 12, 16]. The solutions of such equations can be found in closed form only in special cases. That is why most solution methods for these equations are iterative. Iterative methods are usually studied based on semilocal and local convergence. The semilocal convergence matter is, based on the information around the initial point to give hypotheses ensuring the convergence of the iterative algorithm; while the local one is, based on the information around a solution, to find estimates of the radii of convergence balls as well as error bounds on the distances involved.

© Springer International Publishing Switzerland 2016   163
G.A. Anastassiou and I.K. Argyros, *Intelligent Numerical Methods:*
Applications to Fractional Calculus, Studies in Computational Intelligence 624,
DOI 10.1007/978-3-319-26721-0_10

We introduce the secant-like method defined for each $n = 0, 1, 2, \ldots$ by

$$x_{n+1} = x_n - A(x_n, x_{n-1})^{-1} F(x_n), \tag{10.1.2}$$

where $x_{-1}, x_0 \in D$ are initial points and $A(x, y) \in L(X, Y)$ the space of bounded linear operators from X into Y. There is a plethora on local as well as semilocal convergence theorems for method (10.1.2) provided that the operator A is an approximation to the Fréchet-derivative F' [1, 2, 6–16]. In the present study we do not necessarily assume that operator A is related to F'. This way we expand the applicability of iterative algorithm (10.1.2). Notice that many well known methods are special case of method (10.1.2).

Newton's method: Choose $A(x, x) = F'(x)$ for each $x \in D$.

Secant method: Choose $A(x, y) = [x, y; F]$, where $[x, y; F]$ denotes a divided difference of order one [8, 12, 15].

The so called Newton-like methods and many other methods are special cases of method (10.1.2).

The rest of the chapter is organized as follows. The semilocal as well as the local convergence analysis of method (10.1.2) is given in Sect. 10.2. Some applications from fractional calculus are given in Sect. 10.3.

10.2　Convergence Analysis

We present the main semilocal convergence result for method (10.1.2).

Theorem 10.1 *Let $F : D \subset X \to Y$ be a continuous operator and let $A(x, y) \in L(X, Y)$. Suppose that there exist $x_{-1}, x_0 \in D$, $\eta \geq 0$, $p \geq 1$, a function $g : [0, \infty)^2 \to [0, \infty)$ continuous and nondecreasing such that for each $x, y, z \in D$*

$$A(z, y)^{-1} \in L(Y, X), \tag{10.2.1}$$

$$\max\left\{ \|x_{-1} - x_0\|, \left\| A(x_0, x_{-1})^{-1} F(x_0) \right\| \right\} \leq \eta, \tag{10.2.2}$$

$$\left\| A(z, y)^{-1} (F(z) - F(y) - A(y, x)(z - y)) \right\| \leq g(\|z - y\|, \|y - x\|) \|z - y\|^{p+1}, \tag{10.2.3}$$

$$q := g(\eta, \eta) \eta^p < 1 \tag{10.2.4}$$

and

$$\overline{U}(x_0, r) \subseteq D, \tag{10.2.5}$$

where,

$$r = \frac{\eta}{1-q}. \tag{10.2.6}$$

Then, the sequence $\{x_n\}$ generated by method (10.1.2) is well defined, remains in $\overline{U}(x_0, r)$ for each $n = 0, 1, 2, \ldots$ and converges to some $x^* \in \overline{U}(x_0, r)$ such that

$$\|x_{n+1} - x_n\| \le g(\|x_n - x_{n-1}\|, \|x_{n-1} - x_{n-2}\|)\|x_n - x_{n-1}\|^{p+1}$$

$$\le q\|x_n - x_{n-1}\| \tag{10.2.7}$$

and

$$\|x_n - x^*\| \le \frac{q^n \eta}{1-q}. \tag{10.2.8}$$

Proof The iterate x_1 is well defined by method (10.1.2) for $n = 0$ and (10.2.1). We also have by (10.2.2) and (10.2.6) that
$\|x_1 - x_0\| = \|A(x_0, x_{-1})^{-1} F(x_0)\| \le \eta < r$, so we get that $x_1 \in \overline{U}(x_0, r)$ and x_2 is well defined (by (10.2.5)). Using (10.2.3) and (10.2.4) we get that

$$\|x_2 - x_1\| = \|A(x_1, x_0)^{-1}[F(x_1) - F(x_0) - A(x_0, x_{-1})(x_1 - x_0)]\|$$

$$\le g(\|x_1 - x_0\|, \|x_0 - x_{-1}\|)\|x_1 - x_0\|^{p+1} \le q\|x_1 - x_0\|,$$

which shows (10.2.7) for $n = 1$. Then, we can have that

$$\|x_2 - x_0\| \le \|x_2 - x_1\| + \|x_1 - x_0\| \le q\|x_1 - x_0\| + \|x_1 - x_0\|$$

$$= (1 + q)\|x_1 - x_0\| \le \frac{1 - q^2}{1 - q}\eta < r,$$

so $x_2 \in \overline{U}(x_0, r)$ and x_3 is well defined.

Assuming $\|x_{k+1} - x_k\| \le q\|x_k - x_{k-1}\|$ and $x_{k+1} \in \overline{U}(x_0, r)$ for each $k = 1, 2, \ldots, n$ we get

$$\|x_{k+2} - x_{k+1}\| =$$

$$\|A(x_{k+1}, x_k)^{-1}[F(x_{k+1}) - F(x_k) - A(x_k, x_{k-1})(x_{k+1} - x_k)]\| \le$$

$$g(\|x_{k+1} - x_k\|, \|x_k - x_{k-1}\|)\|x_{k+1} - x_k\|^{p+1} \le$$

$$g(\|x_1 - x_0\|, \|x_0 - x_{-1}\|)\|x_1 - x_0\|^p \|x_{k+1} - x_k\| \le q\|x_{k+1} - x_k\|$$

and

$$\|x_{k+2} - x_0\| \le \|x_{k+2} - x_{k+1}\| + \|x_{k+1} - x_k\| + \cdots + \|x_1 - x_0\|$$

$$\leq \left(q^{k+1} + q^k + \cdots + 1\right) \|x_1 - x_0\| \leq \frac{1 - q^{k+2}}{1 - q} \|x_1 - x_0\|$$

$$< \frac{\eta}{1 - q} = r,$$

which completes the induction for (10.2.7) and $x_{k+2} \in \overline{U}(x_0, r)$. We also have that for $m \geq 0$

$$\|x_{n+m} - x_n\| \leq \|x_{n+m} - x_{n+m-1}\| + \cdots + \|x_{n+1} - x_n\|$$

$$\leq \left(q^{m-1} + q^{m-2} + \cdots + 1\right) \|x_{n+1} - x_n\|$$

$$\leq \frac{1 - q^m}{1 - q} q^n \|x_1 - x_0\|.$$

It follows that $\{x_n\}$ is a complete sequence in a Banach space X and as such it converges to some $x^* \in \overline{U}(x_0, r)$ (since $\overline{U}(x_0, r)$ is a closed set). By letting $m \to \infty$, we obtain (10.2.8). $\qquad\square$

Stronger hypotheses are needed to show that x^* is a solution of equation $F(x) = 0$.

Proposition 10.2 *Let $F : D \subset X \to Y$ be a continuous operator and let $A(x, y) \in L(X, Y)$. Suppose that there exist $x_{-1}, x_0 \in D$, $\eta \geq 0$, $p \geq 1$, $\mu > 0$, a function $g_1 : [0, \infty)^2 \to [0, \infty)$ continuous and nondecreasing such that for each $x, y \in D$*

$$A(x, y)^{-1} \in L(Y, X), \quad \left\|A(x, y)^{-1}\right\| \leq \mu,$$

$$\max \left\{ \|x_{-1} - x_0\|, \left\|A(x_0, x_{-1})^{-1} F(x_0)\right\| \right\} \leq \eta, \qquad (10.2.9)$$

$$\|F(z) - F(y) - A(y, x)(z - y)\| \leq \frac{g_1(\|z - y\|, \|x - y\|)}{\mu} \|z - y\|^{p+1},$$

$$(10.2.10)$$

$$q_1 := g_1(\eta, \eta) \eta^p < 1,$$

and

$$\overline{U}(x_0, r_1) \subseteq D,$$

where,

$$r_1 = \frac{\eta}{1 - q_1}.$$

Then, the conclusions of Theorem 10.1 for sequence $\{x_n\}$ hold with g_1, q_1, r_1, replacing g, q and r, respectively. Moreover, x^* is a solution of the equation $F(x) = 0$.

Proof Notice that

$$\left\| A\left(x_n, x_{n-1}\right)^{-1} \left[F\left(x_n\right) - F\left(x_{n-1}\right) - A\left(x_{n-1}, x_{n-2}\right)\left(x_n - x_{n-1}\right) \right] \right\|$$

$$\leq \left\| A\left(x_n, x_{n-1}\right)^{-1} \right\| \left\| F\left(x_n\right) - F\left(x_{n-1}\right) - A\left(x_{n-1}, x_{n-2}\right)\left(x_n - x_{n-1}\right) \right\|$$

$$\leq g_1\left(\|x_n - x_{n-1}\|, \|x_{n-1} - x_{n-2}\|\right) \|x_n - x_{n-1}\|^{p+1} \leq q_1 \|x_n - x_{n-1}\|.$$

Therefore, the proof of Theorem 10.1 can apply. Then, in view of the estimate

$$\|F\left(x_n\right)\| = \|F\left(x_n\right) - F\left(x_{n-1}\right) - A\left(x_{n-1}, x_{n-2}\right)\left(x_n - x_{n-1}\right)\| \leq$$

$$\frac{g_1\left(\|x_n - x_{n-1}\|\right)}{\mu} \|x_n - x_{n-1}\|^{p+1} \leq q_1 \|x_n - x_{n-1}\|,$$

we deduce by letting $n \to \infty$ that $F\left(x^*\right) = 0$. $\square$

Concerning the uniqueness of the solution x^* we have the following result:

Proposition 10.3 *Under the hypotheses of Proposition 10.2, further suppose that there exists $g_2 : [0, \infty)^2 \to [0, \infty)$ continuous and nondecreasing such that*

$$\|F\left(z\right) - F\left(x\right) - A\left(z, y\right)\left(z - x\right)\| \leq \frac{g_2\left(\|z - x\|, \|y - x\|\right)}{\mu} \|z - x\|^{p+1}$$

$$(10.2.11)$$

and

$$g_2\left(r_1, \eta + r_1\right) r_1^p < 1. \tag{10.2.12}$$

Then, x^* is the only solution of equation $F\left(x\right) = 0$ in $\overline{U}\left(x_0, r_1\right)$.

Proof The existence of the solution $x^* \in \overline{U}\left(x_0, r_1\right)$ has been established in Proposition 10.2. Let $y^* \in \overline{U}\left(x_0, r_1\right)$ with $F\left(y^*\right) = 0$. Then, we have in turn that

$$\left\| x_{n+1} - y^* \right\| = \left\| x_n - y^* - A\left(x_n, x_{n-1}\right)^{-1} F\left(x_n\right) \right\| =$$

$$\left\| A\left(x_n, x_{n-1}\right)^{-1} \left[A\left(x_n, x_{n-1}\right)\left(x_n - y^*\right) - F\left(x_n\right) + F\left(y^*\right) \right] \right\| \leq$$

$$\left\| A\left(x_n, x_{n-1}\right)^{-1} \right\| \left\| F\left(y^*\right) - F\left(x_n\right) - A\left(x_n, x_{n-1}\right)\left(y^* - x_n\right) \right\| \leq$$

$$\mu \frac{g_1\left(\|x_n - y^*\|, \|x_{n-1} - y^*\|\right)}{\mu} \|x_n - y^*\|^{p+1} \leq$$

$$g_2\left(r_1, \eta + r_1\right) r_1^p \|x_n - x^*\| < \|x_n - y^*\|,$$

so we deduce that $\lim\limits_{n \to \infty} x_n = y^*$. But we have that $\lim\limits_{n \to \infty} x_n = x^*$. Hence, we conclude that $x^* = y^*$. $\square$

Next, we present a local convergence analysis for the iterative algorithm (10.1.2).

Proposition 10.4 *Let* $F : D \subset X \to Y$ *be a continuous operator and let* $A(y, x) \in$ $L(X, Y)$. *Suppose that there exist* $x^* \in D$, $p \geq 1$, *a function* $g_3 : [0, \infty)^2 \to [0, \infty)$ *continuous and nondecreasing such that for each* $x, y \in D$

$$F(x^*) = 0, \ A(y, x)^{-1} \in L(Y, X),$$

$$\left\| A(y, x)^{-1} \left[F(x) - F(x^*) - A(y, x)(x - x^*) \right] \right\| \leq$$

$$g_3 \left(\|y - x^*\|, \|x - x^*\| \right) \|y - x^*\|^{p+1}, \tag{10.2.13}$$

and

$$\overline{U}(x^*, r_2) \subseteq D,$$

where r_2 *is the smallest positive solution of equation*

$$h(t) := g_3(t, t) t^p - 1.$$

Then, sequence $\{x_n\}$ *generated by method (10.1.2) for* $x_{-1}, x_0 \in U(x^*, r_2) - \{x^*\}$ *is well defined, remains in* $U(x^*, r_2)$ *for each* $n = 0, 1, 2, \ldots$ *and converges to* x^*. *Moreover, the following estimates hold*

$$\|x_{n+1} - x^*\| \leq g_2 \left(\|x_n - x^*\|, \|x_{n-1} - x^*\| \right) \|x_n - x^*\|^{p+1} < \|x_n - x^*\| < r_2.$$

Proof We have that $h(0) = -1 < 0$ and $h(t) \to +\infty$ as $t \to +\infty$. Then, it follows from the intermediate value theorem that function h has positive zeros. Denote by r_2 the smallest such zero. By hypothesis $x_{-1}, x_0 \in U(x^*, r_2) - \{x^*\}$. Then, we get in turn that

$$\|x_1 - x^*\| = \|x_0 - x^* - A(x_0, x_{-1})^{-1} F(x_0)\| =$$

$$\left\| A(x_0, x_{-1})^{-1} \left[F(x^*) - F(x_0) - A(x_0, x_{-1})(x^* - x_0) \right] \right\| \leq$$

$$g_2 \left(\|x_0 - x^*\|, \|x_{-1} - x^*\| \right) \|x_0 - x^*\|^{p+1} < g_3(r_2, r_2) r_2^p \|x_0 - x^*\| =$$

$$\|x_0 - x^*\| < r_2,$$

which shows that $x_1 \in U(x^*, r_2)$ and x_2 is well defined. By a simple inductive argument as in the preceding estimate we get that

$$\|x_{k+1} - x^*\| = \|x_k - x^* - A(x_k, x_{k-1})^{-1} F(x_k)\| \leq$$

$$\left\| A(x_k, x_{k-1})^{-1} \left[F(x^*) - F(x_k) - A(x_k, x_{k-1})(x^* - x_k) \right] \right\| \leq$$

$$g_2 \left(\left\| x_k - x^* \right\|, \left\| x_{k-1} - x^* \right\| \right) \left\| x_k - x^* \right\|^{p+1} <$$

$$g_3 \left(r_2, r_2 \right) r_2^p \left\| x_k - x^* \right\| = \left\| x_k - x^* \right\| < r_2,$$

which shows $\lim_{k \to \infty} x_k = x^*$ and $x_{k+1} \in U \left(x^*, r_2 \right)$. $\qquad\square$

Remark 10.5 (a) Hypothesis (10.2.3) specializes to Newton-Mysowski-type, if $A(x) = F'(x)$ [8, 12, 15]. However, if F is not Fréchet-differentiable, then our results extend the applicability of iterative algorithm (10.1.2).

(b) Theorem 10.1 has practical value although we do not show that x^* is a solution of equation $F(x) = 0$, since this may be shown in another way.

(c) Hypothesis (10.2.13) can be replaced by the stronger

$$\left\| A(y, x)^{-1} \left[F(x) - F(z) - A(y, x)(x - z) \right] \right\| \leq$$

$$g_3 \left(\| z - y \|, \| z - x \| \right) \| z - y \|^{p+1}.$$

10.3 Applications to Right Fractional Calculus

We present applications of Proposition 10.2.

Let $f : [a, b] \to \mathbb{R}$ such that $f^{(m)} \in L_\infty([a, b])$. The right Caputo fractional derivative of order $\alpha \notin \mathbb{N}$, $\alpha > 0$, $m = \lceil \alpha \rceil$ ($\lceil \cdot \rceil$ ceiling), is defined as follows:

$$\left(D_{b-}^\alpha f \right)(x) := \frac{(-1)^m}{\Gamma(m - \alpha)} \int_x^b (z - x)^{m - \alpha - 1} f^{(m)}(z) \, dz, \tag{10.3.1}$$

$\forall \, x \in [a, b]$, with $D_{b-}^m f(x) := (-1)^m f^{(m)}(x)$, $D_{b-}^0 f := f$, where Γ is the gamma function.

We observe that

$$\left| \left(D_{b-}^\alpha f \right)(x) \right| \leq \frac{1}{\Gamma(m - \alpha)} \int_x^b (z - x)^{m - \alpha - 1} \left| f^{(m)}(z) \right| dz$$

$$\leq \frac{\left\| f^{(m)} \right\|_\infty}{\Gamma(m - \alpha)} \left(\int_x^b (z - x)^{m - \alpha - 1} \, dz \right) = \frac{\left\| f^{(m)} \right\|_\infty}{\Gamma(m - \alpha)} \frac{(b - x)^{m - \alpha}}{m - \alpha} \tag{10.3.2}$$

$$= \frac{\left\| f^{(m)} \right\|_\infty (b - x)^{m - \alpha}}{\Gamma(m - \alpha + 1)}.$$

We have proved that

$$\left|\left(D_{b-}^{\alpha} f\right)(x)\right| \le \frac{\left\|f^{(m)}\right\|_{\infty} (b-x)^{m-\alpha}}{\Gamma(m-\alpha+1)} \le \frac{\left\|f^{(m)}\right\|_{\infty} (b-a)^{m-\alpha}}{\Gamma(m-\alpha+1)}. \tag{10.3.3}$$

Clearly here $\left(D_{b-}^{\alpha} f\right)(b) = 0, 0 < \alpha \notin \mathbb{N}$.

Let $n \in \mathbb{N}$. We denote

$$D_{b-}^{n\alpha} := D_{b-}^{\alpha} D_{b-}^{\alpha} \ldots D_{b-}^{\alpha} \ (n\text{-times}). \tag{10.3.4}$$

The right Riemann-Liouville fractional integral of order α, is defined as follows:

$$\left(I_{b-}^{\alpha} f\right)(x) := \frac{1}{\Gamma(\alpha)} \int_{x}^{b} (z-x)^{\alpha-1} f(z)\, dz, \tag{10.3.5}$$

$\forall\, x \in [a, b], I_{b-}^{0} := I$ (the identity operator).

We denote also

$$I_{b-}^{n\alpha} := I_{b-}^{\alpha} I_{b-}^{\alpha} \ldots I_{b-}^{\alpha} \ (n\text{-times}). \tag{10.3.6}$$

From now on we assume $0 < \alpha \le 1$, that is $m = 1$.

In [4] and Chap. 24, we proved the following right generalized fractional Taylor's formula:

Theorem 10.6 *Suppose that* $D_{b-}^{k\alpha} f \in C([a, b])$, *for* $k = 0, 1, \ldots, n+1$, *where* $0 < \alpha \le 1$. *Then*

$$f(x) = \sum_{i=0}^{n} \frac{(b-x)^{i\alpha}}{\Gamma(i\alpha+1)} \left(D_{b-}^{i\alpha} f\right)(b) + \tag{10.3.7}$$

$$\frac{1}{\Gamma((n+1)\alpha)} \int_{x}^{z} (z-x)^{(n+1)\alpha-1} \left(D_{b-}^{(n+1)\alpha} f\right)(z)\, dz, \quad \forall x \in [a, b].$$

We make

Remark 10.7 In particular, when $f' \in L_{\infty}([a, b])$, $0 < \alpha < 1$, we have that $D_{b-}^{\alpha} f(b) = 0$, also $\left(D_{b-}^{1} f\right)(x) = -f'(x)$, and

$$\left(D_{b-}^{\alpha} f\right)(x) = \frac{-1}{\Gamma(1-\alpha)} \int_{x}^{b} (z-x)^{-\alpha} f'(z)\, dz, \quad \forall x \in [a, b]. \tag{10.3.8}$$

Thus, from (10.3.7) we derive

$$f(x) - f(b) = \sum_{i=2}^{n} \frac{(b-x)^{i\alpha}}{\Gamma(i\alpha+1)} \left(D_{b-}^{i\alpha} f\right)(b) + \tag{10.3.9}$$

$$\frac{1}{\Gamma((n+1)\alpha)} \int_x^b (z-x)^{(n+1)\alpha-1} \left(D_{b-}^{(n+1)\alpha} f\right)(z)\, dz, \quad \forall\, x \in [a,b];\ \ 0 < \alpha < 1.$$

Here we are going to operate more generally. Again we assume $0 < \alpha \le 1$, and $f : [a,b] \to \mathbb{R}$, such that $f' \in C([a,b])$. We define the following right Caputo fractional derivatives:

$$\left(D_{y-}^\alpha f\right)(x) := \frac{-1}{\Gamma(1-\alpha)} \int_x^y (t-x)^{-\alpha} f'(t)\, dt, \tag{10.3.10}$$

for any $x \le y$; $x, y \in [a,b]$, and

$$\left(D_{x-}^\alpha f\right)(y) = \frac{-1}{\Gamma(1-\alpha)} \int_y^x (t-y)^{-\alpha} f'(t)\, dt, \tag{10.3.11}$$

for any $y \le x$; $x, y \in [a,b]$.

Notice $D_{y-}^1 f = -f'$, $D_{x-}^1 f = -f'$, by convention.

Clearly here $D_{y-}^\alpha f$, $D_{x-}^\alpha f$ are continuous functions over $[a,b]$, see [3]. We also make the convention that $\left(D_{y-}^\alpha f\right)(x) = 0$, for $x > y$, and $\left(D_{x-}^\alpha f\right)(y) = 0$, for $y > x$.

Here we assume that

$$D_{y-}^{k\alpha} f, D_{x-}^{k\alpha} f \in C([a,b]), \tag{10.3.12}$$

$k = 0, 1, \ldots, n+1$, $n \in \mathbb{N}$; $\forall\, x, y \in [a,b]$; and $0 < \alpha < 1$.

By (10.3.9) we derive

$$f(x) - f(y) = \sum_{i=2}^n \frac{(y-x)^{i\alpha}}{\Gamma(i\alpha+1)} \left(D_{y-}^{i\alpha} f\right)(y) +$$

$$\frac{1}{\Gamma((n+1)\alpha)} \int_x^y (z-x)^{(n+1)\alpha-1} \left(D_{y-}^{(n+1)\alpha} f\right)(z)\, dz, \tag{10.3.13}$$

$\forall\, x < y$; $x, y \in [a,b]$; $0 < \alpha < 1$, and also it holds

$$f(y) - f(x) = \sum_{i=2}^n \frac{(x-y)^{i\alpha}}{\Gamma(i\alpha+1)} \left(D_{x-}^{i\alpha} f\right)(x) +$$

$$\frac{1}{\Gamma((n+1)\alpha)} \int_y^x (z-y)^{(n+1)\alpha-1} \left(D_{x-}^{(n+1)\alpha} f\right)(z)\, dz, \tag{10.3.14}$$

$\forall\, y < x$; $x, y \in [a,b]$; $0 < \alpha < 1$.

We define the following linear operator

$$(A(f))(x, y) :=$$

$$
\begin{cases}
\sum_{i=2}^{n} \frac{(y-x)^{i\alpha-1}}{\Gamma(i\alpha+1)} \left(D_{y-}^{i\alpha} f\right)(y) - \left(D_{y-}^{(n+1)\alpha} f(x)\right) \frac{(y-x)^{(n+1)\alpha-1}}{\Gamma((n+1)\alpha+1)}, & x < y, \\
\sum_{i=2}^{n} \frac{(x-y)^{i\alpha-1}}{\Gamma(i\alpha+1)} \left(D_{x-}^{i\alpha} f\right)(x) - \left(D_{x-}^{(n+1)\alpha} f(y)\right) \frac{(x-y)^{(n+1)\alpha-1}}{\Gamma((n+1)\alpha+1)}, & x > y, \\
f'(x), & \text{when } x = y,
\end{cases}
\tag{10.3.15}
$$

$\forall\, x, y \in [a, b]; 0 < \alpha < 1$.

We may assume that

$$|(A(f))(x, x) - (A(f))(y, y)| = |f'(x) - f'(y)| \le \Phi |x - y|, \quad \forall x, y \in [a, b],
\tag{10.3.16}$$

with $\Phi > 0$.

We estimate and have:

(i) case $x < y$:

$$|f(x) - f(y) - (A(f))(x, y)(x - y)| =$$

$$|f(y) - f(x) - (A(f))(x, y)(y - x)| = \tag{10.3.17}$$

$$\left| \frac{1}{\Gamma((n+1)\alpha)} \int_{x}^{y} (z - x)^{(n+1)\alpha-1} \left(D_{y-}^{(n+1)\alpha} f\right)(z)\, dz - \right.$$

$$\left. \left(D_{y-}^{(n+1)\alpha} f(x)\right) \frac{(y - x)^{(n+1)\alpha}}{\Gamma((n+1)\alpha + 1)} \right| =$$

$$\frac{1}{\Gamma((n+1)\alpha)} \left| \int_{x}^{y} (z - x)^{(n+1)\alpha-1} \left(D_{y-}^{(n+1)\alpha} f(z) - D_{y-}^{(n+1)\alpha} f(x)\right) dz \right| \le \tag{10.3.18}$$

$$\frac{1}{\Gamma((n+1)\alpha)} \left(\int_{x}^{y} (z - x)^{(n+1)\alpha-1} \left| D_{y-}^{(n+1)\alpha} f(z) - D_{y-}^{(n+1)\alpha} f(x) \right| dz \right)$$

(we assume here that

$$\left| D_{y-}^{(n+1)\alpha} f(z) - D_{y-}^{(n+1)\alpha} f(x) \right| \le \lambda_1 |z - x|, \tag{10.3.19}$$

$\forall\, z, x, y \in [a, b] : y \geq z \geq x;\ \lambda_1 > 0)$

$$\leq \frac{\lambda_1}{\Gamma\left((n+1)\,\alpha\right)} \int_x^y (z - x)^{(n+1)\alpha - 1}\,(z - x)\,dz -$$

$$\frac{\lambda_1}{\Gamma\left((n+1)\,\alpha\right)} \int_x^y (z - x)^{(n+1)\alpha}\,dz = \frac{\lambda_1}{\Gamma\left((n+1)\,\alpha\right)} \frac{(y - x)^{(n+1)\alpha+1}}{\left((n+1)\,\alpha + 1\right)}. \tag{10.3.20}$$

We have proved that

$$|f(x) - f(y) - (A(f))(x, y)(x - y)| \leq \frac{\lambda_1 (y - x)^{(n+1)\alpha+1}}{\Gamma\left((n+1)\,\alpha\right)\left((n+1)\,\alpha + 1\right)}, \tag{10.3.21}$$

for any $x, y \in [a, b] : x < y;\ 0 < \alpha < 1$.

(ii) Case of $x > y$: We have

$$|f(y) - f(x) - (A(f))(x, y)(y - x)| = \tag{10.3.22}$$

$$\left| \frac{1}{\Gamma\left((n+1)\,\alpha\right)} \int_y^x (z - y)^{(n+1)\alpha - 1} \left(D_{x-}^{(n+1)\alpha} f \right)(z)\,dz - \right.$$

$$\left. \left(D_{x-}^{(n+1)\alpha} f(y) \right) \frac{(x - y)^{(n+1)\alpha}}{\Gamma\left((n+1)\,\alpha + 1\right)} \right| =$$

$$\frac{1}{\Gamma\left((n+1)\,\alpha\right)} \left| \int_y^x (z - y)^{(n+1)\alpha - 1} \left(\left(D_{x-}^{(n+1)\alpha} f \right)(z) - \left(D_{x-}^{(n+1)\alpha} f \right)(y) \right) dz \right| \tag{10.3.23}$$

$$\leq \frac{1}{\Gamma\left((n+1)\,\alpha\right)} \int_y^x (z - y)^{(n+1)\alpha - 1} \left| \left(D_{x-}^{(n+1)\alpha} f \right)(z) - \left(D_{x-}^{(n+1)\alpha} f \right)(y) \right| dz$$

(we assume that

$$\left| \left(D_{x-}^{(n+1)\alpha} f \right)(z) - \left(D_{x-}^{(n+1)\alpha} f \right)(y) \right| \leq \lambda_2\,|z - y|, \tag{10.3.24}$$

$\forall\, z, y, x \in [a, b] : x \geq z \geq y;\ \lambda_2 > 0)$

$$\leq \frac{\lambda_2}{\Gamma\left((n+1)\,\alpha\right)} \int_y^x (z - y)^{(n+1)\alpha - 1}\,(z - y)\,dz =$$

$$\frac{\lambda_2}{\Gamma\left((n+1)\,\alpha\right)} \int_y^x (z - y)^{(n+1)\alpha}\,dz = \frac{\lambda_2}{\Gamma\left((n+1)\,\alpha\right)} \frac{(x - y)^{(n+1)\alpha+1}}{\left((n+1)\,\alpha + 1\right)}. \tag{10.3.25}$$

We have proved that

$$|f(x) - f(y) - (A(f))(x, y)(x - y)| \leq \frac{\lambda_2}{\Gamma((n+1)\alpha)} \frac{(x-y)^{(n+1)\alpha+1}}{((n+1)\alpha+1)},$$

(10.3.26)

for any $x, y \in [a, b] : x > y; 0 < \alpha < 1$.

Conclusion 10.8 Let $\lambda = \max(\lambda_1, \lambda_2)$. Then

$$|f(x) - f(y) - (A(f))(x, y)(x - y)| \leq \frac{\lambda}{\Gamma((n+1)\alpha)} \frac{|x-y|^{(n+1)\alpha+1}}{((n+1)\alpha+1)},$$

(10.3.27)

$\forall x, y \in [a, b]$; where $0 < \alpha < 1, n \in \mathbb{N}$.

One may assume that

$$\frac{\lambda}{\Gamma((n+1)\alpha)} < 1.$$

(10.3.28)

Above notice that (10.3.27) is trivial when $x = y$.

Now based on (10.3.16) and (10.3.27), we can apply our numerical methods presented in this chapter to solve $f(x) = 0$.

To have $(n+1)\alpha + 1 \geq 2$, we need to take $1 > \alpha \geq \frac{1}{n+1}$, where $n \in \mathbb{N}$.

Returning back to Proposition 10.2 we see by (10.2.10) and (10.3.27) that crucial estimate (10.2.10) is satisfied, if we choose $p = (i+1)\alpha, i \in \mathbb{N}$ fixed and

$$g_1(s, t) = \frac{\lambda |s - t|^p}{\Gamma(p)(p+1)\mu}.$$

References

1. S. Amat, S. Busquier, Third-order iterative methods under Kantorovich conditions. J. Math. Anal. Appl. **336**, 243–261 (2007)
2. S. Amat, S. Busquier, S. Plaza, Chaotic dynamics of a third-order Newton-like method. J. Math. Anal. Appl. **366**(1), 164–174 (2010)
3. G. Anastassiou, Fractional representation formulae and right fractional inequalities. Math. Comput. Model. **54**(10–12), 3098–3115 (2011)
4. G. Anastassiou, Advanced fractional Taylor's formulae. J. Comput. Anal. Appl. (2016) (to appear)
5. G. Anastassiou, I. Argyros, *A Convergence Analysis for Secant-Like Methods with Applications to Fractional Calculus* (2015) (submitted)
6. I.K. Argyros, Newton-like methods in partially ordered linear spaces. J. Approx. Th. Appl. **9**(1), 1–10 (1993)
7. I.K. Argyros, Results on controlling the residuals of perturbed Newton-like methods on Banach spaces with a convergence structure. Southwest J. Pure Appl. Math. **1**, 32–38 (1995)

8. I.K. Argyros, *Convergence and Applications of Newton-Like Iterations* (Springer, New York, 2008)
9. K. Diethelm, *The Analysis of Fractional Differential Equations*. Lecture Notes in Mathematics, vol. 2004, 1st edn. (Springer, New York, 2010)
10. J.A. Ezquerro, M.A. Hernandez, Newton-like methods of high order and domains of semilocal and global convergence. Appl. Math. Comput. **214**(1), 142–154 (2009)
11. J.A. Ezquerro, J.M. Gutierrez, M.A. Hernandez, N. Romero, M.J. Rubio, The Newton method: from Newton to Kantorovich (Spanish). Gac. R. Soc. Mat. Esp. **13**, 53–76 (2010)
12. L.V. Kantorovich, G.P. Akilov, *Functional Analysis in Normed Spaces* (Pergamon Press, New York, 1964)
13. A.A. Magrenan, Different anomalies in a Jarratt family of iterative root finding methods. Appl. Math. Comput. **233**, 29–38 (2014)
14. A.A. Magrenan, A new tool to study real dynamics: the convergence plane. Appl. Math. Comput. **248**, 215–224 (2014)
15. F.A. Potra, V. Ptak, *Nondiscrete Induction and Iterative Processes* (Pitman Publishing, London, 1984)
16. P.D. Proinov, New general convergence theory for iterative processes and its applications to Newton-Kantorovich type theorems. J. Complexity **26**, 3–42 (2010)

Chapter 11
Secant-Like Methods and Modified
g-Fractional Calculus

We present local and semilocal convergence results for secant-type methods in order
to approximate a locally unique solution of a nonlinear equation in a Banach space
setting. In the last part of the study we present some choices of the operators involved
in fractional calculus where the operators satisfy the convergence conditions. It follows [5].

11.1 Introduction

In this study we are concerned with the problem of approximating a locally unique
solution x^* of the nonlinear equation

$$F(x) = 0, \tag{11.1.1}$$

where F is a continuous operator defined on a subset D of a Banach space X with
values in a Banach space Y.

A lot of problems in Computational Sciences and other disciplines can be brought
in a form like (11.1.1) using Mathematical Modelling [8, 12, 16]. The solutions
of such equations can be found in closed form only in special cases. That is why
most solution methods for these equations are iterative. Iterative methods are usually
studied based on semilocal and local convergence. The semilocal convergence matter
is, based on the information around the initial point to give hypotheses ensuring the
convergence of the iterative algorithm; while the local one is, based on the information
around a solution, to find estimates of the radii of convergence balls as well as error
bounds on the distances involved.

© Springer International Publishing Switzerland 2016   177
G.A. Anastassiou and I.K. Argyros, *Intelligent Numerical Methods:*
Applications to Fractional Calculus, Studies in Computational Intelligence 624,
DOI 10.1007/978-3-319-26721-0_11

We introduce the secant-type method defined for each $n = 0, 1, 2, \ldots$ by

$$x_{n+1} = x_n - A\,(F)\,(x_n, x_{n-1})^{-1}\,F\,(x_n)\,, \tag{11.1.2}$$

where $x_{-1}, x_0 \in D$ are initial points and $A\,(F)\,(x, y) \in L\,(X, Y)$ the space of bounded linear operators from X into Y. There is a plethora on local as well as semilocal convergence theorems for method (11.1.2) provided that the operator A is an approximation to the Fréchet-derivative F' [1, 2, 6–16]. In the present study we do not necessarily assume that operator A is related to F'. This way we expand the applicability of iterative algorithm (11.1.2). Notice that many well known methods are special case of method (11.1.2).

Newton's method: Choose $A\,(F)\,(x, x) = F'\,(x)$ for each $x \in D$.

Secant method: Choose $A\,(F)\,(x, y) = [x, y; F]$, where $[x, y; F]$ denotes a divided difference of order one [8, 12, 15].

The so called Newton-like methods and many other methods are special cases of method (11.1.2).

The rest of the chapter is organized as follows. The semilocal as well as the local convergence analysis of method (11.1.2) is given in Sect. 11.2. Some applications from fractional calculus are given in Sect. 11.3.

11.2 Convergence Analysis

We present the main semilocal convergence result for method (11.1.2).

Theorem 11.1 *Let $F : D \subset X \to Y$ be a continuous operator and let $A\,(F)\,(x, y)$ $\in L\,(X, Y)$. Suppose that there exist $x_{-1}, x_0 \in D$, $\eta \geq 0$, $p \geq 1$, a function $\varphi :$ $[0, \infty)^2 \to [0, \infty)$ continuous and nondecreasing such that for each $x, y, z \in D$*

$$A\,(F)\,(z, y)^{-1} \in L\,(Y, X)\,, \tag{11.2.1}$$

$$\max\left\{\|x_{-1} - x_0\|, \left\|A\,(F)\,(x_0, x_{-1})^{-1}\,F\,(x_0)\right\|\right\} \leq \eta, \tag{11.2.2}$$

$$\left\|A\,(F)\,(z, y)^{-1}\,(F\,(z) - F\,(y) - A\,(F)\,(y, x)\,(z - y))\right\| \leq$$

$$\varphi\,(\|z - y\|, \|y - x\|)\,\|z - y\|^{p+1}\,, \tag{11.2.3}$$

$$q := \varphi\,(\eta, \eta)\,\eta^p < 1 \tag{11.2.4}$$

and

$$\overline{U}\,(x_0, r) \subseteq D, \tag{11.2.5}$$

where,

$$r = \frac{\eta}{1-q}. \tag{11.2.6}$$

Then, the sequence $\{x_n\}$ generated by method (11.1.2) is well defined, remains in $\overline{U}(x_0, r)$ for each $n = 0, 1, 2, \ldots$ and converges to some $x^* \in \overline{U}(x_0, r)$ such that

$$\|x_{n+1} - x_n\| \leq \varphi\left(\|x_n - x_{n-1}\|, \|x_{n-1} - x_{n-2}\|\right) \|x_n - x_{n-1}\|^{p+1}$$

$$\leq q \|x_n - x_{n-1}\| \tag{11.2.7}$$

and

$$\|x_n - x^*\| \leq \frac{q^n \eta}{1-q}. \tag{11.2.8}$$

Proof The iterate x_1 is well defined by method (11.1.2) for $n = 0$ and (11.2.1). We also have by (11.2.2) and (11.2.6) that
$\|x_1 - x_0\| = \|A(F)(x_0, x_{-1})^{-1} F(x_0)\| \leq \eta < r$, so we get that $x_1 \in \overline{U}(x_0, r)$ and x_2 is well defined (by (11.2.5)). Using (11.2.3) and (11.2.4) we get that

$$\|x_2 - x_1\| =$$

$$\left\| A(F)(x_1, x_0)^{-1} \left[F(x_1) - F(x_0) - A(F)(x_0, x_{-1})(x_1 - x_0) \right] \right\|$$

$$\leq \varphi\left(\|x_1 - x_0\|, \|x_0 - x_{-1}\|\right) \|x_1 - x_0\|^{p+1} \leq q \|x_1 - x_0\|,$$

which shows (11.2.7) for $n = 1$. Then, we can have that

$$\|x_2 - x_0\| \leq \|x_2 - x_1\| + \|x_1 - x_0\| \leq q \|x_1 - x_0\| + \|x_1 - x_0\|$$

$$= (1+q)\|x_1 - x_0\| \leq \frac{1-q^2}{1-q}\eta < r,$$

so $x_2 \in \overline{U}(x_0, r)$ and x_3 is well defined.

Assuming $\|x_{k+1} - x_k\| \leq q \|x_k - x_{k-1}\|$ and $x_{k+1} \in \overline{U}(x_0, r)$ for each $k = 1, 2, \ldots, n$ we get

$$\|x_{k+2} - x_{k+1}\| =$$

$$\left\| A(F)(x_{k+1}, x_k)^{-1} \left[F(x_{k+1}) - F(x_k) - A(F)(x_k, x_{k-1})(x_{k+1} - x_k) \right] \right\|$$

$$\leq \varphi\left(\|x_{k+1} - x_k\|, \|x_k - x_{k-1}\|\right) \|x_{k+1} - x_k\|^{p+1}$$

$$\leq \varphi\left(\|x_1 - x_0\|, \|x_0 - x_{-1}\|\right) \|x_1 - x_0\|^p \|x_{k+1} - x_k\| \leq q \|x_{k+1} - x_k\|$$

and

$$\|x_{k+2} - x_0\| \leq \|x_{k+2} - x_{k+1}\| + \|x_{k+1} - x_k\| + \cdots + \|x_1 - x_0\|$$

$$\leq \left(q^{k+1} + q^k + \cdots + 1\right) \|x_1 - x_0\| \leq \frac{1 - q^{k+2}}{1 - q} \|x_1 - x_0\|$$

$$< \frac{\eta}{1 - q} = r,$$

which completes the induction for (11.2.7) and $x_{k+2} \in \overline{U}(x_0, r)$. We also have that for $m \geq 0$

$$\|x_{n+m} - x_n\| \leq \|x_{n+m} - x_{n+m-1}\| + \cdots + \|x_{n+1} - x_n\|$$

$$\leq \left(q^{m-1} + q^{m-2} + \cdots + 1\right) \|x_{n+1} - x_n\|$$

$$\leq \frac{1 - q^m}{1 - q} q^n \|x_1 - x_0\|.$$

It follows that $\{x_n\}$ is a complete sequence in a Banach space X and as such it converges to some $x^* \in \overline{U}(x_0, r)$ (since $\overline{U}(x_0, r)$ is a closed set). By letting $m \to \infty$, we obtain (11.2.8). $\qquad\square$

Stronger hypotheses are needed to show that x^* is a solution of equation $F(x) = 0$.

Proposition 11.2 *Let* $F : D \subset X \to Y$ *be a continuous operator and let* $A(F)(x, y) \in L(X, Y)$. *Suppose that there exist* $x_{-1}, x_0 \in D$, $\eta \geq 0$, $p \geq 1$, $\mu > 0$, *a function* $\varphi_1 : [0, \infty)^2 \to [0, \infty)$ *continuous and nondecreasing such that for each* $x, y \in D$

$$A(F)(x, y)^{-1} \in L(Y, X), \quad \|A(F)(x, y)^{-1}\| \leq \mu,$$

$$\max\left\{\|x_{-1} - x_0\|, \|A(F)(x_0, x_{-1})^{-1} F(x_0)\|\right\} \leq \eta, \qquad (11.2.9)$$

$$\|F(z) - F(y) - A(F)(y, x)(z - y)\| \leq \frac{\varphi_1(\|z - y\|, \|x - y\|)}{\mu} \|z - y\|^{p+1},$$
$$(11.2.10)$$

$$q_1 := \varphi_1(\eta, \eta) \eta^p < 1$$

and

$$\overline{U}(x_0, r_1) \subseteq D,$$

where,

$$r_1 = \frac{\eta}{1 - q_1}.$$

Then, the conclusions of Theorem 11.1 for sequence $\{x_n\}$ hold with φ_1, q_1, r_1, replacing φ, q and r, respectively. Moreover, x^* is a solution of the equation $F(x) = 0$.

Proof Notice that

$$\left\| A(F)(x_n, x_{n-1})^{-1} \left[F(x_n) - F(x_{n-1}) - A(F)(x_{n-1}, x_{n-2})(x_n - x_{n-1}) \right] \right\|$$

$$\leq \left\| A(F)(x_n, x_{n-1})^{-1} \right\| \left\| F(x_n) - F(x_{n-1}) - A(F)(x_{n-1}, x_{n-2})(x_n - x_{n-1}) \right\|$$

$$\leq \varphi_1 \left(\|x_n - x_{n-1}\|, \|x_{n-1} - x_{n-2}\| \right) \|x_n - x_{n-1}\|^{p+1} \leq q_1 \|x_n - x_{n-1}\|.$$

Therefore, the proof of Theorem 11.1 can apply. Then, in view of the estimate

$$\|F(x_n)\| = \|F(x_n) - F(x_{n-1}) - A(F)(x_{n-1}, x_{n-2})(x_n - x_{n-1})\| \leq$$

$$\frac{\varphi_1 \left(\|x_n - x_{n-1}\| \right)}{\mu} \|x_n - x_{n-1}\|^{p+1} \leq q_1 \|x_n - x_{n-1}\|,$$

we deduce by letting $n \to \infty$ that $F(x^*) = 0$. $\qquad\square$

Concerning the uniqueness of the solution x^* we have the following result:

Proposition 11.3 *Under the hypotheses of Proposition 11.2, further suppose that there exists $\varphi_2 : [0, \infty)^2 \to [0, \infty)$ continuous and nondecreasing such that*

$$\|F(z) - F(x) - A(F)(z, y)(z - x)\| \leq \frac{\varphi_2 \left(\|z - x\|, \|y - x\| \right)}{\mu} \|z - x\|^{p+1}$$

$$(11.2.11)$$

and

$$\varphi_2(r_1, \eta + r_1) r_1^p < 1. \qquad\qquad (11.2.12)$$

Then, x^* is the only solution of equation $F(x) = 0$ in $\overline{U}(x_0, r_1)$.

Proof The existence of the solution $x^* \in \overline{U}(x_0, r_1)$ has been established in Proposition 11.2. Let $y^* \in \overline{U}(x_0, r_1)$ with $F(y^*) = 0$. Then, we have in turn that

$$\|x_{n+1} - y^*\| = \|x_n - y^* - A(F)(x_n, x_{n-1})^{-1} F(x_n)\| =$$

$$\left\| A(F)(x_n, x_{n-1})^{-1} \left[A(F)(x_n, x_{n-1})(x_n - y^*) - F(x_n) + F(y^*) \right] \right\| \leq$$

$$\left\| A(F)(x_n, x_{n-1})^{-1} \right\| \left\| F(y^*) - F(x_n) - A(F)(x_n, x_{n-1})(y^* - x_n) \right\| \leq$$

$$\mu \frac{\varphi_1 \left(\|x_n - y^*\|, \|x_{n-1} - y^*\| \right)}{\mu} \|x_n - y^*\|^{p+1} \leq$$

$$\varphi_2(r_1, \eta + r_1) r_1^p \|x_n - x^*\| < \|x_n - y^*\|,$$

so we deduce that $\lim_{n\to\infty} x_n = y^*$. But we have that $\lim_{n\to\infty} x_n = x^*$. Hence, we conclude that $x^* = y^*$. □

Next, we present a local convergence analysis for the iterative algorithm (11.1.2).

Proposition 11.4 *Let* $F : D \subset X \to Y$ *be a continuous operator and let* $A(F)(y, x) \in L(X, Y)$. *Suppose that there exist* $x^* \in D$, $p \geq 1$, *a function* $\varphi_3 : [0, \infty)^2 \to [0, \infty)$ *continuous and nondecreasing such that for each* $x, y \in D$

$$F\left(x^*\right) = 0, \quad A(F)(y, x)^{-1} \in L(Y, X),$$

$$\left\| A(F)(y, x)^{-1} \left[F(x) - F\left(x^*\right) - A(F)(y, x)\left(x - x^*\right) \right] \right\| \leq$$

$$\varphi_3\left(\left\| y - x^*\right\|, \left\| x - x^*\right\|\right) \left\| y - x^*\right\|^{p+1}, \tag{11.2.13}$$

and

$$\overline{U}\left(x^*, r_2\right) \subseteq D,$$

where r_2 *is the smallest positive solution of equation*

$$h(t) := \varphi_3(t, t)\, t^p - 1.$$

Then, sequence $\{x_n\}$ *generated by method (11.1.2) for* $x_{-1}, x_0 \in U(x^*, r_2) - \{x^*\}$ *is well defined, remains in* $U(x^*, r_2)$ *for each* $n = 0, 1, 2, \ldots$ *and converges to* x^*. *Moreover, the following estimates hold*

$$\left\| x_{n+1} - x^*\right\| \leq \varphi_2\left(\left\| x_n - x^*\right\|, \left\| x_{n-1} - x^*\right\|\right) \left\| x_n - x^*\right\|^{p+1} < \left\| x_n - x^*\right\| < r_2.$$

Proof We have that $h(0) = -1 < 0$ and $h(t) \to +\infty$ as $t \to +\infty$. Then, it follows from the intermediate value theorem that function h has positive zeros. Denote by r_2 the smallest such zero. By hypothesis $x_{-1}, x_0 \in U(x^*, r_2) - \{x^*\}$. Then, we get in turn that

$$\left\| x_1 - x^*\right\| = \left\| x_0 - x^* - A(F)(x_0, x_{-1})^{-1} F(x_0)\right\| =$$

$$\left\| A(F)(x_0, x_{-1})^{-1} \left[F\left(x^*\right) - F(x_0) - A(F)(x_0, x_{-1})\left(x^* - x_0\right) \right]\right\| \leq$$

$$\varphi_2\left(\left\| x_0 - x^*\right\|, \left\| x_{-1} - x^*\right\|\right) \left\| x_0 - x^*\right\|^{p+1} < \varphi_3(r_2, r_2)\, r_2^p \left\| x_0 - x^*\right\| =$$

$$\left\| x_0 - x^*\right\| < r_2,$$

which shows that $x_1 \in U(x^*, r_2)$ and x_2 is well defined. By a simple inductive argument as in the preceding estimate we get that

$$\left\| x_{k+1} - x^*\right\| = \left\| x_k - x^* - A(F)(x_k, x_{k-1})^{-1} F(x_k)\right\| \leq$$

$$\left\| A\left(F\right)\left(x_k, x_{k-1}\right)^{-1} \left[F\left(x^*\right) - F\left(x_k\right) - A\left(F\right)\left(x_k, x_{k-1}\right)\left(x^* - x_k\right)\right]\right\| \leq$$

$$\varphi_2\left(\left\| x_k - x^*\right\|, \left\| x_{k-1} - x^*\right\|\right)\left\| x_k - x^*\right\|^{p+1} <$$

$$\varphi_3\left(r_2, r_2\right) r_2^p \left\| x_k - x^*\right\| = \left\| x_k - x^*\right\| < r_2,$$

which shows $\lim_{k \to \infty} x_k = x^*$ and $x_{k+1} \in U\left(x^*, r_2\right)$. $\qquad\square$

Remark 11.5 (a) Hypothesis (11.2.3) specializes to Newton-Mysowski-type, if $A\left(F\right)\left(x\right) = F'\left(x\right)$ [8, 12, 15]. However, if F is not Fréchet-differentiable, then our results extend the applicability of iterative algorithm (11.1.2).

(b) Theorem 11.1 has practical value although we do not show that x^* is a solution of equation $F\left(x\right) = 0$, since this may be shown in another way.

(c) Hypothesis (11.2.13) can be replaced by the stronger

$$\left\| A\left(F\right)\left(y, x\right)^{-1}\left[F\left(x\right) - F\left(z\right) - A\left(F\right)\left(y, x\right)\left(x - z\right)\right]\right\| \leq$$

$$\varphi_3\left(\left\| z - y\right\|, \left\| z - x\right\|\right)\left\| z - y\right\|^{p+1}.$$

11.3 Applications to Modified g-Fractional Calculus

Let $0 < \alpha \leq 1$, $m = \lceil \alpha \rceil = 1$ ($\lceil \cdot \rceil$ ceiling of number), g is strictly increasing and $g \in AC\left(\left[a, b\right]\right)$ (absolutely continuous functions, $f : \left[a, b\right] \to \mathbb{R}$. Assume that $\left(f \circ g^{-1}\right) \in AC\left(\left[g\left(a\right), g\left(b\right)\right]\right)$ (so the above imply that $f \in C\left(\left[a, b\right]\right)$).

Also assume that $\left(f \circ g^{-1}\right)' \circ g \in L_\infty\left(\left[a, b\right]\right)$. In both backgrounds here we follow [4] and Chap. 24.

(I) The right generalized g-fractional derivative of f of order α is defined as follows:

$$\left(D_{b-;g}^\alpha f\right)\left(x\right) := \frac{-1}{\Gamma\left(1 - \alpha\right)} \int_x^b \left(g\left(t\right) - g\left(x\right)\right)^{-\alpha} g'\left(t\right)\left(f \circ g^{-1}\right)'\left(g\left(t\right)\right) dt,$$

(11.3.1)

$a \leq x \leq b$.

If $0 < \alpha < 1$, then $\left(D_{b-;g}^\alpha f\right) \in C\left(\left[a, b\right]\right)$.

Also we define

$$\left(D_{b-;g}^1 f\right)\left(x\right) := -\left(\left(f \circ g^{-1}\right)' \circ g\right)\left(x\right),$$

(11.3.2)

$$\left(D_{b-;g}^0 f\right)\left(x\right) := f\left(x\right), \quad \forall x \in \left[a, b\right].$$

When $g = id$, then

$$D_{b-;g}^{\alpha} f(x) = D_{b-;id}^{\alpha} f(x) = D_{b-}^{\alpha} f(x),\qquad(11.3.3)$$

the usual right Caputo fractional derivative.

Denote by

$$D_{b-;g}^{n\alpha} := D_{b-;g}^{\alpha} D_{b-;g}^{\alpha} \cdots D_{b-;g}^{\alpha} \quad (n \text{ times}), n \in \mathbb{N}. \qquad(11.3.4)$$

We consider the right generalized fractional Riemann-Liouville integral

$$\left(I_{b-;g}^{\alpha} f\right)(x) = \frac{1}{\Gamma(\alpha)} \int_{x}^{b} (g(t) - g(x))^{\alpha-1} g'(t) f(t)\, dt, \quad a \le x \le b. \qquad(11.3.5)$$

Also denote by

$$I_{b-;g}^{n\alpha} := I_{b-;g}^{\alpha} I_{b-;g}^{\alpha} \cdots I_{b-;g}^{\alpha} \quad (n \text{ times}). \qquad(11.3.6)$$

We will be using the following modified g-right generalized Taylor's formula.

Theorem 11.6 ([4]) *Let here* $0 < \alpha \le 1$, $k = 0, 1, \ldots, n + 1$; *and denote* $F_k^b := D_{b-;g}^{k\alpha} f$. *Assume that* $F_k^b \circ g^{-1} \in AC([g(a), g(b)])$, *and* $\left(F_k^b \circ g^{-1}\right)' \circ g \in L_{\infty}([a, b])$, *for all* $k = 0, 1, \ldots, n + 1$. *Then*

$$f(x) - f(b) = \sum_{i=1}^{n} \frac{(g(b) - g(x))^{i\alpha}}{\Gamma(i\alpha + 1)} \left(D_{b-;g}^{i\alpha} f\right)(b) + \qquad(11.3.7)$$

$$\frac{1}{\Gamma((n+1)\alpha)} \int_{x}^{b} (g(t) - g(x))^{(n+1)\alpha-1} g'(t) \left(D_{b-;g}^{(n+1)\alpha} f\right)(t)\, dt,$$

$\forall\, x \in [a, b]$.

Here we are going to operate more generally. We consider $f \in C^1([a, b])$. We define the following right generalized g-fractional derivative:

$$\left(D_{y-;g}^{\alpha} f\right)(x) := \frac{-1}{\Gamma(1-\alpha)} \int_{x}^{y} (g(t) - g(x))^{-\alpha} g'(t) \left(f \circ g^{-1}\right)'(g(t))\, dt, \qquad(11.3.8)$$

all $a \le x \le y$; $y \in [a, b]$,

$$\left(D_{y-;g}^{1} f\right)(x) := -\left(\left(f \circ g^{-1}\right)' \circ g\right)(x), \quad \forall\, x \in [a, b]. \qquad(11.3.9)$$

Similarly we define:

$$\left(D_{x-;y}^{\alpha}f\right)(y) := \frac{-1}{\Gamma(1-\alpha)}\int_{y}^{x}(g(t)-g(y))^{-\alpha}g'(t)\left(f\circ g^{-1}\right)'(g(t))\,dt,$$

$$(11.3.10)$$

all $a \leq y \leq x$; $x \in [a,b]$,

$$\left(D_{x-;g}^{1}f\right)(y) := -\left(\left(f\circ g^{-1}\right)'\circ g\right)(y), \quad \forall\, y \in [a,b].$$

$$(11.3.11)$$

When $0 < \alpha < 1$, $D_{y-;g}^{\alpha}f$ and $D_{x-;g}^{\alpha}f$ are continuous functions on $[a,b]$. Note here that by convention we have that

$$\left(D_{y-;g}^{\alpha}f\right)(x) = 0, \quad \text{for } x > y$$

and

$$\left(D_{x-;g}^{\alpha}f\right)(y) = 0, \quad \text{for } y > x$$

$$(11.3.12)$$

Denote by

$$F_{k}^{y} := D_{y-;g}^{k\alpha}f, \; F_{k}^{x} := D_{x-;g}^{k\alpha}f, \quad \forall\, x, y \in [a,b].$$

$$(11.3.13)$$

We assume that

$$F_{k}^{z}\circ g^{-1} \in AC\left([g(a),g(b)]\right), \text{ and } \left(F_{k}^{z}\circ g^{-1}\right)'\circ g \in L_{\infty}\left([a,b]\right), \quad (11.3.14)$$

$k = 0, 1, \ldots, n+1$; for $z = x, y$; $\forall\, x, y \in [a,b]$; $0 < \alpha < 1$.

We also observe that $(0 < \alpha < 1)$

$$\left|\left(D_{b-;g}^{\alpha}f\right)(x)\right| \leq \frac{1}{\Gamma(1-\alpha)}\int_{x}^{b}(g(t)-g(x))^{-\alpha}g'(t)\left|\left(f\circ g^{-1}\right)'(g(t))\right|dt \leq$$

$$(11.3.15)$$

$$\frac{\left\|\left(f\circ g^{-1}\right)'\circ g\right\|_{\infty,[a,b]}}{\Gamma(1-\alpha)}\int_{x}^{b}(g(t)-g(x))^{-\alpha}g'(t)\,dt =$$

$$\frac{\left\|\left(f\circ g^{-1}\right)'\circ g\right\|_{\infty,[a,b]}}{\Gamma(1-\alpha)}\frac{(g(b)-g(x))^{1-\alpha}}{1-\alpha} =$$

$$\frac{\left\|\left(f\circ g^{-1}\right)'\circ g\right\|_{\infty,[a,b]}}{\Gamma(2-\alpha)}(g(b)-g(x))^{1-\alpha}, \quad \forall\, x \in [a,b].$$

We have proved that

$$\left|\left(D^\alpha_{b-;g}f\right)(x)\right| \le \frac{\left\|\left(f \circ g^{-1}\right)' \circ g\right\|_{\infty,[a,b]}}{\Gamma(2-\alpha)} \left(g(b) - g(x)\right)^{1-\alpha} \qquad (11.3.16)$$

$$\le \frac{\left\|\left(f \circ g^{-1}\right)' \circ g\right\|_{\infty,[a,b]}}{\Gamma(2-\alpha)} \left(g(b) - g(a)\right)^{1-\alpha}, \quad \forall\, x, y \in [a, b].$$

Clearly here we have

$$\left(D^\alpha_{b-;g}f\right)(b) = 0, \quad 0 < \alpha < 1. \qquad (11.3.17)$$

In particular it holds

$$\left(D^\alpha_{x-;g}f\right)(x) = \left(D^\alpha_{y-;g}f\right)(y) = 0, \quad \forall\, x, y \in [a, b];\ 0 < \alpha < 1. \qquad (11.3.18)$$

By (11.3.7) we derive

$$f(x) - f(y) = \sum_{i=2}^{n} \frac{(g(y) - g(x))^{i\alpha}}{\Gamma(i\alpha + 1)} \left(D^{i\alpha}_{y-;g}f\right)(y) + \qquad (11.3.19)$$

$$\frac{1}{\Gamma((n+1)\alpha)} \int_x^y (g(t) - g(x))^{(n+1)\alpha - 1}\, g'(t) \left(D^{(n+1)\alpha}_{y-;g}f\right)(t)\, dt,$$

$\forall\, x < y;\ x, y \in [a, b];\ 0 < \alpha < 1$, and also it holds:

$$f(y) - f(x) = \sum_{i=2}^{n} \frac{(g(x) - g(y))^{i\alpha}}{\Gamma(i\alpha + 1)} \left(D^{i\alpha}_{x-;g}f\right)(x) + \qquad (11.3.20)$$

$$\frac{1}{\Gamma((n+1)\alpha)} \int_y^x (g(t) - g(y))^{(n+1)\alpha - 1}\, g'(t) \left(D^{(n+1)\alpha}_{x-;g}f\right)(t)\, dt,$$

$\forall\, y < x;\ x, y \in [a, b];\ 0 < \alpha < 1$.

We define also the following linear operator

$$(A_1(f))(x, y) :=$$

$$\begin{cases} \sum_{i=2}^{n} \frac{(g(y)-g(x))^{i\alpha-1}}{\Gamma(i\alpha+1)} \left(D^{i\alpha}_{y-;g}f\right)(y) - \left(D^{(n+1)\alpha}_{y-;g}f(x)\right) \frac{(g(y)-g(x))^{(n+1)\alpha-1}}{\Gamma((n+1)\alpha+1)}, & x < y, \\[2mm] \sum_{i=2}^{n} \frac{(g(x)-g(y))^{i\alpha-1}}{\Gamma(i\alpha+1)} \left(D^{i\alpha}_{x-;g}f\right)(x) - \left(D^{(n+1)\alpha}_{x-;g}f(y)\right) \frac{(g(x)-g(y))^{(n+1)\alpha-1}}{\Gamma((n+1)\alpha+1)}, & x > y, \\[2mm] f'(x), & \text{when } x = y, \end{cases}$$

$$\qquad (11.3.21)$$

$\forall\, x, y \in [a, b];\ 0 < \alpha < 1$.

We may assume that

$$|(A_1 (f)) (x, x) - (A_1 (f)) (y, y)| = |f' (x) - f' (y)|$$

$$= |(f' \circ g^{-1}) (g (x)) - (f' \circ g^{-1}) (g (y))| \leq \Phi |g (x) - g (y)|, \qquad (11.3.22)$$

$\forall\, x, y \in [a, b]$; with $\Phi > 0$.

We estimate and have:

(i) case $x < y$:

$$|f (x) - f (y) - (A_1 (f)) (x, y) (g (x) - g (y))| =$$

$$\left| \frac{1}{\Gamma ((n + 1) \alpha)} \int_x^y (g (t) - g (x))^{(n+1)\alpha - 1} g' (t) \left(D_{y-;g}^{(n+1)\alpha} f \right) (t)\, dt - \right.$$

$$\left. \left(D_{y-;g}^{(n+1)\alpha} f (x) \right) \frac{(g (y) - g (x))^{(n+1)\alpha}}{\Gamma ((n + 1) \alpha + 1)} \right| = \qquad (11.3.23)$$

$$\frac{1}{\Gamma ((n + 1) \alpha)} \cdot$$

$$\left| \int_x^y (g (t) - g (x))^{(n+1)\alpha - 1} g' (t) \left(\left(D_{y-;g}^{(n+1)\alpha} f \right) (t) - \left(D_{y-;g}^{(n+1)\alpha} f \right) (x) \right) dt \right| \leq$$

$$\frac{1}{\Gamma ((n + 1) \alpha)} \cdot$$

$$\int_x^y (g (t) - g (x))^{(n+1)\alpha - 1} g' (t) \left| \left(D_{y-;g}^{(n+1)\alpha} f \right) (t) - \left(D_{y-;g}^{(n+1)\alpha} f \right) (x) \right| dt$$

$$\tag{11.3.24}$$

(we assume that

$$\left| \left(D_{y-;g}^{(n+1)\alpha} f \right) (t) - \left(D_{y-;g}^{(n+1)\alpha} f \right) (x) \right| \leq \lambda_1 |g (t) - g (x)|, \qquad (11.3.25)$$

$\forall\, t, x, y \in [a, b] : y \geq t \geq x; \lambda_1 > 0$)

$$\leq \frac{\lambda_1}{\Gamma ((n + 1) \alpha)} \int_x^y (g (t) - g (x))^{(n+1)\alpha - 1} g' (t) (g (t) - g (x))\, dt = \quad (11.3.26)$$

$$\frac{\lambda_1}{\Gamma ((n + 1) \alpha)} \int_x^y (g (t) - g (x))^{(n+1)\alpha} g' (t)\, dt =$$

$$\frac{\lambda_1}{\Gamma((n+1)\alpha)} \frac{(g(y) - g(x))^{(n+1)\alpha+1}}{((n+1)\alpha+1)}. \tag{11.3.27}$$

We have proved that

$$|f(x) - f(y) - (A_1(f))(x, y)(g(x) - g(y))| \le$$

$$\frac{\lambda_1}{\Gamma((n+1)\alpha)} \frac{(g(y) - g(x))^{(n+1)\alpha+1}}{((n+1)\alpha+1)}, \tag{11.3.28}$$

for any $x, y \in [a, b] : x < y; 0 < \alpha < 1$.

(ii) case $x > y$:

$$|f(x) - f(y) - (A_1(f))(x, y)(g(x) - g(y))| =$$

$$|f(y) - f(x) - (A_1(f))(x, y)(g(y) - g(x))| = \tag{11.3.29}$$

$$\left| \frac{1}{\Gamma((n+1)\alpha)} \int_y^x (g(t) - g(y))^{(n+1)\alpha-1} g'(t) \left(D_{x-;g}^{(n+1)\alpha} f \right)(t) \, dt - \right.$$

$$\left. \left(D_{x-;g}^{(n+1)\alpha} f \right)(y) \frac{(g(x) - g(y))^{(n+1)\alpha}}{\Gamma((n+1)\alpha+1)} \right| = \tag{11.3.30}$$

$$\frac{1}{\Gamma((n+1)\alpha)} \cdot$$

$$\left| \int_y^x (g(t) - g(y))^{(n+1)\alpha-1} g'(t) \left(\left(D_{x-;g}^{(n+1)\alpha} f \right)(t) - \left(D_{x-;g}^{(n+1)\alpha} f \right)(y) \right) dt \right| \le$$

$$\frac{1}{\Gamma((n+1)\alpha)} \int_y^x (g(t) - g(y))^{(n+1)\alpha-1} g'(t) \left| D_{x-;g}^{(n+1)\alpha} f(t) - D_{x-;g}^{(n+1)\alpha} f(y) \right| dt$$

(we assume that

$$\left| D_{x-;g}^{(n+1)\alpha} f(t) - D_{x-;g}^{(n+1)\alpha} f(y) \right| \le \lambda_2 |g(t) - g(y)|, \tag{11.3.31}$$

$\forall\, t, y, x \in [a, b] : x \ge t \ge y; \lambda_2 > 0)$

$$\le \frac{\lambda_2}{\Gamma((n+1)\alpha)} \int_y^x (g(t) - g(y))^{(n+1)\alpha-1} g'(t)(g(t) - g(y)) \, dt = \tag{11.3.32}$$

$$\frac{\lambda_2}{\Gamma((n+1)\alpha)} \int_y^x (g(t) - g(y))^{(n+1)\alpha} g'(t) \, dt =$$

$$\frac{\lambda_2}{\Gamma((n+1)\alpha)} \frac{(g(x) - g(y))^{(n+1)\alpha+1}}{((n+1)\alpha+1)}.$$

We have proved that

$$|f(x) - f(y) - (A_1(f))(x,y)(g(x) - g(y))| \leq$$

$$\frac{\lambda_2}{\Gamma((n+1)\alpha)} \frac{(g(x) - g(y))^{(n+1)\alpha+1}}{((n+1)\alpha+1)}, \tag{11.3.33}$$

$\forall\, x, y \in [a,b] : x > y; 0 < \alpha < 1.$

Conclusion 11.7 *Set $\lambda = \max(\lambda_1, \lambda_2)$. We have proved that*

$$|f(x) - f(y) - (A_1(f))(x,y)(g(x) - g(y))| \leq$$

$$\frac{\lambda}{\Gamma((n+1)\alpha)} \frac{|g(x) - g(y)|^{(n+1)\alpha+1}}{((n+1)\alpha+1)}, \tag{11.3.34}$$

$\forall\, x, y \in [a,b]; 0 < \alpha < 1, n \in \mathbb{N}.$

(Notice that (11.3.34) is trivially true when $x = y$.)
One may assume that

$$\frac{\lambda}{\Gamma((n+1)\alpha)} < 1. \tag{11.3.35}$$

Now based on (11.3.22) and (11.3.34), we can apply our numerical methods presented in this chapter to solve $f(x) = 0$.

To have $(n+1)\alpha + 1 \geq 2$, we need to take $1 > \alpha \geq \frac{1}{n+1}$, where $n \in \mathbb{N}$.
Some examples of g follow:

$$g(x) = e^x, \quad x \in [a,b] \subset \mathbb{R},$$
$$g(x) = \sin x,$$
$$g(x) = \tan x, \tag{11.3.36}$$
$$\text{where } x \in \left[-\frac{\pi}{2} + \varepsilon, \frac{\pi}{2} - \varepsilon\right], \ \varepsilon > 0 \text{ small.}$$

Indeed, the above examples of g are strictly increasing and absolutely continuous functions.

(II) The left generalized g-fractional derivative of f of order α is defined as follows:

$$\left(D_{a+;g}^\alpha f\right)(x) = \frac{1}{\Gamma(1-\alpha)} \int_a^x (g(x) - g(t))^{-\alpha} g'(t) \left(f \circ g^{-1}\right)'(g(t)) \, dt, \tag{11.3.37}$$

$\forall\, x \in [a,b].$
If $0 < \alpha < 1$, then $\left(D_{a+;g}^\alpha f\right) \in C([a,b]).$

Also, we define

$$D_{a+;g}^1 f(x) = \left(\left(f \circ g^{-1} \right)' \circ g \right)(x), \tag{11.3.38}$$

$$D_{a+;g}^0 f(x) = f(x), \quad \forall \, x \in [a, b].$$

When $g = id$, then

$$D_{a+;g}^\alpha f = D_{a+;id}^\alpha f = D_{*a}^\alpha f,$$

the usual left Caputo fractional derivative.

Denote by

$$D_{a+;g}^{n\alpha} := D_{a+;g}^\alpha D_{a+;g}^\alpha \cdots D_{a+;g}^\alpha \quad (n \text{ times}), n \in \mathbb{N}. \tag{11.3.39}$$

We consider the left generalized fractional Riemann-Liouville integral

$$\left(I_{a+;g}^\alpha f \right)(x) = \frac{1}{\Gamma(\alpha)} \int_a^x (g(x) - g(t))^{\alpha-1} g'(t) f(t) \, dt, \quad a \le x \le b. \tag{11.3.40}$$

Also denote by

$$I_{a+;g}^{n\alpha} := I_{a+;g}^\alpha I_{a+;g}^\alpha \cdots I_{a+;g}^\alpha \quad (n \text{ times}). \tag{11.3.41}$$

We will be using the following modified g-left generalized Taylor's formula:

Theorem 11.8 ([4]) Let here $0 < \alpha \le 1$, $k = 0, 1, \ldots, n+1$; and denote $G_k^a := D_{a+;g}^{k\alpha} f$. Assume that $G_k^a \circ g^{-1} \in AC([g(a), g(b)])$, and $\left(G_k^a \circ g^{-1} \right)' \circ g \in L_\infty([a, b])$, for all $k = 0, 1, \ldots, n+1$. Then

$$f(x) - f(a) = \sum_{i=1}^n \frac{(g(x) - g(a))^{i\alpha}}{\Gamma(i\alpha + 1)} \left(D_{a+;g}^{i\alpha} f \right)(a) + \tag{11.3.42}$$

$$\frac{1}{\Gamma((n+1)\alpha)} \int_a^x (g(x) - g(t))^{(n+1)\alpha - 1} g'(t) \left(D_{a+;g}^{(n+1)\alpha} f \right)(t) \, dt,$$

$\forall \, x \in [a, b]$.

Here we are going to operate more generally. We consider $f \in C^1([a, b])$. We define the following left generalized g-fractional derivative:

$$\left(D_{y+;g}^\alpha f \right)(x) = \frac{1}{\Gamma(1 - \alpha)} \int_y^x (g(x) - g(t))^{-\alpha} g'(t) \left(f \circ g^{-1} \right)'(g(t)) \, dt, \tag{11.3.43}$$

for any $y \le x \le b$; $x, y \in [a, b]$,

$$\left(D_{y+;g}^1 f \right)(x) = \left(f \circ g^{-1} \right)'(g(x)), \quad \forall \, x \in [a, b]. \tag{11.3.44}$$

Similarly, we define

$$\left(D_{x+;g}^{\alpha}f\right)(y) = \frac{1}{\Gamma(1-\alpha)} \int_{x}^{y} (g(y) - g(t))^{-\alpha} g'(t) \left(f \circ g^{-1}\right)'(g(t)) dt,$$
(11.3.45)

for any $x \le y \le b$; $x, y \in [a, b]$,

$$\left(D_{x+;g}^{1}f\right)(y) = \left(f \circ g^{-1}\right)'(g(y)), \quad \forall y \in [a, b].$$
(11.3.46)

When $0 < \alpha < 1$, $D_{y+;g}^{\alpha}f$ and $D_{x+;g}^{\alpha}f$ are continuous functions on $[a, b]$. Note here that by convention, we have that

$$\left(D_{y+;g}^{\alpha}f\right)(x) = 0, \quad \text{when } x < y,$$
and
$$\left(D_{x+;g}^{\alpha}f\right)(y) = 0, \quad \text{when } y < x.$$
(11.3.47)

Denote by

$$G_{k}^{y} := D_{y+;g}^{k\alpha}f, \quad G_{k}^{x} := D_{x+;g}^{k\alpha}f, \quad \forall x, y \in [a, b].$$
(11.3.48)

We assume that

$$G_{k}^{z} \circ g^{-1} \in AC([g(a), g(b)]), \quad \text{and} \quad \left(G_{k}^{z} \circ g^{-1}\right)' \circ g \in L_{\infty}([a, b]), \quad (11.3.49)$$

$k = 0, 1, \ldots, n + 1$; for $z = y, x$; $\forall x, y \in [a, b]$; $0 < \alpha < 1$.

We also observe that $(0 < \alpha < 1)$

$$\left|\left(D_{a+;g}^{\alpha}f\right)(x)\right| \le \frac{1}{\Gamma(1-\alpha)} \int_{a}^{x} (g(x) - g(t))^{-\alpha} g'(t) \left|\left(f \circ g^{-1}\right)'(g(t))\right| dt \le$$

$$\frac{\left\|\left(f \circ g^{-1}\right)' \circ g\right\|_{\infty,[a,b]}}{\Gamma(1-\alpha)} \int_{a}^{x} (g(x) - g(t))^{-\alpha} g'(t) dt =$$

$$\frac{\left\|\left(f \circ g^{-1}\right)' \circ g\right\|_{\infty,[a,b]}}{\Gamma(1-\alpha)} \frac{(g(x) - g(a))^{1-\alpha}}{1-\alpha} = \quad (11.3.50)$$

$$\frac{\left\|\left(f \circ g^{-1}\right)' \circ g\right\|_{\infty,[a,b]}}{\Gamma(2-\alpha)} (g(x) - g(a))^{1-\alpha}.$$

We have proved that

$$\left|\left(D_{a+;g}^{\alpha}f\right)(x)\right| \le \frac{\left\|\left(f \circ g^{-1}\right)' \circ g\right\|_{\infty,[a,b]}}{\Gamma(2-\alpha)} \left(g(x) - g(a)\right)^{1-\alpha}$$

$$\le \frac{\left\|\left(f \circ g^{-1}\right)' \circ g\right\|_{\infty,[a,b]}}{\Gamma(2-\alpha)} \left(g(b) - g(a)\right)^{1-\alpha}, \quad \forall \, x \in [a,b]. \tag{11.3.51}$$

In particular it holds

$$\left(D_{a+;g}^{\alpha}f\right)(a) = 0, \quad 0 < \alpha < 1, \tag{11.3.52}$$

and

$$\left(D_{y+;g}^{\alpha}f\right)(y) = \left(D_{x+;g}^{\alpha}f\right)(x) = 0, \; \forall \, x,y \in [a,b]; \quad 0 < \alpha < 1. \tag{11.3.53}$$

By (11.3.42) we derive

$$f(x) - f(y) = \sum_{i=2}^{n} \frac{(g(x) - g(y))^{i\alpha}}{\Gamma(i\alpha+1)} \left(D_{y+;g}^{i\alpha}f\right)(y) +$$

$$\frac{1}{\Gamma((n+1)\alpha)} \int_{y}^{x} (g(x) - g(t))^{(n+1)\alpha-1} g'(t) \left(D_{y+;g}^{(n+1)\alpha}f\right)(t)\,dt, \tag{11.3.54}$$

for any $x > y : x, y \in [a,b]; 0 < \alpha < 1$, also it holds

$$f(y) - f(x) = \sum_{i=2}^{n} \frac{(g(y) - g(x))^{i\alpha}}{\Gamma(i\alpha+1)} \left(D_{x+;g}^{i\alpha}f\right)(x) + \tag{11.3.55}$$

$$\frac{1}{\Gamma((n+1)\alpha)} \int_{x}^{y} (g(y) - g(t))^{(n+1)\alpha-1} g'(t) \left(D_{x+;g}^{(n+1)\alpha}f\right)(t)\,dt,$$

for any $y > x : x, y \in [a,b]; 0 < \alpha < 1$.
We define also the following linear operator

$$(A_2(f))(x,y) :=$$

$$\begin{cases} \sum_{i=2}^{n} \frac{(g(x)-g(y))^{i\alpha-1}}{\Gamma(i\alpha+1)} \left(D_{y+;g}^{i\alpha}f\right)(y) + \left(D_{y+;g}^{(n+1)\alpha}f\right)(x) \frac{(g(x)-g(y))^{(n+1)\alpha-1}}{\Gamma((n+1)\alpha+1)}, & x > y, \\[2mm] \sum_{i=2}^{n} \frac{(g(y)-g(x))^{i\alpha-1}}{\Gamma(i\alpha+1)} \left(D_{x+;g}^{i\alpha}f\right)(x) + \left(D_{x+;g}^{(n+1)\alpha}f\right)(y) \frac{(g(y)-g(x))^{(n+1)\alpha-1}}{\Gamma((n+1)\alpha+1)}, & y > x, \\[2mm] f'(x), & \text{when } x = y, \end{cases}$$

$$\tag{11.3.56}$$

$\forall\, x, y \in [a, b]; 0 < \alpha < 1$.

We may assume that

$$|(A_2(f))(x, x) - (A_2(f))(y, y)| = |f'(x) - f'(y)| \tag{11.3.57}$$

$$\leq \Phi^* |g(x) - g(y)|, \ \forall\, x, y \in [a, b];$$

with $\Phi^* > 0$.

We estimate and have

(i) case of $x > y$:

$$|f(x) - f(y) - (A_2(f))(x, y)(g(x) - g(y))| = \tag{11.3.58}$$

$$\left| \frac{1}{\Gamma((n+1)\alpha)} \int_y^x (g(x) - g(t))^{(n+1)\alpha - 1} g'(t) \left(D_{y+;g}^{(n+1)\alpha} f \right)(t)\, dt - \right.$$

$$\left. \left(D_{y+;g}^{(n+1)\alpha} f \right)(x) \frac{(g(x) - g(y))^{(n+1)\alpha}}{\Gamma((n+1)\alpha + 1)} \right| =$$

$$\frac{1}{\Gamma((n+1)\alpha)} \cdot$$

$$\left| \int_y^x (g(x) - g(tx))^{(n+1)\alpha - 1} g'(t) \left(\left(D_{y+;g}^{(n+1)\alpha} f \right)(t) - \left(D_{y+;g}^{(n+1)\alpha} f \right)(x) \right) dt \right| \leq$$

$$\tag{11.3.59}$$

$$\frac{1}{\Gamma((n+1)\alpha)} \cdot$$

$$\int_y^x (g(x) - g(t))^{(n+1)\alpha - 1} g'(t) \left| \left(D_{y+;g}^{(n+1)\alpha} f \right)(t) - \left(D_{y+;g}^{(n+1)\alpha} f \right)(x) \right| dt$$

(we assume here that

$$\left| \left(D_{y+;g}^{(n+1)\alpha} f \right)(t) - \left(D_{y+;g}^{(n+1)\alpha} f \right)(x) \right| \leq \rho_1 |g(t) - g(x)|, \tag{11.3.60}$$

$\forall\, t, x, y \in [a, b] : x \geq t \geq y; \rho_1 > 0)$

$$\leq \frac{\rho_1}{\Gamma((n+1)\alpha)} \int_y^x (g(x) - g(t))^{(n+1)\alpha - 1} g'(t)(g(x) - g(t))\, dt =$$

$$\frac{\rho_1}{\Gamma((n+1)\alpha)} \int_y^x (g(x) - g(t))^{(n+1)\alpha} g'(t)\, dt =$$

$$\frac{\rho_1}{\Gamma((n+1)\alpha)} \frac{(g(x) - g(y))^{(n+1)\alpha+1}}{((n+1)\alpha+1)}. \tag{11.3.61}$$

We have proved that

$$|f(x) - f(y) - (A_2(f))(x, y)(g(x) - g(y))| \le$$

$$\frac{\rho_1}{\Gamma((n+1)\alpha)} \frac{(g(x) - g(y))^{(n+1)\alpha+1}}{((n+1)\alpha+1)}, \tag{11.3.62}$$

$\forall\, x, y \in [a, b] : x > y;\ 0 < \alpha < 1,$
  (ii) case of $y > x$:

$$|f(x) - f(y) - (A_2(f))(x, y)(g(x) - g(y))| = \tag{11.3.63}$$

$$|f(y) - f(x) - (A_2(f))(x, y)(g(y) - g(x))| =$$

$$\left| \frac{1}{\Gamma((n+1)\alpha)} \int_x^y (g(y) - g(t))^{(n+1)\alpha-1} g'(t) \left(D_{x+;g}^{(n+1)\alpha} f \right)(t)\, dt - \right.$$

$$\left. \left(D_{x+;g}^{(n+1)\alpha} f \right)(y) \frac{(g(y) - g(x))^{(n+1)\alpha}}{\Gamma((n+1)\alpha+1)} \right| =$$

$$\frac{1}{\Gamma((n+1)\alpha)} \cdot$$

$$\left| \int_x^y (g(y) - g(t))^{(n+1)\alpha-1} g'(t) \left(\left(D_{x+;g}^{(n+1)\alpha} f \right)(t) - \left(D_{x+;g}^{(n+1)\alpha} f \right)(y) \right) dt \right| \le$$

$$\frac{1}{\Gamma((n+1)\alpha)} \cdot$$

$$\int_x^y (g(y) - g(t))^{(n+1)\alpha-1} g'(t) \left| \left(D_{x+;g}^{(n+1)\alpha} f \right)(t) - \left(D_{x+;g}^{(n+1)\alpha} f \right)(y) \right| dt \tag{11.3.64}$$

(we assume here that

$$\left| \left(D_{x+;g}^{(n+1)\alpha} f \right)(t) - \left(D_{x+;g}^{(n+1)\alpha} f \right)(y) \right| \le \rho_2 |g(t) - g(y)|, \tag{11.3.65}$$

$\forall\, t, y, x \in [a, b] : y \ge t \ge x;\ \rho_2 > 0)$

$$\le \frac{\rho_2}{\Gamma((n+1)\alpha)} \int_x^y (g(y) - g(t))^{(n+1)\alpha-1} g'(t)(g(y) - g(t))\, dt =$$

$$\frac{\rho_2}{\Gamma((n+1)\alpha)} \frac{(g(y) - g(x))^{(n+1)\alpha+1}}{((n+1)\alpha+1)}. \tag{11.3.66}$$

We have proved that

$$|f(x) - f(y) - (A_2(f))(x, y)(g(x) - g(y))| \le$$

$$\frac{\rho_2}{\Gamma((n+1)\alpha)} \frac{(g(y) - g(x))^{(n+1)\alpha+1}}{((n+1)\alpha+1)},$$

$\forall x, y \in [a, b] : y > x; 0 < \alpha < 1.$

Conclusion 11.9 *Set $\rho = \max(\rho_1, \rho_2)$. Then*

$$|f(x) - f(y) - (A_2(f))(x, y)(g(x) - g(y))| \le$$

$$\frac{\rho}{\Gamma((n+1)\alpha)} \frac{|g(x) - g(y)|^{(n+1)\alpha+1}}{((n+1)\alpha+1)}, \tag{11.3.67}$$

$\forall x, y \in [a, b]; 0 < \alpha < 1.$

(Notice (11.3.67) is trivially true when $x = y$.)

One may assume that

$$\frac{\rho}{\Gamma((n+1)\alpha)} < 1. \tag{11.3.68}$$

Now based on (11.3.57) and (11.3.67), we can apply our numerical methods presented in this chapter to solve $f(x) = 0$.

Remark 11.10 (a) Returning back to Conclusion 11.7, we see that Proposition 11.2 can be used, if $g(t) = t$, $F(t) = f(t)$, $A(F)(s, t) = A_1(f)(s, t)$ for each $s, t \in [a, b]$, $p = (i + 1)\alpha, i \in \mathbb{N}$ fixed and

$$\varphi_1(s, t) = \frac{\lambda |s - t|^p}{\Gamma(p)(p + 1)\mu}$$

for each $s, t \in [a, b]$.

(b) According to Conclusion 11.9, as in (a) but we must choose $A(F)(s, t) = A_2(f)(s, t)$ and

$$\varphi_1(s, t) = \frac{\rho |s - t|^p}{\Gamma(p)(p + 1)\mu}$$

for each $s, t \in [a, b]$.

References

1. S. Amat, S. Busquier, Third-order iterative methods under Kantorovich conditions. J. Math. Anal. Appl. **336**, 243–261 (2007)
2. S. Amat, S. Busquier, S. Plaza, Chaotic dynamics of a third-order Newton-like method. J. Math. Anal. Appl. **366**(1), 164–174 (2010)
3. G. Anastassiou, Fractional representation formulae and right fractional inequalities. Math. Comput. Model. **54**(10–12), 3098–3115 (2011)
4. G. Anastassiou, Advanced fractional Taylor's formulae. J. Comput. Anal. Appl. (2016) (to appear)
5. G. Anastassiou, I. Argyros, Semilocal Convergence of Secant-Type Methods with Applications to Modified g-Fractional Calculus (2015) (submitted)
6. I.K. Argyros, Newton-like methods in partially ordered linear spaces. J. Approx. Theory Appl. **9**(1), 1–10 (1993)
7. I.K. Argyros, Results on controlling the residuals of perturbed Newton-like methods on Banach spaces with a convergence structure. Southwest J. Pure Appl. Math. **1**, 32–38 (1995)
8. I.K. Argyros, *Convergence and Applications of Newton-Like Iterations* (Springer, New York, 2008)
9. K. Diethelm, *The Analysis of Fractional Differential Equations*. Lecture Notes in Mathematics, vol. 2004, 1st edn. (Springer, New York, 2010)
10. J.A. Ezquerro, J.M. Gutierrez, M.A. Hernandez, N. Romero, M.J. Rubio, The Newton method: from Newton to Kantorovich (Spanish). Gac. R. Soc. Mat. Esp. **13**, 53–76 (2010)
11. J.A. Ezquerro, M.A. Hernandez, Newton-like methods of high order and domains of semilocal and global convergence. Appl. Math. Comput. **214**(1), 142–154 (2009)
12. L.V. Kantorovich, G.P. Akilov, *Functional Analysis in Normed Spaces* (Pergamon Press, New York, 1964)
13. A.A. Magrenan, Different anomalies in a Jarratt family of iterative root finding methods. Appl. Math. Comput. **233**, 29–38 (2014)
14. A.A. Magrenan, A new tool to study real dynamics: the convergence plane. Appl. Math. Comput. **248**, 215–224 (2014)
15. F.A. Potra, V. Ptak, *Nondiscrete induction and iterative processes* (Pitman Publishing, London, 1984)
16. P.D. Proinov, New general convergence theory for iterative processes and its applications to Newton-Kantorovich type theorems. J. Complexity **26**, 3–42 (2010)

Chapter 12
Secant-Like Algorithms and Generalized Fractional Calculus

We present local and semilocal convergence results for secant-like algorithms in order to approximate a locally unique solution of a nonlinear equation in a Banach space setting. In the last part of the study we present some choices of the operators involved in fractional calculus where the operators satisfy the convergence conditions. It follows [5].

12.1 Introduction

In this study we are concerned with the problem of approximating a locally unique solution x^* of the nonlinear equation

$$F(x) = 0, \qquad (12.1.1)$$

where F is a continuous operator defined on a subset D of a Banach space X with values in a Banach space Y.

A lot of problems in Computational Sciences and other disciplines can be brought in a form like (12.1.1) using Mathematical Modelling [8, 12, 16]. The solutions of such equations can be found in closed form only in special cases. That is why most solution methods for these equations are iterative. Iterative methods are usually studied based on semilocal and local convergence. The semilocal convergence matter is, based on the information around the initial point to give hypotheses ensuring the convergence of the iterative algorithm; while the local one is, based on the information around a solution, to find estimates of the radii of convergence balls as well as error bounds on the distances involved.

We introduce the secant-type method defined for each $n = 0, 1, 2, \ldots$ by

$$x_{n+1} = x_n - A(F)(x_n, x_{n-1})^{-1} F(x_n), \qquad (12.1.2)$$

© Springer International Publishing Switzerland 2016

G.A. Anastassiou and I.K. Argyros, *Intelligent Numerical Methods:*
Applications to Fractional Calculus, Studies in Computational Intelligence 624,
DOI 10.1007/978-3-319-26721-0_12

where $x_{-1}, x_0 \in D$ are initial points and $A(F)(x, y) \in L(X, Y)$ the space of bounded linear operators from X into Y. There is a plethora on local as well as semilocal convergence theorems for method (12.1.2) provided that the operator A is an approximation to the Fréchet-derivative F' [1, 2, 6–16]. In the present study we do not necessarily assume that operator A is related to F'. This way we expand the applicability of iterative algorithm (12.1.2). Notice that many well known methods are special case of method (12.1.2).

Newton's method: Choose $A(F)(x, x) = F'(x)$ for each $x \in D$.

Secant method: Choose $A(F)(x, y) = [x, y; F]$, where $[x, y; F]$ denotes a divided difference of order one [8, 12, 15].

The so called Newton-like algorithms and many other methods are special cases of method (12.1.2).

The rest of the chapter is organized as follows. The semilocal as well as the local convergence analysis of method (12.1.2) is given in Sect. 12.2. Some applications from fractional calculus are given in Sect. 12.3.

12.2   Convergence Analysis

We present the main semilocal convergence result for method (12.1.2).

Theorem 12.1 *Let $F : D \subset X \to Y$ be a continuous operator and let $A(F)$ $(x, y) \in L(X, Y)$. Suppose that there exist $x_{-1}, x_0 \in D$, $\eta \geq 0$, $p \geq 1$, a function $\varphi : [0, \infty)^2 \to [0, \infty)$ continuous and nondecreasing such that for each $x, y, z \in D$*

$$A(F)(z, y)^{-1} \in L(Y, X), \tag{12.2.1}$$

$$\max\left\{ \|x_{-1} - x_0\|, \left\| A(F)(x_0, x_{-1})^{-1} F(x_0) \right\| \right\} \leq \eta, \tag{12.2.2}$$

$$\left\| A(F)(z, y)^{-1} (F(z) - F(y) - A(F)(y, x)(z - y)) \right\| \leq$$

$$\varphi(\|z - y\|, \|y - x\|) \|z - y\|^{p+1}, \tag{12.2.3}$$

$$q := \varphi(\eta, \eta) \eta^p < 1 \tag{12.2.4}$$

and

$$\overline{U}(x_0, r) \subseteq D, \tag{12.2.5}$$

where,

$$r = \frac{\eta}{1 - q}. \tag{12.2.6}$$

Then, the sequence $\{x_n\}$ generated by method (12.1.2) is well defined, remains in $\overline{U}(x_0, r)$ for each $n = 0, 1, 2, \ldots$ and converges to some $x^* \in \overline{U}(x_0, r)$ such that

$$\|x_{n+1} - x_n\| \leq \varphi\left(\|x_n - x_{n-1}\|, \|x_{n-1} - x_{n-2}\|\right) \|x_n - x_{n-1}\|^{p+1}$$

$$\leq q\,\|x_n - x_{n-1}\| \tag{12.2.7}$$

and

$$\|x_n - x^*\| \leq \frac{q^n \eta}{1 - q}. \tag{12.2.8}$$

Proof The iterate x_1 is well defined by method (12.1.2) for $n = 0$ and (12.2.1). We also have by (12.2.2) and (12.2.6) that $\|x_1 - x_0\| = \left\|A(F)(x_0, x_{-1})^{-1} F(x_0)\right\| \leq \eta < r$, so we get that $x_1 \in \overline{U}(x_0, r)$ and x_2 is well defined (by (12.2.5)). Using (12.2.3) and (12.2.4) we get that

$$\|x_2 - x_1\| = \left\|A(F)(x_1, x_0)^{-1}\left[F(x_1) - F(x_0) - A(F)(x_0, x_{-1})(x_1 - x_0)\right]\right\|$$

$$\leq \varphi\left(\|x_1 - x_0\|, \|x_0 - x_{-1}\|\right) \|x_1 - x_0\|^{p+1} \leq q\,\|x_1 - x_0\|,$$

which shows (12.2.7) for $n = 1$. Then, we can have that

$$\|x_2 - x_0\| \leq \|x_2 - x_1\| + \|x_1 - x_0\| \leq q\,\|x_1 - x_0\| + \|x_1 - x_0\|$$

$$= (1 + q)\,\|x_1 - x_0\| \leq \frac{1 - q^2}{1 - q}\,\eta < r,$$

so $x_2 \in \overline{U}(x_0, r)$ and x_3 is well defined.

Assuming $\|x_{k+1} - x_k\| \leq q\,\|x_k - x_{k-1}\|$ and $x_{k+1} \in \overline{U}(x_0, r)$ for each $k = 1, 2, \ldots, n$ we get

$$\|x_{k+2} - x_{k+1}\| =$$

$$\left\|A(F)(x_{k+1}, x_k)^{-1}\left[F(x_{k+1}) - F(x_k) - A(F)(x_k, x_{k-1})(x_{k+1} - x_k)\right]\right\|$$

$$\leq \varphi\left(\|x_{k+1} - x_k\|, \|x_k - x_{k-1}\|\right) \|x_{k+1} - x_k\|^{p+1}$$

$$\leq \varphi\left(\|x_1 - x_0\|, \|x_0 - x_{-1}\|\right) \|x_1 - x_0\|^p \|x_{k+1} - x_k\| \leq q\,\|x_{k+1} - x_k\|$$

and

$$\|x_{k+2} - x_0\| \leq \|x_{k+2} - x_{k+1}\| + \|x_{k+1} - x_k\| + \cdots + \|x_1 - x_0\|$$

$$\leq \left(q^{k+1} + q^k + \cdots + 1\right) \|x_1 - x_0\| \leq \frac{1 - q^{k+2}}{1 - q}\,\|x_1 - x_0\|$$

$$< \frac{\eta}{1 - q} = r,$$

which completes the induction for (12.2.7) and $x_{k+2} \in \overline{U}(x_0, r)$. We also have that for $m \geq 0$

$$\|x_{n+m} - x_n\| \leq \|x_{n+m} - x_{n+m-1}\| + \cdots + \|x_{n+1} - x_n\|$$

$$\leq \left(q^{m-1} + q^{m-2} + \cdots + 1 \right) \|x_{n+1} - x_n\|$$

$$\leq \frac{1 - q^m}{1 - q} q^n \|x_1 - x_0\|.$$

It follows that $\{x_n\}$ is a complete sequence in a Banach space X and as such it converges to some $x^* \in \overline{U}(x_0, r)$ (since $\overline{U}(x_0, r)$ is a closed set). By letting $m \to \infty$, we obtain (12.2.8). $\qquad \square$

Stronger hypotheses are needed to show that x^* is a solution of equation $F(x) = 0$.

Proposition 12.2 *Let* $F : D \subset X \to Y$ *be a continuous operator and let* $A(F)(x, y) \in L(X, Y)$. *Suppose that there exist* $x_{-1}, x_0 \in D$, $\eta \geq 0$, $p \geq 1$, $\mu > 0$, *a function* $\varphi_1 : [0, \infty)^2 \to [0, \infty)$ *continuous and nondecreasing such that for each* $x, y \in D$

$$A(F)(x, y)^{-1} \in L(Y, X), \quad \|A(F)(x, y)^{-1}\| \leq \mu,$$

$$\max \left\{ \|x_{-1} - x_0\|, \|A(F)(x_0, x_{-1})^{-1} F(x_0)\| \right\} \leq \eta, \qquad (12.2.9)$$

$$\|F(z) - F(y) - A(F)(y, x)(z - y)\| \leq \frac{\varphi_1(\|z - y\|, \|x - y\|)}{\mu} \|z - y\|^{p+1},$$
$$(12.2.10)$$

$$q_1 := \varphi_1(\eta, \eta) \eta^p < 1$$

and

$$\overline{U}(x_0, r_1) \subseteq D,$$

where,

$$r_1 = \frac{\eta}{1 - q_1}.$$

Then, the conclusions of Theorem 12.1 for sequence $\{x_n\}$ *hold with* φ_1, q_1, r_1, *replacing* φ, q *and* r, *respectively. Moreover,* x^* *is a solution of the equation* $F(x) = 0$.

Proof Notice that

$$\left\| A(F)(x_n, x_{n-1})^{-1} \left[F(x_n) - F(x_{n-1}) - A(F)(x_{n-1}, x_{n-2})(x_n - x_{n-1}) \right] \right\|$$

$$\leq \left\| A(F)(x_n, x_{n-1})^{-1} \right\| \left\| F(x_n) - F(x_{n-1}) - A(F)(x_{n-1}, x_{n-2})(x_n - x_{n-1}) \right\|$$

$$\leq \varphi_1 (\|x_n - x_{n-1}\|, \|x_{n-1} - x_{n-2}\|) \|x_n - x_{n-1}\|^{p+1} \leq q_1 \|x_n - x_{n-1}\|.$$

Therefore, the proof of Theorem 12.1 can apply. Then, in view of the estimate

$$\|F(x_n)\| = \|F(x_n) - F(x_{n-1}) - A(F)(x_{n-1}, x_{n-2})(x_n - x_{n-1})\| \le$$

$$\frac{\varphi_1(\|x_n - x_{n-1}\|)}{\mu}\|x_n - x_{n-1}\|^{p+1} \le q_1 \|x_n - x_{n-1}\|,$$

we deduce by letting $n \to \infty$ that $F(x^*) = 0$. $\square$

Concerning the uniqueness of the solution x^* we have the following result:

Proposition 12.3 *Under the hypotheses of Proposition 12.2, further suppose that there exists $\varphi_2 : [0, \infty)^2 \to [0, \infty)$ continuous and nondecreasing such that*

$$\|F(z) - F(x) - A(F)(z, y)(z - x)\| \le \frac{\varphi_2(\|z - x\|, \|y - x\|)}{\mu}\|z - x\|^{p+1}$$

$$(12.2.11)$$

and

$$\varphi_2(r_1, \eta + r_1) r_1^p < 1. \tag{12.2.12}$$

Then, x^* is the only solution of equation $F(x) = 0$ in $\overline{U}(x_0, r_1)$.

Proof The existence of the solution $x^* \in \overline{U}(x_0, r_1)$ has been established in Proposition 12.2. Let $y^* \in \overline{U}(x_0, r_1)$ with $F(y^*) = 0$. Then, we have in turn that

$$\|x_{n+1} - y^*\| = \|x_n - y^* - A(F)(x_n, x_{n-1})^{-1} F(x_n)\| =$$

$$\|A(F)(x_n, x_{n-1})^{-1} [A(F)(x_n, x_{n-1})(x_n - y^*) - F(x_n) + F(y^*)]\| \le$$

$$\|A(F)(x_n, x_{n-1})^{-1}\| \|F(y^*) - F(x_n) - A(F)(x_n, x_{n-1})(y^* - x_n)\| \le$$

$$\mu \frac{\varphi_1(\|x_n - y^*\|, \|x_{n-1} - y^*\|)}{\mu}\|x_n - y^*\|^{p+1} \le$$

$$\varphi_2(r_1, \eta + r_1) r_1^p \|x_n - x^*\| < \|x_n - y^*\|,$$

so we deduce that $\lim_{n \to \infty} x_n = y^*$. But we have that $\lim_{n \to \infty} x_n = x^*$. Hence, we conclude that $x^* = y^*$. $\square$

Next, we present a local convergence analysis for the iterative algorithm (12.1.2).

Proposition 12.4 *Let $F : D \subset X \to Y$ be a continuous operator and let $A(F)(y, x) \in L(X, Y)$. Suppose that there exist $x^* \in D$, $p \ge 1$, a function $\varphi_3 : [0, \infty)^2 \to [0, \infty)$ continuous and nondecreasing such that for each $x, y \in D$*

$$F\left(x^*\right) = 0, \quad A\left(F\right)\left(y, x\right)^{-1} \in L\left(Y, X\right),$$

$$\left\| A\left(F\right)\left(y, x\right)^{-1} \left[F\left(x\right) - F\left(x^*\right) - A\left(F\right)\left(y, x\right)\left(x - x^*\right)\right] \right\| \le$$

$$\varphi_3 \left(\left\| y - x^* \right\|, \left\| x - x^* \right\|\right) \left\| y - x^* \right\|^{p+1}, \tag{12.2.13}$$

and

$$\overline{U}\left(x^*, r_2\right) \subseteq D,$$

where r_2 is the smallest positive solution of equation

$$h\left(t\right) := \varphi_3\left(t, t\right) t^p - 1.$$

Then, sequence $\{x_n\}$ generated by method (12.1.2) for $x_{-1}, x_0 \in U\left(x^*, r_2\right) - \{x^*\}$ is well defined, remains in $U\left(x^*, r_2\right)$ for each $n = 0, 1, 2, \ldots$ and converges to x^*. Moreover, the following estimates hold

$$\left\| x_{n+1} - x^* \right\| \le \varphi_2 \left(\left\| x_n - x^* \right\|, \left\| x_{n-1} - x^* \right\|\right) \left\| x_n - x^* \right\|^{p+1} < \left\| x_n - x^* \right\| < r_2.$$

Proof We have that $h\left(0\right) = -1 < 0$ and $h\left(t\right) \to +\infty$ as $t \to +\infty$. Then, it follows from the intermediate value theorem that function h has positive zeros. Denote by r_2 the smallest such zero. By hypothesis $x_{-1}, x_0 \in U\left(x^*, r_2\right) - \{x^*\}$. Then, we get in turn that

$$\left\| x_1 - x^* \right\| = \left\| x_0 - x^* - A\left(F\right)\left(x_0, x_{-1}\right)^{-1} F\left(x_0\right) \right\| =$$

$$\left\| A\left(F\right)\left(x_0, x_{-1}\right)^{-1} \left[F\left(x^*\right) - F\left(x_0\right) - A\left(F\right)\left(x_0, x_{-1}\right)\left(x^* - x_0\right)\right] \right\| \le$$

$$\varphi_2 \left(\left\| x_0 - x^* \right\|, \left\| x_{-1} - x^* \right\|\right) \left\| x_0 - x^* \right\|^{p+1} < \varphi_3\left(r_2, r_2\right) r_2^p \left\| x_0 - x^* \right\| =$$

$$\left\| x_0 - x^* \right\| < r_2,$$

which shows that $x_1 \in U\left(x^*, r_2\right)$ and x_2 is well defined. By a simple inductive argument as in the preceding estimate we get that

$$\left\| x_{k+1} - x^* \right\| = \left\| x_k - x^* - A\left(F\right)\left(x_k, x_{k-1}\right)^{-1} F\left(x_k\right) \right\| \le$$

$$\left\| A\left(F\right)\left(x_k, x_{k-1}\right)^{-1} \left[F\left(x^*\right) - F\left(x_k\right) - A\left(F\right)\left(x_k, x_{k-1}\right)\left(x^* - x_k\right)\right] \right\| \le$$

$$\varphi_2 \left(\left\| x_k - x^* \right\|, \left\| x_{k-1} - x^* \right\|\right) \left\| x_k - x^* \right\|^{p+1} <$$

$$\varphi_3\left(r_2, r_2\right) r_2^p \left\| x_k - x^* \right\| = \left\| x_k - x^* \right\| < r_2,$$

which shows $\lim_{k \to \infty} x_k = x^*$ and $x_{k+1} \in U\left(x^*, r_2\right)$. $\qquad \square$

Remark 12.5 (a) Hypothesis (12.2.3) specializes to Newton-Mysowski-type, if $A(F)(x) = F'(x)$ [8, 12, 15]. However, if F is not Fréchet-differentiable, then our results extend the applicability of iterative algorithm (12.1.2).

(b) Theorem 12.1 has practical value although we do not show that x^* is a solution of equation $F(x) = 0$, since this may be shown in another way.

(c) Hypothesis (12.2.13) can be replaced by the stronger

$$\left\| A(F)(y,x)^{-1} [F(x) - F(z) - A(F)(y,x)(x-z)] \right\| \le$$

$$\varphi_3 \left(\|z - y\|, \|z - x\| \right) \|z - y\|^{p+1}.$$

12.3 Applications to g-Fractional Calculus

Here both backgrounds needed come from [4] and Chap. 24. See also the related [3].

(I) We need:

Definition 12.6 Let $\alpha > 0$, $\lceil \alpha \rceil = n$, $\lceil \cdot \rceil$ the ceiling of the number. Here let $g \in AC([a,b])$ (absolutely continuous functions) and strictly increasing. We assume that $\left(f \circ g^{-1} \right)^{(n)} \circ g \in L_\infty([a,b])$, where $f : [a,b] \to \mathbb{N}$.

We define the left generalized g-fractional derivative of f of order α as follows:

$$\left(D^\alpha_{a+;g} f \right)(x) := \frac{1}{\Gamma(n-\alpha)} \int_a^x (g(x) - g(t))^{n-\alpha-1} g'(t) \left(f \circ g^{-1} \right)^{(n)} (g(t)) \, dt,$$
$$\tag{12.3.1}$$

$a \le x \le b$, where Γ is the gamma function.

If $\alpha \notin \mathbb{N}$, we have that $D^\alpha_{a+;g} f \in C([a,b])$.

We set

$$D^n_{a+;g} f(x) = \left(\left(f \circ g^{-1} \right)^{(n)} \circ g \right)(x), \tag{12.3.2}$$

$$D^0_{a+;g} f(x) = f(x), \quad \forall x \in [a,b].$$

When $g = id$, then

$$\left(D^\alpha_{a+;g} f \right) = \left(D^\alpha_{a+;id} f \right) = \left(D^\alpha_{*a} f \right), \tag{12.3.3}$$

the usual left Caputo fractional derivative.

We will use the following g-left fractional generalized Taylor's formula from [4].

Theorem 12.7 *Let g be strictly increasing function and $g \in AC([a,b])$. We assume that $\left(f \circ g^{-1} \right) \in AC^n([g(a), g(b)])$ (it means $\left(f \circ g^{-1} \right)^{(n-1)} \in AC$ $([g(a), g(b)])$), where $\mathbb{N} \ni n = \lceil \alpha \rceil$, $\alpha > 0$. Also, we assume that $\left(f \circ g^{-1} \right)^{(n)}$ $\circ g \in L_\infty([a,b])$. Then*

$$f(x) - f(a) = \sum_{k=1}^{n-1} \frac{\left(f \circ g^{-1}\right)^{(k)}(g(a))}{k!}(g(x) - g(a))^k + \tag{12.3.4}$$

$$\frac{1}{\Gamma(\alpha)}\int_a^x (g(x) - g(t))^{\alpha-1} g'(t)\left(D_{a+;g}^{\alpha}f\right)(t)\,dt,$$

$\forall x \in [a, b]$.

The remainder of (12.3.4) is a continuous function in $x \in [a, b]$.

Here we are going to operate more generally. We consider $f \in C^n([a, b])$. We define the following left g-fractional derivative of f of order α as follows:

$$\left(D_{y+;g}^{\alpha}f\right)(x) := \frac{1}{\Gamma(n-\alpha)}\int_y^x (g(x) - g(t))^{n-\alpha-1} g'(t)\left(f \circ g^{-1}\right)^{(n)}(g(t))\,dt,$$
$$\tag{12.3.5}$$

for any $a \le y \le x \le b$;

$$D_{y+;g}^n f(x) = \left(\left(f \circ g^{-1}\right)^{(n)} \circ g\right)(x), \quad \forall x, y \in [a, b], \tag{12.3.6}$$

and

$$D_{y+;g}^0 f(x) = f(x), \quad \forall x \in [a, b]. \tag{12.3.7}$$

For $\alpha > 0$, $\alpha \notin \mathbb{N}$, by convention we set that

$$\left(D_{y+;g}^{\alpha}f\right)(x) = 0, \quad \text{for } x < y, \quad \forall x, y \in [a, b]. \tag{12.3.8}$$

Similarly, we define

$$\left(D_{x+;g}^{\alpha}f\right)(y) := \frac{1}{\Gamma(n-\alpha)}\int_x^y (g(y) - g(t))^{n-\alpha-1} g'(t)\left(f \circ g^{-1}\right)^{(n)}(g(t))\,dt,$$
$$\tag{12.3.9}$$

for any $a \le x \le y \le b$;

$$D_{x+;g}^n f(y) = \left(\left(f \circ g^{-1}\right)^{(n)} \circ g\right)(y), \quad \forall x, y \in [a, b], \tag{12.3.10}$$

and

$$D_{x+;g}^0 f(y) = f(y), \quad \forall y \in [a, b]. \tag{12.3.11}$$

For $\alpha > 0$, $\alpha \notin \mathbb{N}$, by convention we set that

$$\left(D_{x+;g}^{\alpha}f\right)(y) = 0, \quad \text{for } y < x, \quad \forall x, y \in [a, b]. \tag{12.3.12}$$

By assuming $\left(f \circ g^{-1}\right)^{(n)} \circ g \in L_\infty\left([a, b]\right)$, we get that

$$\left|\left(D_{a+;g}^\alpha f\right)(x)\right| \leq \frac{1}{\Gamma(n-\alpha)} \int_a^x (g(x) - g(t))^{n-\alpha-1} g'(t) \left|\left(f \circ g^{-1}\right)^{(n)}(g(t))\right| dt$$

$$\tag{12.3.13}$$

$$\leq \frac{\left\|\left(f \circ g^{-1}\right)^{(n)} \circ g\right\|_{\infty,[a,b]}}{\Gamma(n-\alpha)} \int_a^x (g(x) - g(t))^{n-\alpha-1} g'(t) \, dt =$$

$$\frac{\left\|\left(f \circ g^{-1}\right)^{(n)} \circ g\right\|_{\infty,[a,b]}}{\Gamma(n-\alpha+1)} (g(x) - g(a))^{n-\alpha} \leq$$

$$\frac{\left\|\left(f \circ g^{-1}\right)^{(n)} \circ g\right\|_{\infty,[a,b]}}{\Gamma(n-\alpha+1)} (g(b) - g(a))^{n-\alpha}, \quad \forall\, x \in [a, b]. \tag{12.3.14}$$

That is

$$\left(D_{a+;g}^\alpha f\right)(a) = 0, \tag{12.3.15}$$

and

$$\left(D_{y+;g}^\alpha f\right)(y) = \left(D_{x+;g}^\alpha f\right)(x) = 0, \quad \forall\, x, y \in [a, b]. \tag{12.3.16}$$

Thus when $\alpha > 0$, $\alpha \notin \mathbb{N}$, both $D_{y+;g}^\alpha f$, $D_{x+;g}^\alpha f \in C\left([a, b]\right)$.

Notice also, that $\left(f \circ g^{-1}\right) \in AC^n\left([g(x), g(b)]\right)$ and $\left(f \circ g^{-1}\right)^{(n)} \circ g \in L_\infty\left([x, b]\right)$, and of course $g \in AC\left([x, b]\right)$, and strictly incrasing over $[x, b]$, $\forall\, x \in [a, b]$.

Hence, by Theorem 12.7 we obtain

$$f(x) - f(y) = \sum_{k=1}^{n-1} \frac{\left(f \circ g^{-1}\right)^{(k)}(g(y))}{k!} (g(x) - g(y))^k +$$

$$\frac{1}{\Gamma(\alpha)} \int_y^x (g(x) - g(t))^{\alpha-1} g'(t) \left(D_{y+;g}^\alpha f\right)(t) \, dt, \quad \forall\, x \in [y, b], \tag{12.3.17}$$

and

$$f(y) - f(x) = \sum_{k=1}^{n-1} \frac{\left(f \circ g^{-1}\right)^{(k)}(g(x))}{k!} (g(y) - g(x))^k +$$

$$\frac{1}{\Gamma(\alpha)} \int_x^y (g(y) - g(t))^{\alpha-1} g'(t) \left(D_{x+;g}^\alpha f\right)(t) \, dt, \quad \forall\, y \in [x, b], \tag{12.3.18}$$

We define also the following linear operator

$$(A_1 (f)) (x, y) :=$$

$$\begin{cases} \sum_{k=1}^{n-1} \frac{(f \circ g^{-1})^{(k)} (g(y))}{k!} (g(x) - g(y))^{k-1} + \left(D_{y+;g}^\alpha f \right) (x) \frac{(g(x)-g(y))^{\alpha-1}}{\Gamma(\alpha+1)}, & \text{for } x > y, \\ \sum_{k=1}^{n-1} \frac{(f \circ g^{-1})^{(k)} (g(x))}{k!} (g(y) - g(x))^{k-1} + \left(D_{x+;g}^\alpha f \right) (y) \frac{(g(y)-g(x))^{\alpha-1}}{\Gamma(\alpha+1)}, & \text{for } x < y, \\ f^{(n)} (x), & \text{when } x = y, \end{cases}$$

$$\tag{12.3.19}$$

$\forall\, x, y \in [a, b]; \alpha > 0, n = \lceil \alpha \rceil$.

We may assume that

$$|(A_1 (f)) (x, x) - (A_1 (f)) (y, y)| = \left| f^{(n)} (x) - f^{(n)} (y) \right| \tag{12.3.20}$$

$$\left| \left(f^{(n)} \circ g^{-1} \right) (g(x)) - \left(f^{(n)} \circ g^{-1} \right) (g(y)) \right| \le \Phi\, |g(x) - g(y)|, \quad \forall\, x, y \in [a, b];$$

where $\Phi > 0$.

We estimate and have:
(i) case of $x > y$:

$$|f(x) - f(y) - (A_1 (f)) (x, y) (g(x) - g(y))| =$$

$$\left| \frac{1}{\Gamma(\alpha)} \int_y^x (g(x) - g(t))^{\alpha-1} g'(t) \left(D_{y+;g}^\alpha f \right) (t)\, dt - \right.$$

$$\left. \left(D_{y+;g}^\alpha f \right) (x) \frac{(g(x) - g(y))^\alpha}{\Gamma(\alpha + 1)} \right| = \tag{12.3.21}$$

$$\frac{1}{\Gamma(\alpha)} \left| \int_y^x (g(x) - g(t))^{\alpha-1} g'(t) \left(\left(D_{y+;g}^\alpha f \right) (t) - \left(D_{y+;g}^\alpha f \right) (x) \right) dt \right| \le$$

$$\frac{1}{\Gamma(\alpha)} \int_y^x (g(x) - g(t))^{\alpha-1} g'(t) \left| \left(D_{y+;g}^\alpha f \right) (t) - \left(D_{y+;g}^\alpha f \right) (x) \right| dt \tag{12.3.22}$$

(we assume that

$$\left| \left(D_{y+;g}^\alpha f \right) (t) - \left(D_{y+;g}^\alpha f \right) (x) \right| \le \lambda_1 |g(t) - g(x)|, \tag{12.3.23}$$

$\forall\, t, x, y \in [a, b] : x \ge t \ge y; \lambda_1 > 0)$

$$\le \frac{\lambda_1}{\Gamma(\alpha)} \int_y^x (g(x) - g(t))^{\alpha-1} g'(t) (g(x) - g(t)) dt =$$

$$\frac{\lambda_1}{\Gamma(\alpha)} \int_y^x (g(x) - g(t))^{\alpha} g'(t) dt = \frac{\lambda_1}{\Gamma(\alpha)} \frac{(g(x) - g(y))^{\alpha+1}}{(\alpha+1)}. \tag{12.3.24}$$

We have proved that

$$|f(x) - f(y) - (A_1(f))(x, y)(g(x) - g(y))| \le$$

$$\frac{\lambda_1}{\Gamma(\alpha)} \frac{(g(x) - g(y))^{\alpha+1}}{(\alpha+1)}, \tag{12.3.25}$$

$\forall \, x, y \in [a, b] : x > y$.

(ii) case of $y > x$: We have that

$$|f(x) - f(y) - (A_1(f))(x, y)(g(x) - g(y))| = \tag{12.3.26}$$

$$|f(y) - f(x) - (A_1(f))(x, y)(g(y) - g(x))| =$$

$$\left| \frac{1}{\Gamma(\alpha)} \int_x^y (g(y) - g(t))^{\alpha-1} g'(t) \left(D_{x+;g}^{\alpha} f\right)(t) dt - \right.$$

$$\left. \left(D_{x+;g}^{\alpha} f\right)(y) \frac{(g(y) - g(x))^{\alpha}}{\Gamma(\alpha+1)} \right| =$$

$$\frac{1}{\Gamma(\alpha)} \left| \int_x^y (g(y) - g(t))^{\alpha-1} g'(t) \left(\left(D_{x+;g}^{\alpha} f\right)(t) - \left(D_{x+;g}^{\alpha} f\right)(y)\right) dt \right| \le \tag{12.3.27}$$

$$\frac{1}{\Gamma(\alpha)} \int_x^y (g(y) - g(t))^{\alpha-1} g'(t) \left| \left(D_{x+;g}^{\alpha} f\right)(t) - \left(D_{x+;g}^{\alpha} f\right)(y) \right| dt \tag{12.3.28}$$

(we assume here that

$$\left| \left(D_{x+;g}^{\alpha} f\right)(t) - \left(D_{x+;g}^{\alpha} f\right)(y) \right| \le \lambda_2 |g(t) - g(y)|, \tag{12.3.29}$$

$\forall \, t, y, x \in [a, b] : y \ge t \ge x; \lambda_2 > 0$)

$$\le \frac{\lambda_2}{\Gamma(\alpha)} \int_x^y (g(y) - g(t))^{\alpha-1} g'(t) (g(y) - g(t)) dt = \tag{12.3.30}$$

$$\frac{\lambda_2}{\Gamma(\alpha)} \int_x^y (g(y) - g(t))^{\alpha} g'(t) dt = \frac{\lambda_2}{\Gamma(\alpha)} \frac{(g(y) - g(x))^{\alpha+1}}{(\alpha+1)}. \tag{12.3.31}$$

We have proved that

$$|f(x) - f(y) - (A_1(f))(x, y)(g(x) - g(y))| \leq \qquad (12.3.32)$$

$$\frac{\lambda_2}{\Gamma(\alpha)} \frac{(g(y) - g(x))^{\alpha+1}}{(\alpha + 1)}, \quad \forall\, x, y \in [a, b] : y > x.$$

Conclusion 12.8 *Set* $\lambda := \max(\lambda_1, \lambda_2)$. *Then*

$$|f(x) - f(y) - (A_1(f))(x, y)(g(x) - g(y))| \leq$$

$$\frac{\lambda}{\Gamma(\alpha)} \frac{|g(x) - g(y)|^{\alpha+1}}{(\alpha + 1)}, \quad \forall\, x, y \in [a, b]. \qquad (12.3.33)$$

Notice that (12.3.33) is trivially true when $x = y$.
One may assume that

$$\frac{\lambda}{\Gamma(\alpha)} < 1. \qquad (12.3.34)$$

Now based on (12.3.20) and (12.3.33), we can apply our numerical methods presented in this chapter to solve $f(x) = 0$.

(II) In the next background again we use [4]. We need:

Definition 12.9 Let $\alpha > 0$, $\lceil \alpha \rceil = n$, $\lceil \cdot \rceil$ the ceiling of the number. Here let $g \in AC([a, b])$ and strictly increasing. We assume that $\left(f \circ g^{-1}\right)^{(n)} \circ g \in L_\infty([a, b])$, where $f : [a, b] \to \mathbb{R}$.
   We define the right generalized g-fractional derivative of f of order α as follows:

$$\left(D_{b-;g}^\alpha f\right)(x) := \frac{(-1)^n}{\Gamma(n - \alpha)} \int_x^b (g(t) - g(x))^{n-\alpha-1} g'(t) \left(f \circ g^{-1}\right)^{(n)} (g(t))\, dt, \qquad (12.3.35)$$

$\forall\, x \in [a, b]$.
   If $\alpha \notin \mathbb{N}$, we have that $\left(D_{b-;g}^\alpha f\right) \in C([a, b])$.
   We set that

$$D_{b-;g}^n f(x) := (-1)^n \left(\left(f \circ g^{-1}\right)^{(n)} \circ g\right)(x), \qquad (12.3.36)$$

$$D_{b-;g}^0 f(x) = f(x), \quad \forall\, x \in [a, b]. \qquad (12.3.37)$$

When $g = id$, then

$$D_{b-;g}^\alpha f(x) = D_{b-;id}^\alpha f(x) = D_{b-}^\alpha f(x), \qquad (12.3.38)$$

the usual right Caputo fractional derivative.

We will use the following g-right fractional generalized Taylor's formula from [4].

Theorem 12.10 *Let g be strictly increasing function and $y \in AC([a, b])$. We assume that $(f \circ g^{-1}) \in AC^n([g(a), g(b)])$, where $\mathbb{N} \ni n = \lceil \alpha \rceil$, $\alpha > 0$. Also we assume that $(f \circ g^{-1})^{(n)} \circ g \in L_\infty([a, b])$. Then*

$$f(x) - f(b) = \sum_{k=1}^{n-1} \frac{(f \circ g^{-1})^{(k)}(g(b))}{k!} (g(x) - g(b))^k +$$

$$\frac{1}{\Gamma(\alpha)} \int_x^b (g(t) - g(x))^{\alpha-1} g'(t) \left(D^\alpha_{b-;g} f \right)(t) \, dt, \tag{12.3.39}$$

all $a \leq x \leq b$.

The remainder of (12.3.39) is a continuous function in $x \in [a, b]$.

Here we are going to operate more generally. We consider $f \in C^n([a, b])$. We define the following right g-fractional derivative of f of order α as follows:

$$\left(D^\alpha_{y-;g} f \right)(x) = \frac{(-1)^n}{\Gamma(n-\alpha)} \int_x^y (g(t) - g(x))^{n-\alpha-1} g'(t) \left(f \circ g^{-1} \right)^{(n)}(g(t)) \, dt, \tag{12.3.40}$$

$\forall x \in [a, y]$; where $y \in [a, b]$;

$$\left(D^n_{y-;g} f \right)(x) = (-1)^n \left(\left(f \circ g^{-1} \right)^{(n)} \circ g \right)(x), \quad \forall x, y \in [a, b], \tag{12.3.41}$$

$$\left(D^0_{y-;g} f \right)(x) = f(x), \quad \forall x \in [a, b]. \tag{12.3.42}$$

For $\alpha > 0$, $\alpha \notin \mathbb{N}$, by convention we set that

$$\left(D^\alpha_{y-;g} f \right)(x) = 0, \text{ for } x > y, \quad \forall x, y \in [a, b]. \tag{12.3.43}$$

Similarly, we define

$$\left(D^\alpha_{x-;g} f \right)(y) = \frac{(-1)^n}{\Gamma(n-\alpha)} \int_y^x (g(t) - g(y))^{n-\alpha-1} g'(t) \left(f \circ g^{-1} \right)^{(n)}(g(t)) \, dt, \tag{12.3.44}$$

$\forall y \in [a, x]$, where $x \in [a, b]$;

$$\left(D^n_{x-;g} f \right)(y) = (-1)^n \left(\left(f \circ g^{-1} \right)^{(n)} \circ g \right)(y), \quad \forall x, y \in [a, b], \tag{12.3.45}$$

$$\left(D^0_{x-;g} f \right)(y) = f(y), \quad \forall y \in [a, b]. \tag{12.3.46}$$

For $\alpha > 0$, $\alpha \notin \mathbb{N}$, by convention we set that

$$\left(D_{x-;g}^{\alpha} f\right)(y) = 0, \text{ for } y > x, \quad \forall\, x, y \in [a, b].$$ (12.3.47)

By assuming $\left(f \circ g^{-1}\right)^{(n)} \circ g \in L_{\infty}\left([a, b]\right)$, we get that

$$\left|\left(D_{b-;g}^{\alpha} f\right)(x)\right| \leq \frac{\left\|\left(f \circ g^{-1}\right)^{(n)} \circ g\right\|_{\infty,[a,b]}}{\Gamma\left(n - \alpha + 1\right)} \left(g\left(b\right) - g\left(x\right)\right)^{n-\alpha} \leq$$ (12.3.48)

$$\frac{\left\|\left(f \circ g^{-1}\right)^{(n)} \circ g\right\|_{\infty,[a,b]}}{\Gamma\left(n - \alpha + 1\right)} \left(g\left(b\right) - g\left(a\right)\right)^{n-\alpha}, \quad \forall\, x \in [a, b].$$

That is

$$\left(D_{b-;g}^{\alpha} f\right)(b) = 0,$$ (12.3.49)

and

$$\left(D_{y-;g}^{\alpha} f\right)(y) = \left(D_{x-;g}^{\alpha} f\right)(x) = 0, \quad \forall\, x, y \in [a, b].$$ (12.3.50)

Thus when $\alpha > 0$, $\alpha \notin \mathbb{N}$, both $D_{y-;g}^{\alpha} f$, $D_{x-;g}^{\alpha} f \in C\left([a, b]\right)$.

Notice also, that $\left(f \circ g^{-1}\right) \in AC^{n}\left([g\left(a\right), g\left(x\right)]\right)$ and $\left(f \circ g^{-1}\right)^{(n)} \circ g \in L_{\infty}\left([a, x]\right)$, and of course $g \in AC\left([a, x]\right)$, and strictly increasing over $[a, x]$, $\forall\, x \in [a, b]$.

Hence by Theorem 12.10 we obtain

$$f\left(x\right) - f\left(y\right) = \sum_{k=1}^{n-1} \frac{\left(f \circ g^{-1}\right)^{(k)}\left(g\left(y\right)\right)}{k!} \left(g\left(x\right) - g\left(y\right)\right)^{k} +$$

$$\frac{1}{\Gamma\left(\alpha\right)} \int_{x}^{y} \left(g\left(t\right) - g\left(x\right)\right)^{\alpha-1} g'\left(t\right) \left(D_{y-;g}^{\alpha} f\right)\left(t\right) dt, \quad \text{all } a \leq x \leq y \leq b.$$ (12.3.51)

Also, we have

$$f\left(y\right) - f\left(x\right) = \sum_{k=1}^{n-1} \frac{\left(f \circ g^{-1}\right)^{(k)}\left(g\left(x\right)\right)}{k!} \left(g\left(y\right) - g\left(x\right)\right)^{k} +$$

$$\frac{1}{\Gamma\left(\alpha\right)} \int_{y}^{x} \left(g\left(t\right) - g\left(y\right)\right)^{\alpha-1} g'\left(t\right) \left(D_{x-;g}^{\alpha} f\right)\left(t\right) dt, \quad \text{all } a \leq y \leq x \leq b.$$ (12.3.52)

We define also the following linear operator

$$(A_2(f))(x, y) :=$$

$$
\begin{cases}
\sum_{k=1}^{n-1} \frac{(f \circ g^{-1})^{(k)}(g(y))}{k!} (g(x) - g(y))^{k-1} - \left(D_{y-;g}^{\alpha} f\right)(x) \frac{(g(y)-g(x))^{\alpha-1}}{\Gamma(\alpha+1)}, & \text{for } x < y, \\
\sum_{k=1}^{n-1} \frac{(f \circ g^{-1})^{(k)}(g(x))}{k!} (g(y) - g(x))^{k-1} - \left(D_{x-;g}^{\alpha} f\right)(y) \frac{(g(x)-g(y))^{\alpha-1}}{\Gamma(\alpha+1)}, & \text{for } x > y, \\
f^{(n)}(x), & \text{when } x = y,
\end{cases}
$$

$$(12.3.53)$$

$\forall\, x, y \in [a, b]; \alpha > 0, n = \lceil \alpha \rceil$.

We may assume that

$$|(A_2(f))(x, x) - (A_2(f))(y, y)| = \left| f^{(n)}(x) - f^{(n)}(y) \right| \qquad (12.3.54)$$

$$\leq \Phi^* |g(x) - g(y)|, \quad \forall\, x, y \in [a, b];$$

where $\Phi^* > 0$.

We estimate and have

(i) case of $x < y$:

$$|f(x) - f(y) - (A_2(f))(x, y)(g(x) - g(y))| =$$

$$\left| \frac{1}{\Gamma(\alpha)} \int_x^y (g(t) - g(x))^{\alpha-1} g'(t) \left(D_{y-;g}^{\alpha} f\right)(t)\, dt - \right.$$

$$\left. \left(D_{y-;g}^{\alpha} f\right)(x) \frac{(g(y) - g(x))^{\alpha}}{\Gamma(\alpha+1)} \right| = \qquad (12.3.55)$$

$$\frac{1}{\Gamma(\alpha)} \left| \int_x^y (g(t) - g(x))^{\alpha-1} g'(t) \left(\left(D_{y-;g}^{\alpha} f\right)(t) - \left(D_{y-;g}^{\alpha} f\right)(x)\right) dt \right| \leq$$

$$\frac{1}{\Gamma(\alpha)} \int_x^y (g(t) - g(x))^{\alpha-1} g'(t) \left| \left(D_{y-;g}^{\alpha} f\right)(t) - \left(D_{y-;g}^{\alpha} f\right)(x) \right| dt \qquad (12.3.56)$$

(we assume that

$$\left| \left(D_{y-;g}^{\alpha} f\right)(t) - \left(D_{y-;g}^{\alpha} f\right)(x) \right| \leq \rho_1 |g(t) - g(x)|, \qquad (12.3.57)$$

$\forall\, t, x, y \in [a, b] : y \geq t \geq x; \rho_1 > 0)$

$$\leq \frac{\rho_1}{\Gamma(\alpha)} \int_x^y (g(t) - g(x))^{\alpha-1} g'(t) (g(t) - g(x))\, dt =$$

$$\frac{\rho_1}{\Gamma(\alpha)} \int_x^y (g(t) - g(x))^{\alpha} g'(t)\, dt = \frac{\rho_1}{\Gamma(\alpha)} \frac{(g(y) - g(x))^{\alpha+1}}{(\alpha+1)}. \qquad (12.3.58)$$

We have proved that

$$|f(x) - f(y) - (A_2(f))(x, y)(g(x) - g(y))| \leq$$

$$\frac{\rho_1}{\Gamma(\alpha)} \frac{(g(y) - g(x))^{\alpha+1}}{(\alpha+1)}, \tag{12.3.59}$$

$\forall\, x, y \in [a, b] : x < y$.
  (ii) case of $x > y$:

$$|f(x) - f(y) - (A_2(f))(x, y)(g(x) - g(y))| =$$

$$|f(y) - f(x) - (A_2(f))(x, y)(g(y) - g(x))| = \tag{12.3.60}$$

$$|f(y) - f(x) + (A_2(f))(x, y)(g(x) - g(y))| =$$

$$\left| \frac{1}{\Gamma(\alpha)} \int_y^x (g(t) - g(y))^{\alpha-1} g'(t) \left(D_{x-;g}^{\alpha} f\right)(t)\, dt - \right.$$

$$\left. \left(D_{x-;g}^{\alpha} f\right)(y) \frac{(g(x) - g(y))^{\alpha}}{\Gamma(\alpha+1)} \right| =$$

$$\frac{1}{\Gamma(\alpha)} \left| \int_y^x (g(t) - g(y))^{\alpha-1} g'(t) \left(\left(D_{x-;g}^{\alpha} f\right)(t) - \left(D_{x-;g}^{\alpha} f\right)(y)\right) dt \right| \leq$$
$$\tag{12.3.61}$$

$$\frac{1}{\Gamma(\alpha)} \int_y^x (g(t) - g(y))^{\alpha-1} g'(t) \left|\left(D_{x-;g}^{\alpha} f\right)(t) - \left(D_{x-;g}^{\alpha} f\right)(y)\right| dt \tag{12.3.62}$$

(we assume that

$$\left|\left(D_{x-;g}^{\alpha} f\right)(t) - \left(D_{x-;g}^{\alpha} f\right)(y)\right| \leq \rho_2 \,|g(t) - g(y)|, \tag{12.3.63}$$

$\forall\, t, y, x \in [a, b] : x \geq t \geq y;\, \rho_2 > 0$)

$$\leq \frac{\rho_2}{\Gamma(\alpha)} \int_y^x (g(t) - g(y))^{\alpha-1} g'(t) (g(t) - g(y))\, dt =$$

$$\frac{\rho_2}{\Gamma(\alpha)} \int_y^x (g(t) - g(y))^{\alpha} g'(t)\, dt = \tag{12.3.64}$$

$$\frac{\rho_2}{\Gamma(\alpha)} \frac{(g(x) - g(y))^{\alpha+1}}{(\alpha+1)}. \tag{12.3.65}$$

We have proved that

$$|f(x) - f(y) - (A_2(f))(x,y)(g(x) - g(y))| \le$$

$$\frac{\rho_2}{\Gamma(\alpha)} \frac{(g(x) - g(y))^{\alpha+1}}{(\alpha+1)}, \quad \forall\, x, y \in [a, b] : x > y. \tag{12.3.66}$$

Conclusion 12.11 *Set $\rho := \max(\rho_1, \rho_2)$. Then*

$$|f(x) - f(y) - (A_2(f))(x,y)(g(x) - g(y))| \le$$

$$\frac{\rho}{\Gamma(\alpha)} \frac{|g(x) - g(y)|^{\alpha+1}}{(\alpha+1)}, \quad \forall\, x, y \in [a, b]. \tag{12.3.67}$$

Notice that (12.3.67) is trivially true when $x = y$.
One may assume that

$$\frac{\rho}{\Gamma(\alpha)} < 1. \tag{12.3.68}$$

Now based on (12.3.54) and (12.3.67), we can apply our numerical methods presented in this chapter to solve $f(x) = 0$.

In both fractional applications $\alpha + 1 \ge 2$, iff $\alpha \ge 1$.

Also some examples for g follow:

$$g(x) = e^x, \ x \in [a, b] \subset \mathbb{R},$$
$$g(x) = \sin x,$$
$$g(x) = \tan x, \tag{12.3.69}$$
$$\text{where } x \in \left[-\frac{\pi}{2} + \varepsilon, \frac{\pi}{2} - \varepsilon\right], \text{ where } \varepsilon > 0 \text{ small.}$$

Indeed, the above examples of g are strictly increasing and absolutely continuous functions.

Remark 12.12 (a) Returning back to Conclusion 12.8, we see that Proposition 12.2 can be applied, if $p = \alpha$, $g(t) = t$, $F(t) = f(t)$, $A(F)(s,t) = A_1(f)(s,t)$ and

$$\varphi_1(s,t) = \frac{\lambda |s - t|^p}{(\alpha+1)\Gamma(\alpha)\mu}$$

for each $s, t \in [a, b]$.

(b) According to Conclusion 12.11, as in (a) but we must choose $A(F)(s,t) = A_2(f)(s,t)$ and

$$\varphi_1(s,t) = \frac{\rho |s - t|^p}{(\alpha+1)\Gamma(\alpha)\mu}$$

for each $s, t \in [a, b]$.

References

1. S. Amat, S. Busquier, Third-order iterative methods under Kantorovich conditions. J. Math. Anal. Appl. **336**, 243–261 (2007)
2. S. Amat, S. Busquier, S. Plaza, Chaotic dynamics of a third-order Newton-like method. J. Math. Anal. Appl. **366**(1), 164–174 (2010)
3. G. Anastassiou, Fractional representation formulae and right fractional inequalities. Math. Comput. Model. **54**(10–12), 3098–3115 (2011)
4. G. Anastassiou, Advanced fractional Taylor's formulae. J. Comput. Anal. Appl. (2016) (to appear)
5. G. Anastassiou, I. Argyros, *On the Convergence of Secant-Like Algorithms with Applications to Generalized Fractional Calculus* (2015)
6. I.K. Argyros, Newton-like methods in partially ordered linear spaces. J. Approx. Th. Appl. **9**(1), 1–10 (1993)
7. I.K. Argyros, Results on controlling the residuals of perturbed Newton-like methods on Banach spaces with a convergence structure. Southwest J. Pure Appl. Math. **1**, 32–38 (1995)
8. I.K. Argyros, *Convergence and Applications of Newton-Like Iterations* (Springer, New York, 2008)
9. K. Diethelm, in *The Analysis of Fractional Differential Equations, Lecture Notes in Mathematics*, vol. 2004, 1st edn. (Springer, New York, Heidelberg, 2010)
10. J.A. Ezquerro, M.A. Hernandez, Newton-like methods of high order and domains of semilocal and global convergence. Appl. Math. Comput. **214**(1), 142–154 (2009)
11. J.A. Ezquerro, J.M. Gutierrez, M.A. Hernandez, N. Romero, M.J. Rubio, The Newton method: from Newton to Kantorovich (Spanish). Gac. R. Soc. Mat. Esp. **13**, 53–76 (2010)
12. L.V. Kantorovich, G.P. Akilov, *Functional Analysis in Normed Spaces* (Pergamon Press, New York, 1964)
13. A.A. Magrenan, Different anomalies in a Jarratt family of iterative root finding methods. Appl. Math. Comput. **233**, 29–38 (2014)
14. A.A. Magrenan, A new tool to study real dynamics: the convergence plane. Appl. Math. Comput. **248**, 215–224 (2014)
15. F.A. Potra, V. Ptak, *Nondiscrete Induction and Iterative Processes* (Pitman Publishing, London, 1984)
16. P.D. Proinov, New general convergence theory for iterative processes and its applications to Newton-Kantorovich type theorems. J. Complex **26**, 3–42 (2010)

Chapter 13
Secant-Like Methods and Generalized g-Fractional Calculus of Canavati-Type

We present local and semilocal convergence results for secant-like methods in order to approximate a locally unique solution of a nonlinear equation in a Banach space setting. Finally, we present some applications from generalized g-fractional calculus involving Canavati-type functions. It follows [5].

13.1 Introduction

In this study we are concerned with the problem of approximating a locally unique solution x^* of the nonlinear equation

$$F(x) = 0, \tag{13.1.1}$$

where F is a continuous operator defined on a subset D of a Banach space X with values in a Banach space Y.

A lot of problems in Computational Sciences and other disciplines can be brought in a form like (13.1.1) using Mathematical Modelling [8, 12, 16]. The solutions of such equations can be found in closed form only in special cases. That is why most solution methods for these equations are iterative. Iterative methods are usually studied based on semilocal and local convergence. The semilocal convergence matter is, based on the information around the initial point to give hypotheses ensuring the convergence of the iterative algorithm; while the local one is, based on the information around a solution, to find estimates of the radii of convergence balls as well as error bounds on the distances involved.

We introduce the secant-like method defined for each $n = 0, 1, 2, \ldots$ by

$$x_{n+1} = x_n - A(x_n, x_{n-1})^{-1} F(x_n), \tag{13.1.2}$$

© Springer International Publishing Switzerland 2016

G.A. Anastassiou and I.K. Argyros, *Intelligent Numerical Methods:*
Applications to Fractional Calculus, Studies in Computational Intelligence 624,
DOI 10.1007/978-3-319-26721-0_13

where $x_{-1}, x_0 \in D$ are initial points and $A(x, y) \in L(X, Y)$ the space of bounded linear operators from X into Y. There is a plethora on local as well as semilocal convergence theorems for method (13.1.2) provided that the operator A is an approximation to the Fréchet-derivative F' [1, 2, 6–16]. In the present study we do not necessarily assume that operator A is related to F'. This way we expand the applicability of iterative algorithm (13.1.2). Notice that many well known methods are special case of method (13.1.2).

Newton's method: Choose $A(x, x) = F'(x)$ for each $x \in D$.

Secant method: Choose $A(x, y) = [x, y; F]$, $where$ $[x, y; F]$ denotes a divided difference of order one [8, 12, 15].

The so called Newton-like methods and many other methods are special cases of method (13.1.2).

The rest of the chapter is organized as follows. The semilocal as well as the local convergence analysis of method (13.1.2) is given in Sect. 13.2. Some applications from fractional calculus are given in Sect. 13.3.

13.2 Convergence Analysis

We present the main semilocal convergence result for method (13.1.2).

Theorem 13.1 *Let* $F : D \subset X \to Y$ *be a continuous operator and let* $A(x, y) \in L(X, Y)$. *Suppose that there exist* $x_{-1}, x_0 \in D$, $\eta \geq 0$, $p \geq 1$, *a function* $g : [0, \infty)^2 \to [0, \infty)$ *continuous and nondecreasing such that for each* $x, y, z \in D$

$$A(z, y)^{-1} \in L(Y, X), \tag{13.2.1}$$

$$\max \left\{ \|x_{-1} - x_0\|, \left\| A(x_0, x_{-1})^{-1} F(x_0) \right\| \right\} \leq \eta, \tag{13.2.2}$$

$$\left\| A(z, y)^{-1} (F(z) - F(y) - A(y, x)(z - y)) \right\| \leq$$

$$g(\|z - y\|, \|y - x\|) \|z - y\|^{p+1}, \tag{13.2.3}$$

$$q := g(\eta, \eta) \eta^p < 1 \tag{13.2.4}$$

and

$$\overline{U}(x_0, r) \subseteq D, \tag{13.2.5}$$

where,

$$r = \frac{\eta}{1 - q}. \tag{13.2.6}$$

Then, the sequence $\{x_n\}$ generated by method (13.1.2) is well defined, remains in $\overline{U}(x_0, r)$ for each $n = 0, 1, 2, \ldots$ and converges to some $x^* \in \overline{U}(x_0, r)$ such that

$$\|x_{n+1} - x_n\| \leq g(\|x_n - x_{n-1}\|, \|x_{n-1} - x_{n-2}\|) \|x_n - x_{n-1}\|^{p+1}$$

$$\leq q \|x_n - x_{n-1}\| \tag{13.2.7}$$

and

$$\|x_n - x^*\| \leq \frac{q^n \eta}{1 - q}. \tag{13.2.8}$$

Proof The iterate x_1 is well defined by method (13.1.2) for $n = 0$ and (13.2.1). We also have by (13.2.2) and (13.2.6) that
$\|x_1 - x_0\| = \|A(x_0, x_{-1})^{-1} F(x_0)\| \leq \eta < r$, so we get that $x_1 \in \overline{U}(x_0, r)$ and x_2 is well defined (by (13.2.5)). Using (13.2.3) and (13.2.4) we get that

$$\|x_2 - x_1\| = \|A(x_1, x_0)^{-1} [F(x_1) - F(x_0) - A(x_0, x_{-1})(x_1 - x_0)]\|$$

$$\leq g(\|x_1 - x_0\|, \|x_0 - x_{-1}\|) \|x_1 - x_0\|^{p+1} \leq q \|x_1 - x_0\|,$$

which shows (13.2.7) for $n = 1$. Then, we can have that

$$\|x_2 - x_0\| \leq \|x_2 - x_1\| + \|x_1 - x_0\| \leq q \|x_1 - x_0\| + \|x_1 - x_0\|$$

$$= (1 + q) \|x_1 - x_0\| \leq \frac{1 - q^2}{1 - q} \eta < r,$$

so $x_2 \in \overline{U}(x_0, r)$ and x_3 is well defined.

Assuming $\|x_{k+1} - x_k\| \leq q \|x_k - x_{k-1}\|$ and $x_{k+1} \in \overline{U}(x_0, r)$ for each $k = 1, 2, \ldots, n$ we get

$$\|x_{k+2} - x_{k+1}\| =$$

$$\|A(x_{k+1}, x_k)^{-1} [F(x_{k+1}) - F(x_k) - A(x_k, x_{k-1})(x_{k+1} - x_k)]\| \leq$$

$$g(\|x_{k+1} - x_k\|, \|x_k - x_{k-1}\|) \|x_{k+1} - x_k\|^{p+1} \leq$$

$$g(\|x_1 - x_0\|, \|x_0 - x_{-1}\|) \|x_1 - x_0\|^p \|x_{k+1} - x_k\| \leq q \|x_{k+1} - x_k\|$$

and

$$\|x_{k+2} - x_0\| \leq \|x_{k+2} - x_{k+1}\| + \|x_{k+1} - x_k\| + \cdots + \|x_1 - x_0\|$$

$$\leq (q^{k+1} + q^k + \cdots + 1) \|x_1 - x_0\| \leq \frac{1 - q^{k+2}}{1 - q} \|x_1 - x_0\|$$

$$< \frac{\eta}{1 - q} = r,$$

which completes the induction for (13.2.7) and $x_{k+2} \in \overline{U}(x_0, r)$. We also have that for $m \geq 0$

$$\|x_{n+m} - x_n\| \leq \|x_{n+m} - x_{n+m-1}\| + \cdots + \|x_{n+1} - x_n\|$$

$$\leq \left(q^{m-1} + q^{m-2} + \cdots + 1\right) \|x_{n+1} - x_n\|$$

$$\leq \frac{1 - q^m}{1 - q} q^n \|x_1 - x_0\|.$$

It follows that $\{x_n\}$ is a complete sequence in a Banach space X and as such it converges to some $x^* \in \overline{U}(x_0, r)$ (since $\overline{U}(x_0, r)$ is a closed set). By letting $m \to \infty$, we obtain (13.2.8). $\qquad\square$

Stronger hypotheses are needed to show that x^* is a solution of equation $F(x) = 0$.

Proposition 13.2 *Let $F : D \subset X \to Y$ be a continuous operator and let $A(x, y) \in L(X, Y)$. Suppose that there exist $x_{-1}, x_0 \in D$, $\eta \geq 0$, $p \geq 1$, $\mu > 0$, a function $g_1 : [0, \infty)^2 \to [0, \infty)$ continuous and nondecreasing such that for each $x, y \in D$*

$$A(x, y)^{-1} \in L(Y, X), \quad \left\|A(x, y)^{-1}\right\| \leq \mu,$$

$$\max\left\{\|x_{-1} - x_0\|, \left\|A(x_0, x_{-1})^{-1} F(x_0)\right\|\right\} \leq \eta, \qquad (13.2.9)$$

$$\|F(z) - F(y) - A(y, x)(z - y)\| \leq \frac{g_1(\|z - y\|, \|x - y\|)}{\mu} \|z - y\|^{p+1},$$
$$\qquad (13.2.10)$$

$$q_1 := g_1(\eta, \eta) \eta^p < 1$$

and

$$\overline{U}(x_0, r_1) \subseteq D,$$

where,

$$r_1 = \frac{\eta}{1 - q_1}.$$

Then, the conclusions of Theorem 13.1 for sequence $\{x_n\}$ hold with g_1, q_1, r_1, replacing g, q and r, respectively. Moreover, x^* is a solution of the equation $F(x) = 0$.

Proof Notice that

$$\left\|A(x_n, x_{n-1})^{-1} \left[F(x_n) - F(x_{n-1}) - A(x_{n-1}, x_{n-2})(x_n - x_{n-1})\right]\right\|$$

$$\leq \left\|A(x_n, x_{n-1})^{-1}\right\| \|F(x_n) - F(x_{n-1}) - A(x_{n-1}, x_{n-2})(x_n - x_{n-1})\|$$

$$\leq g_1(\|x_n - x_{n-1}\|, \|x_{n-1} - x_{n-2}\|) \|x_n - x_{n-1}\|^{p+1} \leq q_1 \|x_n - x_{n-1}\|.$$

Therefore, the proof of Theorem 13.1 can apply. Then, in view of the estimate

$$\|F(x_n)\| = \|F(x_n) - F(x_{n-1}) - A(x_{n-1}, x_{n-2})(x_n - x_{n-1})\| \le$$

$$\frac{g_1(\|x_n - x_{n-1}\|)}{\mu}\|x_n - x_{n-1}\|^{p+1} \le q_1\|x_n - x_{n-1}\|,$$

we deduce by letting $n \to \infty$ that $F(x^*) = 0$. $\qquad\qquad\square$

Concerning the uniqueness of the solution x^* we have the following result:

Proposition 13.3 *Under the hypotheses of Proposition 13.2, further suppose that there exists $g_2 : [0, \infty)^2 \to [0, \infty)$ continuous and nondecreasing such that*

$$\|F(z) - F(x) - A(z, y)(z - x)\| \le \frac{g_2(\|z - x\|, \|y - x\|)}{\mu}\|z - x\|^{p+1}$$
$$(13.2.11)$$

and

$$g_2(r_1, \eta + r_1)r_1^p < 1. \qquad\qquad (13.2.12)$$

Then, x^* is the only solution of equation $F(x) = 0$ in $\overline{U}(x_0, r_1)$.

Proof The existence of the solution $x^* \in \overline{U}(x_0, r_1)$ has been established in Proposition 13.2. Let $y^* \in \overline{U}(x_0, r_1)$ with $F(y^*) = 0$. Then, we have in turn that

$$\|x_{n+1} - y^*\| = \|x_n - y^* - A(x_n, x_{n-1})^{-1}F(x_n)\| =$$

$$\|A(x_n, x_{n-1})^{-1}[A(x_n, x_{n-1})(x_n - y^*) - F(x_n) + F(y^*)]\| \le$$

$$\|A(x_n, x_{n-1})^{-1}\|\|F(y^*) - F(x_n) - A(x_n, x_{n-1})(y^* - x_n)\| \le$$

$$\mu\frac{g_1(\|x_n - y^*\|, \|x_{n-1} - y^*\|)}{\mu}\|x_n - y^*\|^{p+1} \le$$

$$g_2(r_1, \eta + r_1)r_1^p\|x_n - x^*\| < \|x_n - y^*\|,$$

so we deduce that $\lim_{n \to \infty} x_n = y^*$. But we have that $\lim_{n \to \infty} x_n = x^*$. Hence, we conclude that $x^* = y^*$. $\qquad\qquad\square$

Next, we present a local convergence analysis for the iterative algorithm (13.1.2).

Proposition 13.4 *Let $F : D \subset X \to Y$ be a continuous operator and let $A(y, x) \in L(X, Y)$. Suppose that there exist $x^* \in D$, $p \ge 1$, a function $g_3 : [0, \infty)^2 \to [0, \infty)$ continuous and nondecreasing such that for each $x, y \in D$*

$$F(x^*) = 0, \quad A(y,x)^{-1} \in L(Y,X),$$

$$\left\| A(y,x)^{-1} \left[F(x) - F(x^*) - A(y,x)(x - x^*) \right] \right\| \le$$

$$g_3\left(\|y - x^*\|, \|x - x^*\| \right) \|y - x^*\|^{p+1}, \tag{13.2.13}$$

and

$$\overline{U}(x^*, r_2) \subseteq D,$$

where r_2 is the smallest positive solution of equation

$$h(t) := g_3(t,t) t^p - 1.$$

Then, sequence $\{x_n\}$ generated by method (13.1.2) for $x_{-1}, x_0 \in U(x^*, r_2) - \{x^*\}$ is well defined, remains in $U(x^*, r_2)$ for each $n = 0, 1, 2, \ldots$ and converges to x^*. Moreover, the following estimates hold

$$\|x_{n+1} - x^*\| \le g_2\left(\|x_n - x^*\|, \|x_{n-1} - x^*\| \right) \|x_n - x^*\|^{p+1} < \|x_n - x^*\| < r_2.$$

Proof We have that $h(0) = -1 < 0$ and $h(t) \to +\infty$ as $t \to +\infty$. Then, it follows from the intermediate value theorem that function h has positive zeros. Denote by r_2 the smallest such zero. By hypothesis $x_{-1}, x_0 \in U(x^*, r_2) - \{x^*\}$. Then, we get in turn that

$$\|x_1 - x^*\| = \|x_0 - x^* - A(x_0, x_{-1})^{-1} F(x_0)\| =$$

$$\left\| A(x_0, x_{-1})^{-1} \left[F(x^*) - F(x_0) - A(x_0, x_{-1})(x^* - x_0) \right] \right\| \le$$

$$g_2\left(\|x_0 - x^*\|, \|x_{-1} - x^*\| \right) \|x_0 - x^*\|^{p+1} < g_3(r_2, r_2) r_2^p \|x_0 - x^*\| =$$

$$\|x_0 - x^*\| < r_2,$$

which shows that $x_1 \in U(x^*, r_2)$ and x_2 is well defined. By a simple inductive argument as in the preceding estimate we get that

$$\|x_{k+1} - x^*\| = \|x_k - x^* - A(x_k, x_{k-1})^{-1} F(x_k)\| \le$$

$$\left\| A(x_k, x_{k-1})^{-1} \left[F(x^*) - F(x_k) - A(x_k, x_{k-1})(x^* - x_k) \right] \right\| \le$$

$$g_2\left(\|x_k - x^*\|, \|x_{k-1} - x^*\| \right) \|x_k - x^*\|^{p+1} <$$

$$g_3(r_2, r_2) r_2^p \|x_k - x^*\| = \|x_k - x^*\| < r_2,$$

which shows $\lim_{k \to \infty} x_k = x^*$ and $x_{k+1} \in U(x^*, r_2)$. $\qquad\square$

Remark 13.5 (a) Hypothesis (13.2.3) specializes to Newton-Mysowski-type, if $A(x) = F'(x)$ [8, 12, 15]. However, if F is not Fréchet-differentiable, then our results extend the applicability of iterative algorithm (13.1.2).

(b) Theorem 13.1 has practical value although we do not show that x^* is a solution of equation $F(x) = 0$, since this may be shown in another way.

(c) Hypothesis (13.2.13) can be replaced by the stronger

$$\left\| A(y,x)^{-1} \left[F(x) - F(z) - A(y,x)(x-z) \right] \right\| \le$$

$$g_3 \left(\|z-y\|, \|z-x\| \right) \|z-y\|^{p+1}.$$

13.3 Applications to g-Fractional Calculus of Canavati Type

Here both needed backgrounds come from [4] and Chap. 25.

Let $\nu > 1$, $\nu \notin \mathbb{N}$, with integral part $[\nu] = n \in \mathbb{N}$. Let $g : [a,b] \to \mathbb{R}$ be a strictly increasing function, such that $g \in C^1([a,b])$, $g^{-1} \in C^n([a,b])$, and let $f \in C^n([a,b])$. It clear then we obtain that $(f \circ g^{-1}) \in C^n([g(a), g(b)])$. Let $\alpha := \nu - [\nu] = \nu - n \ (0 < \alpha < 1)$.

(I) Let $h \in C([g(a), g(b)])$, we define the left Riemann-Liouville fractional integral as

$$\left(J_\nu^{z_0} h \right)(z) := \frac{1}{\Gamma(\nu)} \int_{z_0}^{z} (z-t)^{\nu-1} h(t) \, dt, \tag{13.3.1}$$

for $g(a) \le z_0 \le z \le g(b)$, where Γ is the gamma function.

We define the subspace $C_{g(x)}^\nu([g(a), g(b)])$ of $C^n([g(a), g(b)])$, where $x \in [a,b]$:

$$C_{g(x)}^\nu([g(a), g(b)]) :=$$

$$\left\{ h \in C^n([g(a), g(b)]) : J_{1-\alpha}^{g(x)} h^{(n)} \in C^1([g(x), g(b)]) \right\}. \tag{13.3.2}$$

So let $h \in C_{g(x)}^\nu([g(a), g(b)])$; we define the left g-generalized fractional derivative of h of order ν, of Canavati type, over $[g(x), g(b)]$ as

$$D_{g(x)}^\nu h := \left(J_{1-\alpha}^{g(x)} h^{(n)} \right)'. \tag{13.3.3}$$

Clearly, for $h \in C_{g(x)}^\nu([g(a), g(b)])$, there exists

$$\left(D_{g(x)}^\nu h \right)(z) = \frac{1}{\Gamma(1-\alpha)} \frac{d}{dz} \int_{g(x)}^{z} (z-t)^{-\alpha} h^{(n)}(t) \, dt, \tag{13.3.4}$$

for all $g(x) \le z \le g(b)$.

In particular, when $f \circ g^{-1} \in C_{g(x)}^{\nu}([g(a), g(b)])$ we have that

$$
\left(D_{g(x)}^{\nu}\left(f \circ g^{-1}\right)\right)(z) = \frac{1}{\Gamma(1-\alpha)} \frac{d}{dz} \int_{g(x)}^{z} (z-t)^{-\alpha} \left(f \circ g^{-1}\right)^{(n)}(t) \, dt,
$$

(13.3.5)

for all $z : g(x) \le z \le g(b)$.

We have that $D_{g(x)}^{n}\left(f \circ g^{-1}\right) = \left(f \circ g^{-1}\right)^{(n)}$ and $D_{g(x)}^{0}\left(f \circ g^{-1}\right) = f \circ g^{-1}$.

In [4] we proved for $\left(f \circ g^{-1}\right) \in C_{g(x)}^{\nu}([g(a), g(b)])$, where $x \in [a, b]$, (left fractional Taylor's formula) that

$$
f(y) - f(x) = \sum_{k=1}^{n-1} \frac{\left(f \circ g^{-1}\right)^{(k)}(g(x))}{k!} (g(y) - g(x))^{k} +
$$

(13.3.6)

$$
\frac{1}{\Gamma(\nu)} \int_{g(x)}^{g(y)} (g(y) - t)^{\nu-1} \left(D_{g(x)}^{\nu}\left(f \circ g^{-1}\right)\right)(t) \, dt, \quad \text{for all } y \in [a, b] : y \ge x.
$$

Alternatively, for $\left(f \circ g^{-1}\right) \in C_{g(y)}^{\nu}([g(a), g(b)])$, where $y \in [a, b]$, we can write (again left fractional Taylor's formula) that:

$$
f(x) - f(y) = \sum_{k=1}^{n-1} \frac{\left(f \circ g^{-1}\right)^{(k)}(g(y))}{k!} (g(x) - g(y))^{k} +
$$

(13.3.7)

$$
\frac{1}{\Gamma(\nu)} \int_{g(y)}^{g(x)} (g(x) - t)^{\nu-1} \left(D_{g(y)}^{\nu}\left(f \circ g^{-1}\right)\right)(t) \, dt, \quad \text{for all } x \in [a, b] : x \ge y.
$$

Here we consider $f \in C^{n}([a, b])$, such that $\left(f \circ g^{-1}\right) \in C_{g(x)}^{\nu}([g(a), g(b)])$, for every $x \in [a, b]$; which is the same as $\left(f \circ g^{-1}\right) \in C_{g(y)}^{\nu}([g(a), g(b)])$, for every $y \in [a, b]$ (i.e. exchange roles of x and y); we write that as $\left(f \circ g^{-1}\right) \in C_{g+}^{\nu}([g(a), g(b)])$.

We have that

$$
\left(D_{g(y)}^{\nu}\left(f \circ g^{-1}\right)\right)(z) = \frac{1}{\Gamma(1-\alpha)} \frac{d}{dz} \int_{g(y)}^{z} (z-t)^{-\alpha} \left(f \circ g^{-1}\right)^{(n)}(t) \, dt,
$$

(13.3.8)

for all $z : g(y) \le z \le g(b)$.

So here we work with $f \in C^{n}([a, b])$, such that $\left(f \circ g^{-1}\right) \in C_{g+}^{\nu}([g(a), g(b)])$.

We define the left linear fractional operator

$$
(A_1(f))(x,y) := \begin{cases}
\sum_{k=1}^{n-1} \frac{(f\circ g^{-1})^{(k)}(g(x))}{k!}(g(y)-g(x))^{k-1} + \\
\left(D_{g(x)}^{\nu}(f\circ g^{-1})\right)(g(y))\frac{(g(y)-g(x))^{\nu-1}}{\Gamma(\nu+1)}, \quad y > x, \\[2mm]
\sum_{k=1}^{n-1} \frac{(f\circ g^{-1})^{(k)}(g(y))}{k!}(g(x)-g(y))^{k-1} + \\
\left(D_{g(y)}^{\nu}(f\circ g^{-1})\right)(g(x))\frac{(g(x)-g(y))^{\nu-1}}{\Gamma(\nu+1)}, \quad x > y, \\[2mm]
f^{(n)}(x), \quad x = y.
\end{cases}
\tag{13.3.9}
$$

We may assume that

$$
|(A_1(f))(x,x) - (A_1(f))(y,y)| = \left|f^{(n)}(x) - f^{(n)}(y)\right| =
$$

$$
\left|\left(f^{(n)}\circ g^{-1}\right)(g(x)) - \left(f^{(n)}\circ g^{-1}\right)(g(y))\right| \le \Phi |g(x) - g(y)|,
\tag{13.3.10}
$$

where $\Phi > 0$; for any $x, y \in [a,b]$.

We make the following estimations:

(i) case of $y > x$: We have that

$$
|f(y) - f(x) - (A_1(f))(x,y)(g(y) - g(x))| =
$$

$$
\left|\frac{1}{\Gamma(\nu)}\int_{g(x)}^{g(y)}(g(y)-t)^{\nu-1}\left(D_{g(x)}^{\nu}(f\circ g^{-1})\right)(t)\,dt - \right.
$$

$$
\left.\left(D_{g(x)}^{\nu}(f\circ g^{-1})\right)(g(y))\frac{(g(y)-g(x))^{\nu}}{\Gamma(\nu+1)}\right| = \frac{1}{\Gamma(\nu)} \cdot
$$

$$
\left|\int_{g(x)}^{g(y)}(g(y)-t)^{\nu-1}\left(\left(D_{g(x)}^{\nu}(f\circ g^{-1})\right)(t) - \left(D_{g(x)}^{\nu}(f\circ g^{-1})\right)(g(y))\right)dt\right|
\tag{13.3.11}
$$

$$
\le \frac{1}{\Gamma(\nu)} \cdot
$$

$$
\int_{g(x)}^{g(y)}(g(y)-t)^{\nu-1}\left|\left(D_{g(x)}^{\nu}(f\circ g^{-1})\right)(t) - \left(D_{g(x)}^{\nu}(f\circ g^{-1})\right)(g(y))\right|dt
$$

(we assume here that

$$
\left|\left(D_{g(x)}^{\nu}(f\circ g^{-1})\right)(t) - \left(D_{g(x)}^{\nu}(f\circ g^{-1})\right)(g(y))\right| \le \lambda_1 |t - g(y)|,
\tag{13.3.12}
$$

for every $t, g(y), g(x) \in [g(a), g(b)]$ such that $g(y) \ge t \ge g(x)$; $\lambda_1 > 0$)

$$\leq \frac{\lambda_1}{\Gamma(\nu)} \int_{g(x)}^{g(y)} (g(y) - t)^{\nu-1} (g(y) - t) \, dt = \tag{13.3.13}$$

$$\frac{\lambda_1}{\Gamma(\nu)} \int_{g(x)}^{g(y)} (g(y) - t)^{\nu} \, dt = \frac{\lambda_1}{\Gamma(\nu)} \frac{(g(y) - g(x))^{\nu+1}}{(\nu+1)}. \tag{13.3.14}$$

We have proved that

$$|f(y) - f(x) - (A_1(f))(x, y)(g(y) - g(x))| \leq \frac{\lambda_1}{\Gamma(\nu)} \frac{(g(y) - g(x))^{\nu+1}}{(\nu+1)}, \tag{13.3.15}$$

for all $x, y \in [a, b] : y > x$.

(ii) Case of $x > y$: We observe that

$$|f(y) - f(x) - (A_1(f))(x, y)(g(y) - g(x))| =$$

$$|f(x) - f(y) - (A_1(f))(x, y)(g(x) - g(y))| =$$

$$\left| \frac{1}{\Gamma(\nu)} \int_{g(y)}^{g(x)} (g(x) - t)^{\nu-1} \left(D^{\nu}_{g(y)} \left(f \circ g^{-1} \right) \right)(t) \, dt - \right.$$

$$\left. \left(D^{\nu}_{g(y)} \left(f \circ g^{-1} \right) \right)(g(x)) \frac{(g(x) - g(y))^{\nu}}{\Gamma(\nu+1)} \right| = \frac{1}{\Gamma(\nu)}. \tag{13.3.16}$$

$$\left| \int_{g(y)}^{g(x)} (g(x) - t)^{\nu-1} \left(\left(D^{\nu}_{g(y)} \left(f \circ g^{-1} \right) \right)(t) - \left(D^{\nu}_{g(y)} \left(f \circ g^{-1} \right) \right)(g(x)) \right) dt \right|$$

$$\leq \frac{1}{\Gamma(\nu)}.$$

$$\int_{g(y)}^{g(x)} (g(x) - t)^{\nu-1} \left| \left(D^{\nu}_{g(y)} \left(f \circ g^{-1} \right) \right)(t) - \left(D^{\nu}_{g(y)} \left(f \circ g^{-1} \right) \right)(g(x)) \right| dt \tag{13.3.17}$$

(we assume that

$$\left| \left(D^{\nu}_{g(y)} \left(f \circ g^{-1} \right) \right)(t) - \left(D^{\nu}_{g(y)} \left(f \circ g^{-1} \right) \right)(g(x)) \right| \leq \lambda_2 |t - g(x)|, \tag{13.3.18}$$

for all $t, g(x), g(y) \in [g(a), g(b)]$ such that $g(x) \geq t \geq g(y)$; $\lambda_2 > 0$)

$$\leq \frac{\lambda_2}{\Gamma(\nu)} \int_{g(y)}^{g(x)} (g(x) - t)^{\nu-1} (g(x) - t) \, dt = \tag{13.3.19}$$

$$\frac{\lambda_2}{\Gamma(\nu)} \int_{g(y)}^{g(x)} (g(x) - t)^\nu \, dt = \frac{\lambda_2}{\Gamma(\nu)} \frac{(g(x) - g(y))^{\nu+1}}{(\nu + 1)}.$$

We have proved that

$$|f(y) - f(x) - (A_1(f))(x, y)(g(y) - g(x))| \leq \frac{\lambda_2}{\Gamma(\nu)} \frac{(g(x) - g(y))^{\nu+1}}{(\nu + 1)},$$
$$\tag{13.3.20}$$

for any $x, y \in [a, b] : x > y$.

Conclusion 13.6 *Set* $\lambda := \max(\lambda_1, \lambda_2)$. *Then*

$$|f(y) - f(x) - (A_1(f))(x, y)(g(y) - g(x))| \leq \frac{\lambda}{\Gamma(\nu)} \frac{|g(y) - g(x)|^{\nu+1}}{(\nu + 1)},$$
$$\tag{13.3.21}$$

$\forall \, x, y \in [a, b]$ *(the case of $x = y$ is trivially true)*.

We may choose that $\frac{\lambda}{\Gamma(\nu)} < 1$.

Also we notice here that $\nu + 1 > 2$.

(II) Let $h \in C([g(a), g(b)])$, we define the right Riemann-Liouville fractional integral as

$$\left(J_{z_0-}^\nu h\right)(z) := \frac{1}{\Gamma(\nu)} \int_z^{z_0} (t - z)^{\nu-1} h(t) \, dt, \tag{13.3.22}$$

for $g(a) \leq z \leq z_0 \leq g(b)$.

We define the subspace $C_{g(x)-}^\nu([g(a), g(b)])$ of $C^n([g(a), g(b)])$, where $x \in [a, b]$:

$$C_{g(x)-}^\nu([g(a), g(b)]) :=$$

$$\left\{ h \in C^n([g(a), g(b)]) : J_{g(x)-}^{1-\alpha} h^{(n)} \in C^1([g(a), g(x)]) \right\}. \tag{13.3.23}$$

So, let $h \in C_{g(x)-}^\nu([g(a), g(b)])$; we define the right g-generalized fractional derivative of h of order ν, of Canavati type, over $[g(a), g(x)]$ as

$$D_{g(x)-}^\nu h := (-1)^{n-1} \left(J_{g(x)-}^{1-\alpha} h^{(n)}\right)'. \tag{13.3.24}$$

Clearly, for $h \in C_{g(x)-}^\nu([g(a), g(b)])$, there exists

$$\left(D_{g(x)-}^\nu h\right)(z) = \frac{(-1)^{n-1}}{\Gamma(1-\alpha)} \frac{d}{dz} \int_z^{g(x)} (t - z)^{-\alpha} h^{(n)}(t) \, dt, \tag{13.3.25}$$

for all $g(a) \leq z \leq g(x) \leq g(b)$.

In particular, when $f \circ g^{-1} \in C_{g(x)}^{\nu}\left([g\left(a\right), g\left(b\right)]\right)$ we have that

$$\left(D_{g(x)-}^{\nu}\left(f \circ g^{-1}\right)\right)(z) = \frac{(-1)^{n-1}}{\Gamma\left(1-\alpha\right)} \frac{d}{dz} \int_{z}^{g(x)} (t-z)^{-\alpha} \left(f \circ g^{-1}\right)^{(n)}(t)\, dt,$$

(13.3.26)

for all $g\left(a\right) \leq z \leq g\left(x\right) \leq g\left(b\right)$.

We get that

$$\left(D_{g(x)-}^{n}\left(f \circ g^{-1}\right)\right)(z) = (-1)^{n} \left(f \circ g^{-1}\right)^{(n)}(z),$$

(13.3.27)

and

$$\left(D_{g(x)-}^{0}\left(f \circ g^{-1}\right)\right)(z) = \left(f \circ g^{-1}\right)(z),$$

(13.3.28)

for all $z \in [g\left(a\right), g\left(x\right)]$.

In [4] we proved for $\left(f \circ g^{-1}\right) \in C_{g(x)-}^{\nu}\left([g\left(a\right), g\left(b\right)]\right)$, where $x \in [a, b]$, $\nu \geq 1$ (right fractional Taylor's formula) that:

$$f\left(y\right) - f\left(x\right) = \sum_{k=1}^{n-1} \frac{\left(f \circ g^{-1}\right)^{(k)}\left(g\left(x\right)\right)}{k!} \left(g\left(y\right) - g\left(x\right)\right)^{k} +$$

$$\frac{1}{\Gamma\left(\nu\right)} \int_{g(y)}^{g(x)} \left(t - g\left(y\right)\right)^{\nu-1} \left(D_{g(x)-}^{\nu}\left(f \circ g^{-1}\right)\right)(t)\, dt, \quad \text{all } a \leq y \leq x. \quad (13.3.29)$$

Alternatively, for $\left(f \circ g^{-1}\right) \in C_{g(y)-}^{\nu}\left([g\left(a\right), g\left(b\right)]\right)$, where $y \in [a, b]$, $\nu \geq 1$ (again right fractional Taylor's formula) that:

$$f\left(x\right) - f\left(y\right) = \sum_{k=1}^{n-1} \frac{\left(f \circ g^{-1}\right)^{(k)}\left(g\left(y\right)\right)}{k!} \left(g\left(x\right) - g\left(y\right)\right)^{k} +$$

$$\frac{1}{\Gamma\left(\nu\right)} \int_{g(x)}^{g(y)} \left(t - g\left(x\right)\right)^{\nu-1} \left(D_{g(y)-}^{\nu}\left(f \circ g^{-1}\right)\right)(t)\, dt, \quad \text{all } a \leq x \leq y. \quad (13.3.30)$$

Here we consider $f \in C^{n}\left([a, b]\right)$, such that $\left(f \circ g^{-1}\right) \in C_{g(x)-}^{\nu}\left([g\left(a\right), g\left(b\right)]\right)$, for every $x \in [a, b]$; which is the same as $\left(f \circ g^{-1}\right) \in C_{g(y)-}^{\nu}\left([g\left(a\right), g\left(b\right)]\right)$, for every $y \in [a, b]$; (i.e. exchange roles of x and y) we write that as $\left(f \circ g^{-1}\right) \in C_{g-}^{\nu}\left([g\left(a\right), g\left(b\right)]\right)$.

We have that

$$\left(D_{g(y)-}^{\nu}\left(f \circ g^{-1}\right)\right)(z) = \frac{(-1)^{n-1}}{\Gamma\left(1-\alpha\right)} \frac{d}{dz} \int_{z}^{g(y)} (t-z)^{-\alpha} \left(f \circ g^{-1}\right)^{(n)}(t)\, dt,$$

(13.3.31)

for all $g\left(a\right) \leq z \leq g\left(y\right) \leq g\left(b\right)$.

So, here we work with $f \in C^n([a, b])$, such that $(f \circ g^{-1}) \in C^{\nu}_{g-}([g(a), g(b)])$.
We define the right linear fractional operator

$$(A_2(f))(x, y) := \begin{cases} \sum_{k=1}^{n-1} \frac{(f \circ g^{-1})^{(k)}(g(x))}{k!}(g(y) - g(x))^{k-1} - \\ \left(D^{\nu}_{g(x)-}(f \circ g^{-1})\right)(g(y)) \frac{(g(x)-g(y))^{\nu-1}}{\Gamma(\nu+1)}, \quad x > y, \\ \\ \sum_{k=1}^{n-1} \frac{(f \circ g^{-1})^{(k)}(g(y))}{k!}(g(x) - g(y))^{k-1} - \\ \left(D^{\nu}_{g(y)-}(f \circ g^{-1})\right)(g(x)) \frac{(g(y)-g(x))^{\nu-1}}{\Gamma(\nu+1)}, \quad y > x, \\ \\ f^{(n)}(x), \quad x = y. \end{cases} \tag{13.3.32}$$

We may assume that

$$|(A_2(f))(x, x) - (A_2(f))(y, y)| = \left|f^{(n)}(x) - f^{(n)}(y)\right| \leq \Phi^* |g(x) - g(y)|, \tag{13.3.33}$$

where $\Phi^* > 0$; for any $x, y \in [a, b]$.
We make the following estimations:
(i) case of $x > y$: We have that

$$|f(x) - f(y) - (A_2(f))(x, y)(g(x) - g(y))| =$$

$$|f(y) - f(x) - (A_2(f))(x, y)(g(y) - g(x))| = \tag{13.3.34}$$

$$|f(y) - f(x) + (A_2(f))(x, y)(g(x) - g(y))| =$$

$$\left| \frac{1}{\Gamma(\nu)} \int_{g(y)}^{g(x)} (t - g(y))^{\nu-1} \left(D^{\nu}_{g(x)-}(f \circ g^{-1})\right)(t) \, dt - \right.$$

$$\left. \left(D^{\nu}_{g(x)-}(f \circ g^{-1})\right)(g(y)) \frac{(g(x) - g(y))^{\nu}}{\Gamma(\nu+1)} \right| = \frac{1}{\Gamma(\nu)} \cdot \tag{13.3.35}$$

$$\left| \int_{g(y)}^{g(x)} (t - g(y))^{\nu-1} \left(\left(D^{\nu}_{g(x)-}(f \circ g^{-1})\right)(t) - \left(D^{\nu}_{g(x)-}(f \circ g^{-1})\right)(g(y))\right) dt \right|$$

$$\leq \frac{1}{\Gamma(\nu)} \cdot$$

$$\int_{g(y)}^{g(x)} (t - g(y))^{\nu-1} \left|\left(D^{\nu}_{g(x)-}(f \circ g^{-1})\right)(t) - \left(D^{\nu}_{g(x)-}(f \circ g^{-1})\right)(g(y))\right| dt \tag{13.3.36}$$

(we assume here that

$$\left|\left(D_{g(x)-}^{\nu}\left(f \circ g^{-1}\right)\right)(t) - \left(D_{g(x)-}^{\nu}\left(f \circ g^{-1}\right)\right)(g(y))\right| \le \rho_1 \left|t - g(y)\right|, \quad (13.3.37)$$

for every $t, g(y), g(x) \in [g(a), g(b)]$ such that $g(x) \ge t \ge g(y)$; $\rho_1 > 0$)

$$\le \frac{\rho_1}{\Gamma(\nu)} \int_{g(y)}^{g(x)} (t - g(y))^{\nu-1} (t - g(y)) \, dt =$$

$$\frac{\rho_1}{\Gamma(\nu)} \int_{g(y)}^{g(x)} (t - g(y))^{\nu} \, dt = \frac{\rho_1}{\Gamma(\nu)} \frac{(g(x) - g(y))^{\nu+1}}{(\nu+1)}. \qquad (13.3.38)$$

We have proved that

$$|f(x) - f(y) - (A_2(f))(x, y)(g(x) - g(y))| \le \frac{\rho_1}{\Gamma(\nu)} \frac{(g(x) - g(y))^{\nu+1}}{(\nu+1)}, \qquad (13.3.39)$$

$\forall\, x, y \in [a, b] : x > y$.

(ii) Case of $x < y$: We have that

$$|f(x) - f(y) - (A_2(f))(x, y)(g(x) - g(y))| =$$

$$|f(x) - f(y) + (A_2(f))(x, y)(g(y) - g(x))| = \qquad (13.3.40)$$

$$\left| \frac{1}{\Gamma(\nu)} \int_{g(x)}^{g(y)} (t - g(x))^{\nu-1} \left(D_{g(y)-}^{\nu}\left(f \circ g^{-1}\right)\right)(t) \, dt - \right.$$

$$\left. \left(D_{g(y)-}^{\nu}\left(f \circ g^{-1}\right)\right)(g(x)) \frac{(g(y) - g(x))^{\nu}}{\Gamma(\nu+1)} \right| = \frac{1}{\Gamma(\nu)} \cdot$$

$$\left| \int_{g(x)}^{g(y)} (t - g(x))^{\nu-1} \left(\left(D_{g(y)-}^{\nu}\left(f \circ g^{-1}\right)\right)(t) - \left(D_{g(y)-}^{\nu}\left(f \circ g^{-1}\right)\right)(g(x))\right) dt \right|$$

$$\le \frac{1}{\Gamma(\nu)} \cdot$$

$$\int_{g(x)}^{g(y)} (t - g(x))^{\nu-1} \left|\left(D_{g(y)-}^{\nu}\left(f \circ g^{-1}\right)\right)(t) - \left(D_{g(y)-}^{\nu}\left(f \circ g^{-1}\right)\right)(g(x))\right| dt$$

$$(13.3.41)$$

(we assume that

$$\left|\left(D_{g(y)-}^{\nu}\left(f \circ g^{-1}\right)\right)(t) - \left(D_{g(y)-}^{\nu}\left(f \circ g^{-1}\right)\right)(g(x))\right| \le \rho_2 \left|t - g(x)\right|, \quad (13.3.42)$$

for any $t, g(x), g(y) \subset [g(a), g(b)] : g(y) \ge t \ge g(x)$; $\rho_2 > 0$)

$$\leq \frac{\rho_2}{\Gamma(\nu)} \int_{g(x)}^{g(y)} (t - g(x))^{\nu-1} (t - g(x)) \, dt =$$

$$\frac{\rho_2}{\Gamma(\nu)} \int_{g(x)}^{g(y)} (t - g(x))^{\nu} \, dt = \tag{13.3.43}$$

$$\frac{\rho_2}{\Gamma(\nu)} \frac{(g(y) - g(x))^{\nu+1}}{(\nu + 1)}. \tag{13.3.44}$$

We have proved that

$$|f(x) - f(y) - (A_2(f))(x, y)(g(x) - g(y))| \leq \frac{\rho_2}{\Gamma(\nu)} \frac{(g(y) - g(x))^{\nu+1}}{(\nu + 1)}, \tag{13.3.45}$$

$\forall \, x, y \in [a, b] : x < y.$

Conclusion 13.7 *Set* $\rho := \max(\rho_1, \rho_2)$. *Then*

$$|f(x) - f(y) - (A_2(f))(x, y)(g(x) - g(y))| \leq \frac{\rho}{\Gamma(\nu)} \frac{|g(x) - g(y)|^{\nu+1}}{(\nu + 1)}, \tag{13.3.46}$$

$\forall \, x, y \in [a, b]$ *((13.3.46) is trivially true when* $x = y$*)*.

One may choose $\frac{\rho}{\Gamma(\nu)} < 1$.
Here again $\nu + 1 > 2$.

Conclusion 13.8 *Based on (13.3.10) and (13.3.21) of (I), and based on (13.3.33) and (13.3.46) of (II), using our numerical results presented earlier, we can solve numerically* $f(x) = 0$.

Some examples for g follow:

$$g(x) = e^x, \ x \in [a, b] \subset \mathbb{R},$$
$$g(x) = \sin x,$$
$$g(x) = \tan x,$$
where $x \in \left[-\frac{\pi}{2} + \varepsilon, \frac{\pi}{2} - \varepsilon\right]$, with $\varepsilon > 0$ small.

Returning back to Proposition 13.2 we see by (13.2.10) and (13.3.21) that crucial estimate (13.2.10) is satisfied, if we choose $g(x) = x$ for each $x \in [a, b]$, $p = \nu$,

$$g_1(s, t) = \frac{\lambda \, |s - t|^p}{\Gamma(p)(p + 1)\mu}.$$

for each $s, t \in [a, b]$ and $A = A_1(f)$.

Similarly by (13.2.10) and (13.3.46), we must choose $g(x) = x$ for each $x \in [a, b]$, $p = \nu$,

$$g_1(s, t) = \frac{\rho |s - t|^p}{\Gamma(p)(p+1)}$$

for each $s, t \in [a, b]$ and $A = A_2(f)$.

References

1. S. Amat, S. Busquier, Third-order iterative methods under Kantorovich conditions. J. Math. Anal. Appl. **336**, 243–261 (2007)
2. S. Amat, S. Busquier, S. Plaza, Chaotic dynamics of a third-order Newton-like method. J. Math. Anal. Appl. **366**(1), 164–174 (2010)
3. G. Anastassiou, Fractional representation formulae and right fractional inequalities. Math. Comput. Model. **54**(10–12), 3098–3115 (2011)
4. G. Anastassiou, Generalized Canavati type Fractional Taylor's formulae. J. Comput. Anal. Appl. (2016) (to appear)
5. G. Anastassiou, I. Argyros, Generalized g-fractional calculus of Canavati-type and secant-like methods (2015) (submitted)
6. I.K. Argyros, Newton-like methods in partially ordered linear spaces. J. Approx. Theory Appl. **9**(1), 1–10 (1993)
7. I.K. Argyros, Results on controlling the residuals of perturbed Newton-like methods on Banach spaces with a convergence structure. Southwest J. Pure Appl. Math. **1**, 32–38 (1995)
8. I.K. Argyros, *Convergence and Applications of Newton-Like Iterations* (Springer, New York, 2008)
9. K. Diethelm, *The Analysis of Fractional Differential Equations*. Lecture Notes in Mathematics, vol. 2004, 1st edn. (Springer, New York, 2010)
10. J.A. Ezquerro, M.A. Hernandez, Newton-like methods of high order and domains of semilocal and global convergence. Appl. Math. Comput. **214**(1), 142–154 (2009)
11. J.A. Ezquerro, J.M. Gutierrez, M.A. Hernandez, N. Romero, M.J. Rubio, The Newton method: from Newton to Kantorovich (Spanish). Gac. R. Soc. Mat. Esp. **13**, 53–76 (2010)
12. L.V. Kantorovich, G.P. Akilov, *Functional Analysis in Normed Spaces* (Pergamon Press, New York, 1964)
13. A.A. Magrenan, Different anomalies in a Jarratt family of iterative root finding methods. Appl. Math. Comput. **233**, 29–38 (2014)
14. A.A. Magrenan, A new tool to study real dynamics: the convergence plane. Appl. Math. Comput. **248**, 215–224 (2014)
15. F.A. Potra, V. Ptak, *Nondiscrete Induction and Iterative Processes* (Pitman Publ., London, 1984)
16. P.D. Proinov, New general convergence theory for iterative processes and its applications to Newton-Kantorovich type theorems. J. Complex. **26**, 3–42 (2010)

Chapter 14
Iterative Algorithms and Left-Right Caputo Fractional Derivatives

We present a local as well as a semilocal convergence analysis for some iterative algorithms in order to approximate a locally unique solution of a nonlinear equation in a Banach space setting. In the application part of the study, we present some choices of the operators involving the left and right Caputo derivative where the operators satisfy the convergence conditions. It follows [5].

14.1 Introduction

In this study we are concerned with the problem of approximating a locally unique solution x^* of the nonlinear equation

$$F(x) = 0, \tag{14.1.1}$$

where F is a Fréchet-differentiable operator defined on a subset D of a Banach space X with values in a Banach space Y.

A lot of problems in Computational Sciences and other disciplines can be brought in a form like (14.1.1) using Mathematical Modelling [8, 12, 16]. The solutions of such equations can be found in closed form only in special cases. That is why most solution methods for these equations are iterative. Iterative algorithms are usually studied based on semilocal and local convergence. The semilocal convergence matter is, based on the information around the initial point to give hypotheses ensuring the convergence of the iterative algorithm; while the local one is, based on the information around a solution, to find estimates of the radii of convergence balls as well as error bounds on the distances involved.

We introduce the iterative algorithm defined for each $n = 0, 1, 2, \ldots$ by

$$x_{n+1} = x_n - A(x_n)^{-1} F(x_n), \tag{14.1.2}$$

where $x_0 \in D$ is an initial point and $A(x) \in L(X, Y)$ the space of bounded linear operators from X into Y. There is a plethora on local as well as semilocal convergence theorems for iterative algorithm (14.1.2) provided that the operator A is an approximation to the Fréchet-derivative F' [1, 2, 6–16]. Notice that many well known methods are special case of interative algorithm (14.1.2).

Newton's method: Choose $A(x) = F'(x)$ for each $x \in D$.

Steffensen's method: Choose $A(x) = [x, G(x); F]$, where $G : X \to X$ is a known operator and $[x, y; F]$ denotes a divided difference of order one [8, 12, 15].

The so called Newton-like methods and many other methods are special cases of iterative algorithm (14.1.2).

The rest of the chapter is organized as follows. The semilocal as well as the local convergence analysis of iterative algorithm (14.1.2) is given in Sect. 14.2. Some applications from fractional calculus are given in the concluding Sect. 14.3.

14.2 Convergence Analysis

We present the main semilocal convergence result for iterative algorithm (14.1.2).

Theorem 14.1 *Let $F : D \subset X \to Y$ be a Fréchet-differentiable operator and let $A(x) \in L(X, Y)$. Suppose that there exist $x_0 \in D$, $\eta \geq 0$, $p \geq 1$, functions $g_0 : [0, \eta] \to [0, \infty)$, $g_1 : [0, \infty)^2 \to [0, \infty)$ continuous and nondecreasing such that for each $x, y \in D$*

$$A(x)^{-1} \in L(Y, X), \tag{14.2.1}$$

$$\left\| A(x_0)^{-1} F(x_0) \right\| \leq \eta, \tag{14.2.2}$$

$$\left\| A(y)^{-1} \left(F(y) - F(x) - F'(x)(y - x) \right) \right\| \leq g_0 (\|x - y\|) \|x - y\|^{p+1}, \tag{14.2.3}$$

$$\left\| A(y)^{-1} \left(A(x) - F'(x) \right) \right\| \leq g_1 (\|y - x_0\|, \|x - x_0\|). \tag{14.2.4}$$

Moreover, suppose that function $\varphi : [0, \infty) \to \mathbb{R}$ defined by

$$\varphi(t) = \left(1 - \left(g_0(\eta) \eta^p + g_1(t, t) \right) \right) t - \eta \tag{14.2.5}$$

has a smallest positive zero r and

$$\overline{U}(x_0, r) \subseteq D. \tag{14.2.6}$$

Then, the sequence $\{x_n\}$ generated by iterative algorithm (14.1.2) is well defined, remains in $\overline{U}(x_0, r)$ for each $n = 0, 1, 2, \ldots$ and converges to some $x^* \in \overline{U}(x_0, r)$ such that

$$\|x_{n+1} - x_n\| \leq \left[g_0 \left(\|x_n - x_{n-1}\| \right) \|x_n - x_{n-1}\|^p + \right. \quad (14.2.7)$$

$$\left. g_0 \left(\|x_{n-1} - x_0\| \right) \right] \|x_n - x_{n-1}\| \leq q \|x_n - x_{n-1}\|$$

and

$$\|x_n - x^*\| \leq \frac{q^n \eta}{1 - q}, \quad (14.2.8)$$

where

$$q := g_0 (\eta) \eta^p + g_1 (r, r) \in [0, 1).$$

Proof Notice that it follows from (14.2.5) and the definition of r that $q \in [0, 1)$.

The iterate x_1 is well defined by iterative algorithm (14.1.2) for $n = 0$ and (14.2.1) for $x = x_0$. We also have by (14.2.2) that $\|x_1 - x_0\| = \left\| A(x_0)^{-1} F(x_0) \right\| \leq \eta < r$, so we get that $x_1 \in \overline{U}(x_0, r)$ and x_2 is well defined (by (14.2.6)). Using (14.2.3) and (14.2.4) we get that

$$\|x_2 - x_1\| = \left\| A(x_1)^{-1} \left[F(x_1) - F(x_0) - A(x_0)(x_1 - x_0) \right] \right\| \leq$$

$$\left\| A(x_1)^{-1} \left(F(x_1) - F(x_0) - F'(x_0)(x_1 - x_0) \right) \right\| +$$

$$\left\| A(x_1)^{-1} \left(A(x_0) - F'(x_0) \right) \right\| \leq$$

$$\left(g_0 (\eta) \eta^p + g_1 (\eta, 0) \right) \|x_1 - x_0\| \leq q \|x_1 - x_0\|,$$

which shows (14.2.7) for $n = 1$. Then, we can have that

$$\|x_2 - x_0\| \leq \|x_2 - x_1\| + \|x_1 - x_0\| \leq q \|x_1 - x_0\| + \|x_1 - x_0\|$$

$$= (1 + q) \|x_1 - x_0\| \leq \frac{1 - q^2}{1 - q} \eta < r,$$

so $x_2 \in \overline{U}(x_0, r)$ and x_3 is well defined.

Assuming $\|x_{k+1} - x_k\| \leq q \|x_k - x_{k-1}\|$ and $x_{k+1} \in \overline{U}(x_0, r)$ for each $k = 1, 2, \ldots, n$ we get

$$\|x_{k+2} - x_{k+1}\| = \left\| A(x_{k+1})^{-1} \left[F(x_{k+1}) - F(x_k) - A(x_k)(x_{k+1} - x_k) \right] \right\|$$

$$\leq \left\| A(x_{k+1})^{-1} \left(F(x_{k+1}) - F(x_k) - F'(x_k)(x_{k+1} - x_k) \right) \right\|$$

$$+ \left\| A(x_{k+1})^{-1} \left(A(x_k) - F'(x_k) \right) \right\| \leq$$

$$\left[g_0 \left(\| x_{k+1} - x_k \| \right) \| x_{k+1} - x_k \|^p + g_1 \left(\| x_{k+1} - x_0 \|, \| x_k - x_0 \| \right) \right] \| x_{k+1} - x_k \|$$

$$\leq \left[g_0 \left(\eta \right) \eta^p + g_1 \left(r, r \right) \right] \| x_{k+1} - x_k \| \leq q \| x_{k+1} - x_k \|$$

and

$$\| x_{k+2} - x_0 \| \leq \| x_{k+2} - x_{k+1} \| + \| x_{k+1} - x_k \| + \cdots + \| x_1 - x_0 \|$$

$$\leq \left(q^{k+1} + q^k + \cdots + 1 \right) \| x_1 - x_0 \| \leq \frac{1 - q^{k+2}}{1 - q} \| x_1 - x_0 \|$$

$$< \frac{\eta}{1 - q} = r,$$

which completes the induction for (14.2.7) and $x_{k+2} \in \overline{U} \left(x_0, r \right)$. We also have that for $m \geq 0$

$$\| x_{n+m} - x_n \| \leq \| x_{n+m} - x_{n+m-1} \| + \cdots + \| x_{n+1} - x_n \|$$

$$\leq \left(q^{m-1} + q^{m-2} + \cdots + 1 \right) \| x_{n+1} - x_n \|$$

$$\leq \frac{1 - q^m}{1 - q} q^n \| x_1 - x_0 \|.$$

It follows that $\{ x_n \}$ is a complete sequence in a Banach space X and as such it converges to some $x^* \in \overline{U} \left(x_0, r \right)$ (since $\overline{U} \left(x_0, r \right)$ is a closed set). By letting $m \to \infty$, we obtain (14.2.8). □

Stronger hypotheses are needed to show that x^* is a solution of equation $F \left(x \right) = 0$.

Proposition 14.2 *Let $F : D \subset X \to Y$ be a Fréchet-differentiable operator and let $A \left(x \right) \in L \left(X, Y \right)$. Suppose that there exist $x_0 \in D$, $\eta \geq 0$, $p \geq 1$, $\lambda > 0$, functions $g_0 : [0, \eta] \to [0, \infty)$, $g : [0, \infty) \to [0, \infty)$, continuous and nondecreasing such that for each $x, y \in D$*

$$A \left(x \right)^{-1} \in L \left(Y, X \right), \quad \left\| A \left(x \right)^{-1} \right\| \leq \lambda, \quad \left\| A \left(x_0 \right)^{-1} F \left(x_0 \right) \right\| \leq \eta, \qquad (14.2.9)$$

$$\left\| F \left(y \right) - F \left(x \right) - F' \left(x \right) \left(y - x \right) \right\| \leq \frac{g_0 \left(\| x - y \| \right)}{\lambda} \| x - y \|^{p+1}, \qquad (14.2.10)$$

$$\left\| A \left(x \right) - F' \left(x \right) \right\| \leq \frac{g \left(\| x - x_0 \| \right)}{\lambda}. \qquad (14.2.11)$$

Moreover, suppose that function $\psi : [0, \infty) \to \mathbb{R}$ defined by

$$\psi \left(t \right) = \left(1 - \left(g_0 \left(\eta \right) \eta^p + g \left(t \right) \right) \right) t - \eta$$

has a smallest positive zero r_1 *and*

$$\overline{U}(x_0, r_1) \subseteq D,$$

where,

$$r_1 = \frac{\eta}{1 - q_1} \text{ and } q_1 = g_0(\eta)\eta^p + g(r) \in [0, 1].$$

Then, the conclusions of Theorem 14.1 for sequence $\{x_n\}$ *hold with* $\frac{g_0}{\lambda}$, $\frac{g}{\lambda}$, q_1, r_1, *replacing* g_0, g, q *and* r, *respectively. Moreover,* x^* *is a solution of the equation* $F(x) = 0$.

Proof Notice that

$$\left\| A(x_n)^{-1} \left[F(x_n) - F(x_{n-1}) - A(x_{n-1})(x_n - x_{n-1}) \right] \right\|$$

$$\leq \left\| A(x_n)^{-1} \right\| \left\| F(x_n) - F(x_{n-1}) - A(x_{n-1})(x_n - x_{n-1}) \right\|$$

$$\leq \left(g_0(\|x_n - x_{n-1}\|) \|x_n - x_{n-1}\|^p + g(\|x_n - x_0\|) \right) \|x_n - x_{n-1}\|$$

$$\leq q_1 \|x_n - x_{n-1}\|.$$

Therefore, the proof of Theorem 14.1 can apply. Then, in view of the estimate

$$\|F(x_n)\| = \|F(x_n) - F(x_{n-1}) - A(x_{n-1})(x_n - x_{n-1})\| \leq$$

$$\left\| F(x_n) - F(x_{n-1}) - F'(x_{n-1})(x_n - x_{n-1}) \right\| +$$

$$\left\| A(x_n) - F'(x_n) \right\| \|x_n - x_{n-1}\| \leq q_1 \|x_n - x_{n-1}\|,$$

we deduce by letting $n \to \infty$ that $F(x^*) = 0$. □

Concerning the uniqueness of the solution x^* we have the following result:

Proposition 14.3 *Under the hypotheses of Proposition 14.2, further suppose that*

$$g_0(r_1)r_1^p + g(r_1) < 1. \tag{14.2.12}$$

Then, x^* *is the only solution of equation* $F(x) = 0$ *in* $\overline{U}(x_0, r_1)$.

Proof The existence of the solution $x^* \in \overline{U}(x_0, r_1)$ has been established in Proposition 14.2. Let $y^* \in \overline{U}(x_0, r_1)$ with $F(y^*) = 0$. Then, we have in turn that

$$\|x_{n+1} - y^*\| = \|x_n - y^* - A(x_n)^{-1} F(x_n)\| =$$

$$\left\| A(x_n)^{-1} \left[A(x_n)(x_n - y^*) - F(x_n) + F(y^*) \right] \right\| \leq$$

$$\left\| A \left(x_n \right)^{-1} \right\| \left\| F \left(y^* \right) - F \left(x_n \right) - F' \left(x_n \right) \left(y^* - x_n \right) \right\|$$

$$+ \left\| \left(A \left(x_n \right) - F' \left(x_n \right) \right) \left(y^* - x_n \right) \right\| \le$$

$$\left[g_0 \left(\left\| y^* - x_n \right\| \right) \left\| y^* - x_n \right\|^p + g \left(\left\| x_n - x_0 \right\| \right) \right] \left\| y^* - x_n \right\| \le$$

$$\left(g_0 \left(r_1 \right) r_1^p + g \left(r_1 \right) \right) \left\| x_n - y^* \right\| < \left\| x_n - y^* \right\|,$$

so we deduce that $\lim_{n \to \infty} x_n = y^*$. But we have that $\lim_{n \to \infty} x_n = x^*$. Hence, we conclude that $x^* = y^*$. □

Next, we present a local convergence analysis for the iterative algorithm (14.1.2).

Proposition 14.4 Let $F : D \subset X \to Y$ be a Fréchet-differentiable operator and let $A(x) \in L(X, Y)$. Suppose that there exist $x^* \in D$, $p \ge 1$, functions $g_0 : [0, \infty) \to [0, \infty)$, $g : [0, \infty) \to [0, \infty)$ continuous and nondecreasing such that for each $x \in D$

$$F \left(x^* \right) = 0, \ A \left(x \right)^{-1} \in L \left(Y, X \right),$$

$$\left\| A \left(x \right)^{-1} \left[F \left(x \right) - F \left(x^* \right) - F' \left(x \right) \left(x - x^* \right) \right] \right\| \le g_0 \left(\left\| x - x^* \right\| \right) \left\| x - x^* \right\|^{p+1},$$

$$\tag{14.2.13}$$

$$\left\| A \left(x \right)^{-1} \left(A \left(x \right) - F' \left(x \right) \right) \right\| \le g \left(\left\| x - x^* \right\| \right) \tag{14.2.14}$$

and

$$\overline{U} \left(x^*, r_2 \right) \subseteq D,$$

where r_2 is the smallest positive solution of equation

$$h(t) := g_0(t) t^p + g(t) - 1.$$

Then, sequence $\{x_n\}$ generated by algorithm (14.1.2) for $x_0 \in U(x^*, r_2) - \{x^*\}$ is well defined, remains in $U(x^*, r_2)$ for each $n = 0, 1, 2, \ldots$ and converges to x^*. Moreover, the following estimates hold

$$\left\| x_{n+1} - x^* \right\| \le \left(g_0 \left(\left\| x_n - x^* \right\| \right) \left\| x_n - x^* \right\|^p + g \left(\left\| x_n - x^* \right\| \right) \right) \left\| x_n - x^* \right\|$$

$$< \left\| x_n - x^* \right\| < r_2.$$

Proof We have that $h(0) = -1 < 0$ and $h(t) \to +\infty$ as $t \to +\infty$. Then, it follows from the intermediate value theorem that function h has positive zeros. Denote by r_2

the smallest such zero. By hypothesis $x_0 \in U(x^*, r_2) - \{x^*\}$. Then, we get in turn that

$$\|x_1 - x^*\| = \|x_0 - x^* - A(x_0)^{-1} F(x_0)\| =$$

$$\|A(x_0)^{-1} [F(x^*) - F(x_0) - F'(x_0)(x^* - x_0)]\|$$

$$+ \|A(x_0)^{-1} (A(x_0) - F'(x_0))(x^* - x_0)\| \leq$$

$$g_0(\|x_0 - x^*\|) \|x_0 - x^*\|^{p+1} + g(\|x_0 - x^*\|) \|x_0 - x^*\| <$$

$$(h(r_2) + 1) \|x_0 - x^*\| = \|x_0 - x^*\| < r_2,$$

which shows that $x_1 \in U(x^*, r_2)$ and x_2 is well defined. By a simple inductive argument as in the preceding estimate we get that

$$\|x_{k+1} - x^*\| = \|x_k - x^* - A(x_k)^{-1} F(x_k)\| \leq$$

$$\|A(x_k)^{-1} [F(x^*) - F(x_k) - A(x_k)(x^* - x_k)]\| +$$

$$\|A(x_k)^{-1} (A(x_k) - F'(x_k))(x^* - x_k)\| \leq$$

$$g_0(\|x_k - x^*\|) \|x_k - x^*\|^p + g(\|x_k - x^*\|) \|x_k - x^*\| <$$

$$(h(r_2) + 1) \|x_k - x^*\| = \|x_k - x^*\| < r_2,$$

which shows $\lim_{k \to \infty} x_k = x^*$ and $x_{k+1} \in U(x^*, r_2)$. $\square$

Remark 14.5 (a) Hypothesis (14.2.3) specializes to Newton-Mysowski-type, if $A(x) = F'(x)$ [8, 12, 15]. However, if F is not Fréchet-differentiable, then our results extend the applicability of iterative algorithm (14.1.2).

(b) Theorem 14.1 has practical value although we do not show that x^* is a solution of equation $F(x) = 0$, since this may be shown in another way.

(c) Hypothesis (14.2.13) can be replaced by the stronger

$$\|A(x)^{-1} [F(x) - F(y) - A(x)(x - y)]\| \leq g_2(\|x - y\|) \|x - y\|^{p+1}.$$

The preceding results can be extended to hold for two point iterative algorithms defined for each $n = 0, 1, 2, \ldots$ by

$$x_{n+1} = x_n - A(x_n, x_{n-1})^{-1} F(x_n), \tag{14.2.15}$$

where $x_{-1}, x_0 \in D$ are initial points and $A(w, v) \in L(X, Y)$ for each $v, w \in D$. If $A(w, v) = [w, v; F]$, then iterative algorithm (14.2.15) reduces to the popular secant method, where $[w, v; F]$ denotes a divided difference of order one for the operator F. Many other choices for A are also possible [8, 12, 16].

If we simply replace $A(x)$ by $A(y, x)$ in the proof of Proposition 14.2 we arrive at the following semilocal convergence result for iterative algorithm (14.2.15).

Theorem 14.6 *Let* $F : D \subset X \to Y$ *be a Fréchet-differentiable operator and let* $A(y, x) \in L(X, Y)$ *for each* $x, y \in D$. *Suppose that there exist* $x_{-1}, x_0 \in D$, $\eta \geq 0$, $p \geq 1$, $\mu > 0$, *functions* $g_0 : [0, \eta] \to [0, \infty)$, $g_1 : [0, \infty)^2 \to [0, \infty)$ *continuous and nondecreasing such that for each* $x, y \in D$:

$$A(y, x)^{-1} \in L(Y, X), \quad \left\| A(y, x)^{-1} \right\| \leq \mu, \tag{14.2.16}$$

$$\min \left\{ \|x_0 - x_{-1}\|, \left\| A(x_0, x_{-1})^{-1} F(x_0) \right\| \right\} \leq \eta,$$

$$\left\| F(y) - F(x) - F'(y - x) \right\| \leq \frac{g_0(\|x - y\|)}{\mu} \|x - y\|^{p+1}, \tag{14.2.17}$$

$$\left\| A(y, x) - F'(x) \right\| \leq \frac{g_1(\|y - x_0\|, \|x - x_0\|)}{\mu}.$$

Moreover, suppose that function φ *given by (14.2.5) has a smallest positive zero* r *such that*

$$g_0(r) r^p + g_1(r, r) < 1$$

and

$$\overline{U}(x_0, r) \subseteq D,$$

where,

$$r = \frac{\eta}{1 - q}$$

and q *is defined in Theorem 14.1.*

Then, sequence $\{x_n\}$ generated by iterative algorithm (14.2.15) is well defined, remains in $\overline{U}(x_0, r)$ for each $n = 0, 1, 2, \ldots$ and converges to the only solution of equation $F(x) = 0$ in $\overline{U}(x_0, r)$. Moreover, the estimates (14.2.7) and (14.2.8) hold.

Concerning, the local convergence of the iterative algorithm (14.2.15) we obtain the analogous to Proposition 14.4 result.

Proposition 14.7 *Let* $F : D \subset X \to Y$ *be a Fréchet-differentiable operator and let* $A(y, x) \in L(X, Y)$. *Suppose that there exist* $x^* \in D$, $p \geq 1$, *functions* $g_2 : [0, \infty)^2 \to [0, \infty)$, $g_3 : [0, \infty)^2 \to [0, \infty)$ *continuous and nondecreasing such that for each* $x, y \in D$

$$F(x^*) = 0, \ A(y, x)^{-1} \in L(Y, X),$$

$$\left\| A(y, x)^{-1} \left[F(y) - F(x^*) - A(y, x)(y - x^*) \right] \right\| \leq$$

$$g_2 \left(\|y - x^*\|, \|x - x^*\| \right) \|y - x^*\|^{p+1},$$

$$\left\| A(y, x)^{-1} \left(A(y, x) - F'(x) \right) \right\| \le g_3 \left(\|y - x^*\|, \|x - x^*\| \right)$$

and

$$\overline{U}(x^*, r_2) \subseteq D,$$

where r_2 is the smallest positive solution of equation

$$h(t) := g_2(t, t) t^p + g_3(t, t) - 1.$$

Then, sequence $\{x_n\}$ generated by algorithm (14.2.15) for $x_{-1}, x_0 \in U(x^*, r_2) - \{x^*\}$ is well defined, remains in $U(x^*, r_2)$ for each $n = 0, 1, 2, \ldots$ and converges to x^*. Moreover, the following estimates hold

$$\|x_{n+1} - x^*\| \le \left[g_2 \left(\|x_n - x^*\|, \|x_{n-1} - x^*\| \right) \|x_n - x^*\|^p + \right.$$

$$\left. g_3 \left(\|x_n - x^*\|, \|x_{n-1} - x^*\| \right) \right] \|x_n - x^*\| < \|x_n - x^*\| < r_2.$$

14.3   Applications to Fractional Calculus

In this section we apply Proposition 14.2 and iterative algorithm (14.1.2) to fractional calculus for solving $f(x) = 0$.

Let $0 < \alpha < 1$, hence $\lceil \alpha \rceil = 1$, where $\lceil \cdot \rceil$ is the ceiling of the number. Let also $c < a < b < d$, and $f \in C^2([c, d])$, with $f'' \ne 0$.

Clearly we have

$$\left| f'(x) - f'(y) \right| \le \|f''\|_\infty |x - y|, \ \forall \, x, y \in [c, d]. \tag{14.3.1}$$

We notice that

$$f(x) - f(y) = \left(\int_0^1 f'(y + \theta(x - y)) \, d\theta \right) (x - y). \tag{14.3.2}$$

Therefore it holds

$$\left| f(x) - f(y) - f'(x)(x - y) \right| =$$

$$\left| \left(\int_0^1 f'(y + \theta(x - y)) \, d\theta \right) (x - y) - \left(\int_0^1 f'(x) \, d\theta \right) (x - y) \right| =$$

$$\left| \int_0^1 \left(f'(y + \theta(x - y)) - f'(x) \right) d\theta \right| |x - y| \le$$

$$\left(\int_0^1 \left| f'(y + \theta(x - y)) - f'(x) \right| d\theta \right) |x - y| \overset{(14.3.1)}{\leq}$$

$$\|f''\|_\infty \left(\int_0^1 |y + \theta(x - y) - x| \, d\theta \right) |x - y| =$$

$$\|f''\|_\infty \left(\int_0^1 (1 - \theta) |x - y| \, d\theta \right) |x - y| = \qquad (14.3.3)$$

$$\|f''\|_\infty \left(\int_0^1 (1 - \theta) \, d\theta \right) (x - y)^2 =$$

$$\|f''\|_\infty \left(\left. \frac{(1 - \theta)^2}{2} \right|_1^0 \right) (x - y)^2 = \frac{\|f''\|_\infty}{2} (x - y)^2, \ \forall \, x, y \in [c, d].$$

We have proved that

$$\left| f(y) - f(x) - f'(x)(y - x) \right| \leq \frac{\|f''\|_\infty}{2} (y - x)^2, \ \forall \, x, y \in [c, d]. \quad (14.3.4)$$

(I) The left Caputo fractional derivative of f of order $\alpha \in (0, 1)$, anchored at a, is defined as follows:

$$\left(D_{*a}^\alpha f \right)(x) = \frac{1}{\Gamma(1 - \alpha)} \int_a^x (x - t)^{-\alpha} f'(t) \, dt, \forall \, x \in [a, d], \qquad (14.3.5)$$

while $\left(D_{*a}^\alpha f \right)(x) = 0$, for $c \leq x \leq a$.

Next we consider $a < a^* < b$, and $x \in [a^*, b]$, also $x_0 \in (c, a)$.

We define the function

$$A_1(x) := \frac{\Gamma(2 - \alpha)}{(x - a)^{1 - \alpha}} \left(D_{*a}^\alpha f \right)(x), \ \forall \, x \in [a^*, b]. \qquad (14.3.6)$$

Notice that $A_1(a)$ is undefined.

We see that

$$\left| A_1(x) - f'(x) \right| = \left| \frac{\Gamma(2 - \alpha)}{(x - a)^{1 - \alpha}} \left(D_{*a}^\alpha f \right)(x) - f'(x) \right| = \qquad (14.3.7)$$

$$\left| \frac{\Gamma(2 - \alpha)}{(x - a)^{1 - \alpha}} \frac{1}{\Gamma(1 - \alpha)} \int_a^x (x - t)^{-\alpha} f'(t) \, dt - \frac{\Gamma(2 - \alpha)}{(x - a)^{1 - \alpha}} \frac{(x - a)^{1 - \alpha}}{\Gamma(2 - \alpha)} f'(x) \right|$$

$$= \frac{\Gamma(2 - \alpha)}{(x - a)^{1 - \alpha}} \cdot$$

$$\left| \frac{1}{\Gamma(1-\alpha)} \int_a^x (x-t)^{-\alpha} f'(t)\, dt - \frac{1}{\Gamma(1-\alpha)} \int_a^x (x-t)^{-\alpha} f'(x)\, dt \right| =$$

$$\frac{(1-\alpha)}{(x-a)^{1-\alpha}} \left| \int_a^x (x-t)^{-\alpha} \left(f'(t) - f'(x) \right) dt \right| \leq \qquad (14.3.8)$$

$$\frac{(1-\alpha)}{(x-a)^{1-\alpha}} \int_a^x (x-t)^{-\alpha} \left| f'(t) - f'(x) \right| dt \overset{(14.3.1)}{\leq}$$

$$\frac{(1-\alpha) \left\| f'' \right\|_\infty}{(x-a)^{1-\alpha}} \int_a^x (x-t)^{-\alpha} (x-t)\, dt = \frac{(1-\alpha) \left\| f'' \right\|_\infty}{(x-a)^{1-\alpha}} \int_a^x (x-t)^{1-\alpha}\, dt =$$
$$(14.3.9)$$

$$\frac{(1-\alpha) \left\| f'' \right\|_\infty}{(x-a)^{1-\alpha}} \frac{(x-a)^{2-\alpha}}{2-\alpha} = \frac{(1-\alpha)}{(2-\alpha)} \left\| f'' \right\|_\infty (x-a). \qquad (14.3.10)$$

We have proved that

$$\left| A_1(x) - f'(x) \right| \leq \left(\frac{1-\alpha}{2-\alpha} \right) \left\| f'' \right\|_\infty (x-a) \leq \left(\frac{1-\alpha}{2-\alpha} \right) \left\| f'' \right\|_\infty (b-a),$$
$$(14.3.11)$$

$\forall\, x \in [a^*, b]$.

In particular, it holds that

$$\left| A_1(x) - f'(x) \right| \leq \left(\frac{1-\alpha}{2-\alpha} \right) \left\| f'' \right\|_\infty (x - x_0), \qquad (14.3.12)$$

where $x_0 \in (c, a)$, $\forall\, x \in [a^*, b]$.

(II) The right Caputo fractional derivative of f of order $\alpha \in (0, 1)$, anchored at b, is defined as follows:

$$\left(D_{b-}^\alpha f \right)(x) = \frac{-1}{\Gamma(1-\alpha)} \int_x^b (t-x)^{-\alpha} f'(t)\, dt, \forall\, x \in [c, b], \qquad (14.3.13)$$

while $\left(D_{b-}^\alpha f \right)(x) = 0$, for $d \geq x \geq b$.

Next consider $a < b^* < b$, and $x \in [a, b^*]$, also $x_0 \in (b, d)$.

We define the function

$$A_2(x) := -\frac{\Gamma(2-\alpha)}{(b-x)^{1-\alpha}} \left(D_{b-}^\alpha f \right)(x), \ \forall\, x \in \left[a, b^*\right]. \qquad (14.3.14)$$

Notice that $A_2(b)$ is undefined.

We see that

$$\left| A_2(x) - f'(x) \right| = \left| -\frac{\Gamma(2-\alpha)}{(b-x)^{1-\alpha}} \left(D_{b-}^{\alpha} f \right)(x) - f'(x) \right| =$$

$$\left| \frac{\Gamma(2-\alpha)}{(b-x)^{1-\alpha}} \frac{1}{\Gamma(1-\alpha)} \int_x^b (t-x)^{-\alpha} f'(t)\, dt - f'(x) \right| = \qquad (14.3.15)$$

$$\left| \frac{\Gamma(2-\alpha)}{(b-x)^{1-\alpha}} \frac{1}{\Gamma(1-\alpha)} \int_x^b (t-x)^{-\alpha} f'(t)\, dt - \frac{\Gamma(2-\alpha)}{(b-x)^{1-\alpha}} \frac{(b-x)^{1-\alpha}}{\Gamma(2-\alpha)} f'(x) \right|$$

$$= \frac{\Gamma(2-\alpha)}{(b-x)^{1-\alpha}} \cdot$$

$$\left| \frac{1}{\Gamma(1-\alpha)} \int_x^b (t-x)^{-\alpha} f'(t)\, dt - \frac{1}{\Gamma(1-\alpha)} \int_x^b (t-x)^{-\alpha} f'(x)\, dt \right| =$$
$$\qquad (14.3.16)$$

$$\frac{\Gamma(2-\alpha)}{(b-x)^{1-\alpha}} \frac{1}{\Gamma(1-\alpha)} \left| \int_x^b (t-x)^{-\alpha} \left(f'(t) - f'(x) \right) dt \right| \le$$

$$\frac{(1-\alpha)}{(b-x)^{1-\alpha}} \int_x^b (t-x)^{-\alpha} \left| f'(t) - f'(x) \right| dt \overset{(14.3.1)}{\le}$$

$$\frac{(1-\alpha)\, \|f''\|_\infty}{(b-x)^{1-\alpha}} \int_x^b (t-x)^{-\alpha} (t-x)\, dt = \frac{(1-\alpha)\, \|f''\|_\infty}{(b-x)^{1-\alpha}} \int_x^b (t-x)^{1-\alpha}\, dt =$$
$$\qquad (14.3.17)$$

$$\frac{(1-\alpha)\, \|f''\|_\infty}{(b-x)^{1-\alpha}} \frac{(b-x)^{2-\alpha}}{2-\alpha} = \frac{(1-\alpha)\, \|f''\|_\infty}{(2-\alpha)} (b-x).$$

We have proved that

$$\left| A_2(x) - f'(x) \right| \le \left(\frac{1-\alpha}{2-\alpha} \right) \|f''\|_\infty (b-x) \le \left(\frac{1-\alpha}{2-\alpha} \right) \|f''\|_\infty (b-a),$$
$$\qquad (14.3.18)$$

$\forall\, x \in [a, b^*]$.

In particular, it holds that

$$\left| A_2(x) - f'(x) \right| \le \left(\frac{1-\alpha}{2-\alpha} \right) \|f''\|_\infty (x_0 - x), \qquad (14.3.19)$$

where $x_0 \in (b, d)$, $\forall\, x \in [a, b^*]$.

The results of Proposition 14.2 can apply, if we choose $A = A_i, i = 1, 2, p = 1$, $g_0(t) = \frac{\|f\|_\infty}{\lambda}$ and $g_1(t) = \frac{(1-\alpha)\|f''\|_\infty t}{(2-\alpha)\lambda}$ for each $t \in [c, d]$.

References

1. S. Amat, S. Busquier, Third-order iterative methods under Kantorovich conditions. J. Math. Anal. Appl. **336**, 243–261 (2007)
2. S. Amat, S. Busquier, S. Plaza, Chaotic dynamics of a third-order Newton-like method. J. Math. Anal. Appl. **366**(1), 164–174 (2010)
3. G. Anastassiou, *Fractional Differentiation Inequalities* (Springer, New York, 2009)
4. G. Anastassiou, *Intelligent Mathematics: Computational Analysis* (Springer, Heidelberg, 2011)
5. G. Anastassiou, I. Argyros, *A convergence analysis for some iterative algorithms with applications to fractional calculus*, submitted (2015)
6. I.K. Argyros, Newton-like methods in partially ordered linear spaces. J. Approx. Theory Appl. **9**(1), 1–10 (1993)
7. I.K. Argyros, Results on controlling the residuals of perturbed Newton-like methods on Banach spaces with a convergence structure. Southwest J. Pure Appl. Math. **1**, 32–38 (1995)
8. I.K. Argyros, *Convergence and Applications of Newton-Like Iterations* (Springer, New York, 2008)
9. K. Diethelm, The Analysis of Fractional Differential Equations. Lecture Notes in Mathematics, vol. 2004, 1st edn. (Springer, New York, 2010)
10. J.A. Ezquerro, J.M. Gutierrez, M.A. Hernandez, N. Romero, M.J. Rubio, The Newton method: from Newton to Kantorovich (Spanish). Gac. R. Soc. Mat. Esp. **13**, 53–76 (2010)
11. J.A. Ezquerro, M.A. Hernandez, Newton-like methods of high order and domains of semilocal and global convergence. Appl. Math. Comput. **214**(1), 142–154 (2009)
12. L.V. Kantorovich, G.P. Akilov, *Functional Analysis in Normed Spaces* (Pergamon Press, New York, 1964)
13. A.A. Magrenan, Different anomalies in a Jarratt family of iterative root finding methods. Appl. Math. Comput. **233**, 29–38 (2014)
14. A.A. Magrenan, A new tool to study real dynamics: the convergence plane. Appl. Math. Comput. **248**, 215–224 (2014)
15. F.A. Potra, V. Ptak, *Nondiscrete Induction and Iterative Processes* (Pitman, London, 1984)
16. P.D. Proinov, New general convergence theory for iterative processes and its applications to Newton-Kantorovich type theorems. J. Complex. **26**, 3–42 (2010)

Chapter 15
Iterative Methods on Banach Spaces with a Convergence Structure and Fractional Calculus

We present a semilocal convergence for some iterative methods on a Banach space with a convergence structure to locate zeros of operators which are not necessarily Fréchet-differentiable as in earlier studies such as [6–8, 15]. This way we expand the applicability of these methods. If the operator involved is Fréchet-differentiable one approach leads to more precise error estimates on the distances involved than before [8, 15] and under the same hypotheses. Special cases are presented and some examples from fractional calculus. It follows [5].

15.1 Introduction

In this study we are concerned with the problem of locating a locally unique zero x^* of an operator G defined on a convex subset D of a Banach X with values in a Banach space Y. Our results will be presented for the operator F defined by

$$F(x) := JG(x_0 + x), \qquad (15.1.1)$$

where x_0 is an initial point and $J \in L(Y, X)$ the space of bounded linear operators from Y into X.

A lot of real life problems can be formulated like (15.1.1) using Mathematical Modelling [3, 4, 8, 9, 12]. The zeros of F can be found in closed form only in special cases. That is why most solution methods for these problems are usually iterative. There are mainly two types of convergence: semi-local and local convergence. The semi-local convergence case is based on the information around an initial point to find conditions ensuring the convergence of the iterative method; while the local one is based on the information around a solution, to find estimates of the radii of convergence balls [1, 2, 6–8, 10–17].

The most popular methods for approximating a zero of F are undoubtedly the so called Newton-like methods. There is a plethora of local as well as semi-local convergence results for these methods [1, 2, 6–8, 10–17].

© Springer International Publishing Switzerland 2016

G.A. Anastassiou and I.K. Argyros, *Intelligent Numerical Methods:*
Applications to Fractional Calculus, Studies in Computational Intelligence 624,
DOI 10.1007/978-3-319-26721-0_15

In the present study motivated by the works in [6–8, 15] we present a semi-local convergence analysis involving operators F that are not necessarily Fréchet-differentiable (as in [15]). Therefore, we expand the applicability of these methods in this case. We also show that even in the special case of Newton's method (i.e. when F is Fréchet-differentiable) our technique leads to more precise estimates on the distances involved under the same hypotheses as in [15].

The rest of the chapter is organized as follows. To make the paper as selfcontinued as possible, we present some standard concepts on Banach spaces with a convergence structure in Sect. 15.2. The semilocal convergence analysis of Newton-like methods is presented in Sect. 15.3. Special cases and some examples from fractional calculus involving the Caputo fractional derivative are given in the Sects. 15.4 and 15.5.

15.2 Banach Spaces with Convergence Structure

We present some results on Banach spaces with a convergence structure. More details can be found in [6–8, 15] and the references there in.

Definition 15.1 A triple (X, V, E) is a Banach space with convergence structure, if
(i) $(X, \|\cdot\|)$ is a real Banach space.
(ii) $(V, C, \|\cdot\|_V)$ is a real Banach space which is partially ordered by the closed convex cone C. The norm $\|\cdot\|_V$ is assumed to be monotone on C.
(iii) E is a closed convex cone in $X \times V$ such that $\{0\} \times C \subseteq E \subseteq X \times C$.
(iv) The operator $/ \cdot / : D \to C$

$$/x/ := \inf \{q \in C \,|\, (x, q) \in E\}$$

for each $x \in Q$, is well defined, where

$$Q := \{x \in X \,|\, \exists \, q \in E : (x, q) \in E\}.$$

(v) $\|x\| \leq \|/x/\|_V$ for each $x \in Q$.
Notice that it follows by the definition of Q that $Q + Q \subseteq Q$ and for each $\theta > 0$, $\theta Q \subseteq Q$. Define the set

$$U(a) := \{x \in X \,|\, (x, a) \in E\}.$$

Let us provide some examples when $X = \mathbb{R}^k$ equipped with the max-norm [6–8, 15]:
(a) $V = \mathbb{R}$; $E := \left\{(x, q) \in \mathbb{R}^k \times \mathbb{R} \,|\, \|x\|_\infty \leq q\right\}$.
(b) $V = \mathbb{R}^k$; $E := \left\{(x, q) \in \mathbb{R}^k \times \mathbb{R}^k \,|\, |x| \leq q\right\}$.
(c) $V = \mathbb{R}^k$; $E := \left\{(x, q) \in \mathbb{R}^k \times \mathbb{R}^k \,|\, 0 \leq x \leq q\right\}$.
More cases can be found in [8, 15].

Case (a) corresponds to the convergence analysis in a real Banach space; case (b) can be used for componentwise error analysis and case (c) may be used for monotone convergence analysis.

The convergence analysis is considered in the space $X \times V$. If $(x_n, q_n) \in E^{\mathbb{R}}$ is an increasing sequence, then:

$$(x_n, q_n) \leq (x_{n+m}, q_{n+m}) \Rightarrow 0 \leq (x_{n+m} - x_n, q_{n+m} - q_n).$$

Moreover, if $q_n \to q$ $(n \to \infty)$ then, we get: $0 \leq (x_{n+m} - x_n, q - q_n)$. Hence, by (v) of Definition 15.1

$$\|x_{n+m} - x_n\| \leq \|q - q_n\|_V \to 0 \, (n \to \infty).$$

That is we conclude that $\{x_n\}$ is a complete sequence. Set $q_n = w_0 - w_n$, where $\{w_n\} \in C^{\mathbb{R}}$ is a decreasing sequence.

Then, we have that

$$0 \leq (x_{n+m} - x_n, w_n - w_{n+m}) \leq (x_{n+m} - x_n, w_n).$$

Furthermore, if $x_n \to x^*$ $(n \to \infty)$, then we deduce that $/x^* - x_n/ \leq w_n$.

Let $L(X^j)$ denote the space of multilinear, symmetric, bounded operators on a Banach space X, $H : X^j \to X$.

Let also consider an ordered Banach space V:

$$L_+(V^j) := \left\{ L \in L(V^j) \,|\, 0 \leq x_i \Rightarrow 0 \leq L(x_1, x_2, \ldots, x_j) \right\}.$$

Let V_L be an open subset of an ordered Banach space V.

An operator $L \in C^1(V_L \to V)$ is defined to be order convex on an interval $[a, b] \subseteq V_L$, if for each $c, d \in [a, b]$, $c \leq d \Rightarrow L'(d) - L'(c) \in L_+(V)$.

Definition 15.2 The set of bounds for an operator $H \in L(X^j)$ is defined by:

$$B(H) := \left\{ L \in L_+(V^j) \,|\, (x_i, q_i) \in E \Rightarrow \left[H(x_1, \ldots, x_j), L(q_1, \ldots, q_j) \right] \in E \right\}.$$

Lemma 15.3 *Let $H : [0, 1] \to L(X^j)$ and $L : [0, 1] \to L_+(V^j)$ be continuous operators. Then, we have that for each $t \in [0, 1] : L(t) \in B(H(t)) \Rightarrow \int_0^1 L(t) \, dt \in B\left(\int_0^1 H(t) \, dt \right)$.*

Let $T : Y \to Y$ be an operator on a subset Y of a normed space. Denote by $T^n(x)$ the result of n-fold application of T. In particular in case of convergence, we write

$$T^\infty(x) := \lim_{n \to \infty} T^n(x).$$

Next, we define the right inverse:

Definition 15.4 Let $H \in L(X)$ and $u \in X$ be given. Then,

$$H^*u := x^* \Leftrightarrow x^* \in T^\infty(0), T(x) := (I - H)x + u \Leftrightarrow x^* = \sum_{j=0}^{\infty} (I - H)^j u,$$

provided that this limit exists.

Finally, we need two auxiliary results on inequalities in normed spaces and the Banach perturbation Lemma:

Lemma 15.5 *Let $L \in L_+(V)$ and $a, q \in C$ be given such that*

$$Lq + a \leq q \text{ and } L^n q \to 0 \ (n \to \infty).$$

Then, the operator
$$(I - L)^* : [0, a] \to [0, a]$$

is well defined and continuous.

Lemma 15.6 *Let $H \in L(X)$, $L \in B(H)$, $u \in D$ and $q \in C$ be given such that:*

$$Lq + /u/ \leq q \text{ and } L^n q \to 0 \ (n \to \infty).$$

Then, the point given by $x := (I - H)^* u$ is well defined, belongs in D and

$$/x/ \leq (I - L)^* /u/ \leq q.$$

15.3   Semilocal Convergence

We present the semilocal convergence in this section to determine a zero x^* of the operator (15.1.1) under certain conditions denoted by (A).

Let X be a Banach space with convergence structure (X, V, E), where $V = (V, C, \|\cdot\|_V)$, let operators $F : D \to X$ with $D \subseteq X$, $A(\cdot) : D \to L(X)$, $K, L, M : V_L \to V$ with $V_L \subseteq V$, $K_0(\cdot), M(\cdot) : V_L \to L_+(V)$ and a point $a \in C$ be such that the following conditions (A) hold:

(A$_1$) $U(a) \subseteq D$ and $[0, a] \subseteq V_L$.

(A$_2$) $L \leq K \leq M$, $K_0(\cdot) \leq M(\cdot)$.

(A$_3$) $K_0(0) \in B(I - A(0)), (-F(0), K(0))$.

(A$_4$) $K_0(/x/) - K_0(0) \in B(A(0) - A(x))$.

(A$_5$) $K_0(c)(d - c) \leq K(d) - K(c)$ and $M(c)(d - c) \leq M(d) - M(c)$ for each $c, d \in [0, a]$ with $c \leq d$.

(A$_6$) $L\left(/x/+/y-x/\right)-L\left(/x/\right)-M\left(/x/\right)/y-x/\in B\left(F\left(x\right)-F\left(y\right)\right.$
$+A\left(x\right)\left(y-x\right))$.

(A$_7$) $M\left(a\right)a\le a$.

(A$_8$) $M\left(a\right)^n a\to 0\;(n\to\infty)$.

Next, we can show the following semilocal convergence result of Newton-like methods using the preceding notation.

Theorem 15.7 *Suppose that the conditions (A) hold. Then*
(i) the sequences $\{x_n\}$, $\{\delta_n\}$ *defined by*

$$x_0=0,\; x_{n+1}:=x_n+A^*\left(x_n\right)\left(-F\left(x_n\right)\right),$$
$$\delta_0=0,\; \delta_{n+1}:=L\left(\delta_n\right)+M\left(/x_n/\right)\gamma_n,$$

where $\gamma_n:=/x_{n+1}-x_n/$, *are well defined, the sequence* $(x_n,\delta_n)\in(X\times V)^{\mathbb{R}}$ *remains in* $E^{\mathbb{R}}$, *for each* $n=0,1,2,\ldots$, *is monotone, and*

$$\delta_n\le b,\quad for\; each\; n=0,1,2,\ldots,$$

where $b:=M_0^\infty\left(0\right)$ *is the smallest fixed point of* $M\left(\cdot\right)$ *in* $[0,a]$.

(ii) The Newton-like sequence $\{x_n\}$ *is well defined, it remains in* $U\left(a\right)$ *for each* $n=0,1,2,\ldots$, *and converges to a unique zero* x^* *of* F *in* $U\left(a\right)$.

Proof (i) We shall solve the equation

$$q=\left(I-A\left(x_n\right)\right)q+\left(-F\left(x_n\right)\right),\quad for\; each\; n=0,1,2,\ldots. \tag{15.3.1}$$

First notice that the conditions of Theorem 15.7 are satisfied with b replacing a. If $n=1$ in (15.3.1) we get by (A$_2$), (A$_3$), (A$_5$) and (A$_7$) with $q=b$

$$K_0\left(0\right)b+/-F\left(0\right)/\le K\left(b\right)-K\left(0\right)+/-F\left(0\right)/\le K\left(b\right)\le M\left(b\right)b\le b.$$

That is x_1 is well defined and $(x_1,b)\in E$.

We get the estimate

$$x_1=\left(I-A\left(0\right)\right)x_1+\left(-F\left(0\right)\right)$$

so,

$$/x_1/\le K_0\left(0\right)/x_1/+L\left(0\right)\le M\left(0\right)/x_1/+L\left(0\right)=\delta_1$$

and by (A$_2$)

$$\delta_1=M\left(0\right)/x_1/+L\left(0\right)\le M\left(0\right)\left(b\right)+L\left(0\right)\le M\left(b\right)\left(b-0\right)+L\left(0\right)$$

$$\le M\left(b\right)b-M\left(b\right)\left(0\right)+L\left(0\right)\le M\left(b\right)b\le b.$$

Suppose that the sequence is well defined and monotone for $k = 1, 2, \ldots, n$ and $\delta_k \leq b$. Using the induction hypotheses and (A$_6$) we get in turn that

$$/ - F(x_n) / = / - F(x_n) + F(x_{n-1}) + A(x_{n-1})(x_n - x_{n-1}) /$$

$$\leq L(/x_{n-1}/ + \gamma_{n-1}) - L(/x_{n-1}/) - M(/x_{n-1}/)\gamma_{n-1}$$

$$\leq L(\delta_{n-1} + \delta_n - \delta_{n-1}) - L(\delta_{n-1}) - M(/x_{n-1}/)\gamma_{n-1} \qquad (15.3.2)$$

$$= L(\delta_n) - \delta_n. \qquad (15.3.3)$$

By (A$_2$)–(A$_4$) we have the estimate

$$/I - A(x_n) / \leq /I - A(0)/ + /A(0) - A(x_n)/ \leq K_0(0) + K_0(/x_n/) - K_0(0)$$

$$= K_0(/x_n/) \leq M(/x_n/).$$

Then, to solve the Eq. (15.3.1), let $q = b - \delta_n$ to obtain that

$$M(/x_n/)(b - \delta_n) + / - F(x_n)/ + \delta_n \leq M(\delta_n)(b - \delta_n) + L(\delta_n)$$

$$\leq M(b)b \leq b.$$

That is x_{n+1} is well defined by Lemma 15.5 and $\gamma_n \leq b - \delta_n$. Therefore, δ_{n+1} is also well defined and we can have:

$$\delta_{n+1} \leq L(\delta_n) + M(\delta_n)(b - \delta_n) \leq M(b)b \leq b.$$

We also need to show the monotonicity of $(x_n, \delta_n) \leq (x_{n+1}, \delta_{n+1})$:

$$\gamma_n + \delta_n \leq M(/x_n/)\gamma_n + / - F(x_n)/ + \delta_n \leq M(/x_n/)\gamma_n + L(\delta_n) = \delta_{n+1}.$$

The induction is complete and the statement (i) is shown.

(ii) Using induction and the definition of sequence $\{\delta_n\}$ we get $M(0)^n(0) \leq \delta_n \leq b$, which implies $\delta_n \to b$, since $M(0)^n(0) \to b$. It follows from the discussion in Sect. 15.2 that sequence $\{x_n\}$ converges to some $x^* \in U(b)$ (since $U(b)$ is a closed set). By letting $n \to \infty$ in (15.3.3) we deduce that x^* is a zero of F. Let $y^* \in U(a)$ be a zero of F. Then as in [15] we get that

$$/y^* - x_n/ \leq M^n(a)(a) - M^n(0)(0),$$

so, we conclude that $x^* = y^*$. □

Remark 15.8 Concerning a posteriori estimates, we can list a few. It follows from the proof of Theorem 15.7 that

$$/x^* - x_n/ \leq b - \delta_n \leq q - \delta_n,$$

where we can use for q any solution of $M(q)q \leq q$. We can obtain more precise error estimates as in [15] by introducing monotone maps R_n under the (A) conditions as follows:

$$R_n(q) := (I - K_0/x_n/)^* S_n(q) + \gamma_n,$$

where

$$S_n(q) := L(/x_n/ + q) - L(/x_n/) - M(/x_n/)q.$$

Notice that operator S_n is monotone on the interval $I_n := [0, a - /x_n/]$. Suppose that there exists $q_n \in C$ such that $/x_n/ + q_n \leq a$ and

$$S_n(q_n) + M(/x_n/)(q_n - \gamma_n) \leq q_n - \gamma_n.$$

It then follows that operator $R_n : [0, q_n] \to [0, q_n]$ is well defined and monotone by Lemma 15.5 for each $n = 0, 1, 2, \ldots$. A possible choice q_n is $a - \delta_n$. Indeed, this follows from the implications

$$\delta_n + \gamma_n \leq \delta_{n+1} \Rightarrow M(a)(a) - L(\delta_n) - M(/x_n/)\gamma_n \leq a - \delta_n - \gamma_n \Rightarrow$$

$$L(a) - L(\delta_n) - M(/x_n/)\gamma_n \leq a - \delta_n - \gamma_n \Rightarrow$$

$$S_n(a - \delta_n) + M(/x_n/)(a - \delta_n - \gamma_n) \leq a - \delta_n - \gamma_n.$$

The proofs of the next three results are omitted since they follow from the corresponding ones in [6–8, 15] by simply using R_n and S_n instead of

$$\widetilde{R}_n(q) = (I - L'(/x_n/))^* \widetilde{S}_n(q) + c_n,$$

$$\widetilde{S}_n(q) = L(/x_n/ + q) - L(/x_n/) - L'(/x_n/)q,$$

used in the preceding references, where $c_n = \gamma_n$ and for $L \in C^1(V_L \to L)$ being order convex of the interval $[a, b] \subset V_L$. Let us also define the sequence $\{d_n\}$ by

$$d_0 = 0, d_{n+1} = L(d_n) + L'(/x_n/)c_n. \tag{15.3.4}$$

Proposition 15.9 *Suppose that there exists $q \in I_n$ such that $R_n(q) \leq q$. Then, the following hold*

$$\gamma_n \leq R_n(q) =: q_0 \leq q$$

and

$$R_{n+1} (q_0 - \gamma_n) \leq q_0 - \gamma_n.$$

Proposition 15.10 *Suppose that the (A) conditions hold. Moreover, suppose that there exist $q_n \in I_n$ such that $R_n (q_n) \leq q_n$. Then, the sequence $\{p_n\}$ defined by*

$$p_n = q_n, \ p_{n+1} := R_n (p_m) - \gamma_n \quad for \ m \geq n$$

leads to the estimate $/x^* - x_n/ \leq p_m$.

Proposition 15.11 *Suppose that the (A) conditions hold. Then for any $q \in I_n$ satisfying $R_n (q) \leq q$ we have that*

$$/x^* - x_n/ \leq R_n^\infty (0) \leq q.$$

The rest of the results in [6–8, 15] can be generalized along the same framework.

15.4 Special Cases and Examples

Special case: Newton's method

Let us state Theorem 5 from [15] (see also [8]) so we can compare it with Theorem 15.7.

Theorem 15.12 *Suppose that X is a Banach space with convergence structure (X, V, E) with $V = (V, C, \|\cdot\|_V)$, let operator $F \in C^1 (D \to X)$, operator $L \in C^1 (V_L \to V)$ and a point $a \in C$ such that the following conditions hold:*
(h_1) $U (a) \subseteq D$, $[0, a] \subseteq V_L$.
(h_2) L is order convex on $[0, a]$ and satisfies for each $x, y \in U (a)$,

$$/x/ + /y/ \leq a : L' (/x/ + /y/) - L' (/x/) \in B \left(F' (x) - F' (x + y) \right).$$

(h_3) $L' (0) \in B \left(I - F' (0) \right)$, $(-F (0) , L (0)) \in E$.
(h_4) $L (a) \leq a$.
(h_5) $L' (a)^n a \to 0 \ (n \to \infty)$.
Then, the sequence $\{x_n\}$ generated by Newton's method for each $n = 0, 1, 2, \ldots,$

$$x_0 := 0, \ x_{n+1} := x_n + F' (/x_n/)^* (-F (x_n))$$

is well defined, remains in $U (a)$ for each $n = 0, 1, 2, \ldots$ and converges to a unique zero x^* of F in $U (a)$.
Moreover, the following estimates hold

$$/x^* - x_n/ \leq b - d_n,$$

where the sequence $\{d_n\}$ is defined by (15.3.4).

Let $L \in C^1 (V_L \rightarrow V)$, $A(x) = F'(x)$, $K_0 = M = L'$ and $K = L$. Then, Theorem 15.7 reduces to the weaker version of Theorem 15.12 given by:

Theorem 15.13 *Suppose that the hypotheses of Theorem 15.12 but with (h_2) replaced by*
(h_2') *L is order convex on $[0, a]$ and satisfies for each $x, y \in U(a)$, $/x/ + /y/ \le$*
$a : L(/x/ + /y/) - L(/x/) - L'(/x/)/y - x/ \in B(F(x) - F(y)$
$+ F'(x)(y - x))$.
Then, the conclusions of Theorem 15.12 hold.

Remark 15.14 Notice that condition (h_2) implies condition (h_2') but not necessarily vice versa. Hence, Theorem 15.13 is weaker that Theorem 15.12.

Another improvement of Theorem 15.12 can be given as follows:
(h_2^0) there exists: L_0 which is order convex on $[0, a]$ and satisfies for each $/x/ \le$
$a : L_0'(/x/) - L_0'(0) \in B\left(F'(0) - F'(x)\right)$.
Notice however that

$$L_0' \le L' \qquad\qquad (15.4.1)$$

holds in general and $\frac{L'}{L_0}$ can be arbitrarily large [8].

Notice that (h_2^0) is not an additional to (h_2) condition, since in practice the computation of L' requires the computation of L_0' as a special case.

Condition (h_3) can then certainly be replaced by the weaker
(h_3^0) $L_0'(0) \in B\left(I - F'(0)\right)$, $(-F(0), L(0)) \in E$.
Moreover, if

$$L(0)0 \le L(0), \qquad\qquad (15.4.2)$$

then condition (h_3^0) can be replaced by the weaker
(h_3^1) $L_0'(0) \in B\left(I - F'(0)\right)$, $(-F(0), L_0(0)) \in E$.
Define sequence $\{d_n^0\}$ by

$$d_0^0 := 0, \ d_1^0 = L_0\left(d_0^0\right) + L_0'(/x_0/) c_0, \ d_2^0 = L_0\left(d_1^0\right) + L_0'(/x_1/) c_1,$$

$$d_{n+1}^0 := L\left(d_n^0\right) + L_0'(/x_n/) c_n, \quad n = 2, 3, \ldots .$$

Then, we present the following improvement of Theorem 15.12.

Theorem 15.15 *Suppose that the hypotheses of Theorem 15.12 or Theorem 15.13 hold. Then, the conclusions hold with sequence $\{d_n^0\}$ replacing $\{d_n\}$ and L_0, L_0' replacing L, L' in (h_3) and (h_5), respectively. Moreover, we have*

$$d_n^0 \le d_n \quad \text{for each } n = 0, 1, 2, \ldots . \qquad\qquad (15.4.3)$$

Proof Simply notice that the following crucial estimate holds:

$$L'_0 \left(/x_n/\right) \left(b - d_n^0\right) + / - F\left(x_n\right) / + d_n^0$$

$$\leq L'_0 \left(/x_n/\right) \left(b - d_n^0\right) + L\left(d_n^0\right) - L\left(d_{n-1}^0\right) - L'\left(/x_{n-1}/\right) c_{n-1}$$

$$- L'_0 \left(/x_{n-1}/\right) c_{n-1} + L'_0 \left(/x_{n-1}/\right) c_{n-1} + d_n^0$$

$$\leq L'_0 \left(/x_n/\right) \left(b - d_n^0\right) + L'_0 \left(/x_{n-1}/\right) c_{n-1} - L'\left(/x_{n-1}/\right) c_{n-1} + L\left(d_n^0\right)$$

$$\leq L'_0 \left(/x_n/\right) \left(b - d_n^0\right) + L\left(d_n^0\right)$$

$$\leq L'\left(d_n^0\right) \left(b - d_n^0\right) + L\left(d_n^0\right) \leq L\left(b\right) \leq b.$$

Finally, the estimate (15.4.3) follows by the definition of sequences $\{d_n^0\}$, $\{d_n\}$, (15.4.1) and a simply inductive argument.   □

Remark 15.16 Estimate (15.4.3) holds as a strict inequality for $n = 1, 2, \ldots$, if (15.4.1) is a strict inequality. Hence, the error estimates are improved in this case under the hypotheses of Theorem 15.12 or Theorem 15.13. Finally the a posteriori results presented in Sect. 15.3 are also improved in this special case.

15.5   Applications to Fractional Calculus

In this section we apply our numerical method to fractional calculus.

In our cases we take J the identity map, the function G as f, and $x_0 = 0$. We want to solve

$$f\left(x\right) = 0. \tag{15.5.1}$$

(I) Let $1 < \nu < 2$, i.e. $\lceil \nu \rceil = 2$ ($\lceil \cdot \rceil$ ceiling of number); $x, y \in [0, a]$, $a > 0$, and $f \in C^2\left([0, a]\right)$.

We define the following left Caputo fractional derivatives (see [3], p. 270) by

$$\left(D_{*y}^\nu f\right)\left(x\right) := \frac{1}{\Gamma\left(2 - \nu\right)} \int_y^x \left(x - t\right)^{1-\nu} f''\left(t\right) dt, \tag{15.5.2}$$

when $x \geq y$, and

$$\left(D_{*x}^\nu f\right)\left(y\right) := \frac{1}{\Gamma\left(2 - \nu\right)} \int_x^y \left(y - t\right)^{1-\nu} f''\left(t\right) dt, \tag{15.5.3}$$

when $y \geq x$, where Γ is the gamma function.

We define also the linear operator

$$(A_0(f))(x, y) := \begin{cases} f'(y) + (D_{*y}^\nu f)(x) \cdot \frac{(x-y)^{\nu-1}}{\Gamma(\nu+1)}, & x > y, \\ f'(x) + (D_{*x}^\nu f)(y) \cdot \frac{(y-x)^{\nu-1}}{\Gamma(\nu+1)}, & y > x, \\ 0, & x = y. \end{cases} \qquad (15.5.4)$$

When f is increasing and $f \geq 0$, then $(A_0(f))(x, y) \geq 0$.

By left fractional Caputo Taylor's formula (see [9], p. 54 and [3], p. 395) we get that

$$f(x) - f(y) = f'(y)(x - y) + \frac{1}{\Gamma(\nu)} \int_y^x (x - t)^{\nu-1} D_{*y}^\nu f(t)\, dt, \quad \text{for } x > y, \tag{15.5.5}$$

and

$$f(y) - f(x) = f'(x)(y - x) + \frac{1}{\Gamma(\nu)} \int_x^y (y - t)^{\nu-1} D_{*x}^\nu f(t)\, dt, \quad \text{for } x < y, \tag{15.5.6}$$

equivalently, it holds

$$f(x) - f(y) = f'(x)(x - y) - \frac{1}{\Gamma(\nu)} \int_x^y (y - t)^{\nu-1} D_{*x}^\nu f(t)\, dt, \quad \text{for } x < y. \tag{15.5.7}$$

We would like to prove that

$$|f(x) - f(y) - (A_0(f))(x, y) \cdot (x - y)| \leq c \cdot \frac{(x - y)^2}{2}, \tag{15.5.8}$$

for any $x, y \in [0, a], 0 < c < 1$.

When $x = y$ the last condition (15.5.8) is trivial.

We assume $x \neq y$. We distinguish the cases:

(1) $x > y$: We observe that

$$|f(x) - f(y) - (A_0(f))(x, y) \cdot (x - y)| = \tag{15.5.9}$$

$$\left| f'(y)(x - y) + \frac{1}{\Gamma(\nu)} \int_y^x (x - t)^{\nu-1} (D_{*y}^\nu f)(t)\, dt - \right.$$

$$\left. \left(f'(y) + (D_{*y}^\nu f)(x) \cdot \frac{(x - y)^{\nu-1}}{\Gamma(\nu+1)} \right)(x - y) \right| =$$

$$\left| \frac{1}{\Gamma(\nu)} \int_y^x (x - t)^{\nu-1} (D_{*y}^\nu f)(t)\, dt - (D_{*y}^\nu f)(x) \frac{(x - y)^\nu}{\Gamma(\nu+1)} \right| = \tag{15.5.10}$$

$$\left| \frac{1}{\Gamma(\nu)} \int_y^x (x-t)^{\nu-1} \left(D_{*y}^\nu f\right)(t)\, dt - \frac{1}{\Gamma(\nu)} \int_y^x (x-t)^{\nu-1} \left(D_{*y}^\nu f\right)(x)\, dt \right| =$$

$$(15.5.11)$$

$$\frac{1}{\Gamma(\nu)} \left| \int_y^x (x-t)^{\nu-1} \left(\left(D_{*y}^\nu f\right)(t) - \left(D_{*y}^\nu f\right)(x) \right) dt \right| \le$$

$$\frac{1}{\Gamma(\nu)} \int_y^x (x-t)^{\nu-1} \left| \left(D_{*y}^\nu f\right)(t) - \left(D_{*y}^\nu f\right)(x) \right| dt =: (\xi), \qquad (15.5.12)$$

(assume that

$$\left| \left(D_{*y}^\nu f\right)(t) - \left(D_{*y}^\nu f\right)(x) \right| \le \lambda_1 \, |t-x|^{2-\nu}, \qquad (15.5.13)$$

for any $t, x, y \in [0, a] : x \ge t \ge y$, where $\lambda_1 < \Gamma(\nu)$, i.e. $\rho_1 := \frac{\lambda_1}{\Gamma(\nu)} < 1$).
   Therefore

$$(\xi) \le \frac{\lambda_1}{\Gamma(\nu)} \int_y^x (x-t)^{\nu-1} (x-t)^{2-\nu}\, dt \qquad (15.5.14)$$

$$= \frac{\lambda_1}{\Gamma(\nu)} \int_y^x (x-t)\, dt = \frac{\lambda_1}{\Gamma(\nu)} \frac{(x-y)^2}{2} = \rho_1 \frac{(x-y)^2}{2}. \qquad (15.5.15)$$

We have proved that

$$\left| f(x) - f(y) - (A_0(f))(x,y) \cdot (x-y) \right| \le \rho_1 \frac{(x-y)^2}{2}, \qquad (15.5.16)$$

where $0 < \rho_1 < 1$, and $x > y$.
   (2) $x < y$: We observe that

$$\left| f(x) - f(y) - (A_0(f))(x,y) \cdot (x-y) \right| = \qquad (15.5.17)$$

$$\left| f'(x)(x-y) - \frac{1}{\Gamma(\nu)} \int_x^y (y-t)^{\nu-1} D_{*x}^\nu f(t)\, dt - \right.$$

$$\left. \left(f'(x) + \left(D_{*x}^\nu f\right)(y) \cdot \frac{(y-x)^{\nu-1}}{\Gamma(\nu+1)} \right)(x-y) \right| =$$

$$\left| -\frac{1}{\Gamma(\nu)} \int_x^y (y-t)^{\nu-1} D_{*x}^\nu f(t)\, dt + \left(D_{*x}^\nu f\right)(y) \frac{(y-x)^\nu}{\Gamma(\nu+1)} \right| = \qquad (15.5.18)$$

$$\left| \frac{1}{\Gamma(\nu)} \int_x^y (y-t)^{\nu-1} D_{*x}^\nu f(t)\, dt - \left(D_{*x}^\nu f\right)(y) \frac{(y-x)^\nu}{\Gamma(\nu+1)} \right| = \qquad (15.5.19)$$

$$\frac{1}{\Gamma(\nu)} \left| \int_x^y (y-t)^{\nu-1} D_{*x}^\nu f(t)\, dt - \frac{1}{\Gamma(\nu)} \int_x^y (y-t)^{\nu-1} \left(D_{*x}^\nu f\right)(y)\, dt \right| =$$

$$\frac{1}{\Gamma(\nu)} \left| \int_x^y (y-t)^{\nu-1} \left(D_{*x}^\nu f(t) - D_{*x}^\nu f(y) \right) dt \right| \le \qquad (15.5.20)$$

$$\frac{1}{\Gamma(\nu)} \int_x^y (y-t)^{\nu-1} \left| D_{*x}^\nu f(t) - D_{*x}^\nu f(y) \right| dt$$

(by assumption,

$$\left| D_{*x}^\nu f(t) - D_{*x}^\nu f(y) \right| \le \lambda_2 |t-y|^{2-\nu}, \qquad (15.5.21)$$

for any $t, y, x \in [0, a] : y \ge t \ge x$).

$$\le \frac{1}{\Gamma(\nu)} \int_x^y (y-t)^{\nu-1} \lambda_2 |t-y|^{2-\nu} dt$$

$$= \frac{\lambda_2}{\Gamma(\nu)} \int_x^y (y-t)^{\nu-1} (y-t)^{2-\nu} dt \qquad (15.5.22)$$

$$= \frac{\lambda_2}{\Gamma(\nu)} \int_x^y (y-t) dt = \frac{\lambda_2}{\Gamma(\nu)} \frac{(x-y)^2}{2}.$$

Assuming also $\rho_2 := \frac{\lambda_2}{\Gamma(\nu)} < 1$ (i.e. $\lambda_2 < \Gamma(\nu)$), we have proved that

$$|f(x) - f(y) - (A_0(f))(x,y) \cdot (x-y)| \le \rho_2 \frac{(x-y)^2}{2}, \quad \text{for } x < y. \qquad (15.5.23)$$

Conclusion: choosing $\lambda := \max(\lambda_1, \lambda_2)$ and $\rho := \frac{\lambda}{\Gamma(\nu)} < 1$, we have proved that

$$|f(x) - f(y) - (A_0(f))(x,y) \cdot (x-y)| \le \rho \frac{(x-y)^2}{2}, \quad \text{for any } x, y \in [0, a]. \qquad (15.5.24)$$

This is a condition needed to solve numerically $f(x) = 0$.

(II) Let $n - 1 < \nu < n$, $n \in \mathbb{N} - \{1\}$, i.e. $\lceil \nu \rceil = n$; $x, y \in [0, a]$, $a > 0$, and $f \in C^n([0, a])$.

We define the following right Caputo fractional derivatives (see [4], p. 336),

$$D_{x-}^\nu f(y) = \frac{(-1)^n}{\Gamma(n-\nu)} \int_y^x (z-y)^{n-\nu-1} f^{(n)}(z) \, dz, \quad \text{for } y \le x, \qquad (15.5.25)$$

and

$$D_{y-}^\nu f(x) = \frac{(-1)^n}{\Gamma(n-\nu)} \int_x^y (z-x)^{n-\nu-1} f^{(n)}(z) \, dz, \quad \text{for } x \le y. \qquad (15.5.26)$$

By right Caputo fractional Taylor's formula (see [4], p. 341) we have

$$f(x) - f(y) = \sum_{k=1}^{n-1} \frac{f^{(k)}(y)}{k!}(x-y)^k + \frac{1}{\Gamma(\nu)}\int_x^y (z-x)^{\nu-1}\left(D_{y-}^\nu f\right)(z)\,dz,$$

$$(15.5.27)$$

when $x \le y$, and

$$f(y) - f(x) = \sum_{k=1}^{n-1} \frac{f^{(k)}(x)}{k!}(y-x)^k + \frac{1}{\Gamma(\nu)}\int_y^x (z-y)^{\nu-1}\left(D_{x-}^\nu f\right)(z)\,dz,$$

$$(15.5.28)$$

when $x \ge y$.

We define also the linear operator

$$(A_0(f))(x,y) := \begin{cases} \sum_{k=1}^{n-1} \frac{f^{(k)}(x)}{k!}(y-x)^k - \left(D_{x-}^\nu f\right)(y) \cdot \frac{(x-y)^{\nu-1}}{\Gamma(\nu+1)}, & x > y, \\ \sum_{k=1}^{n-1} \frac{f^{(k)}(y)}{k!}(x-y)^k - \left(D_{y-}^\nu f\right)(x) \cdot \frac{(y-x)^{\nu-1}}{\Gamma(\nu+1)}, & y > x, \\ 0, & x = y. \end{cases}$$

$$(15.5.29)$$

When $n = 2$, and f is decreasing and $f \ge 0$, then $(A_0(f))(x,y) \le 0$.

We would like to prove that

$$|f(x) - f(y) - (A_0(f))(x,y) \cdot (x-y)| \le c \cdot \frac{|x-y|^n}{n},$$

$$(15.5.30)$$

for any $x, y \in [0,a], 0 < c < 1$.

When $x = y$ the last condition (15.5.30) is trivial.

We assume $x \ne y$. We distinguish the cases:

(1) $x > y$: We observe that

$$|(f(x) - f(y)) - (A_0(f))(x,y) \cdot (x-y)| =$$

$$(15.5.31)$$

$$|(f(y) - f(x)) - (A_0(f))(x,y) \cdot (y-x)| =$$

$$\left| \left(\sum_{k=1}^{n-1} \frac{f^{(k)}(x)}{k!}(y-x)^k + \frac{1}{\Gamma(\nu)}\int_y^x (z-y)^{\nu-1}\left(D_{x-}^\nu f\right)(z)\,dz \right) - \right.$$

$$\left. \left(\sum_{k=1}^{n-1} \frac{f^{(k)}(x)}{k!}(y-x)^{k-1} - \left(D_{x-}^\nu f\right)(y) \cdot \frac{(x-y)^{\nu-1}}{\Gamma(\nu+1)} \right)(y-x) \right| =$$

$$\left| \frac{1}{\Gamma(\nu)}\int_y^x (z-y)^{\nu-1}\left(D_{x-}^\nu f\right)(z)\,dz + \left(D_{x-}^\nu f\right)(y)\frac{(x-y)^{\nu-1}}{\Gamma(\nu+1)}(y-x) \right| =$$

$$(15.5.32)$$

$$\left| \frac{1}{\Gamma(\nu)}\int_y^x (z-y)^{\nu-1}\left(D_{x-}^\nu f\right)(z)\,dz - \left(D_{x-}^\nu f\right)(y)\frac{(x-y)^\nu}{\Gamma(\nu+1)} \right| =$$

$$\frac{1}{\Gamma(\nu)} \left| \int_y^x (z - y)^{\nu-1} \left(D_{x-}^\nu f\right)(z) \, dz - \int_y^x (z - y)^{\nu-1} \left(D_{x-}^\nu f\right)(y) \, dz \right| =$$

$$\frac{1}{\Gamma(\nu)} \left| \int_y^x (z - y)^{\nu-1} \left(\left(D_{x-}^\nu f\right)(z) - \left(D_{x-}^\nu f\right)(y)\right) dz \right| \leq \quad (15.5.33)$$

$$\frac{1}{\Gamma(\nu)} \int_y^x (z - y)^{\nu-1} \left|\left(D_{x-}^\nu f\right)(z) - \left(D_{x-}^\nu f\right)(y)\right| dz$$

(we assume that

$$\left|\left(D_{x-}^\nu f\right)(z) - \left(D_{x-}^\nu f\right)(y)\right| \leq \lambda_1 |z - y|^{n-\nu}, \quad (15.5.34)$$

$\lambda_1 > 0$, for all $x, z, y \in [0, a]$, with $x \geq z \geq y$)

$$\leq \frac{\lambda_1}{\Gamma(\nu)} \int_y^x (z - y)^{\nu-1} (z - y)^{n-\nu} \, dz = \quad (15.5.35)$$

$$= \frac{\lambda_1}{\Gamma(\nu)} \int_y^x (z - y)^{n-1} \, dz = \frac{\lambda_1}{\Gamma(\nu)} \frac{(x - y)^n}{n}$$

(assume $\lambda_1 < \Gamma(\nu)$, i.e. $\rho_1 := \frac{\lambda_1}{\Gamma(\nu)} < 1$)

$$= \rho_1 \frac{(x - y)^n}{n}.$$

We have proved, when $x > y$, that

$$|f(x) - f(y) - (A_0(f))(x, y) \cdot (x - y)| \leq \rho_1 \frac{(x - y)^n}{n}. \quad (15.5.36)$$

(2) $y > x$: We observe that

$$|f(x) - f(y) - (A_0(f))(x, y) \cdot (x - y)| =$$

$$\left| \left(\sum_{k=1}^{n-1} \frac{f^{(k)}(y)}{k!} (x - y)^k + \frac{1}{\Gamma(\nu)} \int_x^y (z - x)^{\nu-1} \left(D_{y-}^\nu f\right)(z) \, dz \right) - \right.$$

$$\left. \left(\sum_{k=1}^{n-1} \frac{f^{(k)}(y)}{k!} (x - y)^{k-1} - \left(D_{y-}^\nu f\right)(x) \cdot \frac{(y - x)^{\nu-1}}{\Gamma(\nu + 1)} \right) (x - y) \right| = \quad (15.5.37)$$

$$\left| \frac{1}{\Gamma(\nu)} \int_x^y (z - x)^{\nu-1} \left(D_{y-}^\nu f\right)(z) \, dz - \left(D_{y-}^\nu f\right)(x) \frac{(y - x)^\nu}{\Gamma(\nu + 1)} \right| = \quad (15.5.38)$$

$$\left| \frac{1}{\Gamma(\nu)} \int_x^y (z-x)^{\nu-1} \left(D_{y-}^\nu f\right)(z)\, dz - \frac{1}{\Gamma(\nu)} \int_x^y (z-x)^{\nu-1} \left(D_{y-}^\nu f\right)(x)\, dz \right| =$$

$$\frac{1}{\Gamma(\nu)} \left| \int_x^y (z-x)^{\nu-1} \left(\left(D_{y-}^\nu f\right)(z) - \left(D_{y-}^\nu f\right)(x) \right) dz \right| \le \qquad (15.5.39)$$

$$\frac{1}{\Gamma(\nu)} \int_x^y (z-x)^{\nu-1} \left| \left(D_{y-}^\nu f\right)(z) - \left(D_{y-}^\nu f\right)(x) \right| dz$$

(we assume that

$$\left| \left(D_{y-}^\nu f\right)(z) - \left(D_{y-}^\nu f\right)(x) \right| \le \lambda_2 \left| z - x \right|^{n-\nu}, \qquad (15.5.40)$$

$\lambda_2 > 0$, for all $y, z, x \in [0, a]$ with $y \ge z \ge x$)

$$\le \frac{\lambda_2}{\Gamma(\nu)} \int_x^y (z-x)^{\nu-1} (z-x)^{n-\nu}\, dz = \qquad (15.5.41)$$

$$\frac{\lambda_2}{\Gamma(\nu)} \int_x^y (z-x)^{n-1}\, dz = \frac{\lambda_2}{\Gamma(\nu)} \frac{(y-x)^n}{n}.$$

Assume now that $\lambda_2 < \Gamma(\nu)$, that is $\rho_2 := \frac{\lambda_2}{\Gamma(\nu)} < 1$.

We have proved, for $y > x$, that

$$|f(x) - f(y) - (A_0(f))(x, y) \cdot (x - y)| \le \rho_2 \frac{(y-x)^n}{n}. \qquad (15.5.42)$$

Set $\lambda := \max(\lambda_1, \lambda_2)$, and

$$0 < \rho := \frac{\lambda}{\Gamma(\nu)} < 1. \qquad (15.5.43)$$

Conclusion: We have proved that

$$|f(x) - f(y) - (A_0(f))(x, y) \cdot (x - y)| \le \rho \frac{|x - y|^n}{n}, \quad \text{for any } x, y \in [0, a].$$
$$(15.5.44)$$

In the special case of $1 < \nu < 2$, we obtain that

$$|f(x) - f(y) - (A_0(f))(x, y) \cdot (x - y)| \le \rho \frac{(x - y)^2}{2}, \qquad (15.5.45)$$

for any $x, y \in [0, a], 0 < \rho < 1$.

This is a condition needed to solve numerically $f(x) = 0$.

(III) A simple instructive example follows:

Let $f \in C^1([0, a]), a > 0$. We assume that

$$\left| f'(x) - f'(y) \right| \le \lambda \left| x - y \right|, \quad \text{where } 0 < \lambda < 1, \tag{15.5.46}$$

for every $x, y \in [0, a]$. Here we take $A_0(f)(x) := f'(x)$, all $x \in [0, a]$.

We notice that

$$f(x) - f(y) = \left(\int_0^1 f'(y + \theta(x - y)) d\theta \right)(x - y). \tag{15.5.47}$$

Therefore it holds

$$\left| f(x) - f(y) - (A_0(f))(x) \cdot (x - y) \right| = \tag{15.5.48}$$

$$\left| \left(\int_0^1 f'(y + \theta(x - y)) d\theta \right)(x - y) - \left(\int_0^1 f'(x) d\theta \right)(x - y) \right| =$$

$$\left| \int_0^1 \left(f'(y + \theta(x - y)) - f'(x) \right) d\theta \right| |x - y| \le$$

$$\left(\int_0^1 \left| f'(y + \theta(x - y)) - f'(x) \right| d\theta \right) |x - y| \le \tag{15.5.49}$$

$$\lambda \left(\int_0^1 \left| y + \theta(x - y) - x \right| d\theta \right) |x - y| =$$

$$\lambda \left(\int_0^1 (1 - \theta) |x - y| d\theta \right) |x - y| = \tag{15.5.50}$$

$$\lambda \left(\frac{(1 - \theta)^2}{2} \Big|_1^0 \right) |x - y|^2 = \lambda \frac{(x - y)^2}{2},$$

proving that

$$\left| f(x) - f(y) - (A_0(f))(x) \cdot (x - y) \right| \le \lambda \frac{(x - y)^2}{2}, \tag{15.5.51}$$

for all $x, y \in [0, a]$, a condition needed to solve numerically $f(x) = 0$.

Next, we connect the results of this section to a special case of Theorem 15.7 for the real norm $\|\cdot\|_\infty$ as follows:

Define functions F and A by

$$F(x) = Jf(x_0 + x) \text{ and } A(x) = JA_0(x), \tag{15.5.52}$$

where $J = (A_0(0))^{-1}$ and $x_0 = 0$.

Choose $L = K = M$ and $K_0(\cdot) = M(\cdot)$, where $L(t) = \|F(0)\| + \frac{1}{2}\overline{\rho}t^2$ and $K_0(t) = M(t) = L'(t) = \overline{\rho}t$, $\overline{\rho} = \|J\|\rho$.

Then, conditions (A_1)–(A_6) are satisfied. Moreover, condition (A_7) reduces to

$$2\overline{\rho}\|F(0)\| < 1 \qquad\qquad (15.5.53)$$

for

$$a = \frac{1 - \sqrt{1 - 2\overline{\rho}\|F(0)\|}}{\overline{\rho}}.$$

Furthermore, condition (A_8) holds, provided that (15.5.53) is a strict inequality. Hence, we deduce that the conclusions of Theorem 15.7 hold for equation $F(x) \doteq 0$, where F is given by (15.5.52) provided that the Newton-Kantorovich-type condition [8, 12] (15.5.53) holds as a strict inequality.

References

1. S. Amat, S. Busquier, Third-order iterative methods under Kantorovich conditions. J. Math. Anal. Appl. **336**, 243–261 (2007)
2. S. Amat, S. Busquier, S. Plaza, Chaotic dynamics of a third-order Newton-like method. J. Math. Anal. Appl. **366**(1), 164–174 (2010)
3. G. Anastassiou, *Fractional Differentiation Inequalities* (Springer, New York, 2009)
4. G. Anastassiou, *Intelligent Mathematics: Computational Analysis* (Springer, Heidelberg, 2011)
5. G. Anastassiou, I. Argyros, *Convergence for iterative methods on Banach spaces of a convergence structure with applications to fractional calculus*, Sema, accepted (2015)
6. I.K. Argyros, Newton-like methods in partially ordered linear spaces. J. Approx. Theory Appl. **9**(1), 1–10 (1993)
7. I.K. Argyros, Results on controlling the residuals of perturbed Newton-like methods on Banach spaces with a convergence structure. Southwest J. Pure Appl. Math. **1**, 32–38 (1995)
8. I.K. Argyros, *Convergence and Applications of Newton-Like Iterations* (Springer, New York, 2008)
9. K. Diethelm, The Analysis,of Fractional Differential Equations, Lecture Notes in Mathematics, vol. 2004, 1st edn. (Springer, New York, Heidelberg, 2010)
10. J.A. Ezquerro, J.M. Gutierrez, M.A. Hernandez, N. Romero, M.J. Rubio, The Newton method: from Newton to Kantorovich (Spanish). Gac. R. Soc. Mat. Esp. **13**, 53–76 (2010)
11. J.A. Ezquerro, M.A. Hernandez, Newton-like methods of high order and domains of semilocal and global convergence. Appl. Math. Comput. **214**(1), 142–154 (2009)
12. L.V. Kantorovich, G.P. Akilov, *Functional Analysis in Normed Spaces* (Pergamon Press, New York, 1964)
13. A.A. Magrenan, Different anomalies in a Jarratt family of iterative root finding methods. Appl. Math. Comput. **233**, 29–38 (2014)
14. A.A. Magrenan, A new tool to study real dynamics: the convergence plane. Appl. Math. Comput. **248**, 215–224 (2014)
15. P.W. Meyer, A unifying theorem on Newton's method. Numer. Func. Anal. Optim. 13, 5 and 6, 463–473 (1992)
16. F.A. Potra, V. Ptak, *Nondiscrete Induction and Iterative Processes* (Pitman, London, 1984)
17. P.D. Proinov, New general convergence theory for iterative processes and its applications to Newton-Kantorovich type theorems. J. Complex. **26**, 3–42 (2010)

Chapter 16
Inexact Gauss-Newton Method for Singular Equations

A new semi-local convergence analysis of the Gauss-Newton method for solving convex composite optimization problems is presented using the concept of quasi-regularity for an initial point [13, 18, 22, 23, 25]. The convergence analysis is based on a combination of a center-majorant and majorant function. The results extend the applicability of the Gauss-Newton method under the same computational cost as in earlier studies such as [5, 7, 13–43]. In particular, the advantages are: the error estimates on the distances involved are tighter and the convergence ball is at least as large. Numerical examples are also provided in this study. It follows [12].

16.1 Introduction

A lot of problems such as convex inclusion, minimax problems, penalization methods, goal programming, constrained optimization and other problems can be formulated like

$$F(x) = 0, \qquad (16.1.1)$$

where D is open and convex and $F : D \subset \mathbb{R}^j \to \mathbb{R}^m$ is a nonlinear operator with its Fréchet derivative denoted by F'. The solutions of Eq. (16.1.1) can rarely be found in closed form. That is why the solution methods for these equations are usually iterative. In particular, the practice of numerical analysis for finding such solutions is essentially connected to Newton-like methods [2, 6–9, 11, 13, 19, 20, 28, 29, 35, 37]. The study about convergence matter of iterative procedures is usually centered on two types: semilocal and local convergence analysis. The semilocal convergence matter is, based on the information around an initial point, to give criteria ensuring the convergence of iterative procedures; while the local one is, based on the information

© Springer International Publishing Switzerland 2016
G.A. Anastassiou and I.K. Argyros, *Intelligent Numerical Methods:*
Applications to Fractional Calculus, Studies in Computational Intelligence 624,
DOI 10.1007/978-3-319-26721-0_16

around a solution, to find estimates of the radii of convergence balls. A plethora of sufficient conditions for the local as well as the semilocal convergence of Newton-like methods as well as an error analysis for such methods can be found in [11, 19, 20]. In the case $m = j$, the inexact Newton method was defined in [6] by:

$$x_{n+1} = x_n + s_n, \quad F'(x_n)s_n = -F(x_n) + r_n \quad \text{for each} \quad n = 0, 1, 2, \ldots, \quad (16.1.2)$$

where x_0 is an initial point, the residual control r_n satisfy

$$\|r_n\| \leq \lambda_n \|F(x_n)\| \quad \text{for each} \quad n = 0, 1, 2, \ldots, \quad (16.1.3)$$

and $\{\lambda_n\}$ is a sequence of forcing terms such that $0 \leq \lambda_n < 1$. Let x^* be a solution of (16.1.1) such that $F'(x^*)$ is invertible. As shown in [6], if $\lambda_n \leq \lambda < 1$, then, there exists $r > 0$ such that for any initial guess $x_0 \in U(x^*, r) := \{x \in \mathbb{R}^j : \|x - x^*\| < r\}$, the sequence $\{x_n\}$ is well defined and converges to a solution x^* in the norm $\|y\|_* := \|F'(x^*)y\|$, where $\| \cdot \|$ is any norm in $\mathbb{R}^j$. Moreover, the rate of convergence of $\{x_n\}$ to x^* is characterized by the rate of convergence of $\{\lambda_n\}$ to 0. It is worth noting that, in [6], no Lipschitz condition is assumed on the derivative F' to prove that $\{x_n\}$ is well defined and linearly converging. However, no estimate of the convergence radius r is provided. A pointed out by [16] the result of [6] is difficult to apply due to dependence of the norm $\| \cdot \|_*$, which is not computable.

The residual control (16.1.3) is non-affine invariant. The advantages of affine versus non-affine invariant forms have been explained in [20]. That is why, Ypma used in [41] the affine invariant condition of residual control in the form:

$$\|F'(x_n)^{-1}r_n\| \leq \lambda_n \|F'(x_n)^{-1}F(x_n)\| \quad \text{for each} \quad n = 0, 1, 2, \ldots, \quad (16.1.4)$$

to study the local convergence of inexact Newton method (16.1.2). And the radius of convergent result are also obtained.

To study the local convergence of inexact Newton method and inexact Newton-like method (called inexact methods for short below), Morini presented in [32] the following variation for the residual controls:

$$\|P_n r_n\| \leq \lambda_n \|P_n F(x_n)\| \quad \text{for each} \quad n = 0, 1, 2, \ldots, \quad (16.1.5)$$

where $\{P_n\}$ is a sequence of invertible operator from $\mathbb{R}^j$ to $\mathbb{R}^j$ and $\{\lambda_n\}$ is the forcing term. If $P_n = I$ and $P_n = F'(x_n)$ for each n, (16.1.5) reduces to (16.1.3) and (16.1.4), respectively. These methods are linearly convergent under Lipschitz Condition. It is worth nothing that the residual controls (16.1.5) are used in iterative methods if preconditioning is applied and lead to a relaxation on the forcing terms. But we also note that the results obtained in [32] do not provide an estimate of the radius of convergence. This is why Chen and Li [16] obtained the local convergence properties of inexact methods for (16.1.1) under a weak Lipschitz condition, which was first introduced by Wang in [38] to study the local convergence behaviour of Newton's method. The result in [16] easily provides an estimate of convergence ball for the inexact methods. Furthermore, Ferreira and Gonçalves presented in [23] a new local convergence analysis for inexact Newton-like under so-called majorant condition.

Recent attentions are focused on the study of finding zeros of singular nonlinear systems by Gauss-Newton's method, which is defined by

$$x_{n+1} = x_n - F'(x_n)^\dagger F(x_n) \quad \text{for each} \quad n = 0, 1, 2, \ldots, \tag{16.1.6}$$

where $x_0 \in D$ is an initial point and $F'(x_n)^\dagger$ denotes the Moore-Penrose inverse of the linear operator (of matrix) $F'(x_n)$. Shub and Smale extended in [36] the Smale point estimate theory (includes α-theory and γ-theory) to Gauss-Newton's methods for underdetermined analytic systems with surjective derivatives. For overdetermined systems, Dedieu and Schub studied in [18] the local linear convergence properties of Gauss-Newton's for analytic systems with injective derivatives and provided estimates of the radius of convergence balls for Gauss-Newton's method. Dedieu and Kim in [17] generalized both the results of the undetermined case and the overdetermined case to such case where $F'(x)$ is of constant rank (not necessary full rank), which has been improved by some authors in [1, 11, 14, 15, 20, 21].

Recently, several authors have studied the convergence behaviour of inexact versions of Gauss-Newton's method for singular nonlinear systems. For example, Chen [15] employed the ideas of [38] to study the local convergence properties of several inexact Gauss-Newton type methods where a scaled relative residual control is performed at each iteration under weak Lipschitz conditions. Ferreira, Gonçalves and Oliveira presented in their recent paper [26] a local convergence analysis of an inexact version of Gauss-Newton's method for solving nonlinear least squares problems. Moreover, the radius of the convergence balls under the corresponding conditions were estimated in these two papers. The preceding results were improved by Argyros et al. [2–11] using the concept of the center Lipschitz condition (see also (16.2.8) and the numerical examples) under the same computational cost on the parameters and functions involved.

In the present study, we are motivated by the elegant work in [42, 43] and optimization considerations. Using more precise majorant condition and functions, we provide a new local convergence analysis for Gauss-Newton method under the same computational cost and the following advantages: larger radius of convergence; tighter error estimates on the distances $\|x_n - x^*\|$ for each $n = 0, 1, \ldots$ and a clearer relationship between the majorant function (see (16.2.7)) and the associated least squares problems (16.1.1). These advantages are obtained because we use a center-type majorant condition (see (16.2.8)) for the computation of inverses involved which is more precise that the majorant condition used in [21–26, 30, 31, 39–43]. Moreover, these advantages are obtained under the same computational cost, since as we will see in Sects. 16.3 and 16.4, the computation of the majorant function requires the computation of the center-majorant function. Furthermore, these advantages are very important in computational mathematics, since we have a wider choice of initial guesses x_0 and fewer computations to obtain a desired error tolerance on the distances $\|x_n - x^*\|$ for each $n = 0, 1, 2, \ldots$

The rest of this study is organized as follows. In Sect. 16.2, we introduce some preliminary notions and properties of the majorizing function. The main result about the local convergence are stated in Sect. 16.3. In Sect. 16.4, we prove the local convergence results given in Sect. 16.3. Section 16.5 contains the numerical examples and Sect. 16.6 the conclusion of this study.

16.2   Preliminaries

We present some standard results to make the study as selfcontained as possible. More results can be found in [13, 35, 38].

Let $A : \mathbb{R}^j \to \mathbb{R}^m$ be a linear operator (or an $m \times j$ matrix). Recall that an operator (or $j \times m$ matrix) $A^\dagger : \mathbb{R}^m \to \mathbb{R}^j$ is the Moore-Penrose inverse of A if it satisfies the following four equations:

$$A^\dagger A A^\dagger = A^\dagger; \quad A A^\dagger A = A; \quad (A A^\dagger)^* = A A^\dagger; \quad (A^\dagger A) = A^\dagger A,$$

where A^* denotes the adjoint of A. Let $ker\,A$ and $im\,A$ denote the kernel and image of A, respectively. For a subspace E of $\mathbb{R}^j$, we use Π_E to denote the projection onto E. Clearly, we have that

$$A^\dagger A = \Pi_{ker A^\perp} \quad \text{and} \quad A A^\dagger = \Pi_{im A}.$$

In particular, in the case when A is full row rank (or equivalently, when A is surjective), $A A^\dagger = I_{\mathbb{R}^m}$; when A is full column rank (or equivalently, when A is injective), $A^\dagger A = I_{\mathbb{R}^j}$.

The following lemma gives a Banach-type perturbation bound for Moore-Penrose inverse, which is stated in [25].

Lemma 16.1 ([25, Corollary 7.1.1, Corollary 7.1.2]). *Let A and B be $m \times j$ matrices and let $r \le \min\{m, j\}$. Suppose that $rank\,A = r$, $1 \le rank\,B \le A$ and $\|A^\dagger\|\|B - A\| < 1$. Then, $rank\,B = r$ and*

$$\|B^\dagger\| \le \frac{\|A^\dagger\|}{1 - \|A^\dagger\|\|B - A\|}.$$

Also, we need the following useful lemma about elementary convex analysis.

Lemma 16.2 ([25, Proposition 1.3]). *Let $R > 0$. If $\varphi : [0, R] \to \mathbb{R}$ is continuously differentiable and convex, then, the following assertions hold:*

(a) $\dfrac{\varphi(t) - \varphi(\tau t)}{t} \le (1 - \tau)\varphi'(t)$ *for each $t \in (0, R)$ and $\tau \in [0, 1]$.*

(b) $\dfrac{\varphi(u) - \varphi(\tau u)}{u} \le \dfrac{\varphi(v) - \varphi(\tau v)}{v}$ *for each $u, v \in [0, R), u < v$ and $0 \le \tau \le 1$.*

From now on we suppose that the (I) conditions listed below hold.
For a positive real $R \in \mathbb{R}^+$, let

$$\psi : [0, R] \times [0, 1) \times [0, 1) \to \mathbb{R}$$

be a continuous differentiable function of three of its arguments and satisfy the following properties:

(i) $\psi(0, \lambda, \theta) = 0$ *and* $\dfrac{\partial}{\partial t}\psi(t, \lambda, \theta)\Big|_{t=0} = -(1 + \lambda + \theta)$.

(ii) $\dfrac{\partial}{\partial t}\psi(t, \lambda, \theta)$ *is convex and strictly increasing with respect to the argument t.*

For fixed $\lambda, \theta \in [0, 1)$, we write $h_{\lambda,\theta}(t) \triangleq \psi(t, \lambda, \theta)$ for short below. Then the above two properties can be restated as follows.

(iii) $h_{\lambda,\theta}(0) = 0$ *and* $h'_{\lambda,\theta}(0) = -(1 + \lambda + \theta)$.
(iv) $h'_{\lambda,\theta}(t)$ *is convex and strictly increasing.*
 (v) $g : [0, R] \to \mathbb{R}$ *is strictly increasing with $g(0) = 0$.*
(vi) g' *is convex and strictly increasing with $g'(0) = -1$.*
(vii) $g(t) \leq h_{\lambda,\theta}(t)$, $g'(t) \leq h'_{\lambda,\theta}(t)$ *for each $t \in [0, R)$, $\lambda, \theta \in [0, 1]$.*

Define

$$\zeta_0 := \sup\{t \in [0, R) : h'_{0,0}(t) < 0\}, \quad \zeta := \sup\{t \in [0, R) : g'(t) < 0\}, \quad (16.2.1)$$

$$\rho_0 := \sup\left\{t \in [0, \zeta_0) : \left|\frac{h_{\lambda,\theta}(t)}{h'_{0,0}(t)} - t\right| < t\right\},$$

$$\rho = \sup\left\{t \in [0, \zeta) : \left|\frac{h_{\lambda,\theta}(t) - th'_{0,0}(t)}{g'(t)}\right| < t\right\} \qquad (16.2.2)$$

$$\sigma := \sup\{t \in [0, R) : U(x^*, t) \subset D\}. \qquad (16.2.3)$$

The next two lemmas show that the constants ζ and ρ defined in (16.2.1) and (16.2.2), respectively, are positive.

Lemma 16.3 *The constant ζ defined in (16.2.1) is positive and* $\dfrac{th'_{0,0}(t) - h_{\lambda,\theta}(t)}{g'(t)} < 0$ *for each $t \in (0, \zeta)$.*

Proof Since $g'(0) = -1$, there exists $\delta > 0$ such that $g'(t) < 0$ for each $t \in (0, \delta)$. Then, we get $\zeta \geq \delta(> 0)$. We must show that $\dfrac{th'_{0,0}(t) - h_{\lambda,\theta}(t)}{g'(t)} < 0$ for each $t \in (0, \zeta)$. By hypothesis, functions $h'_{\lambda,\theta}$, $g'(t)$ are strictly increasing, then functions $h_{\lambda,\theta}$, $g'(t)$ are strictly convex. It follows from Lemma 16.2 (i) and hypothesis (vii) that

$$\frac{h_{\lambda,\theta}(t) - h_{\lambda,\theta}(0)}{t} < h'_{\lambda,\theta}(t), \quad t \in (0, R).$$

In view of $h_{\lambda,\theta}(0) = 0$ and $g'(t) < 0$ for all $t \in (0, \zeta)$. This together with the last inequality yields the desired inequality. $\square$

Lemma 16.4 *The constant ρ defined in (16.2.2) is positive. Consequently,*
$$\left| \frac{th'_{0,0}(t) - h_{\lambda,\theta}(t)}{g'(t)} \right| < t \text{ for each } t \in (0, \rho).$$

Proof Firstly, by Lemma 16.3, it is clear that $\left(\dfrac{h_{\lambda,\theta}(t)}{th'_{0,0}(t)} - 1 \right) \dfrac{h'_{0,0}(t)}{g'(t)} > 0$ for $t \in$ $(0, \zeta)$. Secondly, we get from Lemma 16.2 (i) that

$$\lim_{t \to 0} \left(\frac{h_{\lambda,\theta}(t)}{th'_{0,0}(t)} - 1 \right) \frac{h'_{0,0}(t)}{g'(t)} = 0.$$

Hence, there exists a $\delta > 0$ such that

$$0 < \left(\frac{h_{\lambda,\theta}(t)}{th'_{0,0}(t)} - 1 \right) \frac{h'_{0,0}(t)}{g'(t)} < 1, \quad t \in (0, \zeta).$$

That is ρ is positive. $\square$

Define

$$r := \min\{\rho, \delta\}, \tag{16.2.4}$$

where ρ and δ are given in (16.2.2) and (16.2.3), respectively. For any starting point $x_0 \in U(x^*, r) \backslash \{x^*\}$, let $\{t_n\}$ be a sequence defined by:

$$t_0 = \|x_0 - x^*\|, \quad t_{n+1} = \left| \left(t_n - \frac{h_{\lambda,\theta}(t_n)}{h'_{0,0}(t_n)} \right) \frac{h'_{0,0}(t_n)}{g'(t_n)} \right| \quad \text{for each} \quad n = 0, 1, 2, \ldots \tag{16.2.5}$$

Lemma 16.5 *The sequence $\{t_n\}$ given by (16.2.5) is well defined, strictly decreasing, remains in $(0, \rho)$ for each $n = 0, 1, 2, \ldots$ and converges to 0.*

Proof Since $0 < t_0 = \|x_0 - x^*\| < r \leq \rho$, using Lemma 16.4, we have that $\{t_n\}$ is well defined, strictly decreasing and remains in $[0, \rho)$ for each $n = 0, 1, 2, \ldots$ Hence, there exists $t^* \in [0, \rho)$ such that $\lim\limits_{n \to +\infty} t_n = t^*$. That is, we have

$$0 \leq t^* = \left(\frac{h_{\lambda,\theta}(t^*)}{h'_{0,0}(t^*)} - t^* \right) \frac{h'_{0,0}(t^*)}{g'(t^*)} < \rho.$$

If $t^* \neq 0$, it follows from Lemma 16.4 that

$$\left(\frac{h_{\lambda,\theta}(t^*)}{h'_{0,0}(t^*)} - t^* \right) \frac{h'_{0,0}(t^*)}{g'(t^*)} < t^*,$$

which is a contradiction. Hence, we conclude that $t_n \to 0$ as $n \to +\infty$. $\square$

If $g(t) = h_{\lambda,\theta}(t)$, then Lemmas 16.3–16.5 reduce to the corresponding ones in [42, 43]. Otherwise, i.e., if $g(t) < h_{\lambda,\theta}(t)$, then our results are better, since

$$\zeta_0 < \zeta \quad \text{and} \quad \rho_0 < \rho.$$

Moreover, the scalar sequence used in [42, 43] is defined by

$$u_0 = \|x_0 - x^*\|, \quad u_{n+1} = \left| u_n - \frac{h_{\lambda,\theta}(u_n)}{h'_{0,0}(u_n)} \right| \quad \text{for each } n = 0, 1, 2, \ldots \quad (16.2.5')$$

Using the properties of the functions $h_{\lambda,\theta}, g$, (16.2.5), (16.2.5') and a simple inductive argument we get that

$$t_0 = u_0, \quad t_1 = u_1, \quad t_n < u_n, \quad t_{n+1} - t_n < u_{n+1} - u_n \quad \text{for each } n = 1, 2, \ldots$$

and

$$t^* \leq u^* = \lim_{n \to +\infty} u_n,$$

which justify the advantages of our approach as claimed in the introduction of this study.

In Sect. 16.3 we shall show that $\{t_n\}$ is a majorizing sequence for $\{x_n\}$.

We state the following modified majorant condition for the convergence of various Newton-like methods in [9–11, 13].

Definition 16.6 Let $r > 0$ be such that $U(x^*, r) \subset D$. Then, F' is said to satisfy the majorant condition on $U(x^*, r)$ if

$$\|F'(x^*)^{\dagger}[F'(x) - F'(x^* + \tau(x - x^*))]\| \leq h'_{\lambda,\theta}(\|x - x^*\|) - h'_{\lambda,\theta}(\tau\|x - x^*\|) \quad (16.2.6)$$

for any $x \in U(x^*, r)$ and $\tau \in [0, 1]$.

In the case when $F'(x^*)$ is not surjective, the information on $im F'(x^*)^{\perp}$ may be lost. This is why the above notion was modified in [42, 43] to suit the case when $F'(x^*)$ is not surjective as follows:

Definition 16.7 Let $r > 0$ be such that $U(x^*, r) \subset D$. Then, f' is said to satisfy the modified majorant condition on $U(x^*, r)$ if

$$\|F'(x^*)^\dagger\| \|F'(x) - F'(x^* + \tau(x - x^*))\| \leq h'_{\lambda,\theta}(\|x - x^*\|) - h'_{\lambda,\theta}(\tau\|x - x^*\|)$$
$$(16.2.7)$$

for any $x \in U(x^*, r)$ and $\tau \in [0, 1]$.

If $\tau = 0$, condition (16.2.7) reduces to

$$\|F'(x^*)^\dagger\| \|F'(x) - F'(x^*)\| \leq h'_{\lambda,\theta}(\|x - x^*\|) - h'_{\lambda,\theta}(0).   (16.2.7')$$

In particular, for $\lambda = \theta = 0$, condition (16.2.7') reduces to

$$\|F'(x^*)^\dagger\| \|F'(x) - F'(x^*)\| \leq h'_{0,0}(\|x - x^*\|) - h'_{0,0}(0).   (16.2.7'')$$

Condition (16.2.7'') is used to produce the Banach-type perturbation Lemmas in [42, 43] for the computation of the upper bounds on the norms $\|F'(x)^\dagger\|$. In this study we use a more flexible function g than $h_{\lambda,\theta}$ function for the same purpose. This way the advantages as stated in the Introduction of this study can be obtained.

In order to achieve these advantages we introduce the following notion [2–11].

Definition 16.8 Let $r > 0$ be such that $U(x^*, r) \subset D$. Then g' is said to satisfy the center-majorant condition on $U(x^*, r)$ if

$$\|F'(x^*)^\dagger\| \|F'(x) - F'(x^*)\| \leq g'(\|x - x^*\|) - g'(0).   (16.2.8)$$

Clearly,

$$g'(t) \leq h'_{\lambda,\theta}(t) \quad \text{for each} \quad t \in [0, R], \quad \lambda, \theta \in [0, 1]   (16.2.9)$$

holds in general and $\dfrac{h'_{\lambda,\theta}(t)}{g'(t)}$ can be arbitrarily large [11].

It is worth noticing that (16.2.8) is not an additional condition to (16.2.7) since in practice the computation of function $h_{\lambda,\theta}$ requires the computation of g as a special case (see also the numerical examples).

16.3   Local Convergence

In this section, we present local convergence for inexact Newton method (16.1.2). Equation (16.1.1) is a surjective-undetermined (resp. injective-overdetermined) system if the number of equations is less (resp. greater) than the number of knowns and $F'(x)$ is of full rank for each $x \in D$. It is well known that, for surjective-underdetermined systems, the fixed points of the Newton operator

$N_F(x) := x - F'(x)^\dagger F(x)$ are the zeros of F, while for injective-overdetermined systems, the fixed points of N_F are the least square solutions of (16.1.1), which, in general, are not necessarily the zeros of F.

Next, we present the local convergence properties of inexact Newton method for general singular systems with constant rank derivatives.

Theorem 16.9 *Let* $F : D \subset \mathbb{R}^j \to \mathbb{R}^m$ *be continuously Fréchet differentiable nonlinear operator, D is open and convex. Suppose that* $F(x^*) = 0$, $F'(x^*) \neq 0$ *and that* F' *satisfies the modified majorant condition (16.2.7) and the center-majorant condition (16.2.8) on* $U(x^*, r)$, *where r is given in (16.2.4). In addition, we assume that* $\operatorname{rank} F'(x) \leq \operatorname{rank} F'(x^*)$ *for any* $x \in U(x^*, r)$ *and that*

$$\|[I_{\mathbb{R}^j} - F'(x)^\dagger F'(x)](x - x^*)\| \leq \theta\|x - x^*\|, \quad x \in U(x^*, r), \qquad (16.3.1)$$

where the constant θ *satisfies* $0 \leq \theta < 1$. *Let sequence* $\{x_n\}$ *be generated by inexact Gauss-Newton method with any initial point* $x_0 \in U(x^*, r)\backslash\{x^*\}$ *and the conditions for the residual* r_n *and the forcing term* λ_n:

$$\|r_n\| \leq \lambda_n\|F(x_n)\|, \quad 0 \leq \lambda_n F'(x_k) \leq \lambda \text{ for each } n = 0, 1, 2, \ldots \qquad (16.3.2)$$

Then, $\{x_n\}$ *converges to a zero* x^* *of* $F'(\cdot)^\dagger F(\cdot)$ *in* $\overline{U}(x^*, r)$. *Moreover, we have the following estimate:*

$$\|x_{n+1} - x^*\| \leq \frac{t_{n+1}}{t_n}\|x_n - x^*\| \text{ for each } n = 0, 1, 2, \ldots, \qquad (16.3.3)$$

where the sequence $\{t_n\}$ *is defined by (16.2.5).*

Remark 16.10 (a) If $g(t) = h_{\lambda,\theta}(t)$, then the results obtained in Theorem 16.9 reduce to the ones given in [42, 43].
(b) If $g(t)$ and $h_{\lambda,\theta}(t)$ are

$$g(t) = h_{\lambda,\theta}(t) = -(1 + \lambda + \theta)t + \int_0^t L(u)(t - u)\, du, \quad t \in [0, R], \quad (16.3.4)$$

then the results obtained in Theorem 16.9 reduce to the one given in [25]. Moreover, if taking $\lambda = 0$ (in this case $\lambda_n = 0$ and $r_n = 0$) in Theorem 16.9, we obtain the local convergence of Newton's method for solving the singular systems, which has been studied by Dedieu and Kim in [17] for analytic singular systems with constant rank derivatives and Li, Xu in [39] and Wang in [38] for some special singular systems with constant rank derivatives.
(c) If $g(t) < h_{\lambda,\theta}(t)$ then the improvements as mentioned in the Introduction of this study we obtained (see also the discussion above and below Definition 16.6)

If $F'(x)$ is full column rank for every $x \in U(x^*, r)$, then we have $F'(x)^\dagger$ $F'(x) = I_{\mathbb{R}^J}$. Thus,

$$\|[I_{\mathbb{R}^m} - F'(x)^\dagger F'(x)](x - x^*)\| = 0,$$

i.e., $\theta = 0$. We immediately have the following corollary:

Corollary 16.11 *Suppose that* $rank F'(x) \le rank F'(x^*)$ *and that*

$$\|[I_{\mathbb{R}^m} - F'^\dagger(x)F'(x)](x - x^*)\| = 0,$$

for any $x \in U(x^*, r)$*. Suppose that* $F(x^*) = 0$*,* $F'(x^*) \ne 0$ *and that* F' *satisfies the modifed majorant condition (16.2.7) and the center-majorant condition (16.2.8). Let sequence* $\{x_n\}$ *be generated by inexact Gauss-Newton method with any initial point* $x_0 \in U(x^*, r)\backslash\{x^*\}$ *and the condition (16.3.2) for the residual* r_n *and the forcing term* λ_n*. Then,* $\{x_n\}$ *converges to a zero* x^* *of* $F'(\cdot)^\dagger F(\cdot)$ *in* $\overline{U}(x^*, r)$*. Moreover, we have the following estimate:*

$$\|x_{n+1} - x^*\| \le \frac{t_{n+1}}{t_n}\|x_n - x^*\| \quad \text{for each} \quad n = 0, 1, 2, \ldots, \tag{16.3.5}$$

where the sequence $\{t_n\}$ *is defined by (16.2.5) for* $\theta = 0$*.*

In the case when $F'(x^*)$ is full row rank, the modified majorant condition (16.2.7) can be replaced by the majorant condition (16.2.6).

Theorem 16.12 *Suppose that* $F(x^*) = 0$*,* $F'(x^*)$ *is full row rank, and that* F' *satisfies the majorant condition (16.2.6) and the center-majorant condition (16.2.8) on* $U(x^*, r)$*, where* r *is given in (16.2.4). In addition, we assume that* $rank F'(x) \le rank F'(x^*)$ *for any* $x \in U(x^*, r)$ *and that condition (16.3.1) holds. Let sequence* $\{x_n\}$ *be generated by inexact Gauss-Newton method with any initial point* $x_0 \in U(x^*, r)\backslash\{x^*\}$ *and the conditions for the residual* r_n *and the forcing term* λ_n*:*

$$\|F'(x^*)^\dagger r_n\| \le \lambda_n \|F'(x^*)^\dagger F(x_n)\|, 0 \le \lambda_n F'(x^*)^\dagger F'(x_n) \le \lambda \tag{16.3.6}$$

for each $n = 0, 1, 2, \ldots$*.*

Then, $\{x_n\}$ converges to a cero ζ of $F(\cdot)$ in $\overline{U}(x^*, r)$. Moreover, we have the following estimate:

$$\|x_{n+1} - x^*\| \le \frac{t_{n+1}}{t_n}\|x_n - x^*\| \quad \text{for each} \quad n = 0, 1, 2, \ldots,$$

where the sentence $\{t_n\}$ *is defined by (16.2.5).*

Remark 16.13 Comments as in Remark 16.10 can follow for this case.

Theorem 16.14 *Suppose that $F(x^*) = 0$, $F'(x^*)$ is full row rank, and that F' satisfies the majorant condition (16.2.6) and the center-majorant condition on $U(x^*, r)$, where r is given in (16.2.4). In addition, we assume that $\operatorname{rank} F'(x) \leq \operatorname{rank} F'(x^*)$ for any $x \in U(x^*, r)$ and that condition (16.3.1) holds. Let sequence $\{x_n\}$ sequence generated by inexact Gauss-Newton method with any initial point $x_0 \in U(x^*, r) \backslash \{x^*\}$ and the conditions for the control residual r_n and the forcing term λ_n:*

$$\|F'(x_n)^\dagger r_n\| \leq \lambda_n \|F'(x_n)^\dagger F(x_n)\|, \quad 0 \leq \lambda_n F'(x_n) \leq \lambda \quad \text{for each} \quad n = 0, 1, 2, \ldots$$
(16.3.7)

Then, $\{x_n\}$ converges to a zero x^* of $f(\cdot)$ in $\overline{U}(x^*, r)$. Moreover, we have the following estimate:

$$\|x_{n+1} - x^*\| \leq \frac{t_{n+1}}{t_n} \|x_k - x^*\| \quad \text{for each} \quad n = 0, 1, 2, \ldots,$$

where sequence $\{t_n\}$ is defined by (16.2.5).

Remark 16.15 In the case when $F'(x^*)$ is invertible in Theorem 16.14, $h_{\lambda,\theta}$ is given by (16.3.4) and $g(t) = -1 + \int_0^t L_0(t)(t - u) \, du$ for each $t \in [0, R]$, we obtain the local convergence results of inexact Gauss-Newton method for nonsingular systems, and the convergence ball r is this case satisfies

$$\frac{\int_0^r L(u) u \, du}{r \left((1 - \lambda) - \int_0^r L_0(u) \, du\right)} \leq 1, \quad \lambda \in [0, 1).$$
(16.3.8)

In particular, if taking $\lambda = 0$, the convergence ball r determined in (16.3.8) reduces to the one given in [38] by Wang and the value r is the optimal radius of the convergence ball when the equality holds. That is our radius is r larger than the one obtained in [38], if $L_0 < L$ (see also the numerical examples). Notice that L is used in [38] for the estimate (16.3.8). Then, we can conclude that vanishing residuals, Theorem 16.14 merges into the theory of Newton's method.

16.4 Proofs

In this section, we prove our main results of local convergence for inexact Gauss-Newton method (16.1.2) given in Sect. 16.3.

16.4.1   Proof of Theorem 16.9

Lemma 16.16 *Suppose that F' satisfies the modified majorant condition on $U(x^*, r)$ and that $\|x^* - x\| < \min\{\rho, x^*\}$, where r, ρ and x^* are defined in (16.2.4), (16.2.2) and (16.2.1), respectively. Then, $\operatorname{rank} F'(x) = \operatorname{rank} F'(x^*)$ and*

$$\|F'(x)^\dagger\| \leq -\frac{\|F'(x^*)^\dagger\|}{g'(\|x - x^*\|)}.$$

Proof Since $g'(0) = -1$, we have

$$\|F'(x^*)^\dagger\| \|F'(x) - F'(x^*)\| \leq g'(\|x - x^*\|) - g'(0) < -g'(0) = 1.$$

It follows from Lemma (16.1) that $\operatorname{rank} F'(x) = \operatorname{rank} F'(x^*)$ and

$$\|F'(x)^\dagger\| \leq \frac{\|F'(x^*)^\dagger\|}{1 - (g'(\|x - x^*\|) - g'(0))} = -\frac{\|F'(x^*)^\dagger\|}{g'(\|x - x^*\|)}.$$

$\square$

Proof of Theorem 16.9 We shall prove by mathematical induction on n that $\{t_n\}$ is the majorizing sequence for $\{x_n\}$, i.e.,

$$\|x^* - x_j\| \leq t_j \quad \text{for each} \quad j = 0, 1, 2, \ldots \tag{16.4.1}$$

Because $t_0 = \|x_0 - x^*\|$; thus (16.4.1) holds for $j = 0$. Suppose that $\|x^* - x_j\| \leq t_j$ for some $j = n \in \mathbb{N}$. For the case $j = n + 1$, we first have that,

$$\begin{aligned} x_{n+1} - x^* &= x_n - x_\dagger^* - F'(x_n)^\dagger[F(x_n) - F(x^*)] + F'(x_n)^\dagger r_n \\ &= F'(x_n)^\dagger[F(x^*) - F(x_n) - F'(x_n)(x_\dagger^* - x_n)] + F'(x_n)^\dagger r_n \\ &\quad + [I_{\mathbb{R}^j} - F'(x_n)^\dagger F'(x_n)](x_n - x^*) \\ &= F'(x_n)^\dagger \int_0^1 [F'(x_n) - F'(x^* + \tau(x_n - x^*))](x_n - x^*) \, d\tau \\ &\quad + F'(x_n)^\dagger r_n + [I_{\mathbb{R}^j} - F'(x_n)^\dagger F'(x_n)](x_n - \varsigma). \end{aligned} \tag{16.4.2}$$

By using the modified majorant condition (16.2.7), Lemma 16.4, the inductive hypothesis (16.4.1) and Lemma 16.2, we obtain in turn that

$$\left\| F'(x_n)^\dagger \int_0^1 [F'(x_n) - F'(x^* + \tau(x_n - x^*))](x_n - x^*) \, d\tau \right\| \leq \tag{16.4.3}$$

$$-\frac{1}{g'(\|x_n - x^*\|)} \int_0^1 \|F'(x^*)^\dagger\| \|F'(x_n) - F'(x^* + \tau(x_n - x^*))\| \|x_n - x^*\| \, d\tau$$

$$= -\frac{1}{g'(\|x_n - x^*\|)} \int_0^1 \frac{h'_{\lambda,0}(\|x_n - x^*\|) - h'_{\lambda,0}(\tau\|x_n - x^*\|)}{\|x_n - x^*\|} \, d\tau \cdot \|x_n - x^*\|^2$$

$$\leq -\frac{1}{g'(t_n)} \int_0^1 \frac{h'_{\lambda,0}(t_n) - h_{\lambda,0}(\tau t_n)}{t_n} \, d\tau \cdot \|x_n - x^*\|^2$$

$$= -\frac{1}{g'(t_n)}(t_n h'_{\lambda,0}(t_n) - h_{\lambda,0}(t_n)) \frac{\|x_n - x^*\|^2}{t_n^2}.$$

In view of (16.3.2),

$$\|F'(x_n)^\dagger r_n\| \leq \|F'(x_n)^\dagger\| \|r_n\| \leq \lambda_n \|F'(x_n)^\dagger\| \|F(x_n)\|. \tag{16.4.4}$$

We have that

$$- F(x_n) = F(x^*) - F(x_n) - F'(x_n)(x^* - x_n) + F'(x_n)(x^* - x_n)$$

$$= \int_0^1 [F'(x_n) - F'(x^* + \tau(x_n - x^*))](x_n - x^*) \, d\tau$$

$$+ F'(x_n)(x^* - x_n). \tag{16.4.5}$$

Then, combining Lemmas 16.2, 16.16, the modified majorant condition (16.2.7), the inductive hypothesis (16.4.1) and the condition (16.3.2), we obtain in turn that

$$\lambda_n \|F'(x_n)^\dagger\| \|F(x_n)\|$$

$$\leq \lambda_n \|F'(x_n)^\dagger\| \int_0^1 \|F'(x_n) - F'(x^* + \tau(x_n - x^*))\| \|x_n - x^*\| \, d\tau$$

$$+ \lambda_n \|F'(x_n)^\dagger\| \|F'(x_n)\| \|x_n - x^*\|$$

$$\leq -\frac{\lambda}{g'(t_n)}(t_n h'_{\lambda,0}(t_n) - h_{\lambda,0}(t_n)) \frac{\|x_n - x^*\|^2}{t_n^2} + \lambda t_n \frac{\|x_n - x^*\|}{t_n}$$

$$\leq \lambda \frac{\lambda t_n + h_{\lambda,0}(t_n)}{g'(t_n)} \frac{\|x_n - x^*\|}{t_n}. \tag{16.4.6}$$

Combining (16.3.1), (16.4.3), (16.4.4) and (16.4.6), we get that

$$\|x_{n+1} - x^*\| \leq$$

$$\left[-\frac{t_n h'_{\lambda,0}(t_n) - h_{\lambda,0}(t_n)}{g'(t_n)} + \lambda \frac{\lambda t_n + h_{\lambda,0}(t_n)}{g'(t_n)} + \theta t_n \right] \frac{\|x_n - x^*\|}{t_n} =$$

$$\left[-t_n + (1 + \lambda)\left(\frac{\lambda t_n}{g'(t_n)} + \frac{h_{\lambda,0}(t_n)}{g'(t_n)} \right) + \theta t_n \right] \frac{\|x_n - x^*\|}{t_n}.$$

But, we have that $-1 < g'(t) < 0$ for any $t \in (0, \rho)$, so

$$(1 + \lambda) \left(\frac{\lambda t_n}{g'(t_n)} + \frac{h_{\lambda,0}(t_n)}{g'(t_n)} \right) + \theta t_n \leq \frac{h_{\lambda,0}(t_n)}{g'(t_n)} + \theta_n$$

$$\leq \frac{h_{\lambda,0}(t_n) - \theta t_n}{g'(t_n)} = \frac{h_{\lambda,\theta}(t_n)}{g'(t_n)}.$$

Using the definition of $\{t_n\}$ given in (16.2.5), we get that

$$\|x_{n+1} - x^*\| \leq \frac{t_{n+1}}{t_n} \|x_n - x^*\|,$$

so we deduce that $\|x_{n+1} - x^*\| \leq t_{n+1}$, which completes the induction. In view of the fact that $\{t_n\}$ converges to 0 (by Lemma 16.5), it follows from (16.4.1) that $\{x_n\}$ converges to x^* and the estimate (16.3.3) holds for all $n \geq 0$. $\square$

16.4.2 Proof of Theorem 16.12

Lemma 16.17 *Suppose that $F(x^*) = 0$, $F'(x^*)$ is full row rank and that F' satisfies the majorant condition (16.2.6) on $U(x^*, r)$. Then, for each $x \in U(x^*, r)$, we have $\mathrm{rank}\, F'(x) = \mathrm{rank}\, F'(x^*)$ and*

$$\|[I_{\mathbb{R}^j} - F'(x^*)^\dagger (F'(x^*) - F'(x))]^{-1}\| \leq -\frac{1}{g'(\|x - x^*\|)}.$$

Proof Since $g'(0) = -1$, we have

$$\|F'(x^*)^\dagger [F'(x) - F'(x^*)]\| \leq g'(\|x - x^*\|) - g'(0) < -g'(0) = 1.$$

It follows from Banach lemma that $[I_{\mathbb{R}^j} - F'(x^*)^\dagger (F'(x^*) - F'(x))]^{-1}$ exists and

$$\|[I_{\mathbb{R}^j} - F'(x^*)^\dagger (F'(x^*) - F'(x))]^{-1}\| \leq -\frac{1}{g'(\|x - x^*\|)}.$$

Since $F'(x^*)$ is full row rank, we have $F'(x^*) F'(x^*)^\dagger = I_{\mathbb{R}^m}$ and

$$F'(x) = F'(x^*)[I_{\mathbb{R}^j} - F'(z^*)^\dagger (F'(x^*) - F'(x))],$$

which implies that $F'(x)$ is full row, i.e., $\mathrm{rank}\, F'(x) = \mathrm{rank}\, F'(x^*)$. $\square$

Proof of Theorem 16.12 Let $\widehat{F} : U(x^*, r) \to \mathbb{R}^m$ be defined by

$$\widehat{F}(x) = F'(x^*)^\dagger \widehat{F}(x), \quad x \in U(x^*, r),$$

with residual $\widehat{r}_k = F'(x^*)^\dagger r_n$. In view of

$$\widehat{F}'(x)^\dagger = [F'(x^*)^\dagger F'(x)]^\dagger = F'(x)^\dagger F'(x^*), \quad x \in U(x^*, r),$$

we have that $\{x_n\}$ coincides with the sequence generated by inexact Gauss-Newton method (16.1.2) for $\widehat{F}$. Moreover, we get that

$$\widehat{F}'(x^*)^\dagger = (F'(x^*)^\dagger F'(x^*))^\dagger = F'(x^*)^\dagger F'(x^*).$$

Consequently,

$$\|\widehat{F}'(x^*)^\dagger \widehat{F}'(x^*)\| = \|F'(x^*)^\dagger F'(x^*)F'(x^*)^\dagger F(x^*)\| = \|F'(x^*)^\dagger F(x^*)\|.$$

Because $\|F'(x^*)^\dagger F(x^*)\| = \|\Pi_{\ker F'(x^*)^\perp}\| = 1$, thus, we have

$$\|\widehat{F}'(x^*)^\dagger\| = \|\widehat{F}'(x^*)^\dagger \widehat{F}'(x^*)\| = 1.$$

Therefore, by (16.2.6), we can obtain that

$$\|\widehat{F}'(x^*)^\dagger\| \|\widehat{F}'(x) - \widehat{F}'(x^* + \tau(x - x^*))\| =$$
$$\|F'(x^*)^\dagger (F'(x) - F'(x^* + \tau(x - x^*)))\| \leq$$
$$h'_{\lambda,\theta}(\|x - x^*\|) - h_{\lambda,\theta}(\tau\|x - x^*\|).$$

Hence, $\widehat{F}$ satisfies the modified majorant condition (16.2.7) on $U(x^*, r)$. Then, Theorem 16.9 is applicable and $\{x_k\}$ converges to x^* follows. Note that, $\widehat{F}'(\cdot)^\dagger \widehat{F}(\cdot) = F'(\cdot)^\dagger F(\cdot)$ and $F(\cdot) = F'(\cdot)F'(\cdot)^\dagger F(\cdot)$. Hence, we conclude that x^* is a zero of F. $\qquad\square$

16.4.3   Proof of Theorem 16.14

Lemma 16.18 *Suppose that $F(x^*) = 0$, $F'(x^*)$ is full row rank and that F' satisfies the majorant condition (16.2.6) on $U(x^*, r)$. Then, we have*

$$\|F'(x)^\dagger F'(x^*)\| \leq -\frac{1}{g'(\|x - x^*\|)} \quad \text{for each} \quad x \in U(x^*, r).$$

Proof Since $F'(x^*)$ is full row rank, we have $F'(x^*)F'(x^*)^\dagger = I_{R^m}$. Then, we get that

$$F'(x)^\dagger F'(x^*)(I_{\mathbb{R}^j} - F'(x^*)^\dagger(F'(x^*) - F'(x^*))) = F'(x)^\dagger F'(x), \quad x \in U(x^*, r).$$

By Lemma 16.17, $I_{\mathbb{R}^j} - F'(x^*)^\dagger(F'(x^*) - F'(x))$ is invertible for any $x \in U(x^*, r)$. Thus, in view of the equality $A^\dagger A = \Pi_{\ker A^\perp}$ for any $m \times j$ matrix A, we obtain that

$$F'(x)^\dagger F'(x^*) = \Pi_{\ker F'(x)^\perp}[I_{\mathbb{R}^j} - F'(x^*)^\dagger(F'(x^*) - F'(x))]^{-1}.$$

Therefore, by Lemma 16.17 we deduce that

$$\|F'(x)^\dagger F'(x^*)\| \le \|\Pi_{\ker F'(x)^\perp}\| \|[I_{\mathbb{R}^j} - F'(x^*)^\dagger(F'(x^*) - F'(x))]^{-1}\|$$

$$\le -\frac{1}{g'(\|x - x^*\|)}. \qquad \square$$

Proof of Theorem 16.14 Using Lemma 16.18, majorant condition (16.2.6) and the residual condition (16.3.7), respectively, instead of Lemma 16.16, modified majorant condition (16.2.7) and condition (16.3.2), one can complete the proof of Theorem 16.14 in an analogous way to the proof of Theorem 16.9. $\qquad \square$

16.5 Numerical Examples

We present some numerical examples, where

$$g(t) < h_{\lambda,\theta}(t) \qquad (16.5.1)$$

and

$$g'(t) < h'_{\lambda,\theta}(t). \qquad (16.5.2)$$

For simplicity we take $F'(x)^\dagger = F'(x)^{-1}$ for each $x \in D$.

Example 16.19 Let $X = Y = (-\infty, +\infty)$ and define function $F : X \to Y$ by

$$F(x) = d_0 x - d_1 \sin(1) + d_1 \sin(e^{d_2 x})$$

where d_0, d_1, d_2 are given real numbers. Then $x^* = 0$. Define functions g and $h_{\lambda,\theta}$ by $g(t) = \frac{L_0}{2}t^2 - t$ and $h_{\lambda,\theta}(t) = \frac{L}{2}t^2 - t$. Then, it can easily be seen that for d_2 sufficiently large and d_1 sufficiently small $\frac{L}{L_0}$ can be arbitrarily large. Hence, (16.5.1) and (16.5.2) hold.

Example 16.20 Let $F(x, y, z) = 0$ be a nonlinear system, where $F : D \subseteq \mathbb{R}^3 \to \mathbb{R}^3$ and $F(x, y, z) = (x, \frac{e - 1}{2}y^2 + y, e^z - 1)$. It is obvious that $(0, 0, 0) = \overline{x}^*$ is a solution of the system.

From F, we deduce

$$F'(\overline{x}) = \begin{pmatrix} 1 & 0 & 0 \\ 0 & (e-1)y & 0 \\ 0 & 0 & e^z \end{pmatrix} \quad \text{and} \quad F'(x^*) = \text{diag}\{1, 1, 1\},$$

where $\overline{x} = (x, y, z)$. Hence, $[F'(\overline{x}^*)]^{-1} = \text{diag}\{1, 1, 1\}$. Moreover, we can define for $L_0 = e - 1 < L = e$, $g(t) = \dfrac{e-1}{2}t^2 - t$ and $h_{\lambda,\theta}(t) = \dfrac{e}{2}t^2 - t$. Then, again (16.5.1) and (16.5.2) hold.

Other examples where (16.5.1) and (16.5.2) are satisfied can be found in [2, 5, 8, 9, 11].

16.6 Conclusion

We expanded the applicability of inexact Gauss-Newton method under a majorant and a center-majorant condition. The advantages of our analysis over earlier works such as [5, 7–43] are also shown under the same computational cost for the functions and constants involved. These advantages include: a large radius of convergence and more precise error estimates on the distances $\|x_{n+1} - x^*\|$ for each $n = 0, 1, 2, \ldots$, leading to a wider choice of initial guesses and computation of less iterates x_n in order to obtain a desired error tolerance. Numerical examples show that the center-function can be smaller than the majorant function.

References

1. S. Amat, S. Busquier, J.M. Gutiérrez, Geometric constructions of iterative functions to solve nonlinear equations. J. Comput. Appl. Math. **157**, 197–205 (2003)
2. I.K. Argyros, *Computational Theory of Iterative Methods*. Studies in Computational Mathematics, vol. 15, ed. by K. Chui, L. Wuytack (Elsevier, New York, 2007)
3. I.K. Argyros, Concerning the convergence of Newton's method and quadratic majorants. J. Appl. Math. Comput. **29**, 391–400 (2009)
4. I.K. Argyros, S. Hilout, On the Gauss-Newton method. J. Appl. Math. 1–14 (2010)
5. I.K. Argyros, S. Hilout, Extending the applicability of the Gauss-Newton method under average Lipschitz-conditions. Numer. Algorithm **58**, 23–52 (2011)
6. I.K. Argyros, A semilocal convergence analysis for directional Newton methods. Math. Comput. AMS **80**, 327–343 (2011)
7. I.K. Argyros, S. Hilout, On the solution of systems of equations with constant rank derivatives. Numer. Algorithm **57**, 235–253 (2011)
8. I.K. Argyros, Y.J. Cho, S, Hilout, *Numerical Methods for Equations and Its Applications* (CRC Press/Taylor and Francis Group, New-York, 2012)
9. I.K. Argyros, S. Hilout, Improved local convergence of Newton's method under weaker majorant condition. J. Comput. Appl. Math. **236**(7), 1892–1902 (2012)

10. I.K. Argyros, S. Hilout, Weaker conditions for the convergence of Newton's method. J. Complex. **28**, 364–387 (2012)
11. I.K. Argyros, S. Hilout, *Computational Methods in Nonlinear Analysis* (World Scientific, New Jersey, 2013)
12. I.K. Argyros, D. Gonzalez, Local Convergence Analysis of Inexact Gauss-Newton Method for Singular Systems of Equations Under Majorant and Condition (2015) (submitted)
13. A. Ben-Israel, T.N.E. Greville, *Generalized Inverses*. CMS Books in Mathematics/Ouvrages de Mathematiques de la SMC, 15. 2nd edn. Theory and Applications (Springer, New York, 2003)
14. J.V. Burke, M.C. Ferris, A Gauss-Newton method for convex composite optimization. Math. Program. Ser. A. **71**, 179–194 (1995)
15. J. Chen, The convergence analysis of inexact Gauss-Newton. Comput. Optim. Appl. **40**, 97–118 (2008)
16. J. Chen, W. Li, Convergence behaviour of inexact Newton methods under weak Lipschitz condition. J. Comput. Appl. Math. **191**, 143–164 (2006)
17. J.P. Dedieu, M.H. Kim, Newton's method for analytic systems of equations with constant rank derivatives. J. Complex. **18**, 187–209 (2002)
18. J.P. Dedieu, M. Shub, Newton's method for overdetermined systems of equations. Math. Comput. **69**, 1099–1115 (2000)
19. R.S. Dembo, S.C. Eisenstat, T. Steihaug, Inexact Newton methods. SIAM J. Numer. Anal. **19**, 400–408 (1982)
20. P. Deuflhard, G. Heindl, Affine invariant convergence theorems for Newton's method and extensions to related methods. SIAM J. Numer. Anal. **16**, 1–10 (1979)
21. O.P. Ferreira, Local convergence of Newton's method in Banach space from the viewpoint of the majorant principle. IMA J. Numer. Anal. **29**, 746–759 (2009)
22. O.P. Ferreira, Local convergence of Newton's method under majorant condition. J. Comput. Appl. Math. **235**, 1515–1522 (2011)
23. O.P. Ferreira, M.L.N. Gonçalves, Local convergence analysis of inexact Newton-like methods under majorant condition. Comput. Optim. Appl. **48**, 1–21 (2011)
24. O.P. Ferreira, M.L.N. Gonçalves, P.R. Oliveira, Local convergence analysis of the Gauss-Newton method under a majorant condition. J. Complex. **27**, 111–125 (2011)
25. O.P. Ferreira, M.L.N. Gonçalves, P.R. Oliveira, Local convergence analysis of inexact Gauss-Newton like method under majorant condition. J. Comput. Appl. Math. **236**, 2487–2498 (2012)
26. M.L.N. Gonçalves, P.R. Oliveira, Convergence of the Gauss-Newton method for a special class of systems of equations under a majorant condition. Optimiz. (2013). doi:10.1080/02331934. 2013.778854
27. W.M. Häussler, A Kantorovich-type convergence analysis for the Gauss-Newton method. Numer. Math. **48**, 119–125 (1986)
28. J.B. Hiriart-Urruty, C. Lemaréchal, *Convex analysis and minimization algorithms* (two volumes). I. Fundamentals. II. Advanced theory and bundle methods, 305 and 306 (Springer, Berlin, 1993)
29. L.V. Kantorovich, G.P. Akilov, *Functional Analysis* (Pergamon Press, Oxford, 1982)
30. C. Li, K.F. Ng, Majorizing functions and convergence of the Gauss-Newton method for convex composite optimization. SIAM J. Optim. **18**(2), 613–692 (2007)
31. C. Li, N. Hu, J. Wang, Convergence behaviour of Gauss-Newton's method and extensions of Smale point estimate theory. J. Complex. **26**, 268–295 (2010)
32. B. Morini, Convergence behaviour of inexact Newton methods. Math. Comput. **68**, 1605–1613 (1999)
33. F.A. Potra, V. Pták, *Nondiscrete Induction and Iterative Processes* Pitman (1994)
34. S.M. Robinson, Extension of Newton's method to nonlinear functions with values in a cone. Numer. Math. **19**, 341–347 (1972)
35. R.T. Rockafellar, *Convex Analysis*. Princeton Mathematical Series, 28. (Princeton University Press, Princeton, 1970)

36. M. Shub, S. Smale, Complexity of Bézout's theorem IV: probability of success extensions. SIAM J. Numer. Anal. **33**, 128–148 (1996)
37. S. Smale, *Newton's Method Estimates From Data at One Point. The Merging of Disciplines: New Directions in Pure, Applied, and Computational Mathematics* (Laramie, Wyo., 1985), pp. 185–196 (Springer, New York, 1986)
38. W. Wang, Convergence of Newton's method and uniqueness of the solution of equations in Banach space. IMA J. Numer. Anal. **20**, 123–134 (2000)
39. X. Xu, C. Li, Convergence of Newton's method for systems of equations with constant rank derivatives. J. Comput. Math. **25**, 705–718 (2007)
40. X. Xu, C. Li, Convergence criterion of Newton's method for singular systems with constant rank derivatives. J. Math. Anal. Appl. **345**, 689–701 (2008)
41. T.J. Ypma, Local convergence of inexact Newton's methods. SIAM J. Numer. Anal. **21**, 583–590 (1984)
42. F. Zhou, An analysis on local convergence of inexact Newton-Gauss method solving singular systems of equations. Sci. World J. (2014). Article ID 752673
43. F. Zhou, On local convergence analysis of inexact Newton method for singular systems of equations under majorant condition. Sci. World J. **2014** (2014). Article ID 498016

Chapter 17
The Asymptotic Mesh Independence Principle

We present a new asymptotic mesh independence principle of Newton's method for discretized nonlinear operator equations. Our hypotheses are weaker than in earlier studies such as [1, 9–13]. This way we extend the applicability of the mesh independence principle which asserts that the behavior of the discretized version is asymptotically the same as that of the original iteration and consequently, the number of steps required by the two processes to converge within a given tolerance is essentially the same. The results apply to solve a boundary value problem that cannot be solved with the earlier hypotheses given in [13]. It follows [8].

17.1 Introduction

In this chapter we are concerned with the problem of approximating a solution x^* of the nonlinear operator equation

$$F(x) = 0, \qquad (17.1.1)$$

where F is a Fréchet-differentiable operator defined on a convex subset D of a Banach space X with values in a Banach space Y. Throughout this study we assume the existence of a unique solution x^* of Eq. (17.1.1).

Many problems in computational Sciences can be written in a form like (17.1.1) using Mathematical Model ling [6, 7, 9, 11, 13]. The solutions of Eq. (17.1.1) can be found in closed form only in special cases. That is why most solution methods for these equations are usually iterative.

The most popular method for generating a sequence approximating x^* is Newton's method which is defined for each $n = 0, 1, 2, \ldots$ by

$$F'(x^k)\Delta x^k = -F(x^k), \quad x^{k+1} = x^k + \Delta x^k \quad \text{for each } k = 0, 1, 2, \ldots \quad (17.1.2)$$

© Springer International Publishing Switzerland 2016

G.A. Anastassiou and I.K. Argyros, *Intelligent Numerical Methods: Applications to Fractional Calculus*, Studies in Computational Intelligence 624, DOI 10.1007/978-3-319-26721-0_17

assuming that the derivatives are invertible and x^0 is an initial point. However, in practice the computation of iterates $\{x^*\}$ is very expensive or even impossible. That is why, we can only solve discretized nonlinear equations defined on finite dimensional spaces on a sequence of successively finer mesh levels such as

$$F_m(x_m) = 0 \quad \text{for each } m = 0, 1, 2, \ldots, \tag{17.1.3}$$

where $F_m : D_m \subset X_m \to Y_m$, D_m is a convex domain and X_m is a finite dimensional subspace of X with values in a finite dimensional subspace Y_m. The operator F_m usually results from a Petrov-Galerkin discretization, such that $F_m(x_m) = r_m F(x_m)$, where $r_m : Y \to Y_m$ is a linear restriction. Then, the corresponding discretized Newton's method for solving Eq. (17.1.3) is defined for each $k = 0, 1, 2, \ldots$ by

$$F'_m(x_m^k) \Delta x_m^k = -F_m(x_m^k), \;\; x_m^{k+1} = x_m^k + \Delta x_m^k. \tag{17.1.4}$$

We have that $F'_m = r_m F'$ and in each Newton step, a system of linear equations must be solved.

The mesh independence principle asserts that the behavior of (17.1.4) is essentially the same as (17.1.2). Many authors have worked on the mesh independence principle under various hypotheses [1–13].

In the present study we are motivated by the work of Weiser et al. in [13]. Their paper presents an affine invariant theory on the asymptotic mesh independence principle of Newton's method. It turns out that their approach is simpler and more intuitive from the algorithmic point of view. Their theory is based on Lipschitz continuity conditions on the operator F' and F'_m. However, there are equations where the Lipschitz continuity conditions do not hold (see e.g. the numerical example at the end of the study). That is why, in the present study we extend the mesh independence principle by considering the more general Hölder continuity conditions on the operators F' and F'_m.

The chapter is organized as follows: In Sect. 17.2 we present the mesh independence principle in the Hölder case, whereas in the concluding Sect. 17.3, we provide a boundary value problem where the Lipschitz conditions given in [13] (or in earlier studies such as [1, 9–12]) do not hold but our Hölder conditions hold.

17.2   The Mesh Independence Principle

We shall show the mesh independence principle in this section under Hölder continuity conditions on the operator F' and F'_m. Let $U(w, \xi)$, $\bar{U}(w, \xi)$ denote the open and closed balls in X, respectively with center $w \in X$ and radius $\xi > 0$.

Next, we present a generalized Newton-Mysovskhi-type semi-local convergence result in affine-invariant form.

Theorem 17.1 *Let* $F : D \subset X \rightarrow Y$ *be a continuously Fréchet-differentiable operator. Suppose: For each* $x \in D$, $F'(x)^{-1} \in L(Y, X)$; *the following holds for collinear* $x, y, z \in D$, $v \in X$ *and some* $L > 0$, $p \in [0, 1]$;

$$\|F'(z)^{-1}(F'(y) - F'(x))v\| \le L\|y - x\|^p\|v\|; \tag{17.2.1}$$

For some $x^0 \in D$ *with* $F(x^0) \neq 0$,

$$h_0 = L\|\Delta x^0\|^p < 1 + p \tag{17.2.2}$$

and

$$\bar{U}(x^0, \rho) \subseteq D, \tag{17.2.3}$$

where

$$\rho = \frac{\|\Delta x^0\|}{1 - \frac{h_0}{1+p}}. \tag{17.2.4}$$

Then, the sequence $\{x^k\}$ *generated by Newton's method (17.1.2) is well defined, remains in* $U(x^0, \rho)$ *for each* $k = 0, 1, 2, \ldots$ *and converges to a unique solution* $x^* \in U(x^0, \rho)$ *of equation (17.1.1). Moreover, the following estimates hold*

$$\|x^{k+1} - x^k\| \le \frac{L}{1 + p}\|x^k - x^{k-1}\|^{1+p}. \tag{17.2.5}$$

Proof We shall show estimate (17.2.5) using Mathematical induction. The point x^1 is well defined, since $F'(x_0)^{-1} \in L(Y, X)$. Using (17.2.3), (17.2.4) and $\|x^1 - x^0\| = \|\Delta x^0\| < \rho$, we get that $x^1 \in \bar{U}(x^0, \rho) \subseteq D$. We also have that x^2 is well defined by Newton's method (17.1.2) for $k = 1$, since $x^1 \in D$ and $F'(x^1)^{-1} \in L(Y, X)$. In view of (17.1.2), we can write

$$F(x^1) = F(x^1) - F(x^0) - F'(x^0)(x^1 - x^0)$$
$$= \int_0^1 [F'(x^0 + \theta(x^1 - x^0)) - F'(x^0)](x^1 - x^0)d\theta. \tag{17.2.6}$$

It follows from (17.2.1) and (17.1.1) that

$$\|x^2 - x^1\| = \|F'(x^1)^{-1}F(x^1)\|$$
$$= \int_0^1 F'(x^1)^{-1}[F'(x^0 + \theta(x^1 - x^0)) - F'(x^0)](x^1 - x^0)d\theta\|$$
$$\le L\int_0^1 \theta^p\|x^1 - x^0\|^{1+p}d\theta = \frac{L}{1 + p}\|x^1 - x^0\|^{1+p}, \tag{17.2.7}$$

which shows (17.2.5) for $k = 1$. We also have by (17.2.2), (17.2.4) and (17.1.1) that

$$
\begin{aligned}
\|x^2 - x^0\| &\leq \|x^2 - x^1\| + \|x^1 - x^0\| \\
&\leq \frac{L}{1+p}\|x^1 - x^0\|^{1+p} + \|x^1 - x^0\| \\
&= [\frac{L}{1+p}\|x^1 - x^0\|^p + 1]\|x^1 - x^0\| \\
&= (\frac{h_0}{1+p} + 1)\|\Delta x^0\| \\
&= \frac{1 - (\frac{h_0}{1+p})^2}{1 - \frac{h_0}{1+p}}\|\Delta x^0\| \\
&< \frac{\|\Delta x^0\|}{1 - \frac{h_0}{1+p}} = \rho,
\end{aligned}
$$

which shows that $x^2 \in U(x^0, \rho)$. Suppose that (17.2.5) holds for each integer $i \leq k$ and $x^{k+1} \in U(x^0, \rho)$. Then, as in (17.2.6) we get that

$$
\begin{aligned}
F(x^{k+1}) &= F(x^{k+1}) - F(x^k) - F'(x^k)(x^{k+1} - x^k) \\
&= \int_0^1 [F'(x^k + \theta(x^{k+1} - x^k)) - F'(x^k)](x^{k+1} - x^k)d\theta
\end{aligned}
$$

so

$$
\begin{aligned}
\|x^{k+2} - x^{k+1}\| &= \|F'(x^{k+1})^{-1}F(x^{k+1})\| \\
&= \int_0^1 F'(x^{k+1})^{-1}[F'(x^k + \theta(x^{k+1} - x^k)) - F'(x^k)](x^{k+1} - x^k)d\theta\| \\
&\leq L\int_0^1 \theta^p \|x^{k+1} - x^k\|^{1+p}d\theta = \frac{L}{1+p}\|x^{k+1} - x^k\|^{1+p},
\end{aligned}
$$

which completes the induction for (17.2.5). We have shown $\|x^2 - x^1\| \leq \frac{h_0}{1+p}\|\Delta x^0\|$. From, the estimate

$$
\begin{aligned}
\|x^3 - x^2\| &\leq \frac{L}{1+p}\|x^2 - x^1\|^{1+p} \\
&\leq \frac{L}{1+p}(\frac{L}{1+p}\|x^1 - x^0\|^{1+p})^{1+p} \\
&\leq \frac{L}{1+p}(\frac{h_0}{1+p}\|x^1 - x^0\|)^{1+p} \\
&= \frac{L}{1+p}\|x^1 - x_0\|^p \|x^1 - x^0\|(\frac{h_0}{1+p})^{1+p} \\
&< (\frac{h_0}{1+p})^{2+p}\|x^1 - x^0\| \leq (\frac{h_0}{1+p})^2\|\Delta x^0\|,
\end{aligned}
$$

(since $(\frac{h_0}{1+p})^p < 1$ by (17.2.2)), similarly, we get that

$$\|x^{k+2} - x^{k+1}\| \le \frac{L}{1+p}\|x^{k+1} - x^k\|^{1+p}$$

$$\vdots$$

$$\le (\frac{h_0}{1+p})^{k+1+p}\|\Delta x^0\|$$

$$\le (\frac{h_0}{1+p})^{k+1}\|\Delta x^0\|. \tag{17.2.8}$$

Then, we get that

$$\|x^{k+2} - x^0\| \le \|x^{k+2} - x^{k+1}\| + \|x^{k+1} - x^k\| + \cdots + \|x^1 - x^0\|$$

$$\le ((\frac{h_0}{1+p})^{k+1} + (\frac{h_0}{1+p})^k + \cdots + 1)\|\Delta x^0\|$$

$$= \frac{1 - (\frac{h_0}{1+p})^{k+2}}{1 - \frac{h_0}{1+p}}\|\Delta x^0\|$$

$$< \frac{\|\Delta x^0\|}{1 - \frac{h_0}{1+p}} = \rho, \tag{17.2.9}$$

which shows $x^{k+2} \in U(x^0, \rho)$.

It follows from (17.1.1) that $\{x^k\}$ is a complete sequence in a Banach space X and as such it converges to some $x^* \in \bar{U}(x^0, \rho)$ (since $\bar{U}(x^0, \rho)$ is a closed set). Then, from the estimate

$$\|x^{k+2} - x^{k+1}\| \le \|F'(x^{k+1})^{-1}F(x^{k+1})\|$$

$$\le (\frac{h_0}{1+p})^{k+1}\|\Delta x^0\| \tag{17.2.10}$$

and the invertability of $F'(x^{k+1})$ we conclude by letting $k \to \infty$ in (17.1.1) that $F(x^*) = 0$. Finally, to show the uniqueness part, let $Q = \int_0^1 F'(y^* + \theta(x^* - y^*))d\theta$ with $F(y^*) = 0$ and $y^* \in \bar{U}(x^0, \rho)$ we have that

$$\|y^* + \theta(x^* - y^*) - x^*\| \le (1 - \theta)\|y^* - x^0\| + \theta\|x^* - x^0\|$$

$$\le (1 - \theta)\rho + \theta\rho = \rho.$$

Hence, $y^* + \theta(x^* - y^*) \in D$ and Q is invertible. Then, from the identity $0 = F(x^*) - F(y^*) = Q(x^* - y^*)$, we deduce that $x^* = y^*$. □

Remark 17.2 If $p = 1$, then Theorem 17.1 merges to the Newton-Mysorskikhi Theorem [9] (see also [6, 7]).

From now on we assume that Theorem 17.1 holds for (17.1.1)–(17.1.4). Notice that in the case of (17.1.4) the analog of (17.2.5) is given by

$$\|x_m^{k+1} - x_m^k\| \leq \frac{L_m}{1+p} \|x_m^k - x_m^{k-1}\|^{1+p}, \tag{17.2.11}$$

where $L_m > 0$ stand for the corresponding affine invariant Hölder parameter. Then, there exist unique discrete solutions x_m^* of Eq. (17.1.3) for each m. The discretization schemes must be chosen such that

$$\lim_{m \to \infty} x_m^* = x^*. \tag{17.2.12}$$

Earlier papers such as [1–7, 10–12] have used the non-affine invariant conditions

$$\|F_m'(x_m)^{-1}\| \leq \beta_m, \ \ \|F_m'(x_m + v_m) - F_m'(x_m)\| \leq \gamma_m \|v_m\| \tag{17.2.13}$$

$$\beta_m \leq \beta, \ \gamma_m \leq \gamma \tag{17.2.14}$$

and have restricted their analyses to smoother subset $W^* \subseteq X$ such that

$$x^*, x^k, \Delta x^k, x^k - x^* \in W^* \tag{17.2.15}$$

for each $k = 0, 1, 2, \ldots$. Notice however that due to the non-invariance conditions (17.2.13) are expressed in terms of operator norms which depend on the relation of the norms in the domain and the image of the operators F_m and F. As already noted in [13], this requirement is not convenient for PDEs, since we may have $\lim_{m \to \infty} \beta_m = \infty$, contradicting assumption (17.2.14). Moreover, assumption (17.2.15) is difficult to verify in many interesting cases. That is why in [13] a new technique was developed using assumption

$$L_m \leq \beta_m \gamma_m \tag{17.2.16}$$

instead of (17.2.13) and (17.2.14). Then, it was shown that L_m is bounded in the limit as long as L is bounded–even if β_m or γ_m are unbounded. Moreover, as in [13] introduce linear projections

$$\Pi_m : X \to X_m \quad \text{for each} \ \ m = 0, 1, 2, \ldots$$

and assume the stability condition

$$q_m = \sup_{x \in W^*, x \neq 0} \frac{\|\Pi_m x\|}{\|x\|} \leq \bar{q} < \infty \tag{17.2.17}$$

Moreover, define

$$\delta_m = \sup_{x \in W^*, x \neq 0} \frac{\|x - \Pi_m x\|}{\|x\|} \quad \text{for each } m = 0, 1, 2, \ldots. \tag{17.2.18}$$

Notice that (17.2.18) is implied by the assumption

$$\lim_{m \to \infty} \delta_m = 0. \tag{17.2.19}$$

We also have that

$$q_m \geq 1 \tag{17.2.20}$$

$$q_m \leq 1 + \delta_m \tag{17.2.21}$$

and

$$\lim_{m \to \infty} q_m = 1. \tag{17.2.22}$$

We need an auxiliary perturbation Lemma.

Lemma 17.3 *Suppose: Newton sequences* $\{x^k\}$, $\{y^k\}$ *given for each* $k = 0, 1, 2, \ldots$ *by*

$$x^{k+1} = x^k + \Delta x^k, \quad y^{k+1} = y^k + \Delta y^k, \tag{17.2.23}$$

where x^0, y^0 *are initial points and* Δx^k, Δy^k *are the corresponding corrections are well-defined; Hölder condition (17.2.1) is satisfied. Then, the following contraction estimate holds:*

$$\|x^{k+1} - y^{k+1}\| \leq L\left(\frac{1}{1+p}\|x^k - y^k\| + \|\Delta y^k\|\right)\|x^k - y^k\|^p. \tag{17.2.24}$$

Proof It is convenient to drop index k. Then, we can write

$$x + \Delta x - y - \Delta y = x - F'(x)^{-1}F(x) - y + F'(y)^{-1}F(y)$$
$$= x - F'(x)^{-1}F(x) + F'(x)^{-1}F(y) - F'(x)^{-1}F(y) - y + F'(y)^{-1}F(y)$$
$$= x - y - F'(x)^{-1}(F(x) - F(y)) + F'(x)^{-1}(F'(y) - F'(x))F'(y)^{-1}F(y)$$
$$= F'(x)^{-1}\left(F'(x)(x - y) - \int_0^1 F'(y + t(x - y))(x - y)dt\right)$$
$$\quad + F'(x)^{-1}(F'(y) - F'(x))\Delta y.$$

Using (17.2.1), we get in turn that

$$\|x^{k+1} - y^{k+1}\| \leq$$

$$\int_0^1 \|F'(x^k)^{-1}(F'(x^k) - F'(y^k + t(x^k - y^k)))(x^k - y^k)\|dt$$

$$+\|F'(x^k)^{-1}(F'(y^k) - F'(x^k))\Delta y^k\|$$

$$\leq \frac{L}{1+p}\|x^k - y^k\|^{1+p} + L\|x^k - y^k\|^p\|\Delta y^k\|,$$

which shows (17.2.24).   □

We also need an auxiliary result about zero of real functions.

Lemma 17.4 *Let* $\mu \geq 1$, $p \in (0, 1]$ *and* $\eta > 0$. *Suppose that*

$$\mu\eta^p < (\frac{p}{1+p})^p. \tag{17.2.25}$$

Then, the scalar function

$$\psi(t) = \frac{\mu}{1+p}t^{1+p} - t + \eta \tag{17.2.26}$$

has two zeros μ_-, μ_+ *such that*

$$\eta < \mu_- < \mu_+ < (\frac{1}{\mu})^{\frac{1}{p}}. \tag{17.2.27}$$

Proof We have that $\psi'(t) = (1+p)(\mu t^p - 1)$. Set $t^* = (\frac{1}{\mu})^{\frac{1}{p}}$. Then, it follows from (17.2.25) that $\psi(t^*) < 0$. Notice also that $\psi(0) = \eta > 0$ and $\psi(\eta) = \frac{\mu}{1+p}\eta^{1+p} > 0$.
□

Using the preceding notation and the two auxiliary results we can show a result connecting the behaviour of the discrete versus the continuous Newton sequences.

Theorem 17.5 *Suppose that for* $x^0 = x_m^0 \in X_m$ *the assumptions of Theorem 17.1 hold. Define for each discrete operator* F_m *and all arguments* $x_m \in D_m := D \cap X_m$

$$F'_m(x^m)\Delta x^m = -F_m(x^m), \quad F(x_m)\Delta x = -F(x_m). \tag{17.2.28}$$

Moreover, suppose that the discretization is such that

$$\|\Delta x^m - \Delta x\| \leq \delta_m \leq \frac{\min\{1, 1+p-h_0\}}{(1+p)L} \tag{17.2.29}$$

for each $x_m \in D_m$. *Furthermore, suppose* $\bar{U}(x^0, \rho_m) \cap X_m \subset D_m$ *for some* $\mu > 0$

$$\rho_m := \frac{\|\Delta x^0\|}{1 - \frac{h_0}{1+p}} + \frac{\frac{1+p}{p}\mu^{\frac{1}{p}}\delta_m}{\min\{1, 1+p-h_0\}}. \tag{17.2.30}$$

Then, the sequence $\{x_m^k\}$ generated by the discrete Newton iterates remains in $U(x^0, \rho_m) \cap X_m$. Moreover, the following estimates hold

$$\|x_m^k - x^k\| \leq \frac{\frac{1+p}{p} \mu^{\frac{1}{p}} \delta_m}{\min\{1, 1 + p - h_0\}} \leq \frac{1}{L} \tag{17.2.31}$$

for each $k = 0, 1, 2, \ldots$ and

$$\lim_{k \to \infty} \sup \|x_m^k - x^k\| \leq \frac{(1+p)}{p} \delta_m. \tag{17.2.32}$$

Proof As in [13] the discrete sequence starting at $x_m^0 = x^{0,0}$ is denoted by $\{x^{k,k}\}$, whereas the continuous Newton sequence starting at $x^0 = x^{0,0}$ is denoted by $\{x^{k,0}\}$. In between, we define more continuous Newton sequences, denoted by $\{x^{i,k}\}$ for each $k = i, i + 1, \ldots$, which start at the discrete Newton iterates $x_m^i = x^{i,i}$ and also run towards x^*. Using induction, we shall show

$$\|x_m^{k-1} - x^0\| < \rho_m, \tag{17.2.33}$$

where ρ_m is defined by (17.2.30). Clearly, (17.2.33) holds for $k = 1$. Let us introduce two majorants

$$L\|\Delta x^k\|^p \leq h_k, \quad \|x_m^k - x^k\| \leq \epsilon_k. \tag{17.2.34}$$

In view of Theorem 17.1 and (17.2.5), we have that

$$h_{k+1} = \lambda h_k^{1+p}, \quad \lambda = \left(\frac{1}{1+p}\right)^p. \tag{17.2.35}$$

In order for us to derive a majorant recursion for ϵ_k, we first need the estimate

$$\|x^{k+1,k+1} - x^{k+1,0}\| \leq \|x^{k+1,k+1} - x^{k+1,k}\| + \|x^{k+1,k} - x^{k+1,0}\|. \tag{17.2.36}$$

In view of the first term in (17.2.36) and (17.2.29) we obtain that

$$\|x^{k+1,k+1} - x^{k+1,k}\| = \|x_m^k + \Delta x_m^k - (x^{k,k} + \Delta x^{k,k})\|$$
$$= \|\Delta x_m^k - \Delta x^{k,k}\| \leq \delta_m. \tag{17.2.37}$$

Using Lemma 17.3, we can obtain for the second term in (17.2.36) that

$$\|x^{k+1,k} - x^{k+1,0}\| \leq L\left(\frac{1}{1+p}\|x_m^{k,k} - x^{k,0}\| + \|\Delta x^{k,0}\|\right)\|x_m^{k,k} - x^{k,0}\|. \tag{17.2.38}$$

Summing up, we get that

$$\|x^{k+1,k} - x^{k+1,0}\| \le \delta_m + \frac{L}{1+p}\epsilon_k^{1+p} + h_k\epsilon_k^p := \epsilon_{k+1}. \tag{17.2.39}$$

Hence, we have that

$$h_{k+1} = \lambda h_k, \quad h_0 = L\|\Delta x^0\|^p,$$

$$\epsilon_{k+1} = \delta_m + \frac{1}{1+p}L\epsilon_k^{1+p} + h_k\epsilon^p, \quad \epsilon_0 = 0. \tag{17.2.40}$$

Then, by combining the majorant equations in (17.1), we get that for $\beta \ge 1$

$$\beta L\epsilon_{k+1} + h_{k+1} = \beta L\delta_m + \frac{\beta L^2\epsilon_k^{1+p}}{1+p} + \beta L h_k\epsilon_k^p + \lambda h_k^p$$

$$= \beta L\delta_m + \frac{\mu}{1+p}(\beta L\epsilon_k + h_k)^{1+p}$$

$$+[\frac{\beta L^2}{1+p}\epsilon_k^{1+p} + \beta L\epsilon_k^p + \lambda h_k^p$$

$$-\frac{\mu}{1+p}(\beta L\epsilon_k + h_k)^{1+p}] \tag{17.2.41}$$

Clearly, the quantity inside the preceding bracket is negative for sufficiently large $\mu > 0$. Hence, the sequence a_k defined by

$$a_{k+1} = \beta L\delta_m + \frac{\mu}{1+p}a_k^{1+p}, \quad a_0 = h_0 \tag{17.2.42}$$

is a majorant for $\beta L\epsilon_k + h_k$. Suppose that

$$\mu(\beta L\delta_m)^p < (\frac{p}{1+p})^p. \tag{17.2.43}$$

Then, according to Lemma 17.4 (see (17.2.25)), (17.2.42) has two equilibrium points a_- and a_+ such that $a_- < a_+ < (\frac{1}{\mu})^{\frac{1}{p}}$. Notice that (17.2.43) certainly holds for

$$1 \le \beta \le \frac{p}{(1+p)L\mu^{\frac{1}{p}}\delta_m}. \tag{17.2.44}$$

The sequence $\{a_k\}$ converges monotonically to a_- for $h_0 < a_+$. We consider two cases $h_0 \le 1$ and $h_0 > 1$. If $h_0 \le 1$, set

$$\beta = \frac{p}{(1+p)L\mu^{\frac{1}{p}}\delta_m}$$

such that $h_0 \le a_- \le 1$. Due to monotonicity the sequence $\{a_k\}$ is bounded above from $a_- \le 1$. Then, we have the upper bound

$$\epsilon_k \le \frac{a_-}{\beta L} \le \frac{1+p}{p}\delta_m \mu^{\frac{1}{p}} \le \frac{1}{L}.$$

Hence, both (17.2.31) and (17.2.32) hold. If $1 < h_0 < 1+p$, choose $\sigma > 0$ sufficiently small and

$$\beta = \frac{h_0(1+p-h_0)}{(1+p+\sigma)L\delta_m}$$

so that $\beta \ge 1$ and $h_0 < a_+$ hold. Due to monotonicity, the sequence $\{a_k\}$ is bounded from above by $a_0 = h_0$. Then, we have

$$\epsilon_k \le \frac{h_0}{\beta L} = \frac{(1+p+\sigma)\delta_m}{1+p-h_0}$$

for sufficiently small $\sigma > 0$, which implies (17.2.31). Then, (17.2.32) follows from $a_k \to a_-$. Moreover, we get that

$$\|x_m^{k+1} - x^0\| \le \|x^{k+1} - x^0\| + \epsilon_{k+1}$$

$$\le \frac{\|\Delta x^0\|}{1 - \frac{h_0}{1+p}} + \frac{\frac{1+p}{p}\mu^{\frac{1}{p}}\delta_m}{\min\{1, 1+p-h_0\}} = \rho_m.$$

$\square$

Remark 17.6 If $p = \mu = 1$. Theorem 17.5 merges to Theorem 2.2 in [13].

Corollary 17.7 *Suppose that the hypotheses of Theorem 17.5 hold. Then, there exists an accumulation point*

$$\hat{x}_m \in \bar{U}(x^*, \frac{1+p}{p}\mu^{\frac{1}{p}}\delta_m) \cap X_m \subset U(x^*, \frac{1}{L}) \cap X_m.$$

This point need not be a solution of the discrete equation $F_m(x_m) = 0$.

Lemma 17.8 *Suppose that for the hypotheses of Theorem 17.1 hold for the operator $F : D \subseteq X \to Y$. Define $u_m \in X_m$, $u \in X$ for each collinear $x_m, y_m, z_m \in X_m$ by*

$$F'(x_m)u = (F'(z_m) - F'(y_m))v_m$$

$$F'_m(x_m)u_m = (F'_m(z_m) - F'_m(y_m))v_m$$

for arbitrary $v_m \in X_m$. Suppose that the discretization method satisfies

$$\|u - u_m\| \le \sigma_m \|z_m - y_m\|^p \|v_m\|.$$

Then, there exist constants

$$L_m \leq L + \sigma_m \tag{17.2.45}$$

such that the affine invariant Hölder condition

$$\|u_m\| \leq L_m \|x_m - y_m\| \|v_m\|$$

holds. We have in turn that

$$\|u_m\| \leq \|u\| + \|u_m - u\| \leq L\|z_m - y_m\|^p \|v_m\|$$
$$+\sigma_m \|z_m - y_m\|^p \|v_m\| = (L + \sigma_m)\|z_m - y_m\|^p \|v_m\|.$$

Corollary 17.9 *Suppose that the hypotheses of Theorem 17.5 and Lemma 17.8 hold for the discrete Newton sequence $\{x_m^k\}$. Then, the sequence $\{x_m^k\}$ converges to a unique discrete solution point $x_m^* \in \bar{U}(x^*, \frac{1+p}{p}\mu^{\frac{1}{p}}\delta_m) \cap X_m \subset U(x^*, \frac{1}{L}) \cap X_m.$*

Proof Simply apply Theorem 17.1 to F_m with starting point $x_m^0 = x^0$ and (17.2.45). $\qquad \square$

Remark 17.10 (a) If $p = \mu = 1$ the last three results merge to the corresponding ones in [13].

(b) If $\lim_{m\to\infty} \delta_m = \lim_{m\to\infty} \sigma_m = 0$, then the convergence speed of the discrete Newton method (17.1.4) is asymptotically the same with the continuous Newton method (17.1.2). Moreover, if initial points x^0 and x_m^0 are chosen, then the number of iterations to achieve a desired error tolerance is nearly the same.

17.3 Numerical Examples

We present a numerical example where the earlier results in [1, 9–13] cannot apply to solve a boundary value problem but our results can apply.

Example 17.11 Let $X = Y = \mathbb{R}^{m-1}$ for a natural integer $m \geq 2$. Let X and Y be equipped with the max-norm $\|x\| = \max_{1\leq i \leq m-1} \|x_i\|$. The corresponding matrix norm is

$$\|R\| = \max_{1\leq i \leq m-1} \sum_{j=1}^{j=m-1} \|r_{i,j}\|$$

for $R = (r_{i,j})_{1\leq i,j\leq m-1}$. We consider the following two point boundary value problem on the interval $[0, 1]$

$$v'' + v^{1+p} = 0 \quad p \in (0, 1]$$
$$v(0) = v(1) = 0. \tag{17.3.1}$$

To discretize the above equation, we divide the interval $[0, 1]$ into m equal parts with length of each part $h = \frac{1}{m}$ and coordinate at each point : $x_i = ih$ with $i = 0, 1, 2, \ldots m$. A second order finite difference discretization of (17.3.1) given by

$$v_i'' = \frac{v_{i-1} - 2v_i + v_{i+1}}{h^2}, \quad i = 1, 2, \ldots, m - 1$$

results in the following set of nonlinear equations

$$F(v) = \begin{cases} v_{i-1} + h^2 v_i^{1+p} - 2v_i + v_{i+1} = 0 \\ for each\ i = 1, 2, \ldots m - 1,\ v_0 = v_m = 0, \end{cases} \quad (17.3.2)$$

where $v = [v_1, v_2, \ldots, v_{m-1}]^T$. The Fréchet-derivative of operator F is given by

$$A = \begin{bmatrix} (1+p)h^2\ v_1^p - 2 & 1 & 0 & \cdots\cdots & & 0 \\ 1 & (1+p)h^2\ v_2^p - 2 & 1 & 0 & \cdots & 0 \\ \vdots & & & & & \vdots \\ 0 & 0 & \cdots & \cdots & 1 & (1+p)h^2\ v_{m-1}^p - 2 \end{bmatrix}$$

Then, if we choose say $p = \frac{1}{2}$, the results in [1, 9–13] cannot apply, but ours can apply, since (17.2.1) is satisfied for $p = \frac{1}{2}$ and $L \geq \frac{3}{2}h^2 \|F'(x)^{-1}\|$.

References

1. E.L. Allgower, K. Böhmer, F.A. Potra, W.C. Rheinboldt, A mesh-independence principle for operator equations and their discretizations. SIAM J. Numer. Anal. **23**, 160–169 (1986)
2. I.K. Argyros, A mesh independence principle for operator equations and their discretizations under mild differentiability conditions. Computing **45**, 265–268 (1990)
3. I.K. Argyros, On a mesh independence principle for operative equations and the Secant method. Acta Math. Hungarica **60**(1–2), 7–19 (1992)
4. I.K. Argyros, The asymptotic mesh independence principle for Newton-Galerkin methods using weak hypotheses on the Fréchet-derivatives. Math. Sci. Res. Hot-line **4**(11), 51–58 (2000)
5. I.K. Argyros, A mesh independence principle for inexact Newton-like methods and their discretizations. Ann. Univ. Sci. Budapest Sect. Comp. **20**, 31–53 (2001)
6. I.K. Argyros, *Convergence and Application of Newton-Like Iterations* (Springer, Berlin, 2008)
7. I.K. Argyros, S. Hilout, *Computational Methods in Nonlinear Analysis* (World Scientific, New Jersey, 2013)
8. I. Argyros, G. Santosh, *The Asymptotic Mesh Independence Principle of Newton's Method Under Weaker Conditions* (2015) (submitted)
9. P. Deuflhard, G. Heindl, Affine invariant convergence theorems for Newton's method and extensions to related methods. SIAM J. Numer. Anal. **16**, 1–10 (1979)
10. P. Deuflhard, F. Potra, Asymptotic mesh independence of Newton-Galerkin methods via a refined Mysovskii theorem. SIAM J. Numer. Anal. **29**, 1395–1412 (1992)

11. C. Kelley, E. Sachs, Mesh independence of Newton-like methods for infinite dimensional problems. J. Intergr. Equ. Appl. **3**, 549–573 (1991)
12. M. Laumen, Newton's mesh independence principle for a class of optimal shape design problems. SIAM J. Control. Optim. **30**, 477–493 (1992)
13. M. Weiser, A. Schiela, P. Deuflhard, Asymptotic mesh independence of Newton's method revisited. SIAM J. Numer. Anal. **42**(5), 1830–1845 (2005)

Chapter 18
Ball Convergence of a Sixth Order Iterative Method

We present a local convergence analysis of a sixth order iterative method for approximate a locally unique solution of an equation defined on the real line. Earlier studies such as [26] have shown convergence of these methods under hypotheses up to the fifth derivative of the function although only the first derivative appears in the method. In this study we expand the applicability of these methods using only hypotheses up to the first derivative of the function. Numerical examples are also presented in this study. It follows [5].

18.1 Introduction

Newton-like methods are famous for approximating a locally unique solution x^* of equation

$$F(x) = 0, \tag{18.1.1}$$

where $F : D \subseteq \mathbb{R} \to \mathbb{R}$ is a differentiable nonlinear function and D is a convex subset of $\mathbb{R}$. These methods are studied based on: semi-local and local convergence [2, 3, 21, 22, 27]. The methods such as Euler's, Halley's, super Halley's, Chebyshev's [2, 3, 7, 8, 10, 17, 23, 27] require the evaluation of the second derivative F'' at each step. To avoid this computation, many authors have used higher order multi-point methods [1, 2, 4, 12, 13, 15, 19, 22, 26, 27].

Newton's method is undoubtedly the most popular method for approximating a locally unique solution x^* provided that the initial point is close enough to the solution. In order to obtain a higher order of convergence Newton-like methods have been studied such as Potra-Ptak, Chebyshev, Cauchy Halley and Ostrowski method. The number of function evaluations per step increases with the order of convergence. In the scalar case the efficiency index [18, 22, 27] $EI = p^{\frac{1}{m}}$ provides

© Springer International Publishing Switzerland 2016

297

G.A. Anastassiou and I.K. Argyros, *Intelligent Numerical Methods: Applications to Fractional Calculus*, Studies in Computational Intelligence 624, DOI 10.1007/978-3-319-26721-0_18

a measure of balance where p is the order of the method and m is the number of function evaluations. According to the Kung-Traub conjecture the convergence of any multi-point method without memory cannot exceed the upper bound 2^{m-1} [22, 27] (called the optimal order). Hence the optimal order for a method with three function evaluations per step is 4. The corresponding efficiency index is $EI = 4^{\frac{1}{3}} = 1.58740...$ which is better than Newtons method which is $EI = 2^{\frac{1}{2}} = 1.414....$ Therefore, the study of new optimal methods of order four is important.

We study the local convergence analysis of three step King-like method with a parameter defined for each $n = 0, 1, 2, \ldots$ by

$$
y_n = x_n - \frac{F(x_n)}{F'(x_n)}
$$

$$
z_n = y_n - \frac{F(y_n)F'(x_n)^{-1}F(x_n)}{F(x_n) - 2F(y_n)}
$$

$$
x_{n+1} = z_n - \frac{(F(x_n) + \alpha F(y_n))F'(x_n)^{-1}F(z_n)}{F(x_n) + (\alpha - 2)F(y_n)}, \tag{18.1.2}
$$

where $x_0 \in D$ is an initial point and $\alpha \in \mathbb{R}$ a parameter. Sharma et al. [26] showed the sixth order of convergence of method (18.1.2) using Taylor expansions and hypotheses reaching up to the fourth derivative of function F although only the first derivative appears in method (18.1.2). These hypotheses limit the applicability of method (18.1.2). As a motivational example, let us define function F on $D = [-\frac{1}{2}, \frac{5}{2}]$ by

$$
F(x) = \begin{cases} x^3 \ln x^2 + x^5 - x^4, & x \neq 0 \\ 0, & x = 0 \end{cases}
$$

Choose $x^* = 1$. We have that

$$
F'(x) = 3x^2 \ln x^2 + 5x^4 - 4x^3 + 2x^2, \quad F'(1) = 3,
$$
$$
F''(x) = 6x \ln x^2 + 20x^3 - 12x^2 + 10x
$$
$$
F'''(x) = 6 \ln x^2 + 60x^2 - 24x + 22.
$$

Then, obviously function F does not have bounded third derivative in X, since $F'''(x)$ is unbounded at $x = 0$ (i.e., unbounded in D). The results in [26] require that all derivatives up to the fourth are bounded. Therefore, the results in [26] cannot be used to show the convergence of method (18.1.2). However, our results can apply (see Example 3.3). Notice that, in-particular there is a plethora of iterative methods for approximating solutions of nonlinear equations defined on $\mathbb{R}$ [2, 2, 4, 6, 8, 12, 13, 15, 19, 22, 26, 27]. These results show that if the initial point x_0 is sufficiently close to the solution x^*, then the sequence $\{x_n\}$ converges to x^*. But how close to the solution x^* the initial guess x_0 should be? These local results give no information on the radius of the convergence ball for the corresponding method. We address this question for method (18.1.2) in Sect. 18.2. The same technique can be used to other methods. In the present study we extend the applicability of these methods by using hypotheses up to the first derivative of function F and contractions. Moreover we avoid Taylor expansions and use instead Lipschitz parameters. Indeed, Taylor expansions and higher order derivatives are needed to obtain the equation of the local error and the order of convergence of the method. Using our technique we find instead the computational order of convergence (COC) or the approximate computational order of convergence that do not require the usage of Taylor expansions or higher order derivatives (see Remark 2.2 part 4). Moreover, using the Lipschitz constants we determine the radius of convergence of method (18.1.2). Notice also that the local error in [26] cannot be used to determine the radius of convergence of method (18.1.2). We do not address the global convergence of the three-step King-like method (18.1.2) in this study. Notice however that the global convergence of King's method (drop the third step of method (18.1.2) to obtain King's method) has not been studied either. This is mainly due to the fact that these methods are considered as special case of Newton-like methods for which there are many results (see e.g. [21]). Therefore, one can simply specialize global convergence results for Newton-like methods to obtain the specific results for method (18.1.2) or King's method.

The chapter is organized as follows. In Sect. 18.2 we present the local convergence analysis. We also provide a radius of convergence, computable error bounds and uniqueness result not given in the earlier studies using Taylor expansions. Special cases and numerical examples are presented in the concluding Sect. 18.3.

18.2 Local Convergence Analysis

We present the local convergence analysis of method (18.1.2) in this section. Let $L_0 > 0, L > 0, M \geq 1$ be given parameters. It is convenient for the local convergence analysis of method (18.1.2) that follows to introduce some scalar functions and parameters. Define functions g_1, p, h_p, q, h_q on the interval $[0, \frac{1}{L_0})$ by

$$g_1(t) = \frac{Lt}{2(1 - L_0 t)},$$

$$p(t) = (\frac{L_0 t}{2} + 2M g_1(t))t,$$

$$h_p(t) = p(t) - 1,$$

$$q(t) = \frac{L_0 t}{2} + |\alpha - 2| M g_1(t),$$

$$h_q(t) = q(t) - 1$$

and parameter r_1 by

$$r_1 = \frac{2}{2L_0 + L}.$$

We have that $h_p(0) = h_q(0) = -1 < 0$ and $h_q(t) \to +\infty$, $h_p(t) \to +\infty$ as $t \to \frac{1}{L_0}^-$. It follows from the intermediate value theorem that functions h_p, h_q have zeros in the interval $(0, \frac{1}{L_0})$. Denote by r_p, r_q the smallest such zeros. Moreover, define functions g_2 and h_2 on the interval $[0, r_p)$ by

$$g_2(t) = [1 + \frac{M^2}{(1 - L_0 t)(1 - p(t))}]g_1(t)$$

and

$$h_2(t) = g_2(t) - 1.$$

We have that $h_2(0) = -1 < 0$ and $h_2(t) \to +\infty$ as $t \to r_p^-$. Denote by r_2 the smallest zero of function h_2 in the interval $(0, r_p)$. Furthermore, define functions g_3 and h_3 on the interval $[0, \min\{r_p, r_q\})$ by

$$g_3(t) = (1 + \frac{M}{1 - L_0 t} + \frac{2M^2 g_1(t)}{(1 - L_0 t)(1 - q(t))})g_2(t)$$

and

$$h_3(t) = g_3(t) - 1.$$

We have that $h_3(0) = -1 < 0$ since $g_1(0) = g_2(0) = 0$ and $h_3(t) \to +\infty$ as $t \to \min\{r_p, r_q\}$. Denote by r_3 the smallest zero of function h_3 in the interval $(0, \min\{r_p, r_q\})$. Set

$$r = \min\{r_1, r_2, r_3\}. \tag{18.2.1}$$

Then, we have that

$$0 < r \leq r_1 < \frac{1}{L_0} \tag{18.2.2}$$

and for each $t \in [0, r)$

$$0 \leq g_1(t) < 1 \tag{18.2.3}$$

$$0 \leq p(t) < 1. \tag{18.2.4}$$

$$0 \leq g_2(t) < 1 \tag{18.2.5}$$

$$0 \leq q(t) < 1. \tag{18.2.6}$$

and

$$0 \leq g_3(t) < 1. \tag{18.2.7}$$

Denote by $U(v, \rho)$, $\bar{U}(v, \rho)$ stand respectively for the open and closed balls in $\mathbb{R}$ with center $v \in \mathbb{R}$ and of radius $\rho > 0$. Next, we present the local convergence analysis of method (18.1.2) using the preceding notation.

Theorem 18.1 *Let* $F : D \subset \mathbb{R} \to \mathbb{R}$ *be a differentiable function. Suppose there exist* $x^* \in D$, $L_0 > 0$, $L > 0$, $M \geq 1$ *such that* $F(x^*) = 0$, $F'(x^*) \neq 0$,

$$|F'(x^*)^{-1}(F'(x) - F'(x^*))| \leq L_0|x - x^*|, \tag{18.2.8}$$

$$|F'(x^*)^{-1}(F'(x) - F'(y))| \leq L|x - y|, \tag{18.2.9}$$

$$|F'(x^*)^{-1}F'(x)| \leq M, \tag{18.2.10}$$

and

$$\bar{U}(x^*, r) \subseteq D; \tag{18.2.11}$$

where the radius of convergence r *is defined by (18.2.1). Then, the sequence* $\{x_n\}$ *generated by method (18.1.2) for* $x_0 \in U(x^*, r) - \{x^*\}$ *is well defined, remains in* $U(x^*, r)$ *for each* $n = 0, 1, 2, \ldots$ *and converges to* x^*. *Moreover, the following estimates hold*

$$|y_n - x^*| \leq g_1(|x_n - x^*|)|x_n - x^*| < |x_n - x^*| < r, \tag{18.2.12}$$

$$|z_n - x^*| \leq g_2(|x_n - x^*|)|x_n - x^*| < |x_n - x^*|, \tag{18.2.13}$$

and

$$|x_{n+1} - x^*| \leq g_3(|x_n - x^*|)|x_n - x^*| < |x_n - x^*|, \tag{18.2.14}$$

where the "g" functions are defined above Theorem 18.1. Furthermore, for $T \in [r, \frac{2}{L_0})$ *the limit point* x^* *is the only solution of equation* $F(x) = 0$ *in* $\bar{U}(x^*, T) \cap D$.

Proof We shall show estimates (18.2.12)–(18.2.14) using mathematical induction. By hypothesis $x_0 \in U(x^*, r) - \{x^*\}$, (18.2.1) and (18.2.8), we have that

$$|F'(x^*)^{-1}(F'(x_0) - F'(x^*))| \leq L_0|x_0 - x^*| < L_0 r < 1. \tag{18.2.15}$$

It follows from (18.2.15) and Banach Lemma on invertible functions [2, 3, 22, 23, 27] that $F'(x_0) \neq 0$ and

$$|F'(x^*)^{-1}F'(x_0)| \leq \frac{1}{1 - L_0|x_0 - x^*|}. \tag{18.2.16}$$

Hence, y_0 is well defined by the first sub-step of method (18.1.2) for $n = 0$. Using (18.2.1), (18.2.3), (18.2.9) and (18.2.16) we get that

$$
\begin{aligned}
|y_0 - x^*| &= |x_0 - x^* - F'(x_0)^{-1} F(x_0)| \\
&\leq |F'(x_0)^{-1} F'(x^*)| |\int_0^1 F'(x^*)^{-1} (F'(x^* + \theta(x_0 - x^*)) \\
&\quad -F'(x_0))(x_0 - x^*) d\theta| \\
&\leq \frac{L|x_0 - x^*|^2}{2(1 - L_0 |x_0 - x^*|)} \\
&= g_1(|x_0 - x^*|)|x_0 - x^*| < |x_0 - x^*| < r,
\end{aligned}
\tag{18.2.17}
$$

which shows (18.2.12) for $n = 0$ and $y_0 \in U(x^*, r)$. We can write that

$$
F(x_0) = F(x_0) - F(x^*) = \int_0^1 F'(x^* + \theta(x_0 - x^*)(x_0 - x^*) d\theta. \tag{18.2.18}
$$

Notice that $|x^* + \theta(x_0 - x^*) - x^*| = \theta|x_0 - x^*| < r$. Hence, we get that $x^* + \theta(x_0 - x^*) \in U(x^*, r)$. Then, by (18.2.10) and (18.2.18), we obtain that

$$
|F'(x^*)^{-1} F(x_0)| \leq M|x_0 - x^*|. \tag{18.2.19}
$$

We also have by (18.2.17) and (18.2.19) (for $y_0 = x_0$) that

$$
|F'(x^*)^{-1} F(y_0) \leq M|y_0 - x^*| \leq M g_1(|x_0 - x^*|)|x_0 - x^*|, \tag{18.2.20}
$$

since $y_0 \in U(x^*, r)$. Next, we shall show $F(x_0) - 2F(y_0) \neq 0$. Using (18.2.1), (18.2.4), (18.2.8), (18.2.20) and the hypothesis $x_0 \neq x^*$, we have in turn that

$$
\begin{aligned}
&|(F'(x^*)(x_0 - x^*))^{-1}[F(x_0) - F'(x^*) - F'(x^*)(x_0 - x^*) - 2F(y_0)| \\
&\leq |x_0 - x^*|^{-1}[|F'(x^*)^{-1}(F(x_0) - F'(x^*) - F'(x^*)(x_0 - x^*))| \\
&\quad +2|F'(x^*)^{-1} F(y_0)| \\
&\leq |x_0 - x^*|^{-1}[\frac{L_0}{2}|x_0 - x^*|^2 + 2M g_1(|x_0 - x^*|)|x_0 - x^*| \\
&= p(|x_0 - x^*|) < p(r) < 1.
\end{aligned}
\tag{18.2.21}
$$

It follows from (18.2.21) that

$$
|(F(x_0) - 2F(y_0))^{-1} F'(x^*)| \leq \frac{1}{|x_0 - x^*|(1 - p(|x_0 - x^*|))}. \tag{18.2.22}
$$

Hence, z_0 and x_1 are is well defined for $n = 0$. Then, using (18.2.1), (18.2.5), (18.2.16), (18.2.17), (18.2.20) and (18.2.22) we get in turn that

$$
|z_0 - x^*| \leq |y_0 - x^*| + \frac{M^2 |y_0 - x^*||x_0 - x^*|}{(1 - L_0|x_0 - x^*|)(1 - p(|x_0 - x^*|))|x_0 - x^*|}
$$

$$
\leq [1 + \frac{M^2}{(1 - L_0|x_0 - x^*|)(1 - p(|x_0 - x^*|))}] g_1(|x_0 - x^*|)|x_0 - x^*|
$$

$$
\leq g_2(|x_0 - x^*|)|x_0 - x^*| < |x_0 - x^*| < r, \tag{18.2.23}
$$

which show (18.2.13) for $n = 0$ and $z_0 \in U(x^*, r)$. Next, we show that $F(x_0) - (\alpha - 2)F(y_0) \neq 0$. Using (18.2.1), (18.2.6), (18.2.8), (18.2.17) and $x_0 \neq x^*$, we obtain in turn that

$$
|(F'(x^*)(x_0 - x^*))^{-1}[F(x_0) - F'(x^*) - F'(x^*)(x_0 - x^*) - (\alpha - 2)F(y_0)|
$$

$$
\leq |x_0 - x^*|^{-1}[|F'(x^*)^{-1}(F(x_0) - F'(x^*) - F'(x^*)(x_0 - x^*))|
$$

$$
+ |\alpha - 2||F'(x^*)^{-1}F(y_0)|
$$

$$
\leq |x_0 - x^*|^{-1}[\frac{L_0}{2}|x_0 - x^*| + |\alpha - 2|M|y_0 - x^*|]
$$

$$
\leq \frac{L_0}{2}|x_0 - x^*| + |\alpha - 2|Mg_1(|x_0 - x^*|)
$$

$$
= p(|x_0 - x^*|) < p(r) < 1. \tag{18.2.24}
$$

It follows from (18.2.24) that

$$
|(F'(x^*)(x_0 - x^*))^{-1}F'(x^*)| \leq \frac{1}{|x_0 - x^*|(1 - q(|x_0 - x^*|))}. \tag{18.2.25}
$$

Hence, x_1 is well defined by the third sub-step of method (18.1.2) for $n = 0$. Then using (18.2.1), (18.2.7),(18.2.16), (18.2.19) (for $z_0 = x_0$), (18.2.23), (18.2.24) and (18.2.25), we obtain in turn that

$$
x_1 - x^* = z_0 - x^* - F'(x_0)^{-1}F(z_0)
$$

$$
+ [1 - \frac{F(x_0) + \alpha F(y_0)}{F(x_0) + (\alpha - 2)F(y_0)}]F'(x_0)^{-1}F(z_0)
$$

so,

$$
|x_1 - x^*| = |z_0 - x^*| + |F'(x_0)^{-1}F'(x^*)||F'(x^*)^{-1}F(z_0)|
$$

$$
+ 2|(F'(x^*)(x_0 - x^*))^{-1}F'(x^*)||F'(x^*)^{-1}F(y_0)||F'(x_0)^{-1}F(x^*||F'(x^*)^{-1}F(z_0)|
$$

$$
\leq |z_0 - x^*| + \frac{M|z_0 - x^*|}{1 - L_0|x_0 - x^*|}
$$

$$+ \frac{2M^2|y_0 - x^*||z_0 - x^*|}{(1 - L_0|x_0 - x^*|)|x_0 - x^*|(1 - q(|x_0 - x^*|))}$$

$$\leq [1 + \frac{M}{1 - L_0|x_0 - x^*|} + \frac{2M^2 g_1(|x_0 - x^*|)}{(1 - L_0|x_0 - x^*|)(1 - q(|x_0 - x^*|))}]|z_0 - x^*|$$

$$= g_3(|x_0 - x^*|)|x_0 - x^*| < |x_0 - x^*| < r, \qquad (18.2.26)$$

which shows (18.2.14) for $n = 0$ and $x_1 \in U(x^*, r)$. By simply replacing x_0, y_0, z_0, x_1 by x_k, y_k, z_k, x_{k+1} in the preceding estimates we arrive at estimates (18.2.12)–(18.2.14). Then, it follows from the estimate $|x_{k+1} - x^*| < |x_k - x^*| < r$, we deduce that $x_{k+1} \in U(x^*, r)$ and $\lim_{k \to \infty} x_k = x^*$. To show the uniqueness part, let $Q = \int_0^1 F'(y^* + \theta(x^* - y^*)d\theta$ for some $y^* \in \bar{U}(x^*, T)$ with $F(y^*) = 0$. Using (18.2.8), we get that

$$|F'(x^*)^{-1}(Q - F'(x^*))| \leq \int_0^1 L_0|y^* + \theta(x^* - y^*) - x^*|d\theta \qquad (18.2.27)$$

$$\leq \int_0^1 L_0(1 - \theta)|x^* - y^*|d\theta \leq \frac{L_0}{2}T < 1.$$

It follows from (18.2.27) and the Banach Lemma on invertible functions that Q is invertible. Finally, from the identity $0 = F(x^*) - F(y^*) = Q(x^* - y^*)$, we deduce that $x^* = y^*$. $\qquad \square$

Remark 18.2   1. In view of (18.2.8) and the estimate

$$|F'(x^*)^{-1}F'(x)| = |F'(x^*)^{-1}(F'(x) - F'(x^*)) + I| \qquad (18.2.28)$$

$$\leq 1 + |F'(x^*)^{-1}(F'(x) - F'(x^*))| \leq 1 + L_0|x - x^*|$$

condition (18.2.10) can be dropped and M can be replaced by

$$M(t) = 1 + L_0 t$$

or

$$M(t) = M = 2,$$

since $t \in [0, \frac{1}{L_0})$.
2. The results obtained here can be used for operators F satisfying autonomous differential equations [2] of the form

$$F'(x) = P(F(x))$$

where P is a continuous operator. Then, since $F'(x^*) = P(F(x^*)) = P(0)$, we can apply the results without actually knowing x^*. For example, let $F(x) = e^x - 1$. Then, we can choose: $P(x) = x + 1$.

3. In [2, 3] we showed that $r_1 = \frac{2}{2L_0+L}$ is the convergence radius of Newton's method:

$$x_{n+1} = x_n - F'(x_n)^{-1} F(x_n) \quad \text{for each} \quad n = 0, 1, 2, \ldots \qquad (18.2.29)$$

under the conditions (18.2.8) and (18.2.9). It follows from the definition of r that the convergence radius r of the method (18.1.2) cannot be larger than the convergence radius r_1 of the second order Newton's method (18.2.29). As already noted in [2, 3] r_1 is at least as large as the convergence radius given by Rheinboldt [25]

$$r_R = \frac{2}{3L}. \qquad (18.2.30)$$

The same value for r_R was given by Traub [27]. In particular, for $L_0 < L$ we have that

$$r_R < r_1$$

and

$$\frac{r_R}{r_1} \to \frac{1}{3} \quad as \quad \frac{L_0}{L} \to 0.$$

That is the radius of convergence r_1 is at most three times larger than Rheinboldt's.

4. It is worth noticing that method (18.1.2) is not changing when we use the conditions of Theorem 18.1 instead of the stronger conditions used in [26]. Moreover, we can compute the computational order of convergence (COC) defined by

$$\xi = \ln \left(\frac{|x_{n+1} - x^*|}{|x_n - x^*|} \right) / \ln \left(\frac{|x_n - x^*|}{|x_{n-1} - x^*|} \right)$$

or the approximate computational order of convergence

$$\xi_1 = \ln \left(\frac{|x_{n+1} - x_n|}{|x_n - x_{n-1}|} \right) / \ln \left(\frac{|x_n - x_{n-1}|}{|x_{n-1} - x_{n-2}|} \right).$$

This way we obtain in practice the order of convergence in a way that avoids the bounds involving estimates using estimates higher than the first Fré chet derivative of operator F.

Table 18.1 Comparison table for radii

Method (18.1.2)	Method in [4]	Method in [6]	r_1	r_R
0.2289	0.1981	0.2016	0.666666667	0.666666667

Table 18.2 Comparison table for radii

Method (18.1.2)	Method in [4]	Method in [6]	r_1	r_R
0.0360	0.0354	0.0355	0.3249	0.2453

18.3   Numerical Examples

We present numerical examples in this section (Table 18.1, 18.2).

Example 18.3   Let $D = (-\infty, +\infty)$. Define function f of D by

$$f(x) = \sin(x). \tag{18.3.1}$$

Then we have for $x^* = 0$ that $L_0 = L = M = 1$. The parameters are

$$r_1 = 0.6667, \ r_p = 0.5858, \ r_q = 0.7192, \ r_2 = 0.3776, \ r_3 = 0.2289 = r.$$

Example 18.4   Let $D = [-1, 1]$. Define function f of D by

$$f(x) = e^x - 1. \tag{18.3.2}$$

Using (18.3.2) and $x^* = 0$, we get that $L_0 = e - 1 < L = e$, $M = 2$. The parameters are

$$r_1 = 0.3249, \ r_p = 0.2916, \ r_q = 0.2843, \ r_2 = 0.09876, \ r_3 = 0.0360 = r.$$

Example 18.5   Returning back to the motivational example at the introduction of this study, we have $L_0 = L = 96.662907$, $M = 2$. The parameters are

$$r_1 = 0.0069, \ r_p = 0.0101, \ r_q = 0.0061, \ r_2 = 0.0025, \ r_3 = 0.0009 = r.$$

References

1. S. Amat, S. Busquier, S. Plaza, Dynamics of the King's and Jarratt iterations. Aequationes. Math. **69**, 212–213 (2005)
2. I. K. Argyros, *Convergence and Application of Newton-Like Iterations* (Springer, Berlin, 2008)

3. I.K. Argyros, S. Hilout, *Computational Methods in Nonlinear Analysis* (World Scientific, New Jersey, 2013)
4. I.K. Argyros, G. Santosh, Ball comparison between two optimal eight-order methods under weak conditions. SeMA Journal Boletin de la Sociedad Espaniola de Matematica Aplicada (2015). doi:10.1007/s40324-015-0035-z
5. I. Argyros, G. Santosh, *Ball Convergence of a Sixth Order Iterative Method With One Parameter For Solving Equations Under Weak Conditions* (2015) (submitted)
6. I.K. Argyros, G. Santosh, Local convergence for an efficient eighth order iterative method with a parameter for solving equations under weak conditions. Int. J. Appl. Comput. Math. doi:10.1007/s40819-015-0078-y
7. I.K. Argyros, D. Chen, Q. Quian, The Jarratt method in Banach space setting. J. Comput. Appl. Math. **51**, 103–106 (1994)
8. I.K. Argyros, G. Santosh, A. Alberto, Magrenan, high convergence order. J. Comput. Appl. Math. **282**, 215–224 (2015)
9. D.K.R. Babajee, M.Z. Dauhoo, An analysis of the properties of Newton's method with third order convergence. Appl. Math. Comput. **183**, 659–684 (2006)
10. V. Candela, A. Marquina, Recurrence relations for rational cubic methods I: the Halley method. Computing **44**, 169–184 (1990)
11. J. Chen, Some new iterative methods with three-order convergence. Appl. Math. Comput. **181**, 1519–1522 (2006)
12. C. Chun, B. Neta, M. Scott, Basins of attraction for optimal eighth order methods to find simple roots of nonlinear equations. Appl. Math. Comput. **227**, 567–592 (2014)
13. A. Cordero, J. Maimo, J. Torregrosa, M.P. Vassileva, P. Vindel, Chaos in King's iterative family. Appl. Math. Lett. **26**, 842–848 (2013)
14. J.A. Ezquerro, M.A. Hernández, New iterations of R-order four with reduced computational cost. BIT Numer. Math. **49**, 325–342 (2009)
15. M. Frontini, E. Sormani, Some variants of Newton's method with third order convergence. Appl. Math. Comput. **140**, 419–426 (2003)
16. V.I. Hasanov, I.G. Ivanov, G. Nedzhibov, A new modification of Newton's method. Acta Math. Eng. **27**, 278–286 (2002)
17. R.F. King, A family of fourth-order methods for nonlinear equations. SIAM. J. Numer. Anal. **10**, 876–879 (1973)
18. H.T. Kung, J.F. Traub, Optimal order of one-point and multipoint iterations. J. Appl. Comput. Math. **21**, 643–651 (1974)
19. M.A. Noor, Some applications of quadrature formulas for solving nonlinear equations. Nonlinear Anal. Forum **12**(1), 91–96 (2007)
20. G. Nedzhibov, *On a few iterative methods for solving nonlinear equations.* Application of Mathematics in Engineering and Economics 28. in *Proceeding of the XXVIII Summer School Sozopol, 2003* (Heron Press, Sofia Bulgaria, 2002)
21. J.M. Ortega, W.C. Rheinbolt, *Iterative Solution of Nonlinear Equations in Several Variables* (Academic Press, New York, 1970)
22. M.S. Petkovic, B. Neta, L. Petkovic, J. Džunič, *Multipoint Methods for Solving Nonlinear Equations* (Elsevier, New York, 2013)
23. F.A. Potra, V. Ptak, *Nondiscrete Induction and Iterative Processes. Research Notes in Mathematics*, vol. 103 (Pitman, Boston, 1984)
24. H. Ren, Q. Wu, W. Bi, New variants of Jarratt method with sixth-order convergence. Numer. Algorithms **52**(4), 585–603 (2009)
25. W.C. Rheinboldt, An adaptive continuation process for solving systems of nonlinear equations. in *Mathematical models and numerical methods*, ed. by A.N.Tikhonov et al. pub.3, vol. 19, pp. 129–142 (Banach Center, Warsaw)
26. J.R. Sharma, R.K. Guha, A family of modified Ostrowski methods with accelerated sixth order convergence. Appl. Math. Comput. **190**, 111–115 (2007)
27. J.F. Traub, *Iterative Methods For the Solution of Equations* (Prentice Hall, Englewood Cliffs, 1964)

Chapter 19
Broyden's Method with Regularly Continuous Divided Differences

In this chapter we provide a semilocal convergence analysis for Broyden's method in a Banach/Hilbert space setting using regularly continuous divided differences. By using: more precise majorizing sequences; the same or weaker hypotheses and the same computational cost as in [7] we provide a new convergence analysis for Broyden's method with the following advantages: larger convergence domain; finer error bounds on the distances involved, and at least as precise information on the location of the solution. It follows [5].

19.1 Introduction

In this study we are concerned with the problem of approximating a locally unique solution x^* of equation

$$F(x) = 0, \qquad (19.1.1)$$

where F is a continuously Fréchet-differentiable operator defined on a convex subset D of a Banach/Hilbert space X with values in a Hilbert space H.

A large number of problems in applied mathematics and also in engineering are solved by finding the solutions of certain equations [1–4, 6–8]. The solution of these equations can rarely be found inclosed form. That is why most solution methods for these equations are iterative.

Methods are usually studied based on: semi–local and local convergence. The semi–local convergence matter is, based on the information around an initial point, to give conditions ensuring the convergence of the iterative procedure; while the local one is, based on the information around a solution, to find a estimates of the radii of convergence balls.

© Springer International Publishing Switzerland 2016
G.A. Anastassiou and I.K. Argyros, *Intelligent Numerical Methods:*
Applications to Fractional Calculus, Studies in Computational Intelligence 624,
DOI 10.1007/978-3-319-26721-0_19

Newton's method

$$x_+ := x - F'(x)^{-1}F(x), \qquad (19.1.2)$$

is undoubtedly the most popular iterative method for generating a sequence approximating x^*. The computation of the inverse $F'(x)^{-1}$ at every step may be very expensive or impossible. That is why Broyden in [6] (for $X = H = \mathbb{R}^m$) replace the inverse Jacobian $F'(x)^{-1}$ by an $m \times m$ matrix A satisfying the equation

$$A(F(x) - F(x_-)) = x - x_-, \qquad (19.1.3)$$

where x_- denotes the iteration preceding the current one x. This way the quasi-Newton methods were introduced [6].

We study the semilocal convergence of Broyden's method defined by

$$x_+ = x - AF(x), \quad A_+ = A - \frac{AF(x_+)\langle A^*AF(x), \cdot \rangle}{\langle A^*AF(x), F(x_+) - F(x) \rangle}, \qquad (19.1.4)$$

where A^* is the adjoint of A and $\langle \cdot, \cdot \rangle$ is the inner product in H.

Semilocal and local convergence results for Broyden's method (19.1.4) and more general Broyden–like methods have already been given in the literature under Lipschitz–type conditions and for smooth operators F. Recently, in the elegant study by A. Galperin [7] the semilocal convergence of Broyden's method (19.1.4) was given for nonsmooth operators using the notion regularly continuous divided differences (RCDD) [1, 4, 8] (to be precised in Definition 19.1). The convergence domain found in [7] is small in general. Hence, it is important to expand this domain without adding hypotheses. This has already be done by us in [1–4] for Newton's method and the Secant method using the notion of the center regularly continuous divided difference (CRCDD) which is always implied by the (RGDD) but not necessarily viceversa. Here, we use this idea to present a new semilocal convergence analysis of Broyden's method with advantages over the work in [7] as already stated in the abstract of this chapter.

19.2   Semilocal Convergence Analysis of Broyden's Method

In the rest of the chapter we use the notation already established in [7].

Let $\underline{h}([x, y|F])$ denote the quantity $\inf_{x,y}\{[x, y|F] : (x, y) \in D^2\}$, and let N be the class of continuous non-decreasing concave functions $\omega : [0, +\infty) \to [0, +\infty)$, such that $\omega(0) = 0$.

We need the definition of RCDD.

Definition 19.1 [7] The $dd[x, y|F]$ is said to be ω-regularly continuous on D (ω-RCDD) if there exist an $\omega \in N$ (call it regularity modulus), and a constant $\underline{h} \in [0, \underline{h}([x, y|F])]$ such that for all $x, y, u, v \in D$

$$\omega^{-1}\Big(\min\{\|\ [x,y|F]\ \|,\ \|\ [u,v|F]\ \|\} - \underline{h} + \|\ [x,y|F] - [u,v|F]\ \|\ \Big)$$

$$-\omega^{-1}\Big(\min\{\|\ [x,y|F]\ \|,\ \|\ [u,v|F]\ \|\} - \underline{h} \Big) \leq \| x-u \| + \| y-v \|.$$

$$(19.2.1)$$

We also say that $dd[x,y|F]$ is regularly continuous on D, if it has a regularity modulus there.

A detailed discussion on the properties of a $dd[x,y|F]$ which is ω-regularly continuous on D is given in [7]. In the same reference a semilocal convergence analysis is provided using only condition (19.2.1). However, in view of condition (19.2.1), for $\overline{x}$ and $\overline{y}$ fixed and all $u, v \in D$, there exist $\omega_0 \in \mathbb{N}$ such that condition (19.2.1) holds with ω_0 replacing function ω. That, the ω_0-CRCDD, is:

$$\omega_0^{-1}\Big(\min\{\|\ [\overline{x},\overline{y}|F]\ \|,\ \|\ [u,v|F]\ \|\} - \underline{h} + \|\ [\overline{x},\overline{y}|F] - [u,v|F]\ \|\ \Big)$$

$$-\omega_0^{-1}\Big(\min\{\|\ [\overline{x},\overline{y}|F]\ \|,\ \|\ [u,v|F]\ \|\} - \underline{h} \Big) \leq \| \overline{x}-u \| + \| \overline{y}-v \|.$$

$$(19.2.2)$$

Clearly,

$$\omega_0(s) \leq \omega(s) \quad \text{for all } s \in [0,+\infty), \tag{19.2.3}$$

holds in general and $\dfrac{\omega(s)}{\omega_0(s)}$ can be arbitrarily large [1–4]. Notice also that in practice the computation of ω requires the computation of ω_0 as a special case. That is (19.2.2) is not an additional hypothesis to (19.2.1).

On the other hand, because of the convexity of ω^{-1}, each ω-regularly continuous dd is also ω-continuous in the sense that

$$\|\ [x,y|F] - [u,v|F]\ \| \leq \omega(\| x-u \| + \| y-v \|) \quad \text{for all } x,y,u,v \in D. \tag{19.2.4}$$

Similar comments can be made for the dd $[x,y|F]$ in connection with function ω_0.

Assume that A_0 is invertible, so that A and F in (19.1.4) can be replaced by their normalizations AA_0^{-1} and A_0F without affecting method (19.1.4). As in [7] we suppose that A and F have already been normalized:

$$A_0 = [x_0, x_{-1}|F] = I.$$

Then, the current approximation (x,A) induces the triple of reals, where

$$\overline{t} := \| x-x_0 \|, \quad \overline{\gamma} := \| x-x_{-1} \| \quad \text{and} \quad \overline{\delta} := \| x_+ - x \|. \tag{19.2.5}$$

From now on the superscript $+$ denotes the non-negative part of real number. That is:

$$r_+ = \max\{r, 0\}.$$

We can have [7]:

$$\bar{t}_+ := \| x_+ - x_0 \| \leq \bar{t} + \bar{\delta},$$

$$\bar{\gamma}_+ := \bar{\delta},$$

and

$$
\begin{aligned}
\bar{\alpha}_+ &:= \omega^{-1}(\| [x_+, x|F] \| - \underline{h}) \\
&\geq \left(\omega^{-1}(1 - \underline{h}) - \| x_+ - x_0 \| - \| x - x_{-1} \| \right)^+ \\
&\geq \left(\omega^{-1}(1 - \underline{h}) - \bar{t}_+ - \bar{t} - \| x_0 - x_{-1} \| \right)^+ .
\end{aligned}
\tag{19.2.6}
$$

It is also convenient for us to introduce notations:

$$\alpha := \omega_0^{-1}(1 - \underline{h}), \quad \bar{\gamma}_0 := \| x_0 - x_{-1} \| \quad \text{and} \quad a := \alpha - \bar{\gamma}_0. \tag{19.2.7}$$

We need the following result relating $\bar{\delta}_{++} = \|x_{++} - x_+\| = \|A_+ F(x_+)\|$ with $(\bar{t}, \bar{\gamma}, \bar{\delta})$.

Lemma 19.2 *Suppose that* $dd[x_1, x_2|F]$ *of F is* ω *-regularly continuous on D. Then, the* $dd[x_1, x_2|F]$ *of F is* ω_0*-regularly continuous on D at a given fixed pair* (x_0, x_{-1}). *If* $\bar{t}_+ + \bar{t} < a$, *then*

$$\bar{\delta}_+ \leq \bar{\delta} \left(\frac{\omega(a - \bar{t}_+ - \bar{t} + \bar{\delta} + \bar{\gamma} - \omega(a - \bar{t}_+ - \bar{t})}{\omega_0(a - \bar{t}_+ - \bar{t})} \right). \tag{19.2.8}$$

Proof $\bar{\delta}_+ \leq \|A_+\| \|F(x_+)\|$. Using the Banach lemma on invertible operators [1–4, 8] we get

$$\|A_+\|^{-1} \geq \|A_0\|^{-1} - \|A_+^{-1} - A_0^{-1}\| \geq 1 - \underline{h} - \|[x_+, x|F] - [x_0, x_{-1}|F]\|, \tag{19.2.9}$$

so, by (19.2.2), we have that

$$\|[x_+, x|F] - [x_0, x_{-1}|F]\| \leq$$

$$
\begin{aligned}
&\omega_0 \left(\min \left\{ \omega_0^{-1}(\|[x_+, x|F]\| - \bar{h}), \omega_0^{-1}(\|[x_0, x_{-1}|F]\| - \bar{h}) \right\} \right. \\
&\left. + \|x_+ - x_0\| + \|x - x_{-1}\| \right) \\
&- \omega_0 \left(\min\{\omega_0^{-1}(\|[x_+, x|F]\| - \underline{h}), \omega_0^{-1}(\|[x_0, x_{-1}|F]\| - \underline{h})\} \right).
\end{aligned}
$$

In view of (19.2.6) (for $\omega_0 = \omega$)

$$\omega_0(\|[x_+, x|F]\| - \underline{h}) \geq (\omega_0^{-1}(1 - \underline{h}) - \|x_+ - x_0\| - \|x - x_{-1}\|)^+$$
$$\geq (\alpha - \bar{t}_+ - \bar{t} - \bar{\gamma}_0)^+. \qquad (19.2.10)$$

By the concavity and monotonicity of ω_0,

$$\|[x_+, x|F] - [x_0, x_{-1}|F]\| \leq$$
$$\omega_0\left(\min\{(\alpha - \bar{t}_+ - \bar{t} - \bar{\gamma}_0)^+, \alpha\} + \bar{t}_+ + \bar{t} + \bar{\gamma}_0\right)$$
$$-\omega_0\left(\min\{(\alpha - \bar{t}_+ - \bar{t} - \bar{\gamma}_0)^+, \alpha\}\right)$$
$$= \omega_0\left((\alpha - \bar{t}_+ - \bar{t} - \bar{\gamma}_0)^+ + \bar{t}_+ + \bar{t} + \bar{\gamma}_0\right)$$
$$-\omega_0\left((\alpha - \bar{t}_+ - \bar{t} - \bar{\gamma}_0)^+\right). \qquad (19.2.11)$$

If this difference $< 1 - \underline{h}$, then it follows from (19.2.9) that

$$\|A_+\| \leq$$

$$\frac{1}{1 - \underline{h} - \omega_0\left((\alpha - \bar{t}_+ - \bar{t} - \bar{\gamma}_0)^+ + \bar{t}_+ + \bar{t} + \bar{\gamma}_0\right) + \omega_0\left((\alpha - \bar{t}_+ - \bar{t} - \bar{\gamma}_0)^+\right)}.$$

Notice that the difference (19.2.11) $< 1 - \underline{h} = \omega_0(\alpha)$ if and only if $\bar{t}_+ + \bar{t} < a$.
Hence, this assumptions implies

$$\|A_+\| \leq \frac{1}{1 - \underline{h} - \omega_0(\alpha) + \omega_0(a - \bar{t}_+ - \bar{t})} = \frac{1}{\omega_0(a - \bar{t}_+ - \bar{t})}. \qquad (19.2.12)$$

Then, as in [7, pp. 48 and 49], we obtain

$$\|F(x_+)\| \leq \bar{\delta}\left(\omega(a - \bar{t}_+ - \bar{t}_+\bar{\delta} + \bar{\gamma}) - \omega(a - \bar{t}_+ - \bar{t})\right) \qquad (19.2.13)$$

which together with (19.2.12) show (19.2.8). $\qquad \square$

Lemma (19.2) motivates us to introduce the following majorant generator $g(t, \gamma, \delta) = (t_+, \gamma_+, \delta_+)$:

$$t_+ := t + \delta, \quad \gamma_+ := \delta,$$

$$\delta_+ := \delta\left(\frac{\omega(a - t_+ - t + \delta + \gamma) - \omega(a - t_+ - t)}{\omega_0(a - t_+ - t)}\right)$$

$$= \delta\left(\frac{\omega(a - 2t + \gamma) - \omega(a - 2t - \delta)}{\omega_0(a - 2t - \delta)}\right) \qquad (19.2.14)$$

We say that the triple $q' = (t', \gamma', \delta')$ majorizes $q = (t, \gamma, \delta)$ (briefly $q \prec q'$) if

$$t \leq t' \quad \& \quad \gamma \leq \gamma' \quad \& \quad \delta \leq \delta'.$$

Lemma (19.2) states that $\overline{q}_+ \prec g(\overline{q})$.

If we begin to fed with the initial triple q_0, the generator iterates producing a majorant sequence as long as the denominator (19.2.14) remains defined:

$$\underset{n}{\&} 2t_n + \delta_n < a. \tag{19.2.15}$$

Remark 19.3 If $\omega_0 = \omega$, then the generator g reduces to the generator $\overline{g}$ given in [7] defined by $\overline{g}(u, \gamma, \delta) = (u_+, \gamma_+, \delta_+)$, $u_+ = u + \theta$, $\overline{\delta}_+ = \theta$,

$$\theta_+ = \theta \left(\frac{\omega(\overline{a} - u_+ - u + \theta + \gamma)}{\omega(\overline{a} - u_+ - u)} - 1 \right) = \theta \left(\frac{\omega(\overline{a} - 2\theta + \gamma)}{\omega(\overline{a} - 2u - \theta)} - 1 \right), \tag{19.2.16}$$

where $\overline{a} = \omega^{-1}(1 - \underline{h}) - \overline{\gamma}_0$. However, if strict inequality holds in (19.2.3), then (19.2.14) generates a more precise majorizing sequence than (19.2.16). That is

$$t < u, \tag{19.2.17}$$

$$\delta_+ < \theta_+ \tag{19.2.18}$$

and

$$t_\infty = \lim_{n \to +\infty} t_n \leq \lim_{n \to +\infty} u_n = u_\infty. \tag{19.2.19}$$

Here, the error bounds are tighter and the information on the location of the solution at least as precise, if we use the generator g instead of the old generator $\overline{g}$ used in [7].

Under condition (19.2.15), we can ensure convergence of the sequence (x_n, A_n) generated by the method (19.1.4) from the starter (x_0, A_0) to a solution of the system

$$F(x) = 0 \quad \& \quad A[x, x|F] = I. \tag{19.2.20}$$

We present the following semilocal convergence result for method (19.1.4).

Theorem 19.4 *If q_0 is such that $\overline{q}_0 \prec q_0$ & $\underset{n}{\&} 2t_n + \delta_n < a$, then sequence $\{x_n\}$ generated by method (19.1.4) is well defined and converges to a solution x_∞ which is the only solution of equation $F(x) = 0$ in $U(x_0, a - t_\infty)$. Moreover the following estimates hold*

$$\|x_{n+1} - x_n\| \leq t_{n+1} - t_n \tag{19.2.21}$$

and

$$\|x_n - x_\infty\| \leq t_\infty - t_n. \tag{19.2.22}$$

Furthermore, the sequence $\{A_n\}$ converges to A_∞ so that (x_∞, A_∞) solve the system (19.2.20).

Proof Simply replace the old generator $\bar{g}$ used in [7, see Lemma 3.2] by the new generator g defined by (19.2.14). □

Remark 19.5 The rest of the results in [7] can be adjusted by switching the generators so we can obtain the advantages as stated in the abstract of this study. However, we leave the details to the motivated reader. Instead, we return to Remark 19.3 and assume that ω_0, ω are linear functions defined by $\omega_0(t) = c_0 t$ and $\omega(t) = ct$ with $c_0 \neq 0$ and $c \neq 0$. Then, the generators g ans $\bar{g}$ provide, respectively the scalar iterations $\{t_n\}$ and $\{u_n\}$ defined by

$$t_{-1} = \gamma_0, \quad t_0 = \delta_0, \quad t_1 = \delta_0 + \|A_0 F(x_0)\|, \quad a = c_0^{-1} - \gamma_0,$$

$$t_{n+2} = t_{n+1} + \frac{(t_{n+1} - t_n)(t_{n+1} - t_{n-1})}{a - (t_{n+1} + t_n)} \tag{19.2.23}$$

and

$$u_{-1} = \gamma_0, \quad u_0 = \delta_0, \quad u_1 = \delta_0 + \|A_0 F(x_0)\|, \quad \bar{a} = c^{-1} - \gamma_0,$$

$$u_{n+2} = u_{n+1} + \frac{(u_{n+1} - u_n)(u_{n+1} - u_{n-1})}{\bar{a} - (u_{n+1} + u_n)}. \tag{19.2.24}$$

Then, we have by (19.2.23) and (19.2.24) that $t_{-1} = u_{-1}$, $t_0 = u_0$, $t_1 = u_1$ and if $c_0 = c$, then $a = \bar{a}$ and $t_n = u_n$. However, if $c_0 < c$, then a simple inductive argument shows that

$$\bar{a} < a, \tag{19.2.25}$$

$$t_n < u_n \quad \text{for each} \quad n = 0, 1, 2, \ldots, \tag{19.2.26}$$

$$t_{n+1} - t_n < u_{n+1} - u_n \quad \text{for each} \quad n = 1, 2, \ldots \tag{19.2.27}$$

and

$$t_\infty \leq u_\infty. \tag{19.2.28}$$

It was shown in [7] that the sufficient convergence condition for sequence $\{u_n\}$ is given by

$$4c^{-1}\delta_0 \leq (c^{-1} - \gamma_0)^2. \tag{19.2.29}$$

Therefore, according to (19.2.25)–(19.2.27), conditions (19.2.29) is also the sufficient convergence conditions for sequence $\{t_n\}$. Notice however that under our new approach the error (19.2.21) and (19.2.22) are tighter and by (19.2.28) the information on the location of the solution x_∞ is also more precise, since $t_\infty - a \leq u_\infty - a$. Moreover, a direct study of sequence $\{t_n\}$ can lead to even weaker sufficient convergence conditions [1–4]. Hence, concluding the error bounds and the information on the location of the solution x_∞ are improved under weaker convergence conditions (if strict inequality holds in (19.2.3)) since the convergence condition in [7] is given by

$$2u_n + \theta_n < \overline{a} \tag{19.2.30}$$

and in this case we have that

$$(19.2.30) \Rightarrow (19.2.15) \tag{19.2.31}$$

but not necesarirly viceversa (unless if $\omega_0 = \omega$).

Examples, where strict inequality holds in (19.2.3) can be found in [1–4].

References

1. I.K. Argyros, in *Computational Theory of Iterative Methods*. Studies in Computational Mathematics, vol. 15 (Elseiver, New York, 2007)
2. I.K. Argyros, On the semiocal convergence of the secant method with regularly continuous divided differences. Commun. Appl. Nonlinear Anal. **19**(2), 55–69 (2012)
3. I.K. Argyros, S. Hilout, Majorizing sequences for iterative methods. J. Comput. Appl. Math. **236**(7), 1947–1960 (2012)
4. I.K. Argyros, S. Hilout, *Computational Methods in Nonlinear Analysis* (World Scientific Publ. Comp., New Jersey, 2013)
5. I. Argyros, D. Gonzales, *Improved Semilocal Convergence of Broyden's Method with Regularity Continuous Divided Differences* (2015) (submitted)
6. C.G. Broyden, A class of methods for solving nonlinear simultaneous equations. Math. Comput. **19**, 577–593 (1965)
7. A.M. Galperin, Broyden's method for operators with regularly continuous divided differences. J. Kor. Math. Soc. **52**(1), 43–65 (2015)
8. L.V. Kantorovich, G.P. Akilov, *Functional Analysis* (Pergamon Press, Oxford, 1982)

Chapter 20
Left General Fractional Monotone Approximation

Here are introduced left general fractional derivatives Caputo style with respect to a base absolutely continuous strictly increasing function g. We give various examples of such fractional derivatives for different g. Let f be p-times continuously differentiable function on $[a, b]$, and let L be a linear left general fractional differential operator such that $L(f)$ is non-negative over a critical closed subinterval I of $[a, b]$. We can find a sequence of polynomials Q_n of degree less-equal n such that $L(Q_n)$ is non-negative over I, furthermore f is approximated uniformly by Q_n over $[a, b]$.

The degree of this constrained approximation is given by an inequality using the first modulus of continuity of $f^{(p)}$. We finish with applications of the main fractional monotone approximation theorem for different g. On the way to prove the main theorem we establish useful related general results. It follows [2].

20.1 Introduction and Preparation

The topic of monotone approximation started in [11] has become a major trend in approximation theory. A typical problem in this subject is: given a positive integer k, approximate a given function whose kth derivative is ≥ 0 by polynomials having this property.

In [4] the authors replaced the kth derivative with a linear differential operator of order k.

Furthermore in [1], the author generalized the result of [4] for linear fractional differential operators.

To describe the motivating result here we need:

Definition 20.1 ([5], p. 50) Let $\alpha > 0$ and $\lceil \alpha \rceil = m$, ($\lceil \cdot \rceil$ ceiling of the number). Consider $f \in C^m([-1, 1])$. We define the left Caputo fractional derivative of f of order α as follows:

© Springer International Publishing Switzerland 2016
G.A. Anastassiou and I.K. Argyros, *Intelligent Numerical Methods:*
Applications to Fractional Calculus, Studies in Computational Intelligence 624,
DOI 10.1007/978-3-319-26721-0_20

$$\left(D_{*-1}^{\alpha}f\right)(x) = \frac{1}{\Gamma(m-\alpha)} \int_{-1}^{x} (x-t)^{m-\alpha-1} f^{(m)}(t)\,dt, \qquad (20.1.1)$$

for any $x \in [-1, 1]$, where Γ is the gamma function $\Gamma(\nu) = \int_0^{\infty} e^{-t} t^{\nu-1} dt, \nu > 0$.
We set

$$D_{*-1}^{0} f(x) = f(x), \qquad (20.1.2)$$

$$D_{*-1}^{m} f(x) = f^{(m)}(x), \quad \forall x \in [-1, 1]. \qquad (20.1.3)$$

We proved:

Theorem 20.2 ([1]) *Let h, k, p be integers, $0 \le h \le k \le p$ and let f be a real function, $f^{(p)}$ continuous in $[-1, 1]$ with modulus of continuity $\omega_1\left(f^{(p)}, \delta\right), \delta > 0$, there. Let $\alpha_j(x)$, $j = h, h+1, \ldots, k$ be real functions, defined and bounded on $[-1, 1]$ and assume for $x \in [0, 1]$ that $\alpha_h(x)$ is either $\ge$ some number $\alpha > 0$ or $\le$ some number $\beta < 0$. Let the real numbers $\alpha_0 = 0 < \alpha_1 \le 1 < \alpha_2 \le 2 < \cdots < \alpha_p \le p$. Here $D_{*-1}^{\alpha_j} f$ stands for the left Caputo fractional derivative of f of order α_j anchored at -1. Consider the linear left fractional differential operator*

$$L := \sum_{j=h}^{k} \alpha_j(x) \left[D_{*-1}^{\alpha_j}\right] \qquad (20.1.4)$$

and suppose, throughout $[0, 1]$,

$$L(f) \ge 0. \qquad (20.1.5)$$

Then, for any $n \in \mathbb{N}$, there exists a real polynomials $Q_n(x)$ of degree $\le n$ such that

$$L(Q_n) \ge 0 \quad throughout \ [0, 1], \qquad (20.1.6)$$

and

$$\max_{-1 \le x \le 1} |f(x) - Q_n(x)| \le Cn^{k-p}\omega_1\left(f^{(p)}, \frac{1}{n}\right), \qquad (20.1.7)$$

where C is independent of n or f.

Notice above that the monotonicity property is only true on $[0, 1]$, see (20.1.5) and (20.1.6). However the approximation property (20.1.7) it is true over the whole interval $[-1, 1]$.

In this chapter we extend Theorem 20.2 to much more general linear left fractional differential operators.

We use a lot here the following generalised fractional integral.

Definition 20.3 (*see also* [8, p. 99]) The left generalised fractional integral of a function f with respect to given function g is defined as follows:

Let $a, b \in \mathbb{R}$, $a < b$, $\alpha > 0$. Here $g \in AC([a, b])$ (absolutely continuous functions) and is strictly increasing, $f \in L_\infty([a, b])$. We set

$$\left(I_{a+;g}^\alpha f\right)(x) = \frac{1}{\Gamma(\alpha)} \int_a^x (g(x) - g(t))^{\alpha-1} g'(t) f(t) dt, \quad x \geq a, \qquad (20.1.8)$$

clearly $\left(I_{a+;g}^\alpha f\right)(a) = 0$.

When g is the identity function id, we get that $I_{a+;id}^\alpha = I_{a+}^\alpha$, the ordinary left Riemann-Liouville fractional integral, where

$$\left(I_{a+}^\alpha f\right)(x) = \frac{1}{\Gamma(\alpha)} \int_a^x (x - t)^{\alpha-1} f(t) dt, \quad x \geq a, \qquad (20.1.9)$$

$\left(I_{a+}^\alpha f\right)(a) = 0$.

When $g(x) = \ln x$ on $[a, b]$, $0 < a < b < \infty$, we get:

Definition 20.4 ([8, p. 110]) Let $0 < a < b < \infty$, $\alpha > 0$. The left Hadamard fractional integral of order α is given by

$$\left(J_{a+}^\alpha f\right)(x) = \frac{1}{\Gamma(\alpha)} \int_a^x \left(\ln \frac{x}{y}\right)^{\alpha-1} \frac{f(y)}{y} dy, \quad x \geq a, \qquad (20.1.10)$$

where $f \in L_\infty([a, b])$.

We mention:

Definition 20.5 The left fractional exponential integral is defined as follows: Let $a, b \in \mathbb{R}$, $a < b$, $\alpha > 0$, $f \in L_\infty([a, b])$. We set

$$\left(I_{a+;e^x}^\alpha f\right)(x) = \frac{1}{\Gamma(\alpha)} \int_a^x \left(e^x - e^t\right)^{\alpha-1} e^t f(t) dt, \quad x \geq a. \qquad (20.1.11)$$

Definition 20.6 Let $a, b \in \mathbb{R}$, $a < b$, $\alpha > 0$, $f \in L_\infty([a, b])$, $A > 1$. We introduce the fractional integral

$$\left(I_{a+;A^x}^\alpha f\right)(x) = \frac{\ln A}{\Gamma(\alpha)} \int_a^x \left(A^x - A^t\right)^{\alpha-1} A^t f(t) dt, \quad x \geq a. \qquad (20.1.12)$$

We also give:

Definition 20.7 Let $\alpha, \sigma > 0$, $0 \leq a < b < \infty$, $f \in L_\infty([a, b])$. We set

$$\left(K_{a+;x^\sigma}^\alpha f\right)(x) = \frac{1}{\Gamma(\alpha)} \int_z^x (x^\sigma - t^\sigma)^{\alpha-1} f(t) \sigma t^{\sigma-1} dt, \quad x \geq a. \qquad (20.1.13)$$

We introduce the following general fractional derivatives.

Definition 20.8 Let $\alpha > 0$ and $\lceil \alpha \rceil = m$, ($\lceil \cdot \rceil$ ceiling of the number). Consider $f \in AC^m([a,b])$ (space of functions f with $f^{(m-1)} \in AC([a,b])$). We define the left general fractional derivative of f of order α as follows

$$\left(D_{*a;g}^{\alpha} f\right)(x) = \frac{1}{\Gamma(m-\alpha)} \int_a^x (g(x) - g(t))^{m-\alpha-1} g'(t) f^{(m)}(t) \, dt, \quad (20.1.14)$$

for any $x \in [a,b]$, where Γ is the gamma function.

We set

$$D_{*\alpha;g}^m f(x) = f^{(m)}(x), \tag{20.1.15}$$

$$D_{*a;g}^0 f(x) = f(x), \quad \forall x \in [a,b]. \tag{20.1.16}$$

When $g = id$, then $D_{*a}^{\alpha} f = D_{*a;id}^{\alpha} f$ is the left Caputo fractional derivative.

So we have the specific general left fractional derivatives.

Definition 20.9

$$D_{*a;\ln x}^{\alpha} f(x) = \frac{1}{\Gamma(m-\alpha)} \int_a^x \left(\ln \frac{x}{y}\right)^{m-\alpha-1} \frac{f^{(m)}(y)}{y} dy, \quad x \geq a > 0, \tag{20.1.17}$$

$$D_{*a;e^x}^{\alpha} f(x) = \frac{1}{\Gamma(m-\alpha)} \int_a^x \left(e^x - e^t\right)^{m-\alpha-1} e^t f^{(m)}(t) \, dt, \quad x \geq a, \tag{20.1.18}$$

and

$$D_{*a;A^x}^{\alpha} f(x) = \frac{\ln A}{\Gamma(m-\alpha)} \int_a^x \left(A^x - A^t\right)^{m-\alpha-1} A^t f^{(m)}(t) \, dt, \quad x \geq a, \tag{20.1.19}$$

$$\left(D_{*a;x^{\sigma}}^{\alpha} f\right)(x) = \frac{1}{\Gamma(m-\alpha)} \int_a^x (x^{\sigma} - t^{\sigma})^{m-\alpha-1} \sigma t^{\sigma-1} f^{(m)}(t) \, dt, \quad x \geq a \geq 0. \tag{20.1.20}$$

We would need a modification of:

Theorem 20.10 (Trigub, [12, 13]) *Let $g \in C^p([-1,1])$, $p \in \mathbb{N}$. Then there exists real polynomial $q_n(x)$ of degree $\leq n$, $x \in [-1,1]$, such that*

$$\max_{-1 \leq x \leq 1} \left| g^{(j)}(x) - q_n^{(j)}(x) \right| \leq R_p n^{j-p} \omega_1 \left(g^{(p)}, \frac{1}{n}\right), \tag{20.1.21}$$

$j = 0, 1, \ldots, p$, *where R_p is independent of n or g.*

We make and need:

Remark 20.11 Here $t \in [-1,1]$, $x \in [a,b]$, $a < b$. Let the map $\varphi : [-1,1] \to [a,b]$ defined by

$$x = \varphi(t) = \left(\frac{b-a}{2}\right)t + \left(\frac{b+a}{2}\right). \qquad (20.1.22)$$

Clearly here φ is $1-1$ and onto map.

We get

$$x' = \varphi'(t) = \frac{b-a}{2}, \qquad (20.1.23)$$

and

$$t = \frac{2x-b-a}{b-a} = 2\left(\frac{x}{b-a}\right) - \left(\frac{b+a}{b-a}\right). \qquad (20.1.24)$$

In fact it holds

$$\varphi(-1) = a, \text{ and } \varphi(1) = b. \qquad (20.1.25)$$

We will prove and use:

Theorem 20.12 *Let $f \in C^p([a,b])$, $p \in \mathbb{N}$. Then there exist real polynomials $Q_n^*(x)$ of degree $\leq n \in \mathbb{N}$, $x \in [a,b]$, such that*

$$\max_{a \leq x \leq b} \left| f^{(j)}(x) - Q_n^{*(j)}(x) \right| \leq R_p \left(\frac{b-a}{2n}\right)^{p-j} \omega_1\left(f^{(p)}, \frac{b-a}{2n}\right), \qquad (20.1.26)$$

$j = 0, 1, \ldots, p$, where R_p is independent of n or g.

Proof We use Theorem 20.10 and Remark 20.11.

Given that $f \in C^p([a,b])$, $x \in [a,b]$, it is clear that

$$g(t) = f\left(\left(\frac{b-a}{2}\right)t + \left(\frac{b+a}{2}\right)\right) \in C^p([-1,1]), \quad t \in [-1,1].$$

We notice that

$$\frac{dg(t)}{dt} = \frac{df\left(\left(\frac{b-a}{2}\right)t + \left(\frac{b+a}{2}\right)\right)}{dt} = f'(x)\left(\frac{b-a}{2}\right), \qquad (20.1.27)$$

thus it holds

$$g'(t) = f'(x)\left(\frac{b-a}{2}\right) = f'\left(\left(\frac{b-a}{2}\right)t + \left(\frac{b+a}{2}\right)\right)\left(\frac{b-a}{2}\right). \qquad (20.1.28)$$

And

$$g''(t) = \frac{df'\left(\left(\frac{b-a}{2}\right)t + \left(\frac{b+a}{2}\right)\right)}{dt} \left(\frac{b-a}{2}\right). \tag{20.1.29}$$

Since as before

$$\frac{df'\left(\left(\frac{b-a}{2}\right)t + \left(\frac{b+a}{2}\right)\right)}{dt} = f''(x)\left(\frac{b-a}{2}\right), \tag{20.1.30}$$

we obtain

$$g''(x) = f''(x)\frac{(b-a)^2}{2^2}. \tag{20.1.31}$$

In general we get

$$g^{(j)}(t) = f^{(j)}(x)\frac{(b-a)^j}{2^j}, \tag{20.1.32}$$

$j = 0, 1, \ldots, p$. Thus we find in detail that $g \in C^p([-1, 1])$. Hence by Theorem 20.10, for any $t \in [-1, 1]$, we have

$$\left|g^{(j)}(t) - q_n^{(j)}(t)\right| \le R_p n^{j-p}\omega_1\left(g^{(p)}, \frac{1}{n}\right), \tag{20.1.33}$$

$j = 0, 1, \ldots, p$, where R_p is independent of n or g.

Notice that

$$q_n^{(j)}(t) \overset{(20.1.24)}{=} q_n^{(j)}\left(\frac{2x-b-a}{b-a}\right), \quad j = 0, 1, \ldots, p. \tag{20.1.34}$$

See that, for $t \in [-1, 1]$, we have

$$q_n(t) = q_n\left(\left(\frac{2}{b-a}\right)x - \left(\frac{b+a}{b-a}\right)\right) =: Q_n^*(x), \quad x \in [a, b], \tag{20.1.35}$$

a polynomial of degree n.

Also it holds

$$Q_n^{*\prime}(x) = \frac{dq_n\left(\left(\frac{2}{b-a}\right)x - \left(\frac{b+a}{b-a}\right)\right)}{dx} = \frac{dq_n(t)}{dt}\frac{dt}{dx} = q_n'(t)\left(\frac{2}{b-a}\right). \tag{20.1.36}$$

That is

$$q_n'(t) = Q_n^{*\prime}(x)\left(\frac{b-a}{2}\right). \tag{20.1.37}$$

Similalry we get

$$Q_n^{*''}(x) = \frac{dQ_n^{*'}(x)}{dx} \overset{(20.1.36)}{=} \frac{dq_n'\left(\left(\frac{2}{b-a}\right)x - \left(\frac{b+a}{b-a}\right)\right)}{dx}\left(\frac{2}{b-a}\right) =$$

$$\frac{dq_n'(t)}{dt}\frac{dt}{dx}\left(\frac{2}{b-a}\right) = q_n''(t)\frac{2^2}{(b-a)^2}. \qquad (20.1.38)$$

Hence

$$q_n''(t) = Q_n^{*''}(x)\frac{(b-a)^2}{2^2}. \qquad (20.1.39)$$

In general it holds

$$q_n^{(j)}(t) = Q_n^{*(j)}(x)\frac{(b-a)^j}{2^j}, \quad j = 0, 1, \ldots, p. \qquad (20.1.40)$$

Thus we have

$$L.H.S.(20.1.33) = \frac{(b-a)^j}{2^j}\left|f^{(j)}(x) - Q_n^{*(j)}(x)\right|, \qquad (20.1.41)$$

$j = 0, 1, \ldots, p, \; x \in [a, b]$.

Next we observe that

$$\omega_1\left(g^{(p)}, \frac{1}{n}\right) = \sup_{\substack{t_1,t_2 \in [-1,1] \\ |t_1-t_2|\le \frac{1}{n}}}\left|g^{(p)}(t_1) - g^{(p)}(t_2)\right| =$$

$$\sup_{\substack{x_1,x_2 \in [a,b] \\ |x_1-x_2|\le \frac{b-a}{2n}}}\frac{(b-a)^p}{2^p}\left|f^{(p)}(x_1) - f^{(p)}(x_2)\right| = \frac{(b-a)^p}{2^p}\omega_1\left(f^{(p)}, \frac{b-a}{2n}\right).$$

$$(20.1.42)$$

An explanation of (20.1.42) follows.

By Remark 20.11 we have that φ is $(1 - 1)$ and onto map, so that for any t_1, $t_2 \in [-1, 1]$ there exist unique $x_1, x_2 \in [a, b]$:

$$t_1 = \left(\frac{2}{b-a}\right)x_1 - \left(\frac{b+a}{b-a}\right),$$

and $\qquad (20.1.43)$

$$t_2 = \left(\frac{2}{b-a}\right)x_2 - \left(\frac{b+a}{b-a}\right).$$

Hence it follows

$$t_1 - t_2 = \left(\frac{2}{b-a}\right)(x_1 - x_2),\qquad(20.1.44)$$

and

$$\frac{1}{n} \geq |t_1 - t_2| = \left(\frac{2}{b-a}\right)|x_1 - x_2|,\qquad(20.1.45)$$

which produces

$$|x_1 - x_2| \leq \frac{b-a}{2n}.\qquad(20.1.46)$$

Finally by (20.1.33) we can find

$$\frac{(b-a)^j}{2^j}\left|f^{(j)}(x) - Q^{*(j)}(x)\right| \leq R_p n^{j-p}\frac{(b-a)^p}{2^p}\omega_1\left(f^{(p)}, \frac{b-a}{2n}\right),\quad(20.1.47)$$

$j = 0, 1, \ldots, p.$
  And it holds

$$\left|f^{(j)}(x) - Q^{*(j)}(x)\right| \leq R_p\frac{(b-a)^{p-j}}{(2n)^{p-j}}\omega_1\left(f^{(p)}, \frac{b-a}{2n}\right),\qquad(20.1.48)$$

for any $x \in [a, b]$, $j = 0, 1, \ldots, p$, proving the claim.   □

We need:

Remark 20.13 Here $g \in AC([a, b])$ (absolutely continuous functions), g is increasing over $[a, b]$, $\alpha > 0$.
  Let $g(a) = c$, $g(b) = d$. We want to calculate

$$I = \int_a^b (g(b) - g(t))^{\alpha-1} g'(t)\, dt.\qquad(20.1.49)$$

Consider the function

$$f(y) = (g(b) - y)^{\alpha-1} = (d - y)^{\alpha-1}, \quad \forall y \in [c, d].\qquad(20.1.50)$$

We have that $f(y) \geq 0$, it may be $+\infty$ when $y = d$ and $0 < \alpha < 1$, but f is measurable on $[c, d]$. By [9], Royden, p. 107, Exercise 13 d, we get that

$$(f \circ g)(t)\, g'(t) = (g(b) - g(t))^{\alpha-1} g'(t)\qquad(20.1.51)$$

is measurable on $[a, b]$, and

$$I = \int_c^d (d - y)^{\alpha-1} \, dy - \frac{(d - c)^{\alpha}}{\alpha} \tag{20.1.52}$$

(notice that $(d - y)^{\alpha-1}$ is Riemann integrable).

That is

$$I = \frac{(g(b) - g(a))^{\alpha}}{\alpha}. \tag{20.1.53}$$

Similarly it holds

$$\int_a^x (g(x) - g(t))^{\alpha-1} \, g'(t) \, dt = \frac{(g(x) - g(a))^{\alpha}}{\alpha}, \quad \forall x \in [a, b]. \tag{20.1.54}$$

Finally we will use:

Theorem 20.14 *Let $\alpha > 0$, $\mathbb{N} \ni m = \lceil \alpha \rceil$, and $f \in C^m([a, b])$. Then $\left(D^{\alpha}_{*a;g} f\right)(x)$ is continuous in $x \in [a, b]$.*

Proof By [3], Apostol, p. 78, we get that g^{-1} exists and it is strictly increasing on $[g(a), g(b)]$. Since g is continuous on $[a, b]$, it implies that g^{-1} is continuous on $[g(a), g(b)]$. Hence $f^{(m)} \circ g^{-1}$ is a continuous function on $[g(a), g(b)]$.

If $\alpha = m \in \mathbb{N}$, then the claim is trivial.

We treat the case of $0 < \alpha < m$.

It holds that

$$\left(D^{\alpha}_{*a;g} f\right)(x) = \frac{1}{\Gamma(m - \alpha)} \int_a^x (g(x) - g(t))^{m-\alpha-1} \, g'(t) \, f^{(m)}(t) \, dt =$$

$$\frac{1}{\Gamma(m - \alpha)} \int_a^x (g(x) - g(t))^{m-\alpha-1} \, g'(t) \left(f^{(m)} \circ g^{-1}\right)(g(t)) \, dt = \tag{20.1.55}$$

$$\frac{1}{\Gamma(m - \alpha)} \int_{g(a)}^{g(x)} (g(x) - z)^{m-\alpha-1} \left(f^{(m)} \circ g^{-1}\right)(z) \, dz.$$

An explanation follows.

The function

$$G(z) = (g(x) - z)^{m-\alpha-1} \left(f^{(m)} \circ g^{-1}\right)(z)$$

is integrable on $[g(a), g(x)]$, and by assumption g is absolutely continuous: $[a, b] \to [g(a), g(b)]$.

Since g is monotone (strictly increasing here) the function

$$(g(x) - g(t))^{m-\alpha-1} g'(t) \left(f^{(m)} \circ g^{-1}\right)(g(t))$$

is integrable on $[a, x]$ (see [7]). Furthermore it holds (see also [7]),

$$\frac{1}{\Gamma(m-\alpha)} \int_{g(a)}^{g(x)} (g(x) - z)^{m-\alpha-1} \left(f^{(m)} \circ g^{-1}\right)(z)\, dz =$$

$$\frac{1}{\Gamma(m-\alpha)} \int_{a}^{x} (g(x) - g(t))^{m-\alpha-1} g'(t) \left(f^{(m)} \circ g^{-1}\right)(g(t))\, dt \qquad (20.1.56)$$

$$= \left(D_{*a;g}^{\alpha} f\right)(x), \quad \forall x \in [a, b].$$

Then, we can write

$$\left(D_{*a;g}^{\alpha} f\right)(x) = \frac{1}{\Gamma(m-\alpha)} \int_{g(a)}^{g(x)} (g(x) - z)^{m-\alpha-1} \left(f^{(m)} \circ g^{-1}\right)(z)\, dz,$$

$$\left(D_{*a;g}^{\alpha} f\right)(y) = \frac{1}{\Gamma(m-\alpha)} \int_{g(a)}^{g(y)} (g(y) - z)^{m-\alpha-1} \left(f^{(m)} \circ g^{-1}\right)(z)\, dz.$$

$$(20.1.57)$$

Here $a \le x \le y \le b$, and $g(a) \le g(x) \le g(y) \le g(b)$, and $0 \le g(x) - g(a) \le g(y) - g(a)$.

Let $\lambda = g(x) - z$, then $z = g(x) - \lambda$. Thus

$$\left(D_{*a;g}^{\alpha} f\right)(x) = \frac{1}{\Gamma(m-\alpha)} \int_{0}^{g(x)-g(a)} \lambda^{m-\alpha-1} \left(f^{(m)} \circ g^{-1}\right)(g(x) - \lambda)\, d\lambda.$$

$$(20.1.58)$$

Clearly, see that $g(a) \le z \le g(x)$, then $-g(a) \ge -z \ge -g(x)$, and $g(x) - g(a) \ge g(x) - z \ge 0$, i.e. $0 \le \lambda \le g(x) - g(a)$.

Similarly

$$\left(D_{*a;g}^{\alpha} f\right)(y) = \frac{1}{\Gamma(m-\alpha)} \int_{0}^{g(y)-g(a)} \lambda^{m-\alpha-1} \left(f^{(m)} \circ g^{-1}\right)(g(y) - \lambda)\, d\lambda.$$

$$(20.1.59)$$

Hence it holds

$$\left(D_{*a;g}^{\alpha} f\right)(y) - \left(D_{*a;g}^{\alpha} f\right)(x) = \frac{1}{\Gamma(m-\alpha)} \cdot$$

$$\left[\int_{0}^{g(x)-g(y)} \lambda^{m-\alpha-1} \left(\left(f^{(m)} \circ g^{-1}\right)(g(y) - \lambda) - \left(f^{(m)} \circ g^{-1}\right)(g(x) - \lambda)\right) d\lambda\right.$$

$$+ \int_{g(x)-g(a)}^{g(y)-g(a)} \lambda^{m-\alpha-1} \left(f^{(m)} \circ g^{-1} \right) \left(g\left(y \right) - \lambda \right) d\lambda \Big]. \tag{20.1.60}$$

Thus we obtain

$$\left| \left(D_{*a;g}^{\alpha} f \right)(y) - \left(D_{*a;g}^{\alpha} f \right)(x) \right| \le \frac{1}{\Gamma\left(m - \alpha \right)} \cdot$$

$$\left[\frac{\left(g\left(x \right) - g\left(a \right) \right)^{m-\alpha}}{m - \alpha} \omega_1 \left(f^{(m)} \circ g^{-1}, |g\left(y \right) - g\left(x \right)| \right) + \tag{20.1.61}$$

$$\frac{\left\| f^{(m)} \circ g^{-1} \right\|_{\infty,[g(a),g(b)]}}{m - \alpha} \left(\left(g\left(y \right) - g\left(a \right) \right)^{m-\alpha} - \left(g\left(x \right) - g\left(a \right) \right)^{m-\alpha} \right) \right] =: (\xi).$$

As $y \to x$, then $g\left(y \right) \to g\left(x \right)$ (since $g \in AC\left([a, b] \right)$). So that $(\xi) \to 0$. As a result

$$\left(D_{*a;g}^{\alpha} f \right)(y) \to \left(D_{*a;g}^{\alpha} f \right)(x), \tag{20.1.62}$$

proving that $\left(D_{*a;g}^{\alpha} f \right)(x)$ is continuous in $x \in [a, b]$. $\qquad\square$

20.2 Main Result

We present:

Theorem 20.15 *Here we assume that $g \in AC\left([a, b] \right)$ and is strictly increasing with $g\left(b \right) - g\left(a \right) > 1$. Let h, k, p be integers, $0 \le h \le k \le p$ and let $f \in C^p\left([a, b] \right)$, $a < b$, with modulus of continuity $\omega_1 \left(f^{(p)}, \delta \right)$, $0 < \delta \le b - a$. Let $\alpha_j\left(x \right)$, $j = h, h + 1, \ldots, k$ be real functions, defined and bounded on $[a, b]$ and assume for $x \in \left[g^{-1}\left(1 + g\left(a \right) \right), b \right]$ that $\alpha_h\left(x \right)$ is either $\ge$ some number $\alpha^* > 0$, or $\le$ some number $\beta^* < 0$. Let the real numbers $\alpha_0 = 0 < \alpha_1 \le 1 < \alpha_2 \le 2 < \cdots < \alpha_p \le p$. Consider the linear left general fractional differential operator*

$$L = \sum_{j=h}^{k} \alpha_j\left(x \right) \left[D_{*a;g}^{\alpha_j} \right], \tag{20.2.63}$$

and suppose, throughout $\left[g^{-1}\left(1 + g\left(a \right) \right), b \right]$,

$$L\left(f \right) \ge 0. \tag{20.2.64}$$

Then, for any $n \in \mathbb{N}$, there exists a real polynomial $Q_n\left(x \right)$ of degree $\le n$ such that

$$L\left(Q_n \right) \ge 0 \quad \text{throughout} \left[g^{-1}\left(1 + g\left(a \right) \right), b \right], \tag{20.2.65}$$

and

$$\max_{x\in[a,b]} |f(x) - Q_n(x)| \leq Cn^{k-p}\omega_1\left(f^{(p)}, \frac{b-a}{2n}\right), \tag{20.2.66}$$

where C is independent of n or f.

Proof of Theorem 20.15.

Here $h, k, p \in \mathbb{Z}_+, 0 \leq h \leq k \leq p$. Let $\alpha_j > 0, j = 1, \ldots, p$, such that $0 < \alpha_1 \leq 1 < \alpha_2 \leq 2 < \alpha_3 \leq 3 \cdots < \cdots < \alpha_p \leq p$. That is $\lceil \alpha_j \rceil = j$, $j = 1, \ldots, p$.

Let $Q_n^*(x)$ be as in Theorem 20.12.

We have that

$$\left(D_{*a;g}^{\alpha_j} f\right)(x) = \frac{1}{\Gamma(j-\alpha_j)} \int_a^x (g(x) - g(t))^{j-\alpha_j-1} g'(t) f^{(j)}(t)\, dt, \tag{20.2.67}$$

and

$$\left(D_{*a;g}^{\alpha_j} Q_n^*\right)(x) = \frac{1}{\Gamma(j-\alpha_j)} \int_a^x (g(x) - g(t))^{j-\alpha_j-1} g'(t) Q_n^{*(j)}(t)\, dt, \tag{20.2.68}$$

$j = 1, \ldots, p$.

Also it holds

$$\left(D_{*a;g}^j f\right)(x) = f^{(j)}(x), \quad \left(D_{*a;g}^j Q_n^*\right)(x) = Q_n^{*(j)}(x), \quad j = 1, \ldots, p. \tag{20.2.69}$$

By [10], we get that there exists g' a.e., and g' is measurable and non-negative.

We notice that

$$\left|\left(D_{*a;g}^{\alpha_j} f\right)(x) - D_{*a;g}^{\alpha_j} Q_n^*(x)\right| =$$

$$\frac{1}{\Gamma(j-\alpha_j)} \left|\int_a^x (g(x) - g(t))^{j-\alpha_j-1} g'(t) \left(f^{(j)}(t) - Q_n^{*(j)}(t)\right) dt\right| \leq$$

$$\frac{1}{\Gamma(j-\alpha_j)} \int_a^x (g(x) - g(t))^{j-\alpha_j-1} g'(t) \left|f^{(j)}(t) - Q_n^{*(j)}(t)\right| dt \overset{(20.1.26)}{\leq}$$

$$\frac{1}{\Gamma(j-\alpha_j)} \left(\int_a^x (g(x) - g(t))^{j-\alpha_j-1} g'(t)\, dt\right) R_p \left(\frac{b-a}{2n}\right)^{p-j} \omega_1\left(f^{(p)}, \frac{b-a}{2n}\right) \tag{20.2.70}$$

$$\overset{(20.1.54)}{=} \frac{(g(x) - g(a))^{j-\alpha_j}}{\Gamma(j-\alpha_j+1)} R_p \left(\frac{b-a}{2n}\right)^{p-j} \omega_1\left(f^{(p)}, \frac{b-a}{2n}\right) \leq$$

$$\frac{(g(b) - g(a))^{j-\alpha_j}}{\Gamma(j-\alpha_j+1)} R_p \left(\frac{b-a}{2n}\right)^{p-j} \omega_1\left(f^{(p)}, \frac{b-a}{2n}\right). \tag{20.2.71}$$

Hence $\forall\, x \in [a, b]$, it holds

$$\left|\left(D_{*a;g}^{\alpha_j}f\right)(x) - D_{*a;g}^{\alpha_j}Q_n^*(x)\right| \le$$

$$\frac{(g\,(b) - g\,(a))^{j-\alpha_j}}{\Gamma\,(j - \alpha_j + 1)}R_p\left(\frac{b-a}{2n}\right)^{p-j}\omega_1\left(f^{(p)}, \frac{b-a}{2n}\right), \qquad (20.2.72)$$

and

$$\max_{x\in[a,b]}\left|D_{*a;g}^{\alpha_j}f(x) - D_{*a;g}^{\alpha_j}Q_n^*(x)\right| \le$$

$$\frac{(g\,(b) - g\,(a))^{j-\alpha_j}}{\Gamma\,(j - \alpha_j + 1)}R_p\left(\frac{b-a}{2n}\right)^{p-j}\omega_1\left(f^{(p)}, \frac{b-a}{2n}\right), \qquad (20.2.73)$$

$j = 0, 1, \ldots, p$.

Above we set $D_{*a;g}^0 f(x) = f(x)$, $D_{*a;g}^0 Q_n^*(x) = Q_n^*(x)$, $\forall\, x \in [a, b]$, and $\alpha_0 = 0$, i.e. $\lceil \alpha_0 \rceil = 0$.

Put

$$s_j = \sup_{a \le x \le b}\left|\alpha_h^{-1}(x)\,\alpha_j(x)\right|, \quad j = h, \ldots, k, \qquad (20.2.74)$$

and

$$\eta_n = R_p\omega_1\left(f^{(p)}, \frac{b-a}{2n}\right)\left(\sum_{j=h}^k s_j\frac{(g\,(b) - g\,(a))^{j-\alpha_j}}{\Gamma\,(j - \alpha_j + 1)}\left(\frac{b-a}{2n}\right)^{p-j}\right). \qquad (20.2.75)$$

I. Suppose, throughout $\left[g^{-1}(1 + g\,(a)), b\right]$, $\alpha_h(x) \ge \alpha^* > 0$. Let $Q_n(x)$ be the real polynomial of degree $\le n$, that corresponds to $(f(x) + \eta_n(h!)^{-1} x^h)$, $x \in [a, b]$, so by Theorem 20.12 and (20.2.73) we get that

$$\max_{x\in[a,b]}\left|D_{*a;g}^{\alpha_j}\left(f(x) + \eta_n(h!)^{-1}x^h\right) - \left(D_{*a;g}^{\alpha_j}Q_n\right)(x)\right| \le \qquad (20.2.76)$$

$$\frac{(g\,(b) - g\,(a))^{j-\alpha_j}}{\Gamma\,(j - \alpha_j + 1)}R_p\left(\frac{b-a}{2n}\right)^{p-j}\omega_1\left(f^{(p)}, \frac{b-a}{2n}\right),$$

$j = 0, 1, \ldots, p$.

In particular $(j = 0)$ holds

$$\max_{x\in[a,b]}\left|\left(f(x) + \eta_n(h!)^{-1}x^h\right) - Q_n(x)\right| \le R_p\left(\frac{b-a}{2n}\right)^p\omega_1\left(f^{(p)}, \frac{b-a}{2n}\right), \qquad (20.2.77)$$

and

$$\max_{x\in[a,b]}|f(x)-Q_n(x)|\le$$

$$\eta_n\,(h!)^{-1}\,(\max\,(|a|,|b|))^h+R_p\left(\frac{b-a}{2n}\right)^p\omega_1\left(f^{(p)},\frac{b-a}{2n}\right)=$$

$$\eta_n\,(h!)^{-1}\max\,\left(|a|^h,|b|^h\right)+R_p\left(\frac{b-a}{2n}\right)^p\omega_1\left(f^{(p)},\frac{b-a}{2n}\right)= \qquad (20.2.78)$$

$$R_p\omega_1\left(f^{(p)},\frac{b-a}{2n}\right)\cdot$$

$$\left(\sum_{j=h}^k s_j\frac{(g(b)-g(a))^{j-\alpha_j}}{\Gamma(j-\alpha_j+1)}\left(\frac{b-a}{2n}\right)^{p-j}\right)(h!)^{-1}\max\,\left(|a|^h,|b|^h\right)$$

$$+R_p\left(\frac{b-a}{2n}\right)^p\omega_1\left(f^{(p)},\frac{b-a}{2n}\right)\le$$

$$R_p\omega_1\left(f^{(p)},\frac{b-a}{2n}\right)n^{k-p}\times$$

$$\left[\left(\sum_{j=h}^k s_j\frac{(g(b)-g(a))^{j-\alpha_j}}{\Gamma(j-\alpha_j+1)}\left(\frac{b-a}{2}\right)^{p-j}\right)(h!)^{-1}\max\,\left(|a|^h,|b|^h\right)+\left(\frac{b-a}{2}\right)^p\right].$$

$$(20.2.79)$$

We have found that

$$\max_{x\in[a,b]}|f(x)-Q_n(x)|\le R_p\left[\left(\frac{b-a}{2}\right)^p+(h!)^{-1}\max\,\left(|a|^h,|b|^h\right)\cdot\right.$$

$$\left.\left(\sum_{j=h}^k s_j\frac{(g(b)-g(a))^{j-\alpha_j}}{\Gamma(j-\alpha_j+1)}\left(\frac{b-a}{2}\right)^{p-j}\right)\right]n^{k-p}\omega_1\left(f^{(p)},\frac{b-a}{2n}\right),$$

$$(20.2.80)$$

proving (20.2.66).

Notice for $j=h+1,\ldots,k$, that

$$\left(D_{*a;g}^{\alpha_j}x^h\right)=\frac{1}{\Gamma(j-\alpha_j)}\int_a^x(g(x)-g(t))^{j-\alpha_j-1}g'(t)\left(t^h\right)^{(j)}dt=0.$$

$$(20.2.81)$$

Here

$$I. = \sum_{j=h}^{k} \alpha_j(x) \left[D_{*a;g}^{\alpha_j} \right],$$

and suppose, throughout $\left[g^{-1}(1 + g(a)), b \right]$, $Lf \geq 0$. So over $g^{-1}(1 + g(a)) \leq x \leq b$, we get

$$\alpha_h^{-1}(x) L(Q_n(x)) \overset{(20.2.81)}{=} \alpha_h^{-1}(x) L(f(x)) + \frac{\eta_n}{h!} \left(D_{*a;g}^{\alpha_h}(x^h) \right) +$$

$$\sum_{j=h}^{k} \alpha_h^{-1}(x) \alpha_j(x) \left[D_{*a;g}^{\alpha_j} Q_n(x) - D_{*a;g}^{\alpha_j} f(x) - \frac{\eta_n}{h!} D_{*a;g}^{\alpha_j} x^h \right] \overset{(20.2.76)}{\geq} \quad (20.2.82)$$

$$\frac{\eta_n}{h!} \left(D_{*a;g}^{\alpha_h}(x^h) \right) - \left(\sum_{j=h}^{k} s_j \frac{(g(b) - g(a))^{j-\alpha_j}}{\Gamma(j - \alpha_j + 1)} \left(\frac{b-a}{2n} \right)^{p-j} \right) R_p \omega_1 \left(f^{(p)}, \frac{b-a}{2n} \right)$$
$$(20.2.83)$$

$$\overset{(20.2.75)}{=} \frac{\eta_n}{h!} \left(D_{*a;g}^{\alpha_h}(x^h) \right) - \eta_n = \eta_n \left(\frac{D_{*a;g}^{\alpha_h}(x^h)}{h!} - 1 \right) = \quad (20.2.84)$$

$$\eta_n \left(\frac{1}{\Gamma(h - \alpha_h) h!} \int_a^x (g(x) - g(t))^{h-\alpha_h-1} g'(t) \left(t^h \right)^{(h)} dt - 1 \right) =$$

$$\eta_n \left(\frac{h!}{h! \Gamma(h - \alpha_h)} \int_a^x (g(x) - g(t))^{h-\alpha_h-1} g'(t) dt - 1 \right) \overset{(20.1.54)}{=}$$

$$\eta_n \left(\frac{(g(x) - g(a))^{h-\alpha_h}}{\Gamma(h - \alpha_h + 1)} - 1 \right) = \quad (20.2.85)$$

$$\eta_n \left(\frac{(g(x) - g(a))^{h-\alpha_h} - \Gamma(h - \alpha_h + 1)}{\Gamma(h - \alpha_h + 1)} \right) \geq$$

$$\eta_n \left(\frac{1 - \Gamma(h - \alpha_h + 1)}{\Gamma(h - \alpha_h + 1)} \right) \geq 0. \quad (20.2.86)$$

Clearly here $g(x) - g(a) \geq 1$.

Hence

$$L(Q_n(x)) \geq 0, \text{ for } x \in \left[g^{-1}(1 + g(a)), b \right]. \quad (20.2.87)$$

A further explanation follows: We know $\Gamma(1) = 1$, $\Gamma(2) = 1$, and Γ is convex and positive on $(0, \infty)$. Here $0 \le h - \alpha_h < 1$ and $1 \le h - \alpha_h + 1 < 2$. Thus

$$\Gamma(h - \alpha_h + 1) \le 1 \text{ and } 1 - \Gamma(h - \alpha_h + 1) \ge 0. \qquad (20.2.88)$$

II. Suppose, throughout $\left[g^{-1}(1 + g(a)), b\right]$, $\alpha_h(x) \le \beta^* < 0$.

Let $Q_n(x)$, $x \in [a, b]$ be a real polynomial of degree $\le n$, according to Theorem 20.12 and (20.2.73), so that

$$\max_{x \in [a,b]} \left| D^{\alpha_j}_{*a;g}\left(f(x) - \eta_n(h!)^{-1} x^h\right) - \left(D^{\alpha_j}_{*a;g} Q_n\right)(x)\right| \le \qquad (20.2.89)$$

$$\frac{(g(b) - g(a))^{j - \alpha_j}}{\Gamma(j - \alpha_j + 1)} R_p \left(\frac{b - a}{2n}\right)^{p-j} \omega_1 \left(f^{(p)}, \frac{b - a}{2n}\right),$$

$j = 0, 1, \ldots, p$.

In particular $(j = 0)$ holds

$$\max_{x \in [a,b]} \left|(f(x) - \eta_n(h!)^{-1} x^h) - Q_n(x)\right| \le R_p \left(\frac{b - a}{2n}\right)^p \omega_1 \left(f^{(p)}, \frac{b - a}{2n}\right),$$
$$\qquad (20.2.90)$$

and

$$\max_{x \in [a,b]} |f(x) - Q_n(x)| \le$$

$$\eta_n(h!)^{-1} (\max(|a|, |b|))^h + R_p \left(\frac{b - a}{2n}\right)^p \omega_1 \left(f^{(p)}, \frac{b - a}{2n}\right) =$$

$$\eta_n(h!)^{-1} \max\left(|a|^h, |b|^h\right) + R_p \left(\frac{b - a}{2n}\right)^p \omega_1 \left(f^{(p)}, \frac{b - a}{2n}\right), \qquad (20.2.91)$$

etc.

We find again that

$$\max_{x \in [a,b]} |f(x) - Q_n(x)| \le R_p \left[\left(\frac{b - a}{2}\right)^p + (h!)^{-1} \max\left(|a|^h, |b|^h\right) \cdot \right.$$

$$\left. \left(\sum_{j=h}^{k} s_j \frac{(g(b) - g(a))^{j - \alpha_j}}{\Gamma(j - \alpha_j + 1)} \left(\frac{b - a}{2}\right)^{p-j}\right)\right] n^{k-p} \omega_1 \left(f^{(p)}, \frac{b - a}{2n}\right), \qquad (20.2.92)$$

reproving (20.2.66).

Here again

$$L = \sum_{j=h}^{k} \alpha_j(x) \left[D_{*a;g}^{\alpha_j} \right],$$

and suppose, throughout $\left[g^{-1}(1+g(a)), b \right]$, $Lf \geq 0$. So over $g^{-1}(1+g(a)) \leq x \leq b$, we get

$$\alpha_h^{-1}(x) L(Q_n(x)) \overset{(20.2.81)}{=} \alpha_h^{-1}(x) L(f(x)) - \frac{\eta_n}{h!} \left(D_{*a;g}^{\alpha_h}(x^h) \right) +$$

$$\sum_{j=h}^{k} \alpha_h^{-1}(x) \alpha_j(x) \left[D_{*a;g}^{\alpha_j} Q_n(x) - D_{*a;g}^{\alpha_j} f(x) + \frac{\eta_n}{h!} D_{*a;g}^{\alpha_j} x^h \right] \overset{(20.2.89)}{\leq} \quad (20.2.93)$$

$$-\frac{\eta_n}{h!} \left(D_{*a;g}^{\alpha_h}(x^h) \right) +$$

$$\left(\sum_{j=h}^{k} s_j \frac{(g(b) - g(a))^{j-\alpha_j}}{\Gamma(j-\alpha_j+1)} \left(\frac{b-a}{2n} \right)^{p-j} \right) R_p \omega_1 \left(f^{(p)}, \frac{b-a}{2n} \right) \quad (20.2.94)$$

$$\overset{(20.2.75)}{=} -\frac{\eta_n}{h!} \left(D_{*a;g}^{\alpha_h}(x^h) \right) + \eta_n = \eta_n \left(1 - \frac{D_{*a;g}^{\alpha_h}(x^h)}{h!} \right) = \quad (20.2.95)$$

$$\eta_n \left(1 - \frac{1}{\Gamma(h-\alpha_h)h!} \int_a^x (g(x) - g(t))^{h-\alpha_h-1} g'(t) \left(t^h \right)^{(h)} dt \right) =$$

$$\eta_n \left(1 - \frac{h!}{h!\Gamma(h-\alpha_h)} \int_a^x (g(x) - g(t))^{h-\alpha_h-1} g'(t) dt \right) \overset{(20.1.54)}{=}$$

$$\eta_n \left(1 - \frac{(g(x) - g(a))^{h-\alpha_h}}{\Gamma(h-\alpha_h+1)} \right) = \quad (20.2.96)$$

$$\eta_n \left(\frac{\Gamma(h-\alpha_h+1) - (g(x) - g(a))^{h-\alpha_h}}{\Gamma(h-\alpha_h+1)} \right) \overset{(20.2.88)}{\leq}$$

$$\eta_n \left(\frac{1 - (g(x) - g(a))^{h-\alpha_h}}{\Gamma(h-\alpha_h+1)} \right) \leq 0. \quad (20.2.97)$$

Hence again

$$L(Q_n(x)) \geq 0, \quad \forall x \in \left[g^{-1}(1+g(a)), b \right].$$

The case of $\alpha_h = h$ is trivially concluded from the above. The proof of the theorem is now over. $\qquad\qquad\qquad\qquad\qquad\qquad\qquad\qquad\qquad\qquad\qquad\qquad\Box$

We make:

Remark 20.16 By Theorem 20.14 we have that $D_{*a;g}^{\alpha_j} f$ are continuous functions, $j = 0, 1, \ldots, p$. Suppose that $\alpha_h (x), \ldots, \alpha_k (x)$ are continuous functions on $[a, b]$, and $L (f) \geq 0$ on $\left[g^{-1} (1 + g (a)), b\right]$ is replaced by $L (f) > 0$ on $\left[g^{-1} (1 + g (a)), b\right]$. Disregard the assumption made in the main theorem on $\alpha_h (x)$. For $n \in \mathbb{N}$, let $Q_n (x)$ be the $Q_n^* (x)$ of Theorem 20.12, and f as in Theorem 20.12 (same as in Theorem 20.15). Then $Q_n (x)$ converges to $f (x)$ at the Jackson rate $\frac{1}{n^{p+1}}$ ([6], p. 18, Theorem VIII) and at the same time, since $L (Q_n)$ converges uniformly to $L (f)$ on $[a, b]$, $L (Q_n) > 0$ on $\left[g^{-1} (1 + g (a)), b\right]$ for all n sufficiently large.

20.3   Applications (to Theorem 20.15)

(1) When $g (x) = \ln x$ on $[a, b]$, $0 < a < b < \infty$.
Here we would assume that $b > ae$, $\alpha_h (x)$ restriction true on $[ae, b]$, and

$$Lf = \sum_{j=h}^{k} \alpha_j (x) \left[D_{*a;\ln x}^{\alpha_j} f\right] \geq 0, \tag{20.3.99}$$

throughout $[ae, b]$.
   Then $L (Q_n) \geq 0$ on $[ae, b]$.
   (2) When $g (x) = e^x$ on $[a, b]$, $a < b < \infty$.
Here we assume that $b > \ln (1 + e^a)$, $\alpha_h (x)$ restriction true on $[\ln (1 + e^a), b]$, and

$$Lf = \sum_{j=h}^{k} \alpha_j (x) \left[D_{*a;e^x}^{\alpha_j} f\right] \geq 0, \tag{20.3.100}$$

throughout $[\ln (1 + e^a), b]$.
   Then $L (Q_n) \geq 0$ on $[\ln (1 + e^a), b]$.
   (3) When, $A > 1$, $g (x) = A^x$ on $[a, b]$, $a < b < \infty$.
Here we assume that $b > \log_A (1 + A^a)$, $\alpha_h (x)$ restriction true on $\left[\log_A (1 + A^a), b\right]$, and

$$Lf = \sum_{j=h}^{k} \alpha_j (x) \left[D_{*a;A^x}^{\alpha_j} f\right] \geq 0, \tag{20.3.101}$$

throughout $\left[\log_A (1 + A^a), b\right]$.
   Then $L (Q_n) \geq 0$ on $\left[\log_A (1 + A^a), b\right]$.

(4) When $\sigma > 0$, $g(x) = x^\sigma$, $0 \le a < b < \infty$.

Here we assume that $b > (1 + a^\sigma)^{\frac{1}{\sigma}}$, $\alpha_h(x)$ restriction true on $\left[(1 + a^\sigma)^{\frac{1}{\sigma}}, b\right]$, and

$$Lf = \sum_{j=h}^{k} \alpha_j(x) \left[D_{*a;x^\sigma}^{\alpha_j} f\right] \ge 0 \tag{20.3.102}$$

throughout $\left[(1 + a^\sigma)^{\frac{1}{\sigma}}, b\right]$.

Then $L(Q_n) \ge 0$ on $\left[(1 + a^\sigma)^{\frac{1}{\sigma}}, b\right]$.

References

1. G.A. Anastassiou, Fractional monotone approximation theory. Indian J. Math. **57**(1), 141–149 (2015)
2. G. Anastassiou, *Left General Fractional Monotone Approximation Theory*, (2015) (submitted)
3. T. Apostol, *Mathematical Analysis* (Addison-Wesley Publ. Co., Reading, Massachusetts, 1969)
4. G.A. Anastassiou, O. Shisha, Monotone approximation with linear differential operators. J. Approx. Theory **44**, 391–393 (1985)
5. K. Diethelm, in *The Analysis of Fractional Differential Equations*. Lecture Notes in Mathematics, vol. 2004, 1st edn. (Spinger, New York, Heidelberg, 2010)
6. D. Jackson, *The Theory of Approximation*, vol XI (American Mathematical Society Colloquium, New York, 1930)
7. R.-Q. Jia, Chapter 3. absolutely continous functions, https://www.ualberta.ca/~rjia/Math418/Notes/Chap.3.pdf
8. A.A. Kilbas, H.M. Srivastava, J.J. Trujillo, in *Theory and Applications of Fractional Differential Equations*. North-Holland Mathematics Studies, vol 204 (Elsevier, New York, 2006)
9. H.L. Royden, *Real Analysis*, 2nd edn. (Macmillan Publishing Co. Inc., New York, 1968)
10. A.R. Schep, Differentiation of monotone functions, https://www.people.math.sc.edu/schep/diffmonotone.pdf
11. O. Shisha, Monotone approximation. Pacific J. Math. **15**, 667–671 (1965)
12. S.A. Teljakovskii, Two theorems on the approximation of functions by algebraic polynomials. Mat. Sb. **70**(112), 252–265 (1966) (Russian); Amer. Math. Soc. Trans. **77**(2), 163–178 (1968)
13. R.M. Trigub, Approximation of functions by polynomials with integer coeficients. Izv. Akad. Nauk SSSR Ser. Mat. **26**, 261–280 (1962) (Russian)

Chapter 21
Right General Fractional Monotone Approximation Theory

Here is introduced a right general fractional derivative Caputo style with respect to a base absolutely continuous strictly increasing function g. We give various examples of such right fractional derivatives for different g. Let f be p-times continuously differentiable function on $[a, b]$, and let L be a linear right general fractional differential operator such that $L(f)$ is non-negative over a critical closed subinterval J of $[a, b]$. We can find a sequence of polynomials Q_n of degree less-equal n such that $L(Q_n)$ is non-negative over J, furthermore f is approximated uniformly by Q_n over $[a, b]$.

The degree of this constrained approximation is given by an inequality using the first modulus of continuity of $f^{(p)}$. We finish we applications of the main right fractional monotone approximation theorem for different g. It follows [3].

21.1 Introduction and Preparation

The topic of monotone approximation started in [12] has become a major trend in approximation theory. A typical problem in this subject is: given a positive integer k, approximate a given function whose kth derivative is ≥ 0 by polynomials having this property.

In [4] the authors replaced the kth derivative with a linear ordinary differential operator of order k.

Furthermore in [1], the author generalized the result of [4] for linear right fractional differential operators.

G.A. Anastassiou and I.K. Argyros, *Intelligent Numerical Methods:*
Applications to Fractional Calculus, Studies in Computational Intelligence 624,
DOI 10.1007/978-3-319-26721-0_21

To describe the motivating result here we need:

Definition 21.1 ([6]) Let $\alpha > 0$ and $\lceil \alpha \rceil = m$, ($\lceil \cdot \rceil$ ceiling of the number). Consider $f \in C^m ([-1, 1])$. We define the right Caputo fractional derivative of f of order α as follows:

$$\left(D_{1-}^{\alpha} f \right)(x) = \frac{(-1)^m}{\Gamma(m - \alpha)} \int_x^1 (t - x)^{m-\alpha-1} f^{(m)}(t)\, dt, \qquad (21.1.1)$$

for any $x \in [-1, 1]$, where Γ is the gamma function $\Gamma(\nu) = \int_0^\infty e^{-t} t^{\nu-1} dt$, $\nu > 0$.
    We set

$$D_{1-}^0 f(x) = f(x), \qquad (21.1.2)$$

$$D_{1-}^m f(x) = (-1)^m f^{(m)}(x), \quad \forall x \in [-1, 1]. \qquad (21.1.3)$$

In [1] we proved:

Theorem 21.2 *Let h, k, p be integers, h is even, $0 \leq h \leq k \leq p$ and let f be a real function, $f^{(p)}$ continuous in $[-1, 1]$ with modulus of continuity $\omega_1 \left(f^{(p)}, \delta \right)$, $\delta > 0$, there. Let $\alpha_j(x)$, $j = h, h + 1, \ldots, k$ be real functions, defined and bounded on $[-1, 1]$ and assume for $x \in [-1, 0]$ that $\alpha_h(x)$ is either $\geq$ some number $\alpha > 0$ or $\leq$ some number $\beta < 0$. Let the real numbers $\alpha_0 = 0 < \alpha_1 < 1 < \alpha_2 < 2 < \cdots < \alpha_p < p$. Here $D_{1-}^{\alpha_j} f$ stands for the right Caputo fractional derivative of f of order α_j anchored at 1. Consider the linear right fractional differential operator*

$$L := \sum_{j=h}^{k} \alpha_j(x) \left[D_{1-}^{\alpha_j} \right] \qquad (21.1.4)$$

and suppose, throughout $[-1, 0]$,

$$L(f) \geq 0. \qquad (21.1.5)$$

Then, for any $n \in \mathbb{N}$, there exists a real polynomial $Q_n(x)$ of degree $\leq n$ such that

$$L(Q_n) \geq 0 \text{ throughout } [-1, 0], \qquad (21.1.6)$$

and

$$\max_{-1 \leq x \leq 1} |f(x) - Q_n(x)| \leq C n^{k-p} \omega_1 \left(f^{(p)}, \frac{1}{n} \right), \qquad (21.1.7)$$

where C is independent of n or f.

Notice above that the monotonicity property is only true on $[-1, 0]$, see (21.1.5) and (21.1.6). However the approximation property (21.1.7) it is true over the whole interval $[-1, 1]$.

In this chapter we extend Theorem 21.2 to much more general linear right fractional differential operators.

We use here the following right generalised fractional integral.

Definition 21.3 (*see also* [9, p. 99]) The right generalised fractional integral of a function f with respect to given function g is defined as follows:

Let $a, b \in \mathbb{R}$, $a < b$, $\alpha > 0$. Here $g \in AC([a, b])$ (absolutely continuous functions) and is strictly increasing, $f \in L_\infty([a, b])$. We set

$$\left(I^\alpha_{b-;g} f\right)(x) = \frac{1}{\Gamma(\alpha)} \int_x^b (g(t) - g(x))^{\alpha-1} g'(t) f(t) \, dt, \quad x \le b, \quad (21.1.8)$$

clearly $\left(I^\alpha_{b-;g} f\right)(b) = 0$.

When g is the identity function id, we get that $I^\alpha_{b-;id} = I^\alpha_{b-}$, the ordinary right Riemann-Liouville fractional integral, where

$$\left(I^\alpha_{b-} f\right)(x) = \frac{1}{\Gamma(\alpha)} \int_x^b (t - x)^{\alpha-1} f(t) \, dt, \quad x \le b, \quad (21.1.9)$$

$\left(I^\alpha_{b-} f\right)(b) = 0$.

When $g(x) = \ln x$ on $[a, b]$, $0 < a < b < \infty$, we get:

Definition 21.4 ([9, p. 110]) Let $0 < a < b < \infty$, $\alpha > 0$. The right Hadamard fractional integral of order α is given by

$$\left(J^\alpha_{b-} f\right)(x) = \frac{1}{\Gamma(\alpha)} \int_x^b \left(\ln \frac{y}{x}\right)^{\alpha-1} \frac{f(y)}{y} \, dy, \quad x \le b, \quad (21.1.10)$$

where $f \in L_\infty([a, b])$.

We mention:

Definition 21.5 The right fractional exponential integral is defined as follows: Let $a, b \in \mathbb{R}$, $a < b$, $\alpha > 0$, $f \in L_\infty([a, b])$. We set

$$\left(I^\alpha_{b-;e^x} f\right)(x) = \frac{1}{\Gamma(\alpha)} \int_x^b (e^t - e^x)^{\alpha-1} e^t f(t) \, dt, \quad x \le b. \quad (21.1.11)$$

Definition 21.6 Let $a, b \in \mathbb{R}$, $a < b$, $\alpha > 0$, $f \in L_\infty([a, b])$, $A > 1$. We introduce the right fractional integral

$$\left(I^\alpha_{b-;A^x} f\right)(x) = \frac{\ln A}{\Gamma(\alpha)} \int_x^b (A^t - A^x)^{\alpha-1} A^t f(t) \, dt, \quad x \le b. \quad (21.1.12)$$

We also give:

Definition 21.7 Let $\alpha, \sigma > 0, 0 \leq a < b < \infty, f \in L_\infty ([a, b])$. We set

$$\left(K_{b-;x^\sigma}^\alpha f\right)(x) = \frac{1}{\Gamma(\alpha)} \int_x^b (t^\sigma - x^\sigma)^{\alpha-1} f(t) \sigma t^{\sigma-1} dt, \quad x \leq b. \qquad (21.1.13)$$

We introduce the following general right fractional derivative.

Definition 21.8 Let $\alpha > 0$ and $\lceil \alpha \rceil = m$, ($\lceil \cdot \rceil$ ceiling of the number). Consider $f \in AC^m ([a, b])$ (space of functions f with $f^{(m-1)} \in AC ([a, b])$). We define the right general fractional derivative of f of order α as follows

$$\left(D_{b-;g}^\alpha f\right)(x) = \frac{(-1)^m}{\Gamma(m - \alpha)} \int_x^b (g(t) - g(x))^{m-\alpha-1} g'(t) f^{(m)}(t) dt, \quad (21.1.14)$$

for any $x \in [a, b]$, where Γ is the gamma function.
    We set

$$D_{b-;g}^m f(x) = (-1)^m f^{(m)}(x), \qquad (21.1.15)$$

$$D_{b-;g}^0 f(x) = f(x), \quad \forall x \in [a, b]. \qquad (21.1.16)$$

When $g = id$, then $D_{b-}^\alpha f = D_{b-;id}^\alpha f$ is the right Caputo fractional derivative.

So we have the specific general right fractional derivatives.

Definition 21.9

$$D_{b-;\ln x}^\alpha f(x) = \frac{(-1)^m}{\Gamma(m - \alpha)} \int_x^b \left(\ln \frac{y}{x}\right)^{m-\alpha-1} \frac{f^{(m)}(y)}{y} dy, \quad 0 < a \leq x \leq b,$$
$$(21.1.17)$$

$$D_{b-;e^x}^\alpha f(x) = \frac{(-1)^m}{\Gamma(m - \alpha)} \int_x^b (e^t - e^x)^{m-\alpha-1} e^t f^{(m)}(t) dt, \quad a \leq x \leq b,$$
$$(21.1.18)$$

and

$$D_{b-;A^x}^\alpha f(x) = \frac{(-1)^m \ln A}{\Gamma(m - \alpha)} \int_x^b (A^t - A^x)^{m-\alpha-1} A^t f^{(m)}(t) dt, \quad a \leq x \leq b,$$
$$(21.1.19)$$

$$\left(D_{b-;x^\sigma}^\alpha f\right)(x) = \frac{(-1)^m}{\Gamma(m - \alpha)} \int_x^b (t^\sigma - x^\sigma)^{m-\alpha-1} \sigma t^{\sigma-1} f^{(m)}(t) dt, \quad 0 \leq a \leq x \leq b.$$
$$(21.1.20)$$

We mention:

Theorem 21.10 (Trigub, [13, 15]) *Let* $g \in C^p$ ([−1, 1]), $p \in \mathbb{N}$. *Then there exists real polynomial* $q_n(x)$ *of degree* $\leq n$, $x \subset [$ 1, 1], *such that*

$$\max_{-1 \leq x \leq 1} \left| g^{(j)}(x) - q_n^{(j)}(x) \right| \leq R_p n^{j-p} \omega_1 \left(g^{(p)}, \frac{1}{n} \right), \qquad (21.1.21)$$

$j = 0, 1, \ldots, p$, *where* R_p *is independent of n or g*.

In [2], based on Theorem 21.10 we proved the following useful here result

Theorem 21.11 *Let* $f \in C^p$ ([a, b]), $p \in \mathbb{N}$. *Then there exist real polynomials* $Q_n^*(x)$ *of degree* $\leq n \in \mathbb{N}$, $x \in [a, b]$, *such that*

$$\max_{a \leq x \leq b} \left| f^{(j)}(x) - Q_n^{*(j)}(x) \right| \leq R_p \left(\frac{b-a}{2n} \right)^{p-j} \omega_1 \left(f^{(p)}, \frac{b-a}{2n} \right), \quad (21.1.22)$$

$j = 0, 1, \ldots, p$, *where* R_p *is independent of n or g*.

Remark 21.12 Here $g \in AC$ ([a, b]) (absolutely continuous functions), g is increasing over $[a, b]$, $\alpha > 0$.

Let $g(a) = c$, $g(b) = d$. We want to calculate

$$I = \int_a^b (g(t) - g(a))^{\alpha-1} g'(t) \, dt. \qquad (21.1.23)$$

Consider the function

$$f(y) = (y - g(a))^{\alpha-1} = (y - c)^{\alpha-1}, \quad \forall \, y \in [c, d]. \qquad (21.1.24)$$

We have that $f(y) \geq 0$, it may be $+\infty$ when $y = c$ and $0 < \alpha < 1$, but f is measurable on $[c, d]$. By [10], Royden, p. 107, Exercise 13d, we get that

$$(f \circ g)(t) g'(t) = (g(t) - g(a))^{\alpha-1} g'(t) \qquad (21.1.25)$$

is measurable on $[a, b]$, and

$$I = \int_c^d (y - c)^{\alpha-1} \, dy = \frac{(d-c)^\alpha}{\alpha} \qquad (21.1.26)$$

(notice that $(y - c)^{\alpha-1}$ is Riemann integrable).

That is

$$I = \frac{(g(b) - g(a))^\alpha}{\alpha}. \qquad (21.1.27)$$

Similarly it holds

$$\int_x^b (g(t) - g(x))^{\alpha-1} g'(t) \, dt = \frac{(g(b) - g(x))^\alpha}{\alpha}, \quad \forall \, x \in [a, b]. \quad (21.1.28)$$

Finally we will use:

Theorem 21.13 *Let* $\alpha > 0$, $\mathbb{N} \ni m = \lceil \alpha \rceil$, *and* $f \in C^m([a, b])$. *Then* $\left(D_{b-;g}^\alpha f \right)(x)$ *is continuous in* $x \in [a, b]$, $-\infty < a < b < \infty$.

Proof By [5], Apostol, p. 78, we get that g^{-1} exists and it is strictly increasing on $[g(a), g(b)]$. Since g is continuous on $[a, b]$, it implies that g^{-1} is continuous on $[g(a), g(b)]$. Hence $f^{(m)} \circ g^{-1}$ is a continuous function on $[g(a), g(b)]$.

If $\alpha = m \in \mathbb{N}$, then the claim is trivial.

We treat the case of $0 < \alpha < m$.

It holds that

$$\left(D_{b-;g}^\alpha f \right)(x) = \frac{(-1)^m}{\Gamma(m-\alpha)} \int_x^b (g(t) - g(x))^{m-\alpha-1} g'(t) f^{(m)}(t) \, dt =$$

$$\frac{(-1)^m}{\Gamma(m-\alpha)} \int_x^b (g(t) - g(x))^{m-\alpha-1} g'(t) \left(f^{(m)} \circ g^{-1} \right)(g(t)) \, dt = \quad (21.1.29)$$

$$\frac{(-1)^m}{\Gamma(m-\alpha)} \int_{g(x)}^{g(b)} (z - g(x))^{m-\alpha-1} \left(f^{(m)} \circ g^{-1} \right)(z) \, dz.$$

An explanation follows.

The function
$$G(z) = (z - g(x))^{m-\alpha-1} \left(f^{(m)} \circ g^{-1} \right)(z)$$

is integrable on $[g(x), g(b)]$, and by assumption g is absolutely continuous: $[a, b] \rightarrow [g(a), g(b)]$.

Since g is monotone (strictly increasing here) the function

$$(g(t) - g(x))^{m-\alpha-1} g'(t) \left(f^{(m)} \circ g^{-1} \right)(g(t))$$

is integrable on $[x, b]$ (see [8]). Furthermore it holds (see also [8]),

$$\frac{(-1)^m}{\Gamma(m-\alpha)} \int_{g(x)}^{g(b)} (z - g(x))^{m-\alpha-1} \left(f^{(m)} \circ g^{-1} \right)(z) \, dz =$$

$$\frac{(-1)^m}{\Gamma(m-\alpha)} \int_x^b (g(t) - g(x))^{m-\alpha-1} g'(t) \left(f^{(m)} \circ g^{-1} \right)(g(t)) \, dt \quad (21.1.30)$$

$$= \left(D_{b-;g}^\alpha f \right)(x), \quad \forall \, x \in [a, b].$$

And we can write

$$\left(D^{\alpha}_{b-;g}f\right)(x) = \frac{(-1)^m}{\Gamma(m-\alpha)} \int_{g(x)}^{g(b)} (z-g(x))^{m-\alpha-1} \left(f^{(m)} \circ g^{-1}\right)(z)\,dz,$$

$$\left(D^{\alpha}_{b-;g}f\right)(y) = \frac{(-1)^m}{\Gamma(m-\alpha)} \int_{g(y)}^{g(b)} (z-g(y))^{m-\alpha-1} \left(f^{(m)} \circ g^{-1}\right)(z)\,dz.$$

$$(21.1.31)$$

Here $a \le y \le x \le b$, and $g(a) \le g(y) \le g(x) \le g(b)$, and $0 \le g(b) - g(x) \le g(b) - g(y)$.

Let $\lambda = z - g(x)$, then $z = g(x) + \lambda$. Thus

$$\left(D^{\alpha}_{b-;g}f\right)(x) = \frac{(-1)^m}{\Gamma(m-\alpha)} \int_{0}^{g(b)-g(x)} \lambda^{m-\alpha-1} \left(f^{(m)} \circ g^{-1}\right)(g(x)+\lambda)\,d\lambda.$$

$$(21.1.32)$$

Clearly, see that $g(x) \le z \le g(b)$, and $0 \le \lambda \le g(b) - g(x)$.

Similarly

$$\left(D^{\alpha}_{b-;g}f\right)(y) = \frac{(-1)^m}{\Gamma(m-\alpha)} \int_{0}^{g(b)-g(y)} \lambda^{m-\alpha-1} \left(f^{(m)} \circ g^{-1}\right)(g(y)+\lambda)\,d\lambda.$$

$$(21.1.33)$$

Hence it holds

$$\left(D^{\alpha}_{b-;g}f\right)(y) - \left(D^{\alpha}_{b-;g}f\right)(x) = \frac{(-1)^m}{\Gamma(m-\alpha)} \cdot$$

$$\left[\int_{0}^{g(b)-g(x)} \lambda^{m-\alpha-1} \left(\left(f^{(m)} \circ g^{-1}\right)(g(y)+\lambda) - \left(f^{(m)} \circ g^{-1}\right)(g(x)+\lambda)\right)\,d\lambda\right.$$

$$\left. + \int_{g(b)-g(x)}^{g(b)-g(y)} \lambda^{m-\alpha-1} \left(f^{(m)} \circ g^{-1}\right)(g(y)+\lambda)\,d\lambda\right].$$

$$(21.1.34)$$

Thus we obtain

$$\left|\left(D^{\alpha}_{b-;g}f\right)(y) - \left(D^{\alpha}_{b-;g}f\right)(x)\right| \le \frac{1}{\Gamma(m-\alpha)} \cdot$$

$$\left[\frac{(g(b)-g(x))^{m-\alpha}}{m-\alpha} \omega_1\left(f^{(m)} \circ g^{-1}, |g(y)-g(x)|\right) + \right.$$

$$(21.1.35)$$

$$\left. \frac{\left\|f^{(m)} \circ g^{-1}\right\|_{\infty,[g(a),g(b)]}}{m-\alpha} \left((g(b)-g(y))^{m-\alpha} - (g(b)-g(x))^{m-\alpha}\right)\right] =: (\xi).$$

As $y \to x$, then $g(y) \to g(x)$ (since $g \in AC([a, b])$). So that $(\xi) \to 0$. As a result

$$\left(D_{b-;g}^{\alpha} f\right)(y) \to \left(D_{b-;g}^{\alpha} f\right)(x), \tag{21.1.36}$$

proving that $\left(D_{b-;g}^{\alpha} f\right)(x)$ is continuous in $x \in [a, b]$. $\qquad\square$

21.2　Main Result

We present:

Theorem 21.14 *Here we assume that $g(b) - g(a) > 1$. Let h, k, p be integers, h is even, $0 \leq h \leq k \leq p$ and let $f \in C^p([a, b])$, $a < b$, with modulus of continuity $\omega_1\left(f^{(p)}, \delta\right)$, $0 < \delta \leq b - a$. Let $\alpha_j(x)$, $j = h, h + 1, \ldots, k$ be real functions, defined and bounded on $[a, b]$ and assume for $x \in \left[a, g^{-1}(g(b) - 1)\right]$ that $\alpha_h(x)$ is either $\geq$ some number $\alpha^* > 0$, or $\leq$ some number $\beta^* < 0$. Let the real numbers $\alpha_0 = 0 < \alpha_1 \leq 1 < \alpha_2 \leq 2 < \cdots < \alpha_p \leq p$. Consider the linear right general fractional differential operator*

$$L = \sum_{j=h}^{k} \alpha_j(x) \left[D_{b-;g}^{\alpha_j}\right], \tag{21.2.1}$$

and suppose, throughout $\left[a, g^{-1}(g(b) - 1)\right]$,

$$L(f) \geq 0. \tag{21.2.2}$$

Then, for any $n \in \mathbb{N}$, there exists a real polynomial $Q_n(x)$ of degree $\leq n$ such that

$$L(Q_n) \geq 0 \text{ throughout } \left[a, g^{-1}(g(b) - 1)\right], \tag{21.2.3}$$

and

$$\max_{x \in [a,b]} |f(x) - Q_n(x)| \leq Cn^{k-p} \omega_1 \left(f^{(p)}, \frac{b-a}{2n}\right), \tag{21.2.4}$$

where C is independent of n or f.

Proof of Theorem 21.14.

Here $h, k, p \in \mathbb{Z}_+$, $0 \leq h \leq k \leq p$. Let $\alpha_j > 0$, $j = 1, \ldots, p$, such that $0 < \alpha_1 \leq 1 < \alpha_2 \leq 2 < \alpha_3 \leq 3 \cdots < \cdots < \alpha_p \leq p$. That is $\lceil \alpha_j \rceil = j$, $j = 1, \ldots, p$.

Let $Q_n^*(x)$ be as in Theorem 21.11.

We have that

$$\left(D_{b-;g}^{\alpha_j} f\right)(x) = \frac{(-1)^j}{\Gamma(j-\alpha_j)} \int_x^b (g(t) - g(x))^{j-\alpha_j-1} g'(t) f^{(j)}(t) dt, \quad (21.2.5)$$

and

$$\left(D_{b-;g}^{\alpha_j} Q_n^*\right)(x) = \frac{(-1)^j}{\Gamma(j-\alpha_j)} \int_x^b (g(t) - g(x))^{j-\alpha_j-1} g'(t) Q_n^{*(j)}(t) dt,$$

$$(21.2.6)$$

$j = 1, \ldots, p$.
Also it holds

$$\left(D_{b-;g}^j f\right)(x) = (-1)^j f^{(j)}(x), \quad \left(D_{b-;g}^j Q_n^*\right)(x) = (-1)^j Q_n^{*(j)}(x), \quad (21.2.7)$$

$j = 1, \ldots, p$.
By [11], we get that there exists g' a.e., and g' is measurable and non-negative.
We notice that

$$\left| \left(D_{b-;g}^{\alpha_j} f\right)(x) - D_{b-;g}^{\alpha_j} Q_n^*(x) \right| =$$

$$\frac{1}{\Gamma(j-\alpha_j)} \left| \int_x^b (g(x) - g(t))^{j-\alpha_j-1} g'(t) \left(f^{(j)}(t) - Q_n^{*(j)}(t)\right) dt \right| \leq$$

$$\frac{1}{\Gamma(j-\alpha_j)} \int_x^b (g(x) - g(t))^{j-\alpha_j-1} g'(t) \left| f^{(j)}(t) - Q_n^{*(j)}(t) \right| dt \overset{(21.1.22)}{\leq}$$

$$\frac{1}{\Gamma(j-\alpha_j)} \left(\int_x^b (g(x) - g(t))^{j-\alpha_j-1} g'(t) dt \right) R_p \left(\frac{b-a}{2n}\right)^{p-j} \omega_1 \left(f^{(p)}, \frac{b-a}{2n}\right)$$

$$\overset{(21.1.28)}{=} \frac{(g(b) - g(x))^{j-\alpha_j}}{\Gamma(j-\alpha_j+1)} R_p \left(\frac{b-a}{2n}\right)^{p-j} \omega_1 \left(f^{(p)}, \frac{b-a}{2n}\right) \leq$$

$$\frac{(g(b) - g(a))^{j-\alpha_j}}{\Gamma(j-\alpha_j+1)} R_p \left(\frac{b-a}{2n}\right)^{p-j} \omega_1 \left(f^{(p)}, \frac{b-a}{2n}\right). \quad (21.2.8)$$

Hence $\forall\, x \in [a, b]$, it holds

$$\left| \left(D_{b-;g}^{\alpha_j} f\right)(x) - D_{b-;g}^{\alpha_j} Q_n^*(x) \right| \leq$$

$$\frac{(g(b) - g(a))^{j-\alpha_j}}{\Gamma(j-\alpha_j+1)} R_p \left(\frac{b-a}{2n}\right)^{p-j} \omega_1 \left(f^{(p)}, \frac{b-a}{2n}\right), \quad (21.2.9)$$

and

$$\max_{x\in[a,b]}\left|D_{b-;g}^{\alpha_j}f(x)-D_{b-;g}^{\alpha_j}Q_n^*(x)\right|\le$$

$$\frac{(g(b)-g(a))^{j-\alpha_j}}{\Gamma(j-\alpha_j+1)}R_p\left(\frac{b-a}{2n}\right)^{p-j}\omega_1\left(f^{(p)},\frac{b-a}{2n}\right),\tag{21.2.10}$$

$j=0,1,\ldots,p$.

Above we set $D_{b-;g}^0f(x)=f(x)$, $D_{b-;g}^0Q_n^*(x)=Q_n^*(x)$, $\forall\,x\in[a,b]$, and $\alpha_0=0$, i.e. $\lceil\alpha_0\rceil=0$.

Put

$$s_j=\sup_{a\le x\le b}\left|\alpha_h^{-1}(x)\alpha_j(x)\right|,\qquad j=h,\ldots,k,\tag{21.2.11}$$

and

$$\eta_n=R_p\omega_1\left(f^{(p)},\frac{b-a}{2n}\right)\left(\sum_{j=h}^k s_j\frac{(g(b)-g(a))^{j-\alpha_j}}{\Gamma(j-\alpha_j+1)}\left(\frac{b-a}{2n}\right)^{p-j}\right).\tag{21.2.12}$$

I. Suppose, throughout $\left[a,g^{-1}(g(b)-1)\right]$, $\alpha_h(x)\ge\alpha^*>0$. Let $Q_n(x)$, $x\in[a,b]$, be a real polynomial of degree $\le n$, according to Theorem 21.11 and (21.2.10), so that

$$\max_{x\in[a,b]}\left|D_{b-;g}^{\alpha_j}\left(f(x)+\eta_n(h!)^{-1}x^h\right)-\left(D_{b-;g}^{\alpha_j}Q_n\right)(x)\right|\le\tag{21.2.13}$$

$$\frac{(g(b)-g(a))^{j-\alpha_j}}{\Gamma(j-\alpha_j+1)}R_p\left(\frac{b-a}{2n}\right)^{p-j}\omega_1\left(f^{(p)},\frac{b-a}{2n}\right),$$

$j=0,1,\ldots,p$.

In particular $(j=0)$ holds

$$\max_{x\in[a,b]}\left|\left(f(x)+\eta_n(h!)^{-1}x^h\right)-Q_n(x)\right|\le R_p\left(\frac{b-a}{2n}\right)^p\omega_1\left(f^{(p)},\frac{b-a}{2n}\right),\tag{21.2.14}$$

and

$$\max_{x\in[a,b]}\left|f(x)-Q_n(x)\right|\le$$

$$\eta_n(h!)^{-1}\left(\max\left(|a|,|b|\right)\right)^h+R_p\left(\frac{b-a}{2n}\right)^p\omega_1\left(f^{(p)},\frac{b-a}{2n}\right)=$$

$$\eta_n(h!)^{-1}\max\left(|a|^h,|b|^h\right)+R_p\left(\frac{b-a}{2n}\right)^p\omega_1\left(f^{(p)},\frac{b-a}{2n}\right)=\tag{21.2.15}$$

$$R_p \omega_1 \left(f^{(p)}, \frac{b-a}{2n} \right) \cdot$$

$$\left(\sum_{j=h}^{k} s_j \frac{(g(b) - g(a))^{j-\alpha_j}}{\Gamma(j - \alpha_j + 1)} \left(\frac{b-a}{2n} \right)^{p-j} \right) (h!)^{-1} \max \left(|a|^h, |b|^h \right)$$

$$+ R_p \left(\frac{b-a}{2n} \right)^p \omega_1 \left(f^{(p)}, \frac{b-a}{2n} \right) \le$$

$$R_p \omega_1 \left(f^{(p)}, \frac{b-a}{2n} \right) n^{k-p} \cdot$$

$$\left[\left(\sum_{j=h}^{k} s_j \frac{(g(b) - g(a))^{j-\alpha_j}}{\Gamma(j - \alpha_j + 1)} \left(\frac{b-a}{2} \right)^{p-j} \right) (h!)^{-1} \max \left(|a|^h, |b|^h \right) + \left(\frac{b-a}{2} \right)^p \right].$$

$$(21.2.16)$$

We have found that

$$\max_{x \in [a,b]} |f(x) - Q_n(x)| \le R_p \left[\left(\frac{b-a}{2} \right)^p + (h!)^{-1} \max \left(|a|^h, |b|^h \right) \cdot \right.$$

$$\left. \left(\sum_{j=h}^{k} s_j \frac{(g(b) - g(a))^{j-\alpha_j}}{\Gamma(j - \alpha_j + 1)} \left(\frac{b-a}{2} \right)^{p-j} \right) \right] n^{k-p} \omega_1 \left(f^{(p)}, \frac{b-a}{2n} \right), \quad (21.2.17)$$

proving (21.2.4).

Notice for $j = h+1, \ldots, k$, that

$$\left(D_{b-;g}^{\alpha_j} x^h \right) = \frac{(-1)^j}{\Gamma(j - \alpha_j)} \int_x^b (g(t) - g(x))^{j-\alpha_j-1} g'(t) \left(t^h \right)^{(j)} dt = 0.$$

$$(21.2.18)$$

Here

$$L = \sum_{j=h}^{k} \alpha_j(x) \left[D_{b-;g}^{\alpha_j} \right],$$

and suppose, throughout $[a, g^{-1}(g(b) - 1)]$, $Lf \ge 0$. So over $a \le x \le g^{-1}(g(b) - 1)$, we get

$$\alpha_h^{-1}(x) L(Q_n(x)) \overset{(21.2.18)}{=} \alpha_h^{-1}(x) L(f(x)) + \frac{\eta_n}{h!} \left(D_{b-;g}^{\alpha_h}(x^h) \right) +$$

$$\sum_{j=h}^{k} \alpha_h^{-1}(x) \alpha_j(x) \left[D_{b-;g}^{\alpha_j} Q_n(x) - D_{b-;g}^{\alpha_j} f(x) - \frac{\eta_n}{h!} D_{b-;g}^{\alpha_j} x^h \right] \overset{(21.2.13)}{\ge} \quad (21.2.19)$$

$$\frac{\eta_n}{h!}\left(D_{b-;g}^{\alpha_h}\left(x^h\right)\right) -$$

$$\left(\sum_{j=h}^{k} s_j \frac{(g\,(b) - g\,(a))^{j-\alpha_j}}{\Gamma\,(j - \alpha_j + 1)}\left(\frac{b-a}{2n}\right)^{p-j}\right) R_p \omega_1\left(f^{(p)}, \frac{b-a}{2n}\right) \overset{(21.2.12)}{=}$$

$$(21.2.20)$$

$$\frac{\eta_n}{h!}\left(D_{b-;g}^{\alpha_h}\left(x^h\right)\right) - \eta_n = \eta_n\left(\frac{D_{b-;g}^{\alpha_h}\left(x^h\right)}{h!} - 1\right) = \qquad (21.2.21)$$

$$\eta_n\left(\frac{1}{\Gamma\,(h - \alpha_h)\,h!}\int_x^b (g\,(t) - g\,(x))^{h-\alpha_h-1}\, g'\,(t)\left(t^h\right)^{(h)}\, dt - 1\right) =$$

$$\eta_n\left(\frac{h!}{h!\Gamma\,(h - \alpha_h)}\int_x^b (g\,(t) - g\,(x))^{h-\alpha_h-1}\, g'\,(t)\, dt - 1\right) \overset{(21.1.28)}{=}$$

$$\eta_n\left(\frac{(g\,(b) - g\,(x))^{h-\alpha_h}}{\Gamma\,(h - \alpha_h + 1)} - 1\right) = \qquad (21.2.22)$$

$$\eta_n\left(\frac{(g\,(b) - g\,(x))^{h-\alpha_h} - \Gamma\,(h - \alpha_h + 1)}{\Gamma\,(h - \alpha_h + 1)}\right) \geq$$

$$\eta_n\left(\frac{1 - \Gamma\,(h - \alpha_h + 1)}{\Gamma\,(h - \alpha_h + 1)}\right) \geq 0. \qquad (21.2.23)$$

Clearly here $g\,(b) - g\,(x) \geq 1$.

Hence

$$L\,(Q_n\,(x)) \geq 0, \quad \text{for } x \in \left[a, g^{-1}\,(g\,(b) - 1)\right]. \qquad (21.2.24)$$

A further explanation follows: We know $\Gamma\,(1) = 1$, $\Gamma\,(2) = 1$, and Γ is convex and positive on $(0, \infty)$. Here $0 \leq h - \alpha_h < 1$ and $1 \leq h - \alpha_h + 1 < 2$. Thus

$$\Gamma\,(h - \alpha_h + 1) \leq 1 \text{ and } 1 - \Gamma\,(h - \alpha_h + 1) \geq 0. \qquad (21.2.25)$$

II. Suppose, throughout $\left[a, g^{-1}\,(g\,(b) - 1)\right]$, $\alpha_h\,(x) \leq \beta^* < 0$.

Let $Q_n\,(x), x \in [a, b]$ be a real polynomial of degree $\leq n$, according to Theorem 21.11 and (21.2.10), so that

$$\max_{x\in[a,b]}\left|D_{b-;g}^{\alpha_j}\left(f\,(x) - \eta_n\,(h!)^{-1}\, x^h\right) - \left(D_{b-;g}^{\alpha_j} Q_n\right)(x)\right| \leq \qquad (21.2.26)$$

$$\frac{(g\,(b) - g\,(a))^{j-\alpha_j}}{\Gamma\,(j - \alpha_j + 1)} R_p\left(\frac{b-a}{2n}\right)^{p-j} \omega_1\left(f^{(p)}, \frac{b-a}{2n}\right),$$

$j = 0, 1, \ldots, p$.

In particular ($j = 0$) holds

$$\max_{x\in[a,b]} \left|(f(x) - \eta_n (h!)^{-1} x^h) - Q_n(x)\right| \leq R_p \left(\frac{b-a}{2n}\right)^p \omega_1 \left(f^{(p)}, \frac{b-a}{2n}\right),$$
(21.2.27)

and

$$\max_{x\in[a,b]} |f(x) - Q_n(x)| \leq$$

$$\eta_n (h!)^{-1} (\max(|a|, |b|))^h + R_p \left(\frac{b-a}{2n}\right)^p \omega_1 \left(f^{(p)}, \frac{b-a}{2n}\right) =$$

$$\eta_n (h!)^{-1} \max(|a|^h, |b|^h) + R_p \left(\frac{b-a}{2n}\right)^p \omega_1 \left(f^{(p)}, \frac{b-a}{2n}\right),$$
(21.2.28)

etc.

We find again that

$$\max_{x\in[a,b]} |f(x) - Q_n(x)| \leq R_p \left[\left(\frac{b-a}{2}\right)^p + (h!)^{-1} \max(|a|^h, |b|^h) \cdot\right.$$

$$\left.\left(\sum_{j=h}^{k} s_j \frac{(g(b) - g(a))^{j-\alpha_j}}{\Gamma(j - \alpha_j + 1)} \left(\frac{b-a}{2}\right)^{p-j}\right)\right] n^{k-p} \omega_1 \left(f^{(p)}, \frac{b-a}{2n}\right),$$
(21.2.29)

reproving (21.2.4).

Here again

$$L = \sum_{j=h}^{k} \alpha_j(x) \left[D_{b-;g}^{\alpha_j}\right],$$

and suppose, throughout $[a, g^{-1}(g(b) - 1)]$, $Lf \geq 0$. So over $a \leq x \leq g^{-1}(g(b) - 1)$, we get

$$\alpha_h^{-1}(x) L(Q_n(x)) \overset{(21.2.54)}{=} \alpha_h^{-1}(x) L(f(x)) - \frac{\eta_n}{h!} \left(D_{b-;g}^{\alpha_h}(x^h)\right) +$$

$$\sum_{j=h}^{k} \alpha_h^{-1}(x) \alpha_j(x) \left[D_{b-;g}^{\alpha_j} Q_n(x) - D_{b-;g}^{\alpha_j} f(x) + \frac{\eta_n}{h!} D_{b-;g}^{\alpha_j} x^h\right] \overset{(21.2.26)}{\leq}$$
(21.2.30)

$$-\frac{\eta_n}{h!} \left(D_{b-;g}^{\alpha_h}(x^h)\right) +$$

$$\left(\sum_{j=h}^{k} s_j \frac{(g(b) - g(a))^{j-\alpha_j}}{\Gamma(j - \alpha_j + 1)} \left(\frac{b-a}{2n} \right)^{p-j} \right) R_p \omega_1 \left(f^{(p)}, \frac{b-a}{2n} \right) \overset{(21.2.12)}{=}$$

$$(21.2.31)$$

$$-\frac{\eta_n}{h!} \left(D_{b-;g}^{\alpha_h} \left(x^h \right) \right) + \eta_n = \eta_n \left(1 - \frac{D_{b-;g}^{\alpha_h} \left(x^h \right)}{h!} \right) = \qquad (21.2.32)$$

$$\eta_n \left(1 - \frac{1}{\Gamma(h - \alpha_h) h!} \int_x^b (g(t) - g(x))^{h-\alpha_h-1} g'(t) \left(t^h \right)^{(h)} dt \right) =$$

$$\eta_n \left(1 - \frac{h!}{h! \Gamma(h - \alpha_h)} \int_x^b (g(t) - g(x))^{h-\alpha_h-1} g'(t) dt \right) \overset{(21.1.28)}{=}$$

$$\eta_n \left(1 - \frac{(g(b) - g(x))^{h-\alpha_h}}{\Gamma(h - \alpha_h + 1)} \right) = \qquad (21.2.33)$$

$$\eta_n \left(\frac{\Gamma(h - \alpha_h + 1) - (g(b) - g(x))^{h-\alpha_h}}{\Gamma(h - \alpha_h + 1)} \right) \overset{(21.2.25)}{\leq}$$

$$\eta_n \left(\frac{1 - (g(b) - g(x))^{h-\alpha_h}}{\Gamma(h - \alpha_h + 1)} \right) \leq 0. \qquad (21.2.34)$$

Hence again

$$L(Q_n(x)) \geq 0, \quad \forall x \in \left[a, g^{-1}(g(b) - 1) \right].$$

The case of $\alpha_h = h$ is trivially concluded from the above. The proof of the theorem is now over. $\square$

We make

Remark 21.15 By Theorem 21.13 we have that $D_{b-;g}^{\alpha_j} f$ are continuous functions, $j = 0, 1, \ldots, p$. Suppose that $\alpha_h(x), \ldots, \alpha_k(x)$ are continuous functions on $[a, b]$, and $L(f) \geq 0$ on $\left[a, g^{-1}(g(b) - 1) \right]$ is replaced by $L(f) > 0$ on $\left[a, g^{-1}(g(b) - 1) \right]$. Disregard the assumption made in the main theorem on $\alpha_h(x)$. For $n \in \mathbb{N}$, let $Q_n(x)$ be the $Q_n^*(x)$ of Theorem 21.11, and f as in Theorem 21.11 (same as in Theorem 21.14). Then $Q_n(x)$ converges to $f(x)$ at the Jackson rate $\frac{1}{n^{p+1}}$ ([7], p. 18, Theorem VIII) and at the same time, since $L(Q_n)$ converges uniformly to $L(f)$ on $[a, b]$, $L(Q_n) > 0$ on $\left[a, g^{-1}(g(b) - 1) \right]$ for all n sufficiently large.

21.3   Applications (to Theorem 21.14)

(1) When $g(x) = \ln x$ on $[a, b]$, $0 < a < b < \infty$.
Here we would assume that $b > ae$, $\alpha_h(x)$ restriction true on $\left[a, \frac{b}{e}\right]$, and

$$Lf = \sum_{j=h}^{k} \alpha_j(x) \left[D_{b-;\ln x}^{\alpha_j} f\right] \geq 0, \qquad (21.3.1)$$

throughout $\left[a, \frac{b}{e}\right]$.
   Then $L(Q_n) \geq 0$ on $\left[a, \frac{b}{e}\right]$.
   (2) When $g(x) = e^x$ on $[a, b]$, $a < b < \infty$.
Here we assume that $b > \ln(1 + e^a)$, $\alpha_h(x)$ restriction true on $\left[a, \ln(e^b - 1)\right]$, and

$$Lf = \sum_{j=h}^{k} \alpha_j(x) \left[D_{b-;e^x}^{\alpha_j} f\right] \geq 0, \qquad (21.3.2)$$

throughout $\left[a, \ln(e^b - 1)\right]$.
   Then $L(Q_n) \geq 0$ on $\left[a, \ln(e^b - 1)\right]$.
   (3) When, $A > 1$, $g(x) = A^x$ on $[a, b]$, $a < b < \infty$.
Here we assume that $b > \log_A(1 + A^a)$, $\alpha_h(x)$ restriction true on $\left[a, \log_A(A^b - 1)\right]$, and

$$Lf = \sum_{j=h}^{k} \alpha_j(x) \left[D_{b-;A^x}^{\alpha_j} f\right] \geq 0, \qquad (21.3.3)$$

throughout $\left[a, \log_A(A^b - 1)\right]$.
   Then $L(Q_n) \geq 0$ on $\left[a, \log_A(A^b - 1)\right]$.
   (4) When $\sigma > 0$, $g(x) = x^\sigma$, $0 \leq a < b < \infty$.
Here we assume that $b > (1 + a^\sigma)^{\frac{1}{\sigma}}$, $\alpha_h(x)$ restriction true on $\left[a, (b^\sigma - 1)^{\frac{1}{\sigma}}\right]$, and

$$Lf = \sum_{j=h}^{k} \alpha_j(x) \left[D_{b-;x^\sigma}^{\alpha_j} f\right] \geq 0 \qquad (21.3.4)$$

throughout $\left[a, (b^\sigma - 1)^{\frac{1}{\sigma}}\right]$.
   Then $L(Q_n) \geq 0$ on $\left[a, (b^\sigma - 1)^{\frac{1}{\sigma}}\right]$.

References

1. G.A. Anastassiou, Right fractional monotone approximation. J. Appl. Funct. Anal. **10**(1–2), 117–124 (2015)
2. G.A. Anastassiou, *Left General Fractional Monotone Approximation Theory* (submitted) (2015)

3. G. Anastassiou, *Right General Fractional Monotone Approximation* (submitted) (2015)
4. G.A. Anastassiou, O. Shisha, Monotone approximation with linear differential operators. J. Approx. Theory **44**, 391–393 (1985)
5. T. Apostol, *Mathematical Analysis* (Addison-Wesley Publ. Co., Reading, 1969)
6. A.M.A. El-Sayed, M. Gaber, On the finite Caputo and finite Riesz derivatives. Electr. J. Theor. Phys. **3**(12), 81–95 (2006)
7. D. Jackson, *The Theory of Approximation*. American Mathematical Society Colloquium, vol. XI (New York, 1930)
8. R.-Q. Jia, *Chapter 3. Absolutely Continous Functions*, https://www.ualberta.ca/~rjia/Math418/ Notes/Chap.3.pdf
9. A.A. Kilbas, H.M. Srivastava, J.J. Trujillo, *Theory and Applications of Fractional Differential Equations*. North-Holland Mathematics Studies, vol. 204 (Elsevier, New York, 2006)
10. H.L. Royden, *Real Analysis*, 2nd edn. (Macmillan Publishing Co., Inc., New York, 1968)
11. A.R. Schep, *Differentiation of Monotone Functions*, http://people.math.sc.edu/schep/ diffmonotone.pdf
12. O. Shisha, Monotone approximation. Pacific J. Math. **15**, 667–671 (1965)
13. S.A. Teljakovskii, Two theorems on the approximation of functions by algebraic polynomials. Mat. Sb. **70**(112), 252–265 (1966)
14. S.A. Teljakovskii, Two theorems on the approximation of functions by algebraic polynomials. Mat. Sb. **70**(112), 252–265 (1966) (Russian); Amer. Math. Soc. Trans. **77**(2), 163–178 (1968)
15. R.M. Trigub, Approximation of functions by polynomials with integer coeficients. Izv. Akad. Nauk SSSR Ser. Mat. **26**, 261–280 (1962) (Russian)

Chapter 22
Left Generalized High Order Fractional Monotone Approximation

Here are used the left general fractional derivatives Caputo style with respect to a base absolutely continuous strictly increasing function g. We mention various examples of such fractional derivatives for different g. Let f be r-times continuously differentiable function on $[a, b]$, and let L be a linear left general fractional differential operator such that $L(f)$ is non-negative over a critical closed subinterval I of $[a, b]$. We can find a sequence of polynomials Q_n of degree less-equal n such that $L(Q_n)$ is non-negative over I, furthermore f is fractionally and simultaneously approximated uniformly by Q_n over $[a, b]$.

The degree of this constrained approximation is given by inequalities using the high order modulus of smoothness of $f^{(r)}$. We finish with applications of the main fractional monotone approximation theorem for different g. It follows [6].

22.1 Introduction

The topic of monotone approximation started in [14] has become a major trend in approximation theory. A typical problem in this subject is: given a positive integer k, approximate a given function whose kth derivative is ≥ 0 by polynomials having this property.

In [5] the authors replaced the kth derivative with a linear differential operator of order k.

Furthermore in [4], the author generalized the result of [5] for linear fractional differential operators.

To describe the motivating result here we need:

Definition 22.1 ([8], p. 50) Let $\alpha > 0$ and $\lceil \alpha \rceil = m$, ($\lceil \cdot \rceil$ ceiling of the number). Consider $f \in C^m([a, b])$, $a < b$. We define the left Caputo fractional derivative of f of order α as follows:

© Springer International Publishing Switzerland 2016

G.A. Anastassiou and I.K. Argyros, *Intelligent Numerical Methods:*
Applications to Fractional Calculus, Studies in Computational Intelligence 624,
DOI 10.1007/978-3-319-26721-0_22

$$\left(D_{*a}^{\alpha} f\right)(x) = \frac{1}{\Gamma(m-\alpha)} \int_{a}^{x} (x-t)^{m-\alpha-1} f^{(m)}(t)\, dt, \qquad (22.1.1)$$

for any $x \in [a, b]$, where Γ is the gamma function $\Gamma(\nu) = \int_{0}^{\infty} e^{-t} t^{\nu-1} dt, \nu > 0$.
    We set

$$D_{*a}^{0} f(x) = f(x), \qquad (22.1.2)$$

$$D_{*a}^{m} f(x) = f^{(m)}(x), \quad \forall\, x \in [a, b]. \qquad (22.1.3)$$

    We proved:

Theorem 22.2 ([4]) *Let h, v, r be integers, $1 \le h \le v \le r$ and let $f \in C^{r}([-1, 1])$, with $f^{(r)}$ having modulus of smoothness $\omega_{s}\left(f^{(r)}, \delta\right)$ there, $s \ge 1$. Let $\alpha_{j}(x)$, $j = h, h+1, \ldots, v$ be real functions, defined and bounded on $[-1, 1]$ and suppose $\alpha_{h}(x)$ is either $\ge \alpha > 0$ or $\le \beta < 0$ on $[0, 1]$. Let the real numbers $\alpha_{0} = 0 < \alpha_{1} \le 1 < \alpha_{2} \le 2 < \cdots < \alpha_{r} \le r$. Here $D_{*-1}^{\alpha_{j}} f$ stands for the left Caputo fractional derivative of f of order α_{j} anchored at -1. Consider the linear left fractional differential operator*

$$L^{*} := \sum_{j=h}^{v} \alpha_{j}(x)\left[D_{*-1}^{\alpha_{j}}\right] \qquad (22.1.4)$$

and suppose, throughout $[0, 1]$,

$$L^{*}(f) \ge 0. \qquad (22.1.5)$$

    Then, for any $n \in \mathbb{N}$ such that $n \ge \max(4(r+1), r+s)$, there exists a real polynomial $Q_{n}(x)$ of degree $\le n$ such that

$$L^{*}(Q_{n}) \ge 0 \quad throughout\ [0, 1], \qquad (22.1.6)$$

and

$$\sup_{-1 \le x \le 1} \left|\left(D_{*-1}^{\alpha_{j}} f\right)(x) - \left(D_{*-1}^{\alpha_{j}} Q_{n}\right)(x)\right| \le$$

$$\frac{2^{j-\alpha_{j}}}{\Gamma(j-\alpha_{j}+1)} \frac{C_{r,s}}{n^{r-j}} \omega_{s}\left(f^{(r)}, \frac{1}{n}\right), \qquad (22.1.7)$$

$j = h+1, \ldots, r$; $C_{r,s}$ *is a constant independent of f and n.*
    Set

$$l_{j} := \sup_{x \in [-1,1]} \left|\alpha_{h}^{-1}(x)\,\alpha_{j}(x)\right|, \quad h \le j \le v. \qquad (22.1.8)$$

When $j = 1, \ldots, h$ we derive

$$\sup_{-1\leq x\leq 1}\left|\left(D_{*-1}^{\alpha_j}f\right)(x)-\left(D_{*-1}^{\alpha_j}Q_n\right)(x)\right|\leq\frac{C_{r,s}}{n^{r-v}}\omega_s\left(f^{(r)},\frac{1}{n}\right)\cdot$$

$$\left[\left(\sum_{\tau=h}^{v}l_{\tau}\frac{2^{\tau-\alpha_{\tau}}}{\Gamma\left(\tau-\alpha_{\tau}+1\right)}\right)\left(\sum_{\lambda=0}^{h-j}\frac{2^{h-\alpha_j-\lambda}}{\lambda!\Gamma\left(h-\alpha_j-\lambda+1\right)}\right)+\frac{2^{j-\alpha_j}}{\Gamma\left(j-\alpha_j+1\right)}\right].$$

$$(22.1.9)$$

Finally it holds

$$\sup_{-1\leq x\leq 1}|f(x)-Q_n(x)|\leq$$

$$\frac{C_{r,s}}{n^{r-v}}\omega_s\left(f^{(r)},\frac{1}{n}\right)\left[\frac{1}{h!}\sum_{\tau=h}^{v}l_{\tau}\frac{2^{\tau-\alpha_{\tau}}}{\Gamma\left(\tau-\alpha_{\tau}+1\right)}+1\right].\qquad(22.1.10)$$

In this chapter we extend Theorem 22.2 to much more general linear left fractional differential operators.

We use a lot here the following generalised fractional integral.

Definition 22.3 (*see also* [11, p. 99]) The left generalised fractional integral of a function f with respect to given function g is defined as follows:

Let $a,b\in\mathbb{R}$, $a<b$, $\alpha>0$. Here $g\in AC([a,b])$ (absolutely continuous functions) and is striclty increasing, $f\in L_{\infty}([a,b])$. We set

$$\left(I_{a+;g}^{\alpha}f\right)(x)=\frac{1}{\Gamma(\alpha)}\int_{a}^{x}(g(x)-g(t))^{\alpha-1}g'(t)f(t)dt,\quad x\geq a,\quad(22.1.11)$$

clearly $\left(I_{a+;g}^{\alpha}f\right)(a)=0$.

When g is the identity function id, we get that $I_{a+;id}^{\alpha}=I_{a+}^{\alpha}$, the ordinary left Riemann-Liouville fractional integral, where

$$\left(I_{a+}^{\alpha}f\right)(x)=\frac{1}{\Gamma(\alpha)}\int_{a}^{x}(x-t)^{\alpha-1}f(t)dt,\quad x\geq a,\qquad(22.1.12)$$

$\left(I_{a+}^{\alpha}f\right)(a)=0$.

When $g(x)=\ln x$ on $[a,b]$, $0<a<b<\infty$, we get:

Definition 22.4 ([11, p. 110]) Let $0<a<b<\infty$, $\alpha>0$. The left Hadamard fractional integral of order α is given by

$$\left(J_{a+}^{\alpha}f\right)(x)=\frac{1}{\Gamma(\alpha)}\int_{a}^{x}\left(\ln\frac{x}{y}\right)^{\alpha-1}\frac{f(y)}{y}dy,\quad x\geq a,\qquad(22.1.13)$$

where $f\in L_{\infty}([a,b])$.

We mention:

Definition 22.5 The left fractional exponential integral is defined as follows: Let $a, b \in \mathbb{R}, a < b, \alpha > 0, f \in L_\infty ([a, b])$. We set

$$\left(I^\alpha_{a+;e^x} f\right)(x) = \frac{1}{\Gamma(\alpha)} \int_a^x \left(e^x - e^t\right)^{\alpha-1} e^t f(t) \, dt, \quad x \geq a. \qquad (22.1.14)$$

Definition 22.6 Let $a, b \in \mathbb{R}, a < b, \alpha > 0, f \in L_\infty ([a, b]), A > 1$. We introduce the fractional integral

$$\left(I^\alpha_{a+;A^x} f\right)(x) = \frac{\ln A}{\Gamma(\alpha)} \int_a^x \left(A^x - A^t\right)^{\alpha-1} A^t f(t) \, dt, \quad x \geq a. \qquad (22.1.15)$$

We also give:

Definition 22.7 Let $\alpha, \sigma > 0, 0 \leq a < b < \infty, f \in L_\infty ([a, b])$. We set

$$\left(K^\alpha_{a+;x^\sigma} f\right)(x) = \frac{1}{\Gamma(\alpha)} \int_a^x (x^\sigma - t^\sigma)^{\alpha-1} f(t) \, \sigma t^{\sigma-1} dt, \quad x \geq a. \qquad (22.1.16)$$

We introduce the following general fractional derivatives:

Definition 22.8 Let $\alpha > 0$ and $\lceil \alpha \rceil = m$. Consider $f \in AC^m ([a, b])$ (space of functions f with $f^{(m-1)} \in AC ([a, b])$). We define the left general fractional derivative of f of order α as follows

$$\left(D^\alpha_{*a;g} f\right)(x) = \frac{1}{\Gamma(m - \alpha)} \int_a^x (g(x) - g(t))^{m-\alpha-1} g'(t) f^{(m)}(t) \, dt, \qquad (22.1.17)$$

for any $x \in [a, b]$.
We set

$$D^m_{*\alpha;g} f(x) = f^{(m)}(x), \qquad (22.1.18)$$

$$D^0_{*a;g} f(x) = f(x) \quad \forall \, x \in [a, b]. \qquad (22.1.19)$$

When $g = id$, then $D^\alpha_{*a} f = D^\alpha_{*a;id} f$ is the left Caputo fractional derivative.

So we have the specific general left fractional derivatives.

Definition 22.9

$$D^\alpha_{*a;\ln x} f(x) = \frac{1}{\Gamma(m - \alpha)} \int_a^x \left(\ln \frac{x}{y}\right)^{m-\alpha-1} \frac{f^{(m)}(y)}{y} dy, \quad x \geq a > 0,$$
$$(22.1.20)$$

$$D^\alpha_{*a;e^x} f(x) = \frac{1}{\Gamma(m - \alpha)} \int_a^x \left(e^x - e^t\right)^{m-\alpha-1} e^t f^{(m)}(t) \, dt, \quad x \geq a, \qquad (22.1.21)$$

and

$$D_{*a;A^x}^\alpha f(x) = \frac{\ln A}{\Gamma(m-\alpha)} \int_a^x \left(A^x - A^t\right)^{m-\alpha-1} A^t f^{(m)}(t)\, dt, \quad x \geq a, \quad (22.1.22)$$

$$\left(D_{*a;x^\sigma}^\alpha f\right)(x) = \frac{1}{\Gamma(m-\alpha)} \int_a^x (x^\sigma - t^\sigma)^{m-\alpha-1} \sigma t^{\sigma-1} f^{(m)}(t)\, dt, \quad x \geq a \geq 0.$$
$$(22.1.23)$$

We need:

Definition 22.10 For $g \in C([-1,1])$ we define

$$\Delta_h^s(g,t) := \sum_{k=0}^s \binom{s}{k} (-1)^{s-k} g(t+kh), \quad (22.1.24)$$

$$A_{sh} := [-1, 1-sh] \ni t. \quad (22.1.25)$$

The s-th modulus of smoothness of g (see [7], p. 44) is defined as

$$\omega_s(g,z) := \sup_{0 < h \leq z} \left\| \Delta_h^s(g,\cdot) \right\|_{\infty, A_{sh}}, \quad z \geq 0. \quad (22.1.26)$$

A similar definition is valid for the arbitrary $f \in C([a,b])$.

In [4] we proved that (see also [9]):

Corollary 22.11 *Let $r \geq 0$ and $s \geq 1$. Then there exists a sequence $Q_n = Q_n^{(r,s)}$ of linear polynomial operators mapping $C^r([-1,1])$ into P_n (space of polynomials of degree $\leq n$), such that for all $g \in C^r([-1,1])$ and all $n \geq \max(4(r+1), r+s)$ we have*

$$\left| g^{(k)}(t) - (Q_n(g))^{(k)}(t) \right| \leq$$

$$\left\| g^{(k)} - (Q_n g)^{(k)} \right\|_{\infty, [-1,1]} \leq \frac{C_{r,s}}{n^{r-k}} \omega_s\left(g^{(r)}, \frac{1}{n}\right), \quad k = 0, 1, \ldots, r, \quad (22.1.27)$$

where $C_{r,s}$ is a constant independent of g and n, for every $t \in [-1,1]$.

We extend the last Corollary from $[-1,1]$ to the arbitrary $[a,b]$. We will establish:

Theorem 22.12 *Let $r \geq 0$ and $s \geq 1$. Then there exists a sequence $Q_n^* = Q_n^{*(r,s)}$ of linear polynomial operators mapping $C^r([a,b])$ into P_n, such that for all $f \in C^r([a,b])$ and all $n \geq \max(4(r+1), r+s)$ we have*

$$\left\| f^{(k)} - (Q_n^*(f))^{(k)} \right\|_{\infty, [a,b]} \leq C_{r,s} \left(\frac{b-a}{2n}\right)^{r-k} \omega_s\left(f^{(r)}, \frac{b-a}{2n}\right), \quad (22.1.28)$$

$k = 0, 1, \ldots, r$, *where the constant $C_{r,s}$ is independent of n or f.*

Proof Let here $t \in [-1, 1]$, $x \in [a, b]$, $a < b$. Let the map $\varphi : [-1, 1] \to [a, b]$ defined by

$$x = \varphi(t) = \left(\frac{b-a}{2}\right) t + \left(\frac{b+a}{2}\right). \tag{22.1.29}$$

Clearly here φ is an $1 - 1$ and onto map.

We get

$$x' = \varphi'(t) = \frac{b-a}{2}, \tag{22.1.30}$$

and

$$t = \frac{2x - b - a}{b - a} = 2\left(\frac{x}{b-a}\right) - \left(\frac{b+a}{b-a}\right).$$

In fact it holds

$$\varphi(-1) = a, \quad \text{and } \varphi(1) = b. \tag{22.1.31}$$

Clearly here it holds

$$g(t) := f\left(\left(\frac{b-a}{2}\right) t + \left(\frac{b+a}{2}\right)\right) \in C^r([-1, 1]), \quad \text{all } t \in [-1, 1]. \tag{22.1.32}$$

We easily get that

$$g^{(k)}(t) = f^{(k)}(x) \frac{(b-a)^k}{2^k}, \quad k = 0, 1, \ldots, r. \tag{22.1.33}$$

Next we apply (22.1.27) to above g: so far we have

$$\left|g^{(k)}(t) - (Q_n(g))^{(k)}(t)\right| = \left|f^{(k)}(x) \frac{(b-a)^k}{2^k} - (Q_n(g))^{(k)}(t)\right| =: (*). \tag{22.1.34}$$

But for $t \in [-1, 1]$, we have that

$$(Q_n(g))(t) = (Q_n(g))\left(\left(\frac{2x}{b-a}\right) - \left(\frac{b+a}{b-a}\right)\right) =: Q_n^*(x), \quad x \in [a, b], \tag{22.1.35}$$

where $Q_n^* \in P_n$.

One can prove easily that

$$(Q_n(g))^{(k)}(t) = \left(Q_n^*\right)^{(k)}(x) \left(\frac{b-a}{2}\right)^k, \quad k = 0, 1, \ldots, r. \tag{22.1.36}$$

Hence it holds

$$(*) = \frac{(b-a)^k}{2^k} \left|f^{(k)}(x) - \left(Q_n^*\right)^{(k)}(x)\right|. \tag{22.1.37}$$

That is

$$\left| g^{(k)}(t) - (Q_n(g))^{(k)}(t) \right| = \frac{(b-a)^k}{2^k} \left| f^{(k)}(x) - (Q_n^*)^{(k)}(x) \right|, \quad k = 0, 1, \ldots, r,$$
(22.1.38)

for any $t \in [-1, 1]$ and the corresponding $x \in [a, b]$.

Furthermore we see that

$$\omega_s \left(g^{(r)}, \frac{1}{n} \right) = \sup_{0 < h \le \frac{1}{n}} \left\| \sum_{k=0}^{s} \binom{s}{k} (-1)^{s-k} g^{(r)}(\cdot + kh) \right\|_{\infty, [-1, 1-sh]} = \quad (22.1.39)$$

$$\left(\frac{b-a}{2} \right)^r .$$

$$\sup_{0 < h \le \frac{1}{n}} \left\| \sum_{k=0}^{s} \binom{s}{k} (-1)^{s-k} f^{(r)} \left(\left(\frac{b-a}{2} \right)(t+kh) + \left(\frac{b+a}{2} \right) \right) \right\|_{t, \infty, [-1, 1-sh]}$$

$$= \left(\frac{b-a}{2} \right)^r \sup_{0 < h^* \le \frac{b-a}{2n}} \left\| \sum_{k=0}^{s} \binom{s}{k} (-1)^{s-k} f^{(r)}(x + kh^*) \right\|_{x, \infty, [a, b-sh^*]}$$

(we denoted $h^* := \left(\frac{b-a}{2} \right) h$)

$$= \left(\frac{b-a}{2} \right)^r \omega_s \left(f^{(r)}, \frac{(b-a)}{2n} \right).$$
(22.1.40)

That is we have proved

$$\omega_s \left(g^{(r)}, \frac{1}{n} \right) = \left(\frac{b-a}{2} \right)^r \omega_s \left(f^{(r)}, \frac{(b-a)}{2n} \right).$$
(22.1.41)

Using (22.1.27), (22.1.38) and (22.1.41), $x \in [a, b]$, we obtain

$$\left(\frac{b-a}{2} \right)^k \left| f^{(k)}(x) - (Q_n^*)^{(k)}(x) \right| \le \frac{C_{r,s}}{n^{r-k}} \left(\frac{b-a}{2} \right)^r \omega_s \left(f^{(r)}, \frac{(b-a)}{2n} \right),$$
(22.1.42)

equivalently it holds

$$\left| f^{(k)}(x) - (Q_n^*)^{(k)}(x) \right| \le \frac{C_{r,s}}{n^{r-k}} \left(\frac{b-a}{2} \right)^{r-k} \omega_s \left(f^{(r)}, \frac{(b-a)}{2n} \right).$$
(22.1.43)

Thus we have proved that

$$\left| f^{(k)}(x) - \left(Q_n^* \right)^{(k)}(x) \right| \le C_{r,s} \left(\frac{b-a}{2n} \right)^{r-k} \omega_s \left(f^{(r)}, \frac{(b-a)}{2n} \right), \qquad (22.1.44)$$

for all $x \in [a, b]$, $k = 0, 1, \ldots, r$.

That is proving (22.1.28). $\qquad\qquad\qquad\qquad\qquad\qquad\qquad\qquad\qquad\qquad$ $\square$

Remark 22.13 Here $g \in AC([a, b])$ (absolutely continuous functions), g is increasing over $[a, b]$, $\alpha > 0$.

Let $g(a) = c$, $g(b) = d$. We want to calculate

$$I = \int_a^b (g(b) - g(t))^{\alpha-1} g'(t)\, dt. \qquad (22.1.45)$$

Consider the function

$$f(y) = (g(b) - y)^{\alpha-1} = (d - y)^{\alpha-1}, \quad \forall\, y \in [c, d]. \qquad (22.1.46)$$

We have that $f(y) \ge 0$, it may be $+\infty$ when $y = d$ and $0 < \alpha < 1$, but f is measurable on $[c, d]$. By [12], Royden, p. 107, Exercise 13d, we get that

$$(f \circ g)(t)\, g'(t) = (g(b) - g(t))^{\alpha-1} g'(t) \qquad (22.1.47)$$

is measurable on $[a, b]$, and

$$I = \int_c^d (d - y)^{\alpha-1}\, dy = \frac{(d - c)^\alpha}{\alpha} \qquad (22.1.48)$$

(notice that $(d - y)^{\alpha-1}$ is Riemann integrable).

That is

$$I = \frac{(g(b) - g(a))^\alpha}{\alpha}. \qquad (22.1.49)$$

Similarly it holds

$$\int_a^x (g(x) - g(t))^{\alpha-1} g'(t)\, dt = \frac{(g(x) - g(a))^\alpha}{\alpha}, \quad \forall\, x \in [a, b]. \qquad (22.1.50)$$

We use:

Theorem 22.14 *Let* $r > 0$, $a < b$, $F \in L_\infty([a, b])$, $g \in AC([a, b])$ *and* g *is strictly increasing.*

Consider

$$G(s) := \int_a^s (g(s) - g(t))^{r-1} g'(t) F(t) \, dt, \quad \text{for all } s \in [a, b]. \qquad (22.1.51)$$

Then $G \in C([a, b])$.

Proof There exists a Borel measurable function $F^* : [a, b] \to \mathbb{R}$ such that $F = F^*$, a.e., and it holds

$$G(s) = \int_a^s (g(s) - g(t))^{r-1} g'(t) F^*(t) \, dt, \quad \text{for all } s \in [a, b]. \qquad (22.1.52)$$

Notice that

$$\|F^*\|_\infty = \|F\|_\infty < \infty. \qquad (22.1.53)$$

We can write

$$G(s) = \int_a^s (g(s) - g(t))^{r-1} g'(t) \left(F^* \circ g^{-1} \right) (g(t)) \, dt, \quad \text{for all } s \in [a, b]. \qquad (22.1.54)$$

By [3], we get that

$$\left\| F^* \circ g^{-1} \right\|_{\infty, [g(a), g(b)]} \le \|F\|_\infty. \qquad (22.1.55)$$

Next we consider the function $(g(s) - z)^{r-1} \left(F^* \circ g^{-1} \right) (z)$, where $z \in [g(a), g(s)]$, the last is integrable over $[g(a), g(s)]$.

Since g is monotone, by [10], $(g(s) - g(t))^{r-1} g'(t) \left(F^* \circ g^{-1} \right) (g(t))$ is integrable on $[a, s]$.

Furthermore, again by [10], it holds

$$\lambda(g(s)) := \int_{g(a)}^{g(s)} (g(s) - z)^{r-1} \left(F^* \circ g^{-1} \right) (z) \, dz = \qquad (22.1.56)$$

$$\int_a^s (g(s) - g(t))^{r-1} g'(t) \left(F^* \circ g^{-1} \right) (g(t)) \, dt =$$

$$\int_a^s (g(s) - g(t))^{r-1} g'(t) F^*(t) \, dt = \qquad (22.1.57)$$

$$\int_a^s (g(s) - g(t))^{r-1} g'(t) F(t) \, dt = G(s).$$

That is

$$G\left(s\right) = \int_{g(a)}^{g(s)} \left(g\left(s\right) - z\right)^{r-1} \left(F^* \circ g^{-1}\right)\left(z\right) dz = \lambda\left(g\left(s\right)\right), \quad \text{all } s \in [a, b].$$

(22.1.58)

By [2], p. 388, we have the function

$$\lambda\left(y\right) = \int_{g(a)}^{y} \left(y - z\right)^{r-1} \left(F^* \circ g^{-1}\right)\left(z\right) dz, \qquad (22.1.59)$$

is continuous in y over $[g\left(a\right), g\left(b\right)]$.

Let now $s_n, s \in [a, b] : s_n \to s$, then $g\left(s_n\right) \to g\left(s\right)$, and $\lambda\left(g\left(s_n\right)\right) \to \lambda\left(g\left(s\right)\right)$, that is $G\left(s_n\right) \to G\left(s\right)$, as $n \to \infty$, proving the continuity of G. $\qquad \square$

We need:

Corollary 22.15 *Let* $g \in AC\left([a, b]\right)$ *and* g *is strictly increasing. Let* $\alpha > 0$, $\alpha \notin \mathbb{N}$, *and* $\lceil \alpha \rceil = m$, $f \in AC^m\left([a, b]\right)$ *with* $f^{(m)} \in L_\infty\left([a, b]\right)$. *Then* $\left(D_{*a;g}^{\alpha} f\right) \in C\left([a, b]\right)$.

Proof By Theorem 22.14. $\qquad \square$

22.2   Main Result

We present:

Theorem 22.16 *Here we assume that* $g\left(b\right) - g\left(a\right) > 1$. *Let* h, v, r *be integers,* $1 \leq h \leq v \leq r$ *and let* $f \in C^r\left([a, b]\right)$, $a < b$, *with* $f^{(r)}$ *having modulus of smoothness* $\omega_s\left(f^{(r)}, \delta\right)$ *there,* $s \geq 1$. *Let* $\alpha_j\left(x\right)$, $j = h, h + 1, \ldots, v$ *be real functions, defined and bounded on* $[a, b]$ *and suppose for* $x \in \left[g^{-1}\left(1 + g\left(a\right)\right), b\right]$ *that* $\alpha_h\left(x\right)$ *is either* $\geq \alpha^* > 0$ *or* $\leq \beta^* < 0$. *Let the real numbers* $\alpha_0 = 0 < \alpha_1 \leq 1 < \alpha_2 \leq 2 < \cdots < \alpha_r \leq r$. *Here* $D_{*a;g}^{\alpha_j} f$ *stands for the left general fractional derivative of* f *of order* α_j *anchored at* a. *Consider the linear left general fractional differential operator*

$$L := \sum_{j=h}^{v} \alpha_j\left(x\right) \left[D_{*a;g}^{\alpha_j}\right] \qquad (22.2.1)$$

and suppose, throughout $\left[g^{-1}\left(1 + g\left(a\right)\right), b\right]$,

$$L\left(f\right) \geq 0. \qquad (22.2.2)$$

Then, for any $n \in \mathbb{N}$ such that $n \geq \max(4(r+1), r+s)$, there exists a real polynomial $Q_n(x)$ of degree $\leq n$ such that

$$L(Q_n) \geq 0 \quad \text{throughout} \quad \left[g^{-1}(1+g(a)), b\right], \tag{22.2.3}$$

and

$$\sup_{a \leq x \leq b} \left|\left(D^{\alpha_j}_{*a;g} f\right)(x) - \left(D^{\alpha_j}_{*a;g} Q_n\right)(x)\right| \leq$$

$$\frac{(g(b) - g(a))^{j-\alpha_j}}{\Gamma(j - \alpha_j + 1)} C_{r,s} \left(\frac{b-a}{2n}\right)^{r-j} \omega_s \left(f^{(r)}, \frac{b-a}{2n}\right), \tag{22.2.4}$$

$j = h+1, \ldots, r$; $C_{r,s}$ *is a constant independent of f and n.*
   Set

$$l_j := \sup_{x \in [a,b]} \left|\alpha_h^{-1}(x) \alpha_j(x)\right|, \quad h \leq j \leq v. \tag{22.2.5}$$

When $j = 1, \ldots, h$ we derive

$$\max_{a \leq x \leq b} \left|D^{\alpha_j}_{*a;g} f - D^{\alpha_j}_{*a;g} Q_n\right| \leq \frac{C_{r,s}}{n^{r-v}} \omega_s \left(f^{(r)}, \frac{b-a}{2n}\right) \frac{(g(b) - g(a))^{j-\alpha_j}}{\Gamma(j - \alpha_j + 1)} \cdot$$

$$\left[\left(\sum_{j^*=h}^{v} s_{j^*} \frac{(g(b) - g(a))^{j^*-\alpha_{j^*}}}{\Gamma(j^* - \alpha_{j^*} + 1)} \left(\frac{b-a}{2}\right)^{r-j^*}\right) \frac{(\max(|a|, |b|))^{h-j}}{(h-j)!} + \left(\frac{b-a}{2}\right)^{r-j}\right]. \tag{22.2.6}$$

Finally it holds

$$\sup_{a \leq x \leq b} |f(x) - Q_n(x)| \leq \frac{C_{r,s}}{n^{r-v}} \omega_s \left(f^{(r)}, \frac{b-a}{2n}\right) \cdot \tag{22.2.7}$$

$$\left[\left(\frac{b-a}{2}\right)^r + (h!)^{-1} \max\left(|a|^h, |b|^h\right) \left(\sum_{j=h}^{v} s_j \frac{(g(b) - g(a))^{j-\alpha_j}}{\Gamma(j - \alpha_j + 1)} \left(\frac{b-a}{2}\right)^{r-j}\right)\right].$$

Proof of Theorem 22.16.
   Here $h, v, r \in \mathbb{Z}_+$, $0 \leq h \leq v \leq r$. Let $\alpha_j > 0$, $j = 1, \ldots, r$, such that $0 < \alpha_1 \leq 1 < \alpha_2 \leq 2 < \alpha_3 \leq 3 \cdots < \cdots < \alpha_r \leq r$. That is $\lceil \alpha_j \rceil = j$, $j = 1, \ldots, r$.
   Let $Q_n^*(x)$ be as in Theorem 22.12.
   We have that

$$\left(D^{\alpha_j}_{*a;g} f\right)(x) = \frac{1}{\Gamma(j - \alpha_j)} \int_a^x (g(x) - g(t))^{j-\alpha_j-1} g'(t) f^{(j)}(t) \, dt, \tag{22.2.8}$$

and

$$\left(D^{\alpha_j}_{*a;g}Q^*_n\right)(x) = \frac{1}{\Gamma(j-\alpha_j)}\int_a^x (g(x)-g(t))^{j-\alpha_j-1}\,g'(t)\,Q^{*(j)}_n(t)\,dt,$$

$$(22.2.9)$$

$j = 1,\ldots,r.$

Also it holds

$$\left(D^j_{*a;g}f\right)(x) = f^{(j)}(x),\quad \left(D^j_{*a;g}Q^*_n\right)(x) = Q^{*(j)}_n(x),\quad j = 1,\ldots,r.$$

$$(22.2.10)$$

By [13], we get that there exists g' a.e., and g' is measurable and non-negative.

We notice that

$$\left|\left(D^{\alpha_j}_{*a;g}f\right)(x) - D^{\alpha_j}_{*a;g}Q^*_n(x)\right| =$$

$$\frac{1}{\Gamma(j-\alpha_j)}\left|\int_a^x (g(x)-g(t))^{j-\alpha_j-1}\,g'(t)\left(f^{(j)}(t) - Q^{*(j)}_n(t)\right)dt\right| \le$$

$$\frac{1}{\Gamma(j-\alpha_j)}\int_a^x (g(x)-g(t))^{j-\alpha_j-1}\,g'(t)\left|f^{(j)}(t) - Q^{*(j)}_n(t)\right|dt \overset{(22.1.28)}{\le}$$

$$\frac{1}{\Gamma(j-\alpha_j)}\left(\int_a^x (g(x)-g(t))^{j-\alpha_j-1}\,g'(t)\,dt\right)C_{r,s}\left(\frac{b-a}{2n}\right)^{r-j}\omega_s\left(f^{(r)},\frac{b-a}{2n}\right)$$

$$(22.2.11)$$

$$\overset{(22.1.50)}{=}\frac{(g(x)-g(a))^{j-\alpha_j}}{\Gamma(j-\alpha_j+1)}C_{r,s}\left(\frac{b-a}{2n}\right)^{r-j}\omega_s\left(f^{(r)},\frac{b-a}{2n}\right) \le$$

$$\frac{(g(b)-g(a))^{j-\alpha_j}}{\Gamma(j-\alpha_j+1)}C_{r,s}\left(\frac{b-a}{2n}\right)^{r-j}\omega_s\left(f^{(r)},\frac{b-a}{2n}\right).\qquad (22.2.12)$$

Hence $\forall\, x \in [a,b]$, it holds

$$\left|\left(D^{\alpha_j}_{*a;g}f\right)(x) - D^{\alpha_j}_{*a;g}Q^*_n(x)\right| \le$$

$$\frac{(g(b)-g(a))^{j-\alpha_j}}{\Gamma(j-\alpha_j+1)}C_{r,s}\left(\frac{b-a}{2n}\right)^{r-j}\omega_s\left(f^{(r)},\frac{b-a}{2n}\right),\qquad (22.2.13)$$

and

$$\max_{x\in[a,b]}\left|D^{\alpha_j}_{*a;g}f(x) - D^{\alpha_j}_{*a;g}Q^*_n(x)\right| \le$$

$$\frac{(g(b)-g(a))^{j-\alpha_j}}{\Gamma(j-\alpha_j+1)}C_{r,s}\left(\frac{b-a}{2n}\right)^{r-j}\omega_s\left(f^{(r)},\frac{b-a}{2n}\right),\qquad (22.2.14)$$

$j = 0,1,\ldots,r.$

Above we set $D^0_{*a;g} f(x) = f(x)$, $D^0_{*a;g} Q^*_n(x) = Q^*_n(x)$, $\forall\, x \in [a, b]$, and $\alpha_0 = 0$, i.e. $\lceil \alpha_0 \rceil = 0$.

Put

$$s_j = \sup_{a \le x \le b} \left| \alpha_h^{-1}(\lambda)\, \alpha_j(\lambda) \right|, \quad j = h, \dots, v, \tag{22.2.15}$$

and

$$\eta_n = C_{r,s} \omega_s \left(f^{(r)}, \frac{b-a}{2n} \right) \left(\sum_{j=h}^{v} s_j \frac{(g(b) - g(a))^{j - \alpha_j}}{\Gamma(j - \alpha_j + 1)} \left(\frac{b-a}{2n} \right)^{r-j} \right). \tag{22.2.16}$$

I. Suppose, throughout $\left[g^{-1}(1 + g(a)), b \right]$, $\alpha_h(x) \ge \alpha^* > 0$. Let $Q_n(x)$, $x \in [a, b]$, be a real polynomial of degree $\le n$, according to Theorem 22.12 and (22.2.14), so that

$$\max_{x \in [a,b]} \left| D^{\alpha_j}_{*a;g} \left(f(x) + \eta_n (h!)^{-1} x^h \right) - \left(D^{\alpha_j}_{*a;g} Q_n \right)(x) \right| \le \tag{22.2.17}$$

$$\frac{(g(b) - g(a))^{j - \alpha_j}}{\Gamma(j - \alpha_j + 1)} C_{r,s} \left(\frac{b-a}{2n} \right)^{r-j} \omega_s \left(f^{(r)}, \frac{b-a}{2n} \right),$$

$j = 0, 1, \dots, r$.

In particular ($j = 0$) holds

$$\max_{x \in [a,b]} \left| \left(f(x) + \eta_n (h!)^{-1} x^h \right) - Q_n(x) \right| \le C_{r,s} \left(\frac{b-a}{2n} \right)^r \omega_s \left(f^{(r)}, \frac{b-a}{2n} \right), \tag{22.2.18}$$

and

$$\max_{x \in [a,b]} \left| f(x) - Q_n(x) \right| \le$$

$$\eta_n (h!)^{-1} (\max(|a|, |b|))^h + C_{r,s} \left(\frac{b-a}{2n} \right)^r \omega_s \left(f^{(r)}, \frac{b-a}{2n} \right) =$$

$$\eta_n (h!)^{-1} \max \left(|a|^h, |b|^h \right) + C_{r,s} \left(\frac{b-a}{2n} \right)^r \omega_s \left(f^{(r)}, \frac{b-a}{2n} \right) = \tag{22.2.19}$$

$$C_{r,s} \omega_s \left(f^{(r)}, \frac{b-a}{2n} \right).$$

$$\left(\sum_{j=h}^{v} s_j \frac{(g(b) - g(a))^{j - \alpha_j}}{\Gamma(j - \alpha_j + 1)} \left(\frac{b-a}{2n} \right)^{r-j} \right) (h!)^{-1} \max \left(|a|^h, |b|^h \right)$$

$$+C_{r,s}\left(\frac{b-a}{2n}\right)^{r}\omega_{s}\left(f^{(r)},\frac{b-a}{2n}\right)\leq$$

$$C_{r,s}\omega_{s}\left(f^{(r)},\frac{b-a}{2n}\right)n^{\upsilon-r}\cdot$$

$$\left[\left(\sum_{j=h}^{\upsilon}s_{j}\frac{(g(b)-g(a))^{j-\alpha_{j}}}{\Gamma(j-\alpha_{j}+1)}\left(\frac{b-a}{2}\right)^{r-j}\right)(h!)^{-1}\max\left(|a|^{h},|b|^{h}\right)+\left(\frac{b-a}{2}\right)^{r}\right].$$

$$(22.2.20)$$

We have found that

$$\max_{x\in[a,b]}|f(x)-Q_{n}(x)|\leq C_{r,s}\left[\left(\frac{b-a}{2}\right)^{r}+(h!)^{-1}\max\left(|a|^{h},|b|^{h}\right)\cdot\right.$$

$$\left.\left(\sum_{j=h}^{\upsilon}s_{j}\frac{(g(b)-g(a))^{j-\alpha_{j}}}{\Gamma(j-\alpha_{j}+1)}\left(\frac{b-a}{2}\right)^{r-j}\right)\right]n^{\upsilon-r}\omega_{s}\left(f^{(r)},\frac{b-a}{2n}\right),\quad(22.2.21)$$

proving (22.2.7).

Notice for $j=h+1,\ldots,\upsilon$, that

$$\left(D_{*a;g}^{\alpha_{j}}x^{h}\right)=\frac{1}{\Gamma(j-\alpha_{j})}\int_{a}^{x}(g(x)-g(t))^{j-\alpha_{j}-1}g'(t)\left(t^{h}\right)^{(j)}dt=0.$$

$$(22.2.22)$$

Hence inequality (22.2.4) is obvious.

When $j=1,\ldots,h$, from (22.2.17) we get

$$\left\|D_{*a;g}^{\alpha_{j}}f-D_{*a;g}^{\alpha_{j}}Q_{n}\right\|_{\infty,[a,b]}\leq\eta_{n}\frac{\left\|D_{*a;g}^{\alpha_{j}}\left(x^{h}\right)\right\|_{\infty,[a,b]}}{h!}+\qquad(22.2.23)$$

$$\frac{(g(b)-g(a))^{j-\alpha_{j}}}{\Gamma(j-\alpha_{j}+1)}C_{r,s}\left(\frac{b-a}{2n}\right)^{r-j}\omega_{s}\left(f^{(r)},\frac{b-a}{2n}\right)=$$

$$C_{r,s}\omega_{s}\left(f^{(r)},\frac{b-a}{2n}\right)\left(\sum_{j^{*}=h}^{\upsilon}s_{j^{*}}\frac{(g(b)-g(a))^{j^{*}-\alpha_{j^{*}}}}{\Gamma(j^{*}-\alpha_{j^{*}}+1)}\left(\frac{b-a}{2}\right)^{r-j^{*}}\frac{1}{n^{r-j^{*}}}\right)$$

$$\cdot\frac{\left\|D_{*a;g}^{\alpha_{j}}\left(x^{h}\right)\right\|_{\infty,[a,b]}}{h!}$$

$$+ \frac{(g(b) - g(a))^{j-\alpha_j}}{\Gamma(j - \alpha_j + 1)} C_{r,s} \left(\frac{b-a}{2}\right)^{r-j} \omega_s \left(f^{(r)}, \frac{b-a}{2n}\right) \frac{1}{n^{r-j}} \leq \quad (22.2.24)$$

$$\frac{C_{r,s}}{n^{r-v}} \omega_s \left(f^{(r)}, \frac{b-a}{2n}\right) \left[\left(\sum_{j^*=h}^{v} s_{j^*} \frac{(g(b) - g(a))^{j^*-\alpha_{j^*}}}{\Gamma(j^* - \alpha_{j^*} + 1)} \left(\frac{b-a}{2}\right)^{r-j^*} \right) \right.$$

$$\cdot \frac{\left\| D_{*a;g}^{\alpha_j}(x^h) \right\|_{\infty,[a,b]}}{h!}$$

$$\left. + \frac{(g(b) - g(a))^{j-\alpha_j}}{\Gamma(j - \alpha_j + 1)} \left(\frac{b-a}{2}\right)^{r-j} \right] =: (\psi). \quad (22.2.25)$$

But we have $(j = 1, \ldots, h)$

$$\frac{D_{*a;g}^{\alpha_j}(x^h)}{h!} = \frac{1}{\Gamma(j - \alpha_j)} \int_a^x (g(x) - g(t))^{j-\alpha_j-1} g'(t) \frac{t^{h-j}}{(h-j)!} dt. \quad (22.2.26)$$

Hence

$$\frac{\left| D_{*a;g}^{\alpha_j}(x^h) \right|}{h!} \leq \frac{1}{\Gamma(j - \alpha_j)(h-j)!} \int_a^x (g(x) - g(t))^{j-\alpha_j-1} g'(t) |t|^{h-j} dt \leq$$

$$\frac{1}{\Gamma(j - \alpha_j)(h-j)!} \left(\int_a^x (g(x) - g(t))^{j-\alpha_j-1} g'(t) dt \right) (\max(|a|, |b|))^{h-j} =$$

$$\quad (22.2.27)$$

$$\frac{(\max(|a|, |b|))^{h-j}}{\Gamma(j - \alpha_j)(h-j)!} \frac{(g(x) - g(a))^{j-\alpha_j}}{j - \alpha_j} =$$

$$\frac{(\max(|a|, |b|))^{h-j}}{\Gamma(j - \alpha_j + 1)(h-j)!} (g(x) - g(a))^{j-\alpha_j} \leq$$

$$\frac{(\max(|a|, |b|))^{h-j}}{\Gamma(j - \alpha_j + 1)(h-j)!} (g(b) - g(a))^{j-\alpha_j}. \quad (22.2.28)$$

That is

$$\frac{\left\| D_{*a;g}^{\alpha_j}(x^h) \right\|_{\infty,[a,b]}}{h!} \leq \frac{(\max(|a|, |b|))^{h-j}}{\Gamma(j - \alpha_j + 1)(h-j)!} (g(b) - g(a))^{j-\alpha_j}. \quad (22.2.29)$$

Therefore we have

$$(\psi) \le \frac{C_{r,s}}{n^{r-v}}\omega_s\left(f^{(r)}, \frac{b-a}{2n}\right)\left[\left(\sum_{j^*=h}^{v} s_{j^*}\frac{(g(b)-g(a))^{j^*-\alpha_{j^*}}}{\Gamma(j^*-\alpha_{j^*}+1)}\left(\frac{b-a}{2}\right)^{r-j^*}\right)\right.$$

$$\frac{(\max(|a|,|b|))^{h-j}}{\Gamma(j-\alpha_j+1)(h-j)!}(g(b)-g(a))^{j-\alpha_j} + \frac{(g(b)-g(a))^{j-\alpha_j}}{\Gamma(j-\alpha_j+1)}\left(\frac{b-a}{2}\right)^{r-j}\right] \tag{22.2.30}$$

$$= \frac{C_{r,s}}{n^{r-v}}\omega_s\left(f^{(r)}, \frac{b-a}{2n}\right)\frac{(g(b)-g(a))^{j-\alpha_j}}{\Gamma(j-\alpha_j+1)}\cdot$$

$$\left[\left(\sum_{j^*=h}^{v} s_{j^*}\frac{(g(b)-g(a))^{j^*-\alpha_{j^*}}}{\Gamma(j^*-\alpha_{j^*}+1)}\left(\frac{b-a}{2}\right)^{r-j^*}\right)\frac{(\max(|a|,|b|))^{h-j}}{(h-j)!}\right. \tag{22.2.31}$$

$$\left. +\left(\frac{b-a}{2}\right)^{r-j}\right],$$

proving (22.2.6).

Here

$$L = \sum_{j=h}^{v}\alpha_j(x)\left[D_{*a;g}^{\alpha_j}\right],$$

and suppose, throughout $\left[g^{-1}(1+g(a)), b\right]$, $Lf \ge 0$. So over $g^{-1}(1+g(a)) \le x \le b$, we get

$$\alpha_h^{-1}(x) L (Q_n(x)) \overset{(22.2.22)}{=} \alpha_h^{-1}(x) L (f(x)) + \frac{\eta_n}{h!}\left(D_{*a;g}^{\alpha_h}\left(x^h\right)\right) +$$

$$\sum_{j=h}^{v}\alpha_h^{-1}(x)\alpha_j(x)\left[D_{*a;g}^{\alpha_j}Q_n(x) - D_{*a;g}^{\alpha_j}f(x) - \frac{\eta_n}{h!}D_{*a;g}^{\alpha_j}x^h\right] \overset{(22.2.17)}{\ge} \tag{22.2.32}$$

$$\frac{\eta_n}{h!}\left(D_{*a;g}^{\alpha_h}\left(x^h\right)\right) -$$

$$\left(\sum_{j=h}^{v}s_j\frac{(g(b)-g(a))^{j-\alpha_j}}{\Gamma(j-\alpha_j+1)}\left(\frac{b-a}{2}\right)^{r-j}\right)C_{r,s}\omega_s\left(f^{(r)}, \frac{b-a}{2n}\right) \overset{(22.2.16)}{=} \tag{22.2.33}$$

$$\frac{\eta_n}{h!}\left(D_{*a;g}^{\alpha_h}\left(x^h\right)\right) - \eta_n = \eta_n\left(\frac{D_{*a;g}^{\alpha_h}\left(x^h\right)}{h!} - 1\right) = \tag{22.2.34}$$

$$\eta_n \left(\frac{1}{\Gamma (h - \alpha_h) \, h!} \int_a^x (g(x) - g(t))^{h - \alpha_h - 1} \, g'(t) \left(t^h \right)^{(h)} dt - 1 \right) =$$

$$\eta_n \left(\frac{h!}{\Gamma (h - \alpha_h)} \int_a^x (g(x) - g(t))^{h - \alpha_h - 1} \, g'(t) \, dt - 1 \right) \overset{(22.2.50)}{=}$$

$$\eta_n \left(\frac{(g(x) - g(a))^{h - \alpha_h}}{\Gamma (h - \alpha_h + 1)} - 1 \right) = \tag{22.2.35}$$

$$\eta_n \left(\frac{(g(x) - g(a))^{h - \alpha_h} - \Gamma (h - \alpha_h + 1)}{\Gamma (h - \alpha_h + 1)} \right) \geq$$

$$\eta_n \left(\frac{1 - \Gamma (h - \alpha_h + 1)}{\Gamma (h - \alpha_h + 1)} \right) \geq 0. \tag{22.2.36}$$

Clearly here $g(x) - g(a) \geq 1$.

Hence

$$L(Q_n(x)) \geq 0, \quad \text{for } x \in \left[g^{-1}(1 + g(a)), b \right]. \tag{22.2.37}$$

A further explanation follows: We know $\Gamma(1) = 1$, $\Gamma(2) = 1$, and Γ is convex and positive on $(0, \infty)$. Here $0 \leq h - \alpha_h < 1$ and $1 \leq h - \alpha_h + 1 < 2$. Thus

$$\Gamma (h - \alpha_h + 1) \leq 1 \quad \text{and} \quad 1 - \Gamma (h - \alpha_h + 1) \geq 0. \tag{22.2.38}$$

II. Suppose, throughout $\left[g^{-1}(1 + g(a)), b \right]$, $\alpha_h(x) \leq \beta^* < 0$.

Let $Q_n(x)$, $x \in [a, b]$ be a real polynomial of degree $\leq n$, according to Theorem 22.12 and (22.2.14), so that

$$\max_{x \in [a,b]} \left| D_{*a;g}^{\alpha_j} \left(f(x) - \eta_n (h!)^{-1} x^h \right) - \left(D_{*a;g}^{\alpha_j} Q_n \right) (x) \right| \leq \tag{22.2.39}$$

$$\frac{(g(b) - g(a))^{j - \alpha_j}}{\Gamma (j - \alpha_j + 1)} C_{r,s} \left(\frac{b - a}{2n} \right)^{r - j} \omega_s \left(f^{(r)}, \frac{b - a}{2n} \right),$$

$j = 0, 1, \ldots, r$.

In particular ($j = 0$) holds

$$\max_{x \in [a,b]} \left| \left(f(x) - \eta_n (h!)^{-1} x^h \right) - Q_n(x) \right| \leq C_{r,s} \left(\frac{b - a}{2n} \right)^r \omega_s \left(f^{(r)}, \frac{b - a}{2n} \right), \tag{22.2.40}$$

and

$$\max_{x \in [a,b]} |f(x) - Q_n(x)| \leq$$

$$\eta_n\,(h!)^{-1}\,(\max\,(|a|\,,\,|b|))^h + C_{r,s}\left(\frac{b-a}{2n}\right)^r \omega_s\left(f^{(r)},\frac{b-a}{2n}\right) =$$

$$\eta_n\,(h!)^{-1}\max\left(|a|^h\,,\,|b|^h\right) + C_{r,s}\left(\frac{b-a}{2n}\right)^r \omega_s\left(f^{(r)},\frac{b-a}{2n}\right), \qquad (22.2.41)$$

etc.

So using triangle's inequality on (22.2.39) and similar reasoning as in the first part of the proof we establish again (22.2.4), (22.2.6) and (22.2.7).

Here again

$$L = \sum_{j=h}^{v} \alpha_j\,(x)\left[D_{*a;g}^{\alpha_j}\right],$$

and suppose, throughout $\left[g^{-1}\,(1+g\,(a))\,,\,b\right]$, $Lf \geq 0$. So over $g^{-1}\,(1+g\,(a)) \leq x \leq b$, we get

$$\alpha_h^{-1}\,(x)\,L\,(Q_n\,(x)) \overset{(22.2.22)}{=} \alpha_h^{-1}\,(x)\,L\,(f\,(x)) - \frac{\eta_n}{h!}\left(D_{*a;g}^{\alpha_h}\left(x^h\right)\right) +$$

$$\sum_{j=h}^{v}\alpha_h^{-1}\,(x)\,\alpha_j\,(x)\left[D_{*a;g}^{\alpha_j}Q_n\,(x) - D_{*a;g}^{\alpha_j}f\,(x) + \frac{\eta_n}{h!}D_{*a;g}^{\alpha_j}x^h\right] \overset{(22.2.39)}{\leq} \quad (22.2.42)$$

$$-\frac{\eta_n}{h!}\left(D_{*a;g}^{\alpha_h}\left(x^h\right)\right) +$$

$$\left(\sum_{j=h}^{v}s_j\frac{(g\,(b)-g\,(a))^{j-\alpha_j}}{\Gamma\,(j-\alpha_j+1)}\left(\frac{b-a}{2n}\right)^{r-j}\right)C_{r,s}\omega_s\left(f^{(r)},\frac{b-a}{2n}\right) \overset{(22.2.16)}{=}$$

$$(22.2.43)$$

$$-\frac{\eta_n}{h!}\left(D_{*a;g}^{\alpha_h}\left(x^h\right)\right) + \eta_n = \eta_n\left(1 - \frac{D_{*a;g}^{\alpha_h}\left(x^h\right)}{h!}\right) = \qquad (22.2.44)$$

$$\eta_n\left(1 - \frac{1}{\Gamma\,(h-\alpha_h)\,h!}\int_a^x (g\,(x)-g\,(t))^{h-\alpha_h-1}\,g'\,(t)\left(t^h\right)^{(h)}dt\right) =$$

$$\eta_n\left(1 - \frac{h!}{\Gamma\,(h-\alpha_h)}\int_a^x (g\,(x)-g\,(t))^{h-\alpha_h-1}\,g'\,(t)\,dt\right) \overset{(22.1.50)}{=}$$

$$\eta_n\left(1 - \frac{(g\,(x)-g\,(a))^{h-\alpha_h}}{\Gamma\,(h-\alpha_h+1)}\right) = \qquad (22.2.45)$$

$$\eta_n\left(\frac{\Gamma\,(h-\alpha_h+1) - (g\,(x)-g\,(a))^{h-\alpha_h}}{\Gamma\,(h-\alpha_h+1)}\right) \overset{(22.2.38)}{\leq}$$

$$\eta_n \left(\frac{1 - (g(x) - g(a))^{h-\alpha_h}}{\Gamma(h - \alpha_h + 1)} \right) \le 0. \tag{22.2.46}$$

Hence again

$$L(Q_n(x)) \ge 0, \quad \forall\, x \in \left[g^{-1}(1 + g(a)), b \right].$$

The case of $\alpha_h = h$ is trivially concluded from the above. The proof of the theorem is now over. $\qquad\square$

We make:

Remark 22.17 By Corollary 22.15 we have that $D_{*a;g}^{\alpha_j} f$ are continuous functions, $j = 0, 1, \ldots, r$. Suppose that $\alpha_h(x), \ldots, \alpha_u(x)$ are continuous functions on $[a, b]$, and $L(f) \ge 0$ on $\left[g^{-1}(1 + g(a)), b \right]$ is replaced by $L(f) > 0$ on $\left[g^{-1}(1 + g(a)), b \right]$. Disregard the assumption made in the main theorem on $\alpha_h(x)$. For $n \in \mathbb{N}$, let $Q_n(x)$ be the $Q_n^*(x)$ of Theorem 22.12, and f as in Theorem 22.12 (same as in Theorem 22.16). Then $D_{*a;g}^{\alpha_j} Q_n^*$ converges uniformly to $D_{*a;g}^{\alpha_j} f$ at a higher rate given by inequality (22.2.14), in particular for $h \le j \le v$. Moreover, because $L(Q_n^*)$ converges uniformly to $L(f)$ on $[a, b]$, $L(Q_n^*) > 0$ on $\left[g^{-1}(1 + g(a)), b \right]$ for sufficiently large n.

22.3 Applications (to Theorem 22.16)

(1) When $g(x) = \ln x$ on $[a, b]$, $0 < a < b < \infty$.
Here we need $b > ae$, and the restriction are on $[ae, b]$.
(2) When $g(x) = e^x$ on $[a, b]$, $a < b < \infty$.
Here we need $b > \ln(1 + e^a)$, and the restriction are on $[\ln(1 + e^a), b]$.
(3) When, $A > 1$, $g(x) = A^x$ on $[a, b]$, $a < b < \infty$.
Here we need $b > \log_A(1 + A^a)$, and the restriction are on $\left[\log_A(1 + A^a), b \right]$.
(4) When $\sigma > 0$, $g(x) = x^\sigma$, $0 \le a < b < \infty$.
Here we need $b > (1 + a^\sigma)^{\frac{1}{\sigma}}$, and the restriction are on $\left[(1 + a^\sigma)^{\frac{1}{\sigma}}, b \right]$.

References

1. G.A. Anastassiou, Higher order monotone approximation with linear differential operators. Indian J. Pure Appl. Math. **24**(4), 263–266 (1993)
2. G.A. Anastassiou, *Fractional Differentiation Inequalities* (Springer, New York, 2009)
3. G.A. Anastassiou, The reduction method in fractional calculus and fractional Ostrowski type inequalities. Indian J. Math. **56**(3), 333–357 (2014)
4. G.A. Anastassiou, Univariate left fractional polynomial high order monotone approximation. Bull. Korean Math. Soc. **52**(2), 593–601 (2015)
5. G.A. Anastassiou, O. Shisha, Monotone approximation with linear differential operators. J. Approx. Theory **44**, 391–393 (1985)

6. G. Anastassiou, *Univariate Left General High Order Fractional Monotone Approximation* (submitted) (2015)
7. R.A. DeVore, G.G. Lorentz, *Constructive Approximation* (Springer, New York, 1993)
8. K. Diethelm, *The Analysis of Fractional Differential Equations*. Lecture Notes in Mathematics, vol. 2004, 1st edn. (Spinger, New York, 2010)
9. H.H. Gonska, E. Hinnemann, Pointwise estimated for approximation by algebraic polynomials. Acta Math. Hungar. **46**, 243–254 (1985)
10. R.-Q. Jia, *Chapter 3. Absolutely Continuous Functions*, https://www.ualberta.ca/~rjia/Math418/Notes/Chap.3.pdf
11. A.A. Kilbas, H.M. Srivastava, J.J. Trujillo, *Theory and Applications of Fractional Differential Equations*. North-Holland Mathematics Studies, vol. 204 (Elsevier, New York, 2006)
12. H.L. Royden, *Real Analysis*, 2nd edn. (Macmillan Publishing Co., Inc., New York, 1968)
13. A.R. Schep, *Differentiation of Monotone Functions*, http://people.math.sc.edu/schep/diffmonotone.pdf
14. O. Shisha, Monotone approximation. Pacific J. Math. **15**, 667–671 (1965)

Chapter 23
Right Generalized High Order Fractional Monotone Approximation

Here are applied the right general fractional derivatives Caputo type with respect to a base absolutely continuous strictly increasing function g. We mention various examples of such right fractional derivatives for different g. Let f be r-times continuously differentiable function on $[a, b]$, and let L be a linear right general fractional differential operator such that $L(f)$ is non-negative over a critical closed subinterval J of $[a, b]$. We can find a sequence of polynomials Q_n of degree less-equal n such that $L(Q_n)$ is non-negative over J, furthermore f is right fractionally and simultaneously approximated uniformly by Q_n over $[a, b]$.

The degree of this constrained approximation is given by inequalities employing the high order modulus of smoothness of $f^{(r)}$. We end chapter with applications of the main right fractional monotone approximation theorem for different g. It follows [4].

23.1 Introduction

The topic of monotone approximation started in [13] has become a major trend in approximation theory. A typical problem in this subject is: given a positive integer k, approximate a given function whose kth derivative is ≥ 0 by polynomials having this property.

In [7] the authors replaced the kth derivative with a linear differential operator of order k.

Furthermore in [3], the author generalized the result of [7] for linear right fractional differential operators.

© Springer International Publishing Switzerland 2016
G.A. Anastassiou and I.K. Argyros, *Intelligent Numerical Methods:*
Applications to Fractional Calculus, Studies in Computational Intelligence 624,
DOI 10.1007/978-3-319-26721-0_23

To describe the motivating result here we need:

Definition 23.1 ([3]) Let $\alpha > 0$ and $\lceil \alpha \rceil = m$, ($\lceil \cdot \rceil$ ceiling of the number). Consider $f \in C^m ([-1, 1])$. The right Caputo fractional derivative of f of order α anchored at 1 is given by

$$\left(D_{1-}^{\alpha} f \right)(x) = \frac{(-1)^m}{\Gamma (m - \alpha)} \int_x^1 (t - x)^{m-\alpha-1} f^{(m)} (t) \, dt, \tag{23.1.1}$$

for any $x \in [-1, 1]$, where Γ is the gamma function.

In particular

$$D_{1-}^0 f (x) = f (x), \tag{23.1.2}$$

$$D_{1-}^m f (x) = (-1)^m f^{(m)} (x), \quad \forall \, x \in [-1, 1]. \tag{23.1.3}$$

Here ω_s stands for the modulus of smoothness, see [8], p. 44.

We have proved:

Theorem 23.2 ([3]) *Let* h, v, r *be integers,* h *is even,* $1 \leq h \leq v \leq r$ *and let* $f \in C^r ([-1, 1])$, *with* $f^{(r)}$ *having modulus of smoothness* $\omega_s \left(f^{(r)}, \delta \right)$ *there,* $s \geq 1$. *Let* $\alpha_j (x)$, $j = h, h + 1, \ldots, v$ *be real functions, defined and bounded on* $[-1, 1]$ *and suppose* $\alpha_h (x)$ *is either* $\geq \alpha > 0$ *or* $\leq \beta < 0$ *on* $[-1, 0]$. *Let the real numbers* $\alpha_0 = 0 < \alpha_1 \leq 1 < \alpha_2 \leq 2 < \cdots < \alpha_r \leq r$. *Consider the linear right fractional differential operator*

$$L^* := \sum_{j=h}^{v} \alpha_j (x) \left[D_{1-}^{\alpha_j} \right] \tag{23.1.4}$$

and suppose, throughout $[-1, 0]$,

$$L^* (f) \geq 0. \tag{23.1.5}$$

Then, for any $n \in \mathbb{N}$ *such that* $n \geq \max \left(4 (r + 1), r + s \right)$, *there exists a real polynomial* Q_n *of degree* $\leq n$ *such that*

$$L^* (Q_n) \geq 0 \text{ throughout } [-1, 0], \tag{23.1.6}$$

and

$$\sup_{-1 \leq x \leq 1} \left| \left(D_{1-}^{\alpha_j} f \right)(x) - \left(D_{1-}^{\alpha_j} Q_n \right)(x) \right| \leq$$

$$\frac{2^{j-\alpha_j}}{\Gamma (j - \alpha_j + 1)} \frac{C_{r,s}}{n^{r-j}} \omega_s \left(f^{(r)}, \frac{1}{n} \right), \tag{23.1.7}$$

$j = h + 1, \ldots, r$; $C_{r,s}$ *is a constant independent of* f *and* n.

Set

$$l_j := \sup_{x \in [-1,1]} \left| \alpha_h^{-1}(x) \, \alpha_j(x) \right|, \quad h \le j \le v. \tag{23.1.8}$$

When $j = 1, \ldots, h$ we derive

$$\sup_{-1 \le x \le 1} \left| \left(D_{1-}^{\alpha_j} f \right)(x) - \left(D_{1-}^{\alpha_j} Q_n \right)(x) \right| \le \frac{C_{r,s}}{n^{r-v}} \omega_s \left(f^{(r)}, \frac{1}{n} \right) \cdot$$

$$\left[\left(\sum_{\tau=h}^{v} l_\tau \frac{2^{\tau - \alpha_\tau}}{\Gamma(\tau - \alpha_\tau + 1)} \right) \left(\sum_{\lambda=0}^{h-j} \frac{2^{h - \alpha_j - \lambda}}{\lambda! \Gamma(h - \alpha_j - \lambda + 1)} \right) + \frac{2^{j - \alpha_j}}{\Gamma(j - \alpha_j + 1)} \right]. \tag{23.1.9}$$

Finally it holds

$$\sup_{-1 \le x \le 1} \left| f(x) - Q_n(x) \right| \le$$

$$\frac{C_{r,s}}{n^{r-v}} \omega_s \left(f^{(r)}, \frac{1}{n} \right) \left[\frac{1}{h!} \sum_{\tau=h}^{v} l_\tau \frac{2^{\tau - \alpha_\tau}}{\Gamma(\tau - \alpha_\tau + 1)} + 1 \right]. \tag{23.1.10}$$

Notice above that the monotonicity property is only true on $[-1, 0]$, see (23.1.5) and (23.1.6). However the approximation properties (23.1.7), (23.1.9) and (23.1.10), are true over the whole interval $[-1, 1]$.

In this chapter we extend Theorem 23.2 to much more general linear right fractional differential operators.

We use here the following right generalised fractional integral:

Definition 23.3 (*see also* [10, p. 99]) The right generalised fractional integral of a function f with respect to given function g is defined as follows:

Let $a, b \in \mathbb{R}$, $a < b$, $\alpha > 0$. Here $g \in AC([a, b])$ (absolutely continuous functions) and is striclty increasing, $f \in L_\infty([a, b])$. We set

$$\left(I_{b-;g}^{\alpha} f \right)(x) = \frac{1}{\Gamma(\alpha)} \int_x^b (g(t) - g(x))^{\alpha - 1} g'(t) f(t) \, dt, \quad x \le b, \tag{23.1.11}$$

clearly $\left(I_{b-;g}^{\alpha} f \right)(b) = 0$.

When g is the identity function id, we get that $I_{b-;id}^{\alpha} = I_{b-}^{\alpha}$, the ordinary right Riemann-Liouville fractional integral, where

$$\left(I_{b-}^{\alpha} f \right)(x) = \frac{1}{\Gamma(\alpha)} \int_x^b (t - x)^{\alpha - 1} f(t) \, dt, \quad x \le b, \tag{23.1.12}$$

$\left(I_{b-}^{\alpha} f \right)(b) = 0$.

When $g(x) = \ln x$ on $[a, b]$, $0 < a < b < \infty$, we get:

Definition 23.4 ([10, p. 110]) Let $0 < a < b < \infty$, $\alpha > 0$. The right Hadamard fractional integral of order α is given by

$$\left(J_{b-}^{\alpha} f\right)(x) = \frac{1}{\Gamma(\alpha)} \int_{x}^{b} \left(\ln \frac{y}{x}\right)^{\alpha-1} \frac{f(y)}{y} dy, \quad x \leq b, \qquad (23.1.13)$$

where $f \in L_{\infty}([a, b])$.

We mention:

Definition 23.5 ([6]) The right fractional exponential integral is defined as follows: Let $a, b \in \mathbb{R}$, $a < b$, $\alpha > 0$, $f \in L_{\infty}([a, b])$. We set

$$\left(I_{b-;e^{x}}^{\alpha} f\right)(x) = \frac{1}{\Gamma(\alpha)} \int_{x}^{b} \left(e^{t} - e^{x}\right)^{\alpha-1} e^{t} f(t) dt, \quad x \leq b. \qquad (23.1.14)$$

Definition 23.6 ([6]) Let $a, b \in \mathbb{R}$, $a < b$, $\alpha > 0$, $f \in L_{\infty}([a, b])$, $A > 1$. We introduce the right fractional integral

$$\left(I_{b-;A^{x}}^{\alpha} f\right)(x) = \frac{\ln A}{\Gamma(\alpha)} \int_{x}^{b} \left(A^{t} - A^{x}\right)^{\alpha-1} A^{t} f(t) dt, \quad x \leq b. \qquad (23.1.15)$$

We also give:

Definition 23.7 ([6]) Let $\alpha, \sigma > 0$, $0 \leq a < b < \infty$, $f \in L_{\infty}([a, b])$. We set

$$\left(K_{b-;x^{\sigma}}^{\alpha} f\right)(x) = \frac{1}{\Gamma(\alpha)} \int_{x}^{b} \left(t^{\sigma} - x^{\sigma}\right)^{\alpha-1} f(t) \sigma t^{\sigma-1} dt, \quad x \leq b. \qquad (23.1.16)$$

We use the following general right fractional derivatives:

Definition 23.8 ([6]) Let $\alpha > 0$ and $\lceil \alpha \rceil = m$. Consider $f \in AC^{m}([a, b])$ (space of functions f with $f^{(m-1)} \in AC([a, b])$). We define the right general fractional derivative of f of order α as follows

$$\left(D_{b-;g}^{\alpha} f\right)(x) = \frac{(-1)^{m}}{\Gamma(m-\alpha)} \int_{x}^{b} (g(t) - g(x))^{m-\alpha-1} g'(t) f^{(m)}(t) dt, \qquad (23.1.17)$$

for any $x \in [a, b]$, where Γ is the gamma function.

We set

$$D_{b-;g}^{m} f(x) = (-1)^{m} f^{(m)}(x), \qquad (23.1.18)$$

$$D_{b-;g}^{0} f(x) = f(x), \quad \forall x \in [a, b]. \qquad (23.1.19)$$

When $g = id$, then $D_{b-}^{\alpha} f = D_{b-;id}^{\alpha} f$ is the right Caputo fractional derivative.

So we have the specific general right fractional derivatives:

Definition 23.9 ([6])

$$D_{b-;\ln x}^{\alpha} f(x) = \frac{(-1)^m}{\Gamma(m-\alpha)} \int_x^b \left(\ln \frac{y}{x}\right)^{m-\alpha-1} \frac{f^{(m)}(y)}{y} dy, \ 0 < a \leq x \leq b,$$

$$(23.1.20)$$

$$D_{b-;e^x}^{\alpha} f(x) = \frac{(-1)^m}{\Gamma(m-\alpha)} \int_x^b \left(e^t - e^x\right)^{m-\alpha-1} e^t f^{(m)}(t) dt, \ a \leq x \leq b,$$

$$(23.1.21)$$

and

$$D_{b-;A^x}^{\alpha} f(x) = \frac{(-1)^m \ln A}{\Gamma(m-\alpha)} \int_x^b \left(A^t - A^x\right)^{m-\alpha-1} A^t f^{(m)}(t) dt, \ a \leq x \leq b,$$

$$(23.1.22)$$

$$\left(D_{b-;x^\sigma}^{\alpha} f\right)(x) = \frac{(-1)^m}{\Gamma(m-\alpha)} \int_x^b \left(t^\sigma - x^\sigma\right)^{m-\alpha-1} \sigma t^{\sigma-1} f^{(m)}(t) dt, \ 0 \leq a \leq x \leq b.$$

$$(23.1.23)$$

We mention:

Theorem 23.10 ([5]) *Let $r \geq 0$ and $s \geq 1$. Then there exists a sequence $Q_n^* = Q_n^{*(r,s)}$ of linear polynomial operators mapping $C^r([a, b])$ into P_n, such that for all $f \in C^r([a, b])$ and all $n \geq \max(4(r+1), r+s)$ we have*

$$\left\| f^{(k)} - \left(Q_n^*(f)\right)^{(k)} \right\|_{\infty,[a,b]} \leq C_{r,s} \left(\frac{b-a}{2n}\right)^{r-k} \omega_s \left(f^{(r)}, \frac{b-a}{2n}\right), \quad (23.1.24)$$

$k = 0, 1, \ldots, r$, *where the constant $C_{r,s}$ is independent of n or f.*

Remark 23.11 Here $g \in AC([a, b])$ (absolutely continuous functions), g is increasing over $[a, b]$, $\alpha > 0$.

Let $g(a) = c$, $g(b) = d$. We want to calculate

$$I = \int_a^b (g(t) - g(a))^{\alpha-1} g'(t) dt. \quad (23.1.25)$$

Consider the function

$$f(y) = (y - g(a))^{\alpha-1} = (y - c)^{\alpha-1}, \forall y \in [c, d]. \quad (23.1.26)$$

We have that $f(y) \geq 0$, it may be $+\infty$ when $y = c$ and $0 < \alpha < 1$, but f is measurable on $[c, d]$. By [11], Royden, p. 107, Exercise 13 d, we get that

$$(f \circ g)(t) g'(t) = (g(t) - g(a))^{\alpha-1} g'(t) \quad (23.1.27)$$

is measurable on $[a, b]$, and

$$I = \int_c^d (y - c)^{\alpha - 1} \, dy = \frac{(d - c)^\alpha}{\alpha} \tag{23.1.28}$$

(notice that $(y - c)^{\alpha - 1}$ is Riemann integrable).

That is

$$I = \frac{(g(b) - g(a))^\alpha}{\alpha}. \tag{23.1.29}$$

Similarly it holds

$$\int_x^b (g(t) - g(x))^{\alpha - 1} \, g'(t) \, dt = \frac{(g(b) - g(x))^\alpha}{\alpha}, \quad \forall \, x \in [a, b]. \tag{23.1.30}$$

We need:

Theorem 23.12 ([1]) *Let $r > 0$, $F \in L_\infty([a, b])$ and*

$$G(s) := \int_s^b (t - s)^{r-1} \, F(t) \, dt, \quad all \, s \in [a, b]. \tag{23.1.31}$$

Then $G \in C([a, b])$ (absolutely continuous functions) for $r \geq 1$ and $G \in C([a, b])$, when $r \in (0, 1)$.

We use:

Theorem 23.13 *Let $r > 0$, $a < b$, $F \in L_\infty([a, b])$, $g \in AC([a, b])$ and g is strictly increasing.*

Consider

$$B(s) := \int_s^b (g(t) - g(s))^{r-1} \, g'(t) \, F(t) \, dt, \quad for \, all \, s \in [a, b]. \tag{23.1.32}$$

Then $B \in C([a, b])$.

Proof There exists a Borel measurable function $F^* : [a, b] \to \mathbb{R}$ such that $F = F^*$, a.e., and it holds

$$B(s) = \int_s^b (g(t) - g(s))^{r-1} \, g'(t) \, F^*(t) \, dt, \quad for \, all \, s \in [a, b]. \tag{23.1.33}$$

Notice that

$$\|F^*\|_\infty = \|F\|_\infty < \infty. \tag{23.1.34}$$

We can write

$$B(s) = \int_s^b (g(t) - g(s))^{r-1} g'(t) \left(F^* \cap g^{-1}\right)(g(t)) \, dt, \text{ for all } s \in [a, b].$$

(23.1.35)

By [2], we get that

$$\left\| F^* \circ g^{-1} \right\|_{\infty, [g(a), g(b)]} \le \|F\|_\infty.$$

(23.1.36)

Next we consider the function $(z - g(s))^{r-1} \left(F^* \circ g^{-1}\right)(z)$, where $z \in [g(s), g(b)]$, the last is integrable over $[g(s), g(b)]$.

Since g is monotone, by [9], $(g(t) - g(s))^{r-1} g'(t) \left(F^* \circ g^{-1}\right)(g(t))$ is integrable on $[s, b]$.

Furthermore, again by [9], it holds

$$\rho(g(s)) := \int_{g(s)}^{g(b)} (z - g(s))^{r-1} \left(F^* \circ g^{-1}\right)(z) \, dz =$$

(23.1.37)

$$\int_s^b (g(t) - g(s))^{r-1} g'(t) \left(F^* \circ g^{-1}\right)(g(t)) \, dt =$$

$$\int_s^b (g(t) - g(s))^{r-1} g'(t) \, F^*(t) \, dt =$$

(23.1.38)

$$\int_s^b (g(t) - g(s))^{r-1} g'(t) \, F(t) \, dt = B(s), \quad \text{all } s \in [a, b].$$

That is

$$B(s) = \int_{g(s)}^{g(b)} (z - g(s))^{r-1} \left(F^* \circ g^{-1}\right)(z) \, dz = \rho(g(s)), \quad \text{all } s \in [a, b].$$

(23.1.39)

By Theorem 23.12 we have the function

$$\rho(y) = \int_y^{g(b)} (z - y)^{r-1} \left(F^* \circ g^{-1}\right)(z) \, dz,$$

(23.1.40)

is continuous in y over $[g(a), g(b)]$.

Let now $s_n, s \in [a, b] : s_n \to s$, then $g(s_n) \to g(s)$, and $\rho(g(s_n)) \to \rho(g(s))$, that is $B(s_n) \to B(s)$, as $n \to \infty$, proving the continuity of B. $\qquad\square$

We need:

Corollary 23.14 *Let $g \in AC([a, b])$ and g is strictly increasing. Let $\alpha > 0$, $\alpha \notin \mathbb{N}$, and $\lceil \alpha \rceil = m$, $f \in AC^m([a, b])$ with $f^{(m)} \in L_\infty([a, b])$. Then $\left(D_{b-;g}^{\alpha} f \right) \in C([a, b])$.*

Proof By Theorem 23.13. $\square$

23.2   Main Result

We present:

Theorem 23.15 *Here we assume that $g(b) - g(a) > 1$. Let h, v, r be integers, h is even, $1 \le h \le v \le r$ and let $f \in C^r([a, b])$, $a < b$, with $f^{(r)}$ having modulus of smoothness $\omega_s \left(f^{(r)}, \delta \right)$ there, $s \ge 1$. Let $\alpha_j(x)$, $j = h, h+1, \ldots, v$ be real functions, defined and bounded on $[a, b]$ and suppose for $x \in \left[a, g^{-1}(g(b) - 1) \right]$ that $\alpha_h(x)$ is either $\ge \alpha^* > 0$ or $\le \beta^* < 0$. Let the real numbers $\alpha_0 = 0 < \alpha_1 \le 1 < \alpha_2 \le 2 < \cdots < \alpha_r \le r$. Here $D_{b-;g}^{\alpha_j} f$ stands for the right general fractional derivative of f of order α_j anchored at b. Consider the linear right general fractional differential operator*

$$L := \sum_{j=h}^{v} \alpha_j(x) \left[D_{b-;g}^{\alpha_j} \right] \tag{23.2.1}$$

and suppose, throughout $\left[a, g^{-1}(g(b) - 1) \right]$,

$$L(f) \ge 0. \tag{23.2.2}$$

Then, for any $n \in \mathbb{N}$ such that $n \ge \max(4(r+1), r+s)$, there exists a real polynomial $Q_n(x)$ of degree $\le n$ such that

$$L(Q_n) \ge 0 \text{ throughout } \left[a, g^{-1}(g(b) - 1) \right], \tag{23.2.3}$$

and

$$\sup_{a \le x \le b} \left| \left(D_{b-;g}^{\alpha_j} f \right)(x) - \left(D_{b-;g}^{\alpha_j} Q_n \right)(x) \right| \le$$

$$\frac{(g(b) - g(a))^{j - \alpha_j}}{\Gamma(j - \alpha_j + 1)} C_{r,s} \left(\frac{b-a}{2n} \right)^{r-j} \omega_s \left(f^{(r)}, \frac{b-a}{2n} \right), \tag{23.2.4}$$

$j = h+1, \ldots, r$; $C_{r,s}$ is a constant independent of f and n.
     Set

$$l_j := \sup_{x \in [a,b]} \left| \alpha_h^{-1}(x) \alpha_j(x) \right|, \ h \le j \le v. \tag{23.2.5}$$

When $j = 1, \ldots, h$ *we derive*

$$\max_{a \leq x \leq b} \left| D^{\alpha_j}_{b-;g} f - D^{\alpha_j}_{b-;g} Q_n \right| \leq \frac{C_{r,s}}{n^{r-v}} \omega_s \left(f^{(r)}, \frac{b-a}{2n} \right) \frac{(g(b) - g(a))^{j-\alpha_j}}{\Gamma(j - \alpha_j + 1)} \cdot$$

$$\left[\left(\sum_{j^*=h}^{v} s_{j^*} \frac{(g(b) - g(a))^{j^*-\alpha_{j^*}}}{\Gamma(j^* - \alpha_{j^*} + 1)} \left(\frac{b-a}{2} \right)^{r-j^*} \right) \frac{(\max(|a|, |b|))^{h-j}}{(h-j)!} + \left(\frac{b-a}{2} \right)^{r-j} \right].$$

$$(23.2.6)$$

Finally it holds

$$\sup_{a \leq x \leq b} |f(x) - Q_n(x)| \leq \frac{C_{r,s}}{n^{r-v}} \omega_s \left(f^{(r)}, \frac{b-a}{2n} \right). \qquad (23.2.7)$$

$$\left[\left(\frac{b-a}{2} \right)^r + (h!)^{-1} \max \left(|a|^h, |b|^h \right) \left(\sum_{j=h}^{v} s_j \frac{(g(b) - g(a))^{j-\alpha_j}}{\Gamma(j - \alpha_j + 1)} \left(\frac{b-a}{2} \right)^{r-j} \right) \right].$$

Proof of Theorem 23.15.

Here $h, u, r \in \mathbb{Z}_+$, $0 \leq h \leq u \leq r$. Let $\alpha_j > 0$, $j = 1, \ldots, r$, such that $0 < \alpha_1 \leq 1 < \alpha_2 \leq 2 < \alpha_3 \leq 3 \cdots < \cdots < \alpha_r \leq r$. That is $\lceil \alpha_j \rceil = j$, $j = 1, \ldots, r$.

Let $Q_n^*(x)$ be as in Theorem 23.10.

We have that

$$\left(D^{\alpha_j}_{b-;g} f \right)(x) = \frac{(-1)^j}{\Gamma(j - \alpha_j)} \int_x^b (g(t) - g(x))^{j-\alpha_j-1} g'(t) f^{(j)}(t) \, dt, \quad (23.2.8)$$

and

$$\left(D^{\alpha_j}_{b-;g} Q_n^* \right)(x) = \frac{(-1)^j}{\Gamma(j - \alpha_j)} \int_x^b (g(t) - g(x))^{j-\alpha_j-1} g'(t) Q_n^{*(j)}(t) \, dt,$$

$$(23.2.9)$$

$j = 1, \ldots, r$.

Also it holds

$$\left(D^j_{b-;g} f \right)(x) = (-1)^j f^{(j)}(x), \quad \left(D^j_{b-;g} Q_n^* \right)(x) = (-1)^j Q_n^{*(j)}(x), \quad (23.2.10)$$

$j = 1, \ldots, r$.

By [12], we get that there exists g' a.e., and g' is measurable and non-negative.

We notice that

$$\left| \left(D^{\alpha_j}_{b-;g} f \right)(x) - D^{\alpha_j}_{b-;g} Q_n^*(x) \right| =$$

$$\frac{1}{\Gamma\left(j-\alpha_j\right)}\left|\int_x^b \left(g\left(t\right)-g\left(x\right)\right)^{j-\alpha_j-1} g'\left(t\right) \left(f^{(j)}\left(t\right)-Q_n^{*(j)}\left(t\right)\right) dt\right| \le$$

$$\frac{1}{\Gamma\left(j-\alpha_j\right)}\int_x^b \left(g\left(t\right)-g\left(x\right)\right)^{j-\alpha_j-1} g'\left(t\right) \left|f^{(j)}\left(t\right)-Q_n^{*(j)}\left(t\right)\right| dt \overset{(23.1.24)}{\le}$$

$$\frac{1}{\Gamma\left(j-\alpha_j\right)} \cdot$$

$$\left(\int_x^b \left(g\left(t\right)-g\left(x\right)\right)^{j-\alpha_j-1} g'\left(t\right) dt\right) C_{r,s} \left(\frac{b-a}{2n}\right)^{r-j} \omega_s \left(f^{(r)}, \frac{b-a}{2n}\right) \overset{(23.1.30)}{=}$$

$$\text{(23.2.11)}$$

$$\frac{\left(g\left(b\right)-g\left(x\right)\right)^{j-\alpha_j}}{\Gamma\left(j-\alpha_j+1\right)} C_{r,s} \left(\frac{b-a}{2n}\right)^{r-j} \omega_s \left(f^{(r)}, \frac{b-a}{2n}\right) \le$$

$$\frac{\left(g\left(b\right)-g\left(a\right)\right)^{j-\alpha_j}}{\Gamma\left(j-\alpha_j+1\right)} C_{r,s} \left(\frac{b-a}{2n}\right)^{r-j} \omega_s \left(f^{(r)}, \frac{b-a}{2n}\right). \tag{23.2.12}$$

Hence $\forall\, x \in [a, b]$, it holds

$$\left|\left(D_{b-;g}^{\alpha_j} f\right)\left(x\right) - D_{b-;g}^{\alpha_j} Q_n^*\left(x\right)\right| \le$$

$$\frac{\left(g\left(b\right)-g\left(a\right)\right)^{j-\alpha_j}}{\Gamma\left(j-\alpha_j+1\right)} C_{r,s} \left(\frac{b-a}{2n}\right)^{r-j} \omega_s \left(f^{(r)}, \frac{b-a}{2n}\right), \tag{23.2.13}$$

and

$$\max_{x\in[a,b]} \left|D_{b-;g}^{\alpha_j} f\left(x\right) - D_{b-;g}^{\alpha_j} Q_n^*\left(x\right)\right| \le$$

$$\frac{\left(g\left(b\right)-g\left(a\right)\right)^{j-\alpha_j}}{\Gamma\left(j-\alpha_j+1\right)} C_{r,s} \left(\frac{b-a}{2n}\right)^{r-j} \omega_s \left(f^{(r)}, \frac{b-a}{2n}\right), \tag{23.2.14}$$

$j = 0, 1, \ldots, r$.

Above we set $D_{b-;g}^0 f\left(x\right) = f\left(x\right)$, $D_{b-;g}^0 Q_n^*\left(x\right) = Q_n^*\left(x\right)$, $\forall\, x \in [a, b]$, and $\alpha_0 = 0$, i.e. $\lceil\alpha_0\rceil = 0$.

Put

$$s_j = \sup_{a\le x\le b} \left|\alpha_h^{-1}\left(x\right) \alpha_j\left(x\right)\right|, \quad j = h, \ldots, v, \tag{23.2.15}$$

and

$$\eta_n = C_{r,s}\omega_s\left(f^{(r)},\frac{b-a}{2n}\right)\left(\sum_{j=h}^{v}s_j\frac{(g(b)-g(a))^{j-\alpha_j}}{\Gamma(j-\alpha_j+1)}\left(\frac{b-a}{2n}\right)^{r-j}\right).$$

(23.2.16)

I. Suppose, throughout $\left[a,g^{-1}(g(b)-1)\right]$, $\alpha_h(x)\geq\alpha^*>0$. Let $Q_n(x)$, $x\in[a,b]$, be a real polynomial of degree $\leq n$, according to Theorem 23.10 and (23.2.14), so that

$$\max_{x\in[a,b]}\left|D_{b-;g}^{\alpha_j}\left(f(x)+\eta_n(h!)^{-1}x^h\right)-\left(D_{b-;g}^{\alpha_j}Q_n\right)(x)\right|\leq$$

(23.2.17)

$$\frac{(g(b)-g(a))^{j-\alpha_j}}{\Gamma(j-\alpha_j+1)}C_{r,s}\left(\frac{b-a}{2n}\right)^{r-j}\omega_s\left(f^{(r)},\frac{b-a}{2n}\right),$$

$j=0,1,\ldots,r$.

In particular ($j=0$) holds

$$\max_{x\in[a,b]}\left|\left(f(x)+\eta_n(h!)^{-1}x^h\right)-Q_n(x)\right|\leq C_{r,s}\left(\frac{b-a}{2n}\right)^r\omega_s\left(f^{(r)},\frac{b-a}{2n}\right),$$

(23.2.18)

and

$$\max_{x\in[a,b]}|f(x)-Q_n(x)|\leq$$

$$\eta_n(h!)^{-1}(\max(|a|,|b|))^h+C_{r,s}\left(\frac{b-a}{2n}\right)^r\omega_s\left(f^{(r)},\frac{b-a}{2n}\right)=$$

$$\eta_n(h!)^{-1}\max(|a|^h,|b|^h)+C_{r,s}\left(\frac{b-a}{2n}\right)^r\omega_s\left(f^{(r)},\frac{b-a}{2n}\right)=$$

(23.2.19)

$$C_{r,s}\omega_s\left(f^{(r)},\frac{b-a}{2n}\right)\left(\sum_{j=h}^{v}s_j\frac{(g(b)-g(a))^{j-\alpha_j}}{\Gamma(j-\alpha_j+1)}\left(\frac{b-a}{2n}\right)^{r-j}\right).$$

$$(h!)^{-1}\max(|a|^h,|b|^h)+C_{r,s}\left(\frac{b-a}{2n}\right)^r\omega_s\left(f^{(r)},\frac{b-a}{2n}\right)\leq$$

$$C_{r,s}\omega_s\left(f^{(r)},\frac{b-a}{2n}\right)n^{v-r}.$$

$$\left[\left(\sum_{j=h}^{v}s_j\frac{(g(b)-g(a))^{j-\alpha_j}}{\Gamma(j-\alpha_j+1)}\left(\frac{b-a}{2}\right)^{r-j}\right)(h!)^{-1}\max\left(|a|^h,|b|^h\right)+\left(\frac{b-a}{2}\right)^r\right].$$

(23.2.20)

We have found that

$$\max_{x \in [a,b]} |f(x) - Q_n(x)| \le C_{r,s} \left[\left(\frac{b-a}{2} \right)^r + (h!)^{-1} \max \left(|a|^h, |b|^h \right) \cdot \right.$$

$$\left. \left(\sum_{j=h}^{v} s_j \frac{(g(b) - g(a))^{j-\alpha_j}}{\Gamma(j-\alpha_j+1)} \left(\frac{b-a}{2} \right)^{r-j} \right) \right] n^{v-r} \omega_s \left(f^{(r)}, \frac{b-a}{2n} \right), \quad (23.2.21)$$

proving (23.2.7).

Notice for $j = h+1, \ldots, v$, that

$$\left(D_{b-;g}^{\alpha_j} x^h \right) = \frac{(-1)^j}{\Gamma(j-\alpha_j)} \int_x^b (g(t) - g(x))^{j-\alpha_j-1} g'(t) \left(t^h \right)^{(j)} dt = 0.$$
$$(23.2.22)$$

Hence inequality (23.2.4) is obvious.

When $j = 1, \ldots, h$, from (23.2.17) we get

$$\left\| D_{b-;g}^{\alpha_j} f - D_{b-;g}^{\alpha_j} Q_n \right\|_{\infty,[a,b]} \le \eta_n \frac{\left\| D_{b-;g}^{\alpha_j} \left(x^h \right) \right\|_{\infty,[a,b]}}{h!} + \quad (23.2.23)$$

$$\frac{(g(b) - g(a))^{j-\alpha_j}}{\Gamma(j-\alpha_j+1)} C_{r,s} \left(\frac{b-a}{2n} \right)^{r-j} \omega_s \left(f^{(r)}, \frac{b-a}{2n} \right) \overset{(23.2.16)}{=}$$

$$C_{r,s} \omega_s \left(f^{(r)}, \frac{b-a}{2n} \right).$$

$$\left(\sum_{j^*=h}^{v} s_{j^*} \frac{(g(b) - g(a))^{j^*-\alpha_{j^*}}}{\Gamma(j^*-\alpha_{j^*}+1)} \left(\frac{b-a}{2} \right)^{r-j^*} \frac{1}{n^{r-j^*}} \right) \frac{\left\| D_{b-;g}^{\alpha_j} \left(x^h \right) \right\|_{\infty,[a,b]}}{h!}$$

$$+ \frac{(g(b) - g(a))^{j-\alpha_j}}{\Gamma(j-\alpha_j+1)} C_{r,s} \left(\frac{b-a}{2} \right)^{r-j} \omega_s \left(f^{(r)}, \frac{b-a}{2n} \right) \frac{1}{n^{r-j}} \le \quad (23.2.24)$$

$$\frac{C_{r,s}}{n^{r-v}} \omega_s \left(f^{(r)}, \frac{b-a}{2n} \right).$$

$$\left[\left(\sum_{j^*=h}^{v} s_{j^*} \frac{(g(b) - g(a))^{j^*-\alpha_{j^*}}}{\Gamma(j^*-\alpha_{j^*}+1)} \left(\frac{b-a}{2} \right)^{r-j^*} \right) \frac{\left\| D_{b-;g}^{\alpha_j} \left(x^h \right) \right\|_{\infty,[a,b]}}{h!} \right.$$

$$\left. + \frac{(g(b) - g(a))^{j-\alpha_j}}{\Gamma(j-\alpha_j+1)} \left(\frac{b-a}{2} \right)^{r-j} \right] =: (\phi). \quad (23.2.25)$$

But we have ($j = 1, \ldots, h$)

$$\frac{D_{b-;g}^{\alpha_j}\left(x^h\right)}{h!} = \frac{(-1)^j}{\Gamma\left(j - \alpha_j\right)} \int_x^b \left(g\left(t\right) - g\left(x\right)\right)^{j - \alpha_j - 1} g'\left(t\right) \frac{t^{h-j}}{(h - j)!} dt. \quad (23.2.26)$$

Hence

$$\frac{\left|D_{b-;g}^{\alpha_j}\left(x^h\right)\right|}{h!} \leq \frac{1}{\Gamma\left(j - \alpha_j\right)(h - j)!} \int_x^b \left(g\left(t\right) - g\left(x\right)\right)^{j - \alpha_j - 1} g'\left(t\right) |t|^{h-j} dt \leq$$

$$\frac{1}{\Gamma\left(j - \alpha_j\right)(h - j)!} \left(\int_x^b \left(g\left(t\right) - g\left(x\right)\right)^{j - \alpha_j - 1} g'\left(t\right) dt\right) \left(\max\left(|a|, |b|\right)\right)^{h-j}$$

$$\quad (23.2.27)$$

$$\overset{(23.1.30)}{=} \frac{\left(\max\left(|a|, |b|\right)\right)^{h-j}}{\Gamma\left(j - \alpha_j\right)(h - j)!} \frac{\left(g\left(b\right) - g\left(x\right)\right)^{j - \alpha_j}}{j - \alpha_j} =$$

$$\frac{\left(\max\left(|a|, |b|\right)\right)^{h-j}}{\Gamma\left(j - \alpha_j + 1\right)(h - j)!} \left(g\left(b\right) - g\left(x\right)\right)^{j - \alpha_j} \leq$$

$$\frac{\left(\max\left(|a|, |b|\right)\right)^{h-j}}{\Gamma\left(j - \alpha_j + 1\right)(h - j)!} \left(g\left(b\right) - g\left(a\right)\right)^{j - \alpha_j}. \quad (23.2.28)$$

That is

$$\frac{\left\|D_{b-;g}^{\alpha_j}\left(x^h\right)\right\|_{\infty,[a,b]}}{h!} \leq \frac{\left(\max\left(|a|, |b|\right)\right)^{h-j}}{\Gamma\left(j - \alpha_j + 1\right)(h - j)!} \left(g\left(b\right) - g\left(a\right)\right)^{j - \alpha_j}. \quad (23.2.29)$$

Therefore we have

$$(\phi) \leq \frac{C_{r,s}}{n^{r-v}} \omega_s\left(f^{(r)}, \frac{b-a}{2n}\right) \left[\left(\sum_{j^*=h}^{v} s_{j^*} \frac{\left(g\left(b\right) - g\left(a\right)\right)^{j^* - \alpha_{j^*}}}{\Gamma\left(j^* - \alpha_{j^*} + 1\right)} \left(\frac{b-a}{2}\right)^{r-j^*}\right)\right.$$

$$\frac{\left(\max\left(|a|, |b|\right)\right)^{h-j}}{\Gamma\left(j - \alpha_j + 1\right)(h - j)!} \left(g\left(b\right) - g\left(a\right)\right)^{j - \alpha_j} + \left.\frac{\left(g\left(b\right) - g\left(a\right)\right)^{j - \alpha_j}}{\Gamma\left(j - \alpha_j + 1\right)} \left(\frac{b-a}{2}\right)^{r-j}\right]$$

$$\quad (23.2.30)$$

$$= \frac{C_{r,s}}{n^{r-v}} \omega_s\left(f^{(r)}, \frac{b-a}{2n}\right) \frac{\left(g\left(b\right) - g\left(a\right)\right)^{j - \alpha_j}}{\Gamma\left(j - \alpha_j + 1\right)}.$$

$$\left[\left(\sum_{j^*=h}^{v} s_{j^*} \frac{(g(b) - g(a))^{j^* - \alpha_{j^*}}}{\Gamma(j^* - \alpha_{j^*} + 1)} \left(\frac{b-a}{2}\right)^{r-j^*}\right) \frac{(\max(|a|, |b|))^{h-j}}{(h-j)!} \right. \tag{23.2.31}$$

$$\left. + \left(\frac{b-a}{2}\right)^{r-j}\right],$$

proving (23.2.6).

Here

$$L = \sum_{j=h}^{v} \alpha_j(x) \left[D_{b-;g}^{\alpha_j}\right],$$

and suppose, throughout $\left[a, g^{-1}(g(b) - 1)\right]$, $Lf \geq 0$. So over $a \leq x \leq g^{-1}$ $(g(b) - 1)$, we get

$$\alpha_h^{-1}(x) L(Q_n(x)) \overset{(23.2.27)}{=} \alpha_h^{-1}(x) L(f(x)) + \frac{\eta_n}{h!}\left(D_{b-;g}^{\alpha_h}(x^h)\right) +$$

$$\sum_{j=h}^{v} \alpha_h^{-1}(x) \alpha_j(x) \left[D_{b-;g}^{\alpha_j} Q_n(x) - D_{b-;g}^{\alpha_j} f(x) - \frac{\eta_n}{h!} D_{b-;g}^{\alpha_j} x^h\right] \overset{(23.2.17)}{\geq} \tag{23.2.32}$$

$$\frac{\eta_n}{h!}\left(D_{b-;g}^{\alpha_h}(x^h)\right) -$$

$$\left(\sum_{j=h}^{v} s_j \frac{(g(b) - g(a))^{j-\alpha_j}}{\Gamma(j - \alpha_j + 1)} \left(\frac{b-a}{2n}\right)^{r-j}\right) C_{r,s} \omega_s\left(f^{(r)}, \frac{b-a}{2n}\right) \tag{23.2.33}$$

$$\overset{(23.2.16)}{=} \frac{\eta_n}{h!}\left(D_{b-;g}^{\alpha_h}(x^h)\right) - \eta_n = \eta_n\left(\frac{D_{b-;g}^{\alpha_h}(x^h)}{h!} - 1\right) = \tag{23.2.34}$$

$$\eta_n\left(\frac{1}{\Gamma(h - \alpha_h) h!} \int_x^b (g(t) - g(x))^{h-\alpha_h-1} g'(t) (t^h)^{(h)} dt - 1\right) =$$

$$\eta_n\left(\frac{1}{\Gamma(h - \alpha_h)} \int_x^b (g(t) - g(x))^{h-\alpha_h-1} g'(t) dt - 1\right) \overset{(23.1.30)}{=}$$

$$\eta_n\left(\frac{(g(b) - g(x))^{h-\alpha_h}}{\Gamma(h - \alpha_h + 1)} - 1\right) = \tag{23.2.35}$$

$$\eta_n\left(\frac{(g(b) - g(x))^{h-\alpha_h} - \Gamma(h - \alpha_h + 1)}{\Gamma(h - \alpha_h + 1)}\right) \geq$$

$$\eta_n \left(\frac{1 - \Gamma (h - \alpha_h + 1)}{\Gamma (h - \alpha_h + 1)} \right) \geq 0. \qquad (23.2.36)$$

Clearly here $g(b) - g(x) \geq 1$.

Hence

$$L(Q_n(x)) \geq 0, \text{ for } x \in \left[a, g^{-1} (g(b) - 1) \right]. \qquad (23.2.37)$$

A further explanation follows: We know $\Gamma(1) = 1$, $\Gamma(2) = 1$, and Γ is convex and positive on $(0, \infty)$. Here $0 \leq h - \alpha_h < 1$ and $1 \leq h - \alpha_h + 1 < 2$. Thus

$$\Gamma(h - \alpha_h + 1) \leq 1 \text{ and } 1 - \Gamma(h - \alpha_h + 1) \geq 0. \qquad (23.2.38)$$

II. Suppose, throughout $\left[a, g^{-1} (g(b) - 1) \right]$, $\alpha_h(x) \leq \beta^* < 0$.

Let $Q_n(x)$, $x \in [a, b]$ be a real polynomial of degree $\leq n$, according to Theorem 23.10 and (23.2.14), so that

$$\max_{x \in [a,b]} \left| D_{b-;g}^{\alpha_j} \left(f(x) - \eta_n (h!)^{-1} x^h \right) - \left(D_{b-;g}^{\alpha_j} Q_n \right)(x) \right| \leq \qquad (23.2.39)$$

$$\frac{(g(b) - g(a))^{j-\alpha_j}}{\Gamma(j - \alpha_j + 1)} C_{r,s} \left(\frac{b-a}{2n} \right)^{r-j} \omega_s \left(f^{(r)}, \frac{b-a}{2n} \right),$$

$j = 0, 1, \ldots, r$.

In particular ($j = 0$) holds

$$\max_{x \in [a,b]} \left| \left(f(x) - \eta_n (h!)^{-1} x^h \right) - Q_n(x) \right| \leq C_{r,s} \left(\frac{b-a}{2n} \right)^r \omega_s \left(f^{(r)}, \frac{b-a}{2n} \right),$$
$$(23.2.40)$$

and

$$\max_{x \in [a,b]} |f(x) - Q_n(x)| \leq$$

$$\eta_n (h!)^{-1} (\max(|a|, |b|))^h + C_{r,s} \left(\frac{b-a}{2n} \right)^r \omega_s \left(f^{(r)}, \frac{b-a}{2n} \right) =$$

$$\eta_n (h!)^{-1} \max \left(|a|^h, |b|^h \right) + C_{r,s} \left(\frac{b-a}{2n} \right)^r \omega_s \left(f^{(r)}, \frac{b-a}{2n} \right), \qquad (23.2.41)$$

etc.

So using triangle's inequality on (23.2.39) and similar reasoning as in the first part of the proof we establish again (23.2.4), (23.2.6) and (23.2.7).

Here again

$$L = \sum_{j=h}^{v} \alpha_j(x) \left[D_{b-;g}^{\alpha_j} \right],$$

and suppose, throughout $\left[a, g^{-1}\left(g\left(b\right) - 1\right)\right]$, $Lf \geq 0$. So over $a \leq x \leq g^{-1}$ $\left(g\left(b\right) - 1\right)$, we get

$$\alpha_h^{-1}\left(x\right) L\left(Q_n\left(x\right)\right) \overset{(23.2.22)}{=} \alpha_h^{-1}\left(x\right) L\left(f\left(x\right)\right) - \frac{\eta_n}{h!}\left(D_{b-;g}^{\alpha_h}\left(x^h\right)\right) +$$

$$\sum_{j=h}^{v} \alpha_h^{-1}\left(x\right) \alpha_j\left(x\right) \left[D_{b-;g}^{\alpha_j} Q_n\left(x\right) - D_{b-;g}^{\alpha_j} f\left(x\right) + \frac{\eta_n}{h!} D_{b-;g}^{\alpha_j} x^h\right] \overset{(23.2.39)}{\leq} \quad (23.2.42)$$

$$-\frac{\eta_n}{h!}\left(D_{b-;g}^{\alpha_h}\left(x^h\right)\right) +$$

$$\left(\sum_{j=h}^{v} s_j \frac{\left(g\left(b\right) - g\left(a\right)\right)^{j-\alpha_j}}{\Gamma\left(j - \alpha_j + 1\right)} \left(\frac{b-a}{2n}\right)^{r-j}\right) C_{r,s} \omega_s \left(f^{(r)}, \frac{b-a}{2n}\right) \overset{(23.2.16)}{=}$$

$$(23.2.43)$$

$$-\frac{\eta_n}{h!}\left(D_{b-;g}^{\alpha_h}\left(x^h\right)\right) + \eta_n = \eta_n\left(1 - \frac{D_{b-;g}^{\alpha_h}\left(x^h\right)}{h!}\right) = \quad (23.2.44)$$

$$\eta_n\left(1 - \frac{1}{\Gamma\left(h - \alpha_h\right) h!} \int_x^b \left(g\left(t\right) - g\left(x\right)\right)^{h-\alpha_h-1} g'\left(t\right) \left(t^h\right)^{(h)} dt\right) =$$

$$\eta_n\left(1 - \frac{1}{\Gamma\left(h - \alpha_h\right)} \int_x^b \left(g\left(t\right) - g\left(x\right)\right)^{h-\alpha_h-1} g'\left(t\right) dt\right) \overset{(23.1.30)}{=}$$

$$\eta_n\left(1 - \frac{\left(g\left(b\right) - g\left(x\right)\right)^{h-\alpha_h}}{\Gamma\left(h - \alpha_h + 1\right)}\right) = \quad (23.2.45)$$

$$\eta_n\left(\frac{\Gamma\left(h - \alpha_h + 1\right) - \left(g\left(b\right) - g\left(x\right)\right)^{h-\alpha_h}}{\Gamma\left(h - \alpha_h + 1\right)}\right) \overset{(23.2.38)}{\leq}$$

$$\eta_n\left(\frac{1 - \left(g\left(b\right) - g\left(x\right)\right)^{h-\alpha_h}}{\Gamma\left(h - \alpha_h + 1\right)}\right) \leq 0. \quad (23.2.46)$$

Hence again
$$L\left(Q_n\left(x\right)\right) \geq 0, \quad \forall x \in \left[a, g^{-1}\left(g\left(b\right) - 1\right)\right].$$

The case of $\alpha_h = h$ is trivially concluded from the above. The proof of the theorem is now over. $\qquad\square$

We make:

Remark 23.16 By Corollary 23.14 we have that $D_{b-;g}^{\alpha_j} f$ are continuous functions, $j = 0, 1, \ldots, r$. Suppose that $\alpha_h\left(x\right), \ldots, \alpha_v\left(x\right)$ are continuous functions on $[a, b]$, and $L(f) \geq 0$ on $\left[a, g^{-1}(g(b) - 1)\right]$ is replaced by $L(f) > 0$ on $\left[a, g^{-1}(g(b) - 1)\right]$. Disregard the assumption made in the main theorem on $\alpha_h(x)$. For $n \subset \mathbb{N}$, let $Q_n(x)$ be the $Q_n^*(x)$ of Theorem 23.10, and f as in Theorem 23.10 (same as in Theorem 23.15). Then $D_{b-;g}^{\alpha_j} Q_n^*$ converges uniformly to $D_{b-;g}^{\alpha_j} f$ at a higher rate given by inequality (23.2.14), in particular for $h \leq j \leq v$. Moreover, because $L\left(Q_n^*\right)$ converges uniformly to $L(f)$ on $[a, b]$, $L\left(Q_n^*\right) > 0$ on $\left[a, g^{-1}(g(b) - 1)\right]$ for sufficiently large n.

23.3 Applications (to Theorem 23.15)

(1) When $g(x) = \ln x$ on $[a, b]$, $0 < a < b < \infty$:
Here we need $ae < b$, and the restriction are on $\left[a, \frac{b}{e}\right]$.
(2) When $g(x) = e^x$ on $[a, b]$, $a < b < \infty$:
Here we need $b > \ln(1 + e^a)$, and the restriction are on $\left[a, \ln\left(e^b - 1\right)\right]$.
(3) When, $A > 1$, $g(x) = A^x$ on $[a, b]$, $a < b < \infty$:
Here we need $b > \log_A(1 + A^a)$, and the restriction are $\left[a, \log_A\left(A^b - 1\right)\right]$.
(4) When $\sigma > 0$, $g(x) = x^\sigma$, $0 \leq a < b < \infty$:
Here we need $b > (1 + a^\sigma)^{\frac{1}{\sigma}}$, and the restriction are on $\left[a, (b^\sigma - 1)^{\frac{1}{\sigma}}\right]$.

References

1. G.A. Anastassiou, Fractional representation formulae and right fractional inequalities. Math. Comput. Model. **54**(11–12), 3098–3115 (2011)
2. G.A. Anastassiou, The reduction method in fractional calculus and fractional Ostrowski type inequalities. Indian J. Math. **56**(3), 333–357 (2014)
3. G.A. Anastassiou, *Univariate Right Fractional Polynomial High order Monotone Approximation*, Demonstratio Mathematica (2014)
4. G. Anastassiou, *Univariate right general high order fractional monotone approximation theory*, Panam. Math. J. (2015)
5. G.A. Anastassiou, *Univariate left General High order Fractional Monotone Approximation*, submitted for publication (2015)
6. G.A. Anastassiou, *Right General Fractional Monotone Approximation*, submitted for publication (2015)
7. G.A. Anastassiou, O. Shisha, Monotone approximation with linear differential operators. J. Approx. Theory **44**, 391–393 (1985)
8. R.A. DeVore, G.G. Lorentz, *Constructive Approximation* (Springer, New York, 1993)
9. Rong-Qing Jia, *Chapter 3. Absolutely Continuous Functions*, https://www.ualberta.ca/~rjia/Math418/Notes/Chap.3.pdf
10. A.A. Kilbas, H.M. Srivastava, J.J. Trujillo, *Theory and Applications of Fractional Differential Equations*, vol 204, North-Holland Mathematics Studies (Elsevier, New York, 2006)
11. H.L. Royden, *Real Analysis*, 2nd edn. (Macmillan Publishing Co. Inc., New York, 1968)
12. A.R. Schep, *Differentiation of Monotone Functions*, https://people.math.sc.edu/schep/diffmonotone.pdf
13. O. Shisha, Monotone approximation. Pac. J. Math. **15**, 667–671 (1965)

Chapter 24
Advanced Fractional Taylor's Formulae

Here are presented five new advanced fractional Taylor's formulae under as weak as possible assumptions. It follows [6].

24.1 Introduction

In [3] we proved:

Theorem 24.1 *Let* $f, f', \ldots, f^{(n)}; g, g'$ *be continuous functions from* $[a, b]$ *(or* $[b, a]$*) into* $\mathbb{R}$, $n \in \mathbb{N}$. *Assume that* $(g^{-1})^{(k)}$, $k = 0, 1, \ldots, n$, *are continuous functions. Then it holds*

$$f(b) = f(a) + \sum_{k=1}^{n-1} \frac{\left(f \circ g^{-1}\right)^{(k)}(g(a))}{k!} (g(b) - g(a))^k + R_n(a, b), \quad (24.1.1)$$

where

$$R_n(a, b) := \frac{1}{(n-1)!} \int_a^b (g(b) - g(s))^{n-1} \left(f \circ g^{-1}\right)^{(n)} (g(s)) g'(s) \, ds \quad (24.1.2)$$

$$= \frac{1}{(n-1)!} \int_{g(a)}^{g(b)} (g(b) - t)^{n-1} \left(f \circ g^{-1}\right)^{(n)} (t) \, dt.$$

Remark 24.2 Let g be strictly increasing and $g \in AC([a, b])$ (absolutely continuous functions). Set $g([a, b]) = [c, d]$, where $c, d \in \mathbb{R}$, i.e. $g(a) = c, g(b) = d$, and call $l := f \circ g^{-1}$.

Assume that $l \in AC^n([c, d])$ (i.e. $l^{(n-1)} \in AC([c, d])$).

[Obviously here it is implied that $f \in C([a, b])$.]

© Springer International Publishing Switzerland 2016
G.A. Anastassiou and I.K. Argyros, *Intelligent Numerical Methods:*
Applications to Fractional Calculus, Studies in Computational Intelligence 624,
DOI 10.1007/978-3-319-26721-0_24

Furthermore assume that $\left(f \circ g^{-1}\right)^{(n)} \in L_\infty\left([c, d]\right)$. [By this very last assumption, the function $(g(b) - t)^{n-1}\left(f \circ g^{-1}\right)^{(n)}(t)$ is integrable over $[c, d]$. Since $g \in AC([a, b])$ and it is increasing, by [10] the function $(g(b) - g(s))^{n-1}\left(f \circ g^{-1}\right)^{(n)}$ $(g(s))\, g'(s)$ is integrable on $[a, b]$, and again by [10], (24.1.2) is valid in this general setting.] Clearly (24.1.1) is now valid under these general assumptions.

24.2 Results

We need:

Lemma 24.3 *Let g be strictly increasing and $g \in AC([a, b])$. Assume that $\left(f \circ g^{-1}\right)^{(m)}$ is Lebesgue measurable function over $[c, d]$. Then*

$$\left\|\left(f \circ g^{-1}\right)^{(m)}\right\|_{\infty,[c,d]} \le \left\|\left(f \circ g^{-1}\right)^{(m)} \circ g\right\|_{\infty,[a,b]}, \qquad (24.2.1)$$

where $\left(f \circ g^{-1}\right)^{(m)} \circ g \in L_\infty\left([a, b]\right)$.

Proof We observe by definition of $\|\cdot\|_\infty$ that:

$$\left\|\left(f \circ g^{-1}\right)^{(m)} \circ g\right\|_{\infty,[a,b]} = \qquad (24.2.2)$$

$$\inf\left\{M : m\left\{t \in [a, b] : \left|\left(\left(f \circ g^{-1}\right)^{(m)} \circ g\right)(t)\right| > M\right\} = 0\right\},$$

where m is the Lebesgue measure.

Because g is absolutely continuous and strictly increasing function on $[a, b]$, by [12], p. 108, Exercise 14, we get that

$$m\left\{z \in [c, d] : \left|\left(f \circ g^{-1}\right)^{(m)}(z)\right| > M\right\} =$$

$$m\left\{g(t) \in [c, d] : \left|\left(f \circ g^{-1}\right)^{(m)}(g(t))\right| > M\right\} =$$

$$m\left(g\left(\left\{t \in [a, b] : \left|\left(f \circ g^{-1}\right)^{(m)}(g(t))\right| > M\right\}\right)\right) = 0,$$

given that

$$m\left\{t \in [a, b] : \left|\left(\left(f \circ g^{-1}\right)^{(m)} \circ g\right)(t)\right| > M\right\} = 0.$$

Therefore each M of (24.2.2) fulfills

$$M \in \left\{ L : m \left\{ z \in [c, d] : \left| \left(f \circ g^{-1} \right)^{(m)} (z) \right| > L \right\} = 0 \right\}. \qquad (24.2.3)$$

The last implies (24.2.1). □

We give:

Definition 24.4 (*see also* [11, p. 99]) The left and right fractional integrals, respectively, of a function f with respect to given function g are defined as follows:

Let $a, b \in \mathbb{R}$, $a < b$, $\alpha > 0$. Here $g \in AC([a, b])$ and is strictly increasing, $f \in L_\infty([a, b])$. We set

$$\left(I^\alpha_{a+;g} f \right) (x) = \frac{1}{\Gamma(\alpha)} \int_a^x (g(x) - g(t))^{\alpha-1} g'(t) f(t) \, dt, \quad x \geq a, \qquad (24.2.4)$$

where Γ is the gamma function, clearly $\left(I^\alpha_{a+;g} f \right) (a) = 0$, $I^0_{a+;g} f := f$ and

$$\left(I^\alpha_{b-;g} f \right) (x) = \frac{1}{\Gamma(\alpha)} \int_x^b (g(t) - g(x))^{\alpha-1} g'(t) f(t) \, dt, \quad x \leq b, \qquad (24.2.5)$$

clearly $\left(I^\alpha_{b-;g} f \right) (b) = 0$, $I^0_{b-;g} f := f$.

When g is the identity function id, we get that $I^\alpha_{a+;id} = I^\alpha_{a+}$, and $I^\alpha_{b-;id} = I^\alpha_{b-}$, the ordinary left and right Riemann-Liouville fractional integrals, where

$$\left(I^\alpha_{a+} f \right) (x) = \frac{1}{\Gamma(\alpha)} \int_a^x (x - t)^{\alpha-1} f(t) \, dt, \quad x \geq a, \qquad (24.2.6)$$

$\left(I^\alpha_{a+} f \right) (a) = 0$ andsss

$$\left(I^\alpha_{b-} f \right) (x) = \frac{1}{\Gamma(\alpha)} \int_x^b (t - x)^{\alpha-1} f(t) \, dt, \quad x \leq b, \qquad (24.2.7)$$

$\left(I^\alpha_{b-} f \right) (b) = 0$.

In [5], we proved:

Lemma 24.5 *Let* $g \in AC([a, b])$ *which is strictly increasing and* f *Borel measurable in* $L_\infty([a, b])$. *Then* $f \circ g^{-1}$ *is Lebesgue measurable, and*

$$\|f\|_{\infty,[a,b]} \geq \left\| f \circ g^{-1} \right\|_{\infty,[g(a),g(b)]}, \qquad (24.2.8)$$

i.e. $\left(f \circ g^{-1} \right) \in L_\infty([g(a), g(b)])$.

If additionally $g^{-1} \in AC([g(a), g(b)])$, then

$$\|f\|_{\infty,[a,b]} = \|f \circ g^{-1}\|_{\infty,[g(a),g(b)]}. \tag{24.2.9}$$

Remark 24.6 We proved ([5]) that

$$\left(I_{a+;g}^{\alpha}f\right)(x) = \left(I_{g(a)+}^{\alpha}\left(f \circ g^{-1}\right)\right)(g(x)), \quad x \geq a \tag{24.2.10}$$

and

$$\left(I_{b-;g}^{\alpha}f\right)(x) = \left(I_{g(b)-}^{\alpha}\left(f \circ g^{-1}\right)\right)(g(x)), \quad x \leq b. \tag{24.2.11}$$

It is well known that, if f is a Lebesgue measurable function, then there exists f^* a Borel measurable function, such that $f = f^*$, a.e. Also it holds $\|f\|_{\infty} = \|f^*\|_{\infty}$, and $\int \ldots f \ldots dx = \int \ldots f^* \ldots dx$.

Of course a Borel measurable function is a Lebesgue measurable function.

Thus, by Lemma 24.5, we get

$$\|f\|_{\infty,[a,b]} = \|f^*\|_{\infty,[a,b]} \geq \|f^* \circ g^{-1}\|_{\infty,[g(a),g(b)]}. \tag{24.2.12}$$

We observe the following:

Let $\alpha, \beta > 0$, then

$$\left(I_{a+;g}^{\beta}\left(I_{a+;g}^{\alpha}f\right)\right)(x) = \left(I_{a+;g}^{\beta}\left(I_{a+;g}^{\alpha}f^*\right)\right)(x) =$$

$$I_{g(a)+}^{\beta}\left(\left(I_{a+;g}^{\alpha}f^*\right) \circ g^{-1}\right)(g(x)) = I_{g(a)+}^{\beta}\left(I_{g(a)+}^{\alpha}\left(f^* \circ g^{-1}\right) \circ g \circ g^{-1}\right)(g(x)) \tag{24.2.13}$$

$$= \left(I_{g(a)+}^{\beta}I_{g(a)+}^{\alpha}\left(f^* \circ g^{-1}\right)\right)(g(x)) \overset{(by\ [8],\,p.14)}{=}$$

$$\left(I_{g(a)+}^{\beta+\alpha}f^* \circ g^{-1}\right)(g(x)) = \left(I_{a+;g}^{\beta+\alpha}f^*\right)(x) = \left(I_{a+;g}^{\beta+\alpha}f\right)(x) \text{ a.e.}$$

The last is true for all x, if $\alpha + \beta \geq 1$ or $f \in C([a,b])$.

We have proved the semigroup composition property

$$\left(I_{a+;g}^{\alpha}I_{a+;g}^{\beta}f\right)(x) = \left(I_{a+;g}^{\alpha+\beta}f\right)(x) = \left(I_{a+;g}^{\beta}I_{a+;g}^{\alpha}f\right)(x), \quad x \geq a, \tag{24.2.14}$$

a.e., which is true for all x, if $\alpha + \beta \geq 1$ or $f \in C([a,b])$.

Similarly we get

$$\left(I_{b-;g}^{\beta}\left(I_{b-;g}^{\alpha}f\right)\right)(x) = \left(I_{b-;g}^{\beta}\left(I_{b-;g}^{\alpha}f^*\right)\right)(x) =$$

$$I_{g(b)-}^{\beta}\left(\left(I_{b-;g}^{\alpha}f^*\right) \circ g^{-1}\right)(g(x)) = I_{g(b)-}^{\beta}\left(I_{g(b)-}^{\alpha}\left(f^* \circ g^{-1}\right) \circ g \circ g^{-1}\right)(g(x)) \tag{24.2.15}$$

$$= I_{g(b)-}^{\beta} \left(I_{g(b)-}^{\alpha} \left(f^* \circ g^{-1} \right) \right) (g(x)) \overset{\text{(by [1])}}{=}$$

$$\left(I_{g(b)-}^{\beta+\alpha} \left(f^* \circ g^{-1} \right) \right) (g(x)) = \left(I_{b-;g}^{\beta+\alpha} f^* \right) (x) = \left(I_{b-;g}^{\beta+\alpha} f \right) (x) \text{ a.e.},$$

true for all $x \in [a, b]$, if $\alpha + \beta \geq 1$ or $f \in C([a, b])$.

We have proved the semigroup property that

$$\left(I_{b-;g}^{\alpha} I_{b-;g}^{\beta} f \right) (x) = \left(I_{b-;g}^{\alpha+\beta} f \right) (x) = \left(I_{b-;g}^{\beta} I_{b-;g}^{\alpha} f \right) (x), \text{ a.e., } x \leq b, \quad (24.2.16)$$

which is true for all $x \in [a, b]$, if $\alpha + \beta \geq 1$ or $f \in C([a, b])$.

From now on without loss of generality, within integrals we may assume that $f = f^*$, and we mean that $f = f^*$, a.e.

We make:

Definition 24.7 Let $\alpha > 0$, $\lceil \alpha \rceil = n$, $\lceil \cdot \rceil$ the ceiling of the number. Again here $g \in AC([a, b])$ and strictly increasing. We assume that $\left(f \circ g^{-1} \right)^{(n)} \circ g \in L_\infty([a, b])$. We define the left generalized g-fractional derivative of f of order α as follows:

$$\left(D_{a+;g}^{\alpha} f \right) (x) := \frac{1}{\Gamma(n-\alpha)} \int_a^x (g(x) - g(t))^{n-\alpha-1} g'(t) \left(f \circ g^{-1} \right)^{(n)} (g(t)) \, dt,$$
$$(24.2.17)$$

$x \geq a$.

If $\alpha \notin \mathbb{N}$, by [7], we have that $D_{a+;g}^{\alpha} f \in C([a, b])$.

We see that

$$\left(I_{a+;g}^{n-\alpha} \left(\left(f \circ g^{-1} \right)^{(n)} \circ g \right) \right) (x) = \left(D_{a+;g}^{\alpha} f \right) (x), \quad x \geq a. \quad (24.2.18)$$

We set

$$D_{a+;g}^{n} f (x) := \left(\left(f \circ g^{-1} \right)^{(n)} \circ g \right) (x), \quad (24.2.19)$$

$$D_{a+;g}^{0} f (x) = f(x), \quad \forall x \in [a, b]. \quad (24.2.20)$$

When $g = id$, then

$$D_{a+;g}^{\alpha} f = D_{a+;id}^{\alpha} f = D_{*a}^{\alpha} f, \quad (24.2.21)$$

the usual left Caputo fractional derivative.

We make:

Remark 24.8 Under the assumption that $\left(f \circ g^{-1} \right)^{(n)} \circ g \in L_\infty([a, b])$, which could be considered as Borel measurable within integrals, we obtain

$$\left(I^{\alpha}_{a+;g} D^{\alpha}_{a+;g} f\right)(x) = \left(I^{\alpha}_{a+;g}\left(I^{n-\alpha}_{a+;g}\left(\left(f \circ g^{-1}\right)^{(n)} \circ g\right)\right)\right)(x) =$$

$$\left(I^{\alpha+n-\alpha}_{a+;g}\left(\left(f \circ g^{-1}\right)^{(n)} \circ g\right)\right)(x) = I^n_{a+;g}\left(\left(f \circ g^{-1}\right)^{(n)} \circ g\right)(x) = \quad (24.2.22)$$

$$\frac{1}{(n-1)!}\int_a^x (g(x) - g(t))^{n-1} g'(t)\left(\left(f \circ g^{-1}\right)^{(n)} \circ g\right)(t)\,dt.$$

We have proved that

$$\left(I^{\alpha}_{a+;g} D^{\alpha}_{a+;g} f\right)(x) = \frac{1}{(n-1)!}\int_a^x (g(x) - g(t))^{n-1} g'(t)\left(f \circ g^{-1}\right)^{(n)}(g(t))\,dt$$

$$\qquad\qquad\qquad (24.2.23)$$

$$= R_n(a, x), \quad \forall x \geq a,$$

see (24.1.2).

But also it holds

$$R_n(a, x) = \left(I^{\alpha}_{a+;g} D^{\alpha}_{a+;g} f\right)(x) = \qquad (24.2.24)$$

$$\frac{1}{\Gamma(\alpha)}\int_a^x (g(x) - g(t))^{\alpha-1} g'(t)\left(D^{\alpha}_{a+;g} f\right)(t)\,dt, \quad x \geq a.$$

We have proved the following g-left fractional generalized Taylor's formula:

Theorem 24.9 *Let g be strictly increasing function and $g \in AC([a, b])$. We assume that $\left(f \circ g^{-1}\right) \in AC^n([g(a), g(b)])$, where $\mathbb{N} \ni n = \lceil \alpha \rceil, \alpha > 0$. Also we assume that $\left(f \circ g^{-1}\right)^{(n)} \circ g \in L_{\infty}([a, b])$. Then*

$$f(x) = f(a) + \sum_{k=1}^{n-1} \frac{\left(f \circ g^{-1}\right)^{(k)}(g(a))}{k!} (g(x) - g(a))^k +$$

$$\frac{1}{\Gamma(\alpha)}\int_a^x (g(x) - g(t))^{\alpha-1} g'(t)\left(D^{\alpha}_{a+;g} f\right)(t)\,dt, \quad \forall x \in [a, b]. \quad (24.2.25)$$

Calling $R_n(a, x)$ the remainder of (24.2.25), we get that

$$R_n(a, x) = \frac{1}{\Gamma(\alpha)}\int_{g(a)}^{g(x)} (g(x) - z)^{\alpha-1}\left(\left(D^{\alpha}_{a+;g} f\right) \circ g^{-1}\right)(z)\,dz, \quad \forall x \in [a, b]. \quad (24.2.26)$$

Remark 24.10 By [7], $R_n(a, x)$ is a continuous function in $x \in [a, b]$. Also, by [10], change of variable in Lebesgue integrals, (24.2.26) is valid.

By [3] we have:

Theorem 24.11 Let $f, f', \ldots, f^{(n)}$; g, g' be continuous from $[a, b]$ into $\mathbb{R}$, $n \in \mathbb{N}$. Assume that $\left(g^{-1}\right)^{(k)}$, $k = 0, 1, \ldots, n$, are continuous. Then

$$f(x) = f(b) + \sum_{k=1}^{n-1} \frac{\left(f \circ g^{-1}\right)^{(k)}(g(b))}{k!} (g(x) - g(b))^k + R_n(b, x), \quad (24.2.27)$$

where

$$R_n(b, x) := \frac{1}{(n-1)!} \int_b^x (g(x) - g(s))^{n-1} \left(f \circ g^{-1}\right)^{(n)}(g(s)) g'(s) \, ds$$
$$(24.2.28)$$

$$= \frac{1}{(n-1)!} \int_{g(b)}^{g(x)} (g(x) - t)^{n-1} \left(f \circ g^{-1}\right)^{(n)}(t) \, dt, \quad \forall x \in [a, b]. \quad (24.2.29)$$

Notice that (24.2.27)–(24.2.29) are valid under more general weaker assumptions, as follows: g is strictly increasing and $g \in AC([a, b])$, $\left(f \circ g^{-1}\right) \in AC^n([g(a), g(b)])$, and $\left(f \circ g^{-1}\right)^{(n)} \in L_\infty([g(a), g(b)])$.

We make:

Definition 24.12 Here we assume that $\left(f \circ g^{-1}\right)^{(n)} \circ g \in L_\infty([a, b])$, where $\mathbb{N} \ni n = \lceil \alpha \rceil$, $\alpha > 0$. We define the right generalized g-fractional derivative of f of order α as follows:

$$\left(D_{b-;g}^\alpha f\right)(x) := \frac{(-1)^n}{\Gamma(n-\alpha)} \int_x^b (g(t) - g(x))^{n-\alpha-1} g'(t) \left(f \circ g^{-1}\right)^{(n)}(g(t)) \, dt,$$
$$(24.2.30)$$

all $x \in [a, b]$.

If $\alpha \notin \mathbb{N}$, by [8], we get that $\left(D_{b-;g}^\alpha f\right) \in C([a, b])$.

We see that

$$I_{b-;g}^{n-\alpha}\left((-1)^n \left(f \circ g^{-1}\right)^{(n)} \circ g\right)(x) = \left(D_{b-;g}^\alpha f\right)(x), \quad a \le x \le b. \quad (24.2.31)$$

We set

$$D_{b-;g}^n f(x) = (-1)^n \left(\left(f \circ g^{-1}\right)^{(n)} \circ g\right)(x), \quad (24.2.32)$$

$$D_{b-;g}^0 f(x) = f(x), \quad \forall x \in [a, b].$$

When $g = id$, then

$$D_{b-;g}^\alpha f(x) = D_{b-;id}^\alpha f(x) = D_{b-}^\alpha f, \quad (24.2.33)$$

the usual right Caputo fractional derivative.

We make:

Remark 24.13 Furthermore it holds

$$\left(I^{\alpha}_{b-;g}D^{\alpha}_{b-;g}f\right)(x) = \left(I^{\alpha}_{b-;g}I^{n-\alpha}_{b-;g}\left((-1)^n\left(f\circ g^{-1}\right)^{(n)}\circ g\right)\right)(x) =$$

$$\left(I^n_{b-;g}\left((-1)^n\left(f\circ g^{-1}\right)^{(n)}\circ g\right)\right)(x) = (-1)^n\left(I^n_{b-;g}\left(\left(f\circ g^{-1}\right)^{(n)}\circ g\right)\right)(x)$$

$$(24.2.34)$$

$$= \frac{(-1)^n}{(n-1)!}\int_x^b (g(t)-g(x))^{n-1}\,g'(t)\left(\left(f\circ g^{-1}\right)^{(n)}\circ g\right)(t)\,dt =$$

$$\frac{(-1)^{2n}}{(n-1)!}\int_b^x (g(x)-g(t))^{n-1}\,g'(t)\left(\left(f\circ g^{-1}\right)^{(n)}\circ g\right)(t)\,dt =$$

$$\frac{1}{(n-1)!}\int_b^x (g(x)-g(t))^{n-1}\,g'(t)\left(\left(f\circ g^{-1}\right)^{(n)}\circ g\right)(t)\,dt = R_n(b,x),$$

$$(24.2.35)$$

as in (24.2.28).

That is

$$R_n(b,x) = \left(I^{\alpha}_{b-;g}D^{\alpha}_{b-;g}f\right)(x) =$$

$$\frac{1}{\Gamma(\alpha)}\int_x^b (g(t)-g(x))^{\alpha-1}\,g'(t)\left(D^{\alpha}_{b-;g}f\right)(t)\,dt, \quad \text{all } a \le x \le b. \quad (24.2.36)$$

We have proved the g-right generalized fractional Taylor's formula:

Theorem 24.14 *Let g be strictly increasing function and $g \in AC([a,b])$. We assume that $\left(f\circ g^{-1}\right) \in AC^n([g(a),g(b)])$, where $\mathbb{N} \ni n = \lceil\alpha\rceil$, $\alpha > 0$. Also we assume that $\left(f\circ g^{-1}\right)^{(n)}\circ g \in L_\infty([a,b])$. Then*

$$f(x) = f(b) + \sum_{k=1}^{n-1}\frac{\left(f\circ g^{-1}\right)^{(k)}(g(b))}{k!}(g(x)-g(b))^k +$$

$$\frac{1}{\Gamma(\alpha)}\int_x^b (g(t)-g(x))^{\alpha-1}\,g'(t)\left(D^{\alpha}_{b-;g}f\right)(t)\,dt, \quad \text{all } a \le x \le b. \quad (24.2.37)$$

Calling $R_n(b,x)$ the remainder in (24.2.37), we get that

$$R_n(b,x) = \frac{1}{\Gamma(\alpha)}\int_{g(x)}^{g(b)}(z-g(x))^{\alpha-1}\left(\left(D^{\alpha}_{b-;g}f\right)\circ g^{-1}\right)(z)\,dz, \quad \forall x \in [a,b].$$

$$(24.2.38)$$

Remark 24.15 By [8], $R_n(b,x)$ is a continuous function in $x \in [a,b]$. Also, by [10], change of variable in Lebesgue integrals, (24.2.38) is valid.

Basics 24.16 *The right Riemann-Liouville fractional integral of order* $\alpha > 0$, $f \in L_1 ([a, b])$, $a < b$, *is defined as follows:*

$$I_{b-}^{\alpha} f (x) := \frac{1}{\Gamma (\alpha)} \int_x^b (z - x)^{\alpha-1} f (z) \, dz, \quad \forall x \in [a, b]. \qquad (24.2.39)$$

$$I_{b-}^0 := I \ (the \ identity \ operator).$$

Let $\alpha, \beta \geq 0$, $f \in L_1 ([a, b])$. *Then, by [1], we have*

$$I_{b-}^{\alpha} I_{b-}^{\beta} f = I_{b-}^{\alpha+\beta} f = I_{b-}^{\beta} I_{b-}^{\alpha} f, \qquad (24.2.40)$$

valid a.e. on $[a, b]$. *If* $f \in C ([a, b])$ *or* $\alpha + \beta \geq 1$, *then the last identity is true on all of* $[a, b]$.

The right Caputo fractional derivative of order $\alpha > 0$, $m = \lceil \alpha \rceil$, $f \in AC^m ([a, b])$ *is defined as follows:*

$$D_{b-}^{\alpha} f (x) := (-1)^m I_{b-}^{m-\alpha} f^{(m)} (x), \qquad (24.2.41)$$

that is

$$D_{b-}^{\alpha} f (x) = \frac{(-1)^m}{\Gamma (m - \alpha)} \int_x^b (z - x)^{m-\alpha-1} f^{(m)} (z) \, dz, \quad \forall x \in [a, b], \quad (24.2.42)$$

with $D_{b-}^m f (x) := (-1)^m f^{(m)} (x)$.

By [1], we have the following right fractional Taylor's formula:
Let $f \in AC^m ([a, b])$, $x \in [a, b]$, $\alpha > 0$, $m = \lceil \alpha \rceil$, then

$$f (x) - \sum_{k=0}^{m-1} \frac{f^{(k)} (b)}{k!} (x - b)^k = \frac{1}{\Gamma (\alpha)} \int_x^b (z - x)^{\alpha-1} D_{b-}^{\alpha} f (z) \, dz = \quad (24.2.43)$$

$$\left(I_{b-}^{\alpha} D_{b-}^{\alpha} f \right) (x) = (-1)^m \left(I_{b-}^{\alpha} I_{b-}^{m-\alpha} f^{(m)} \right) (x) = (-1)^m \left(I_{b-}^m f^{(m)} \right) (x) =$$

$$(-1)^m \frac{1}{(m - 1)!} \int_x^b (z - x)^{m-1} f^{(m)} (z) \, dz =$$

$$(-1)^m \frac{(-1)^m}{(m - 1)!} \int_b^x (x - z)^{m-1} f^{(m)} (z) \, dz = \qquad (24.2.44)$$

$$\frac{1}{(m - 1)!} \int_b^x (x - z)^{m-1} f^{(m)} (z) \, dz.$$

That is

$$\left(I_{b-}^{\alpha} D_{b-}^{\alpha} f\right)(x) = (-1)^m \left(I_{b-}^m f^{(m)}\right)(x) =$$

$$f(x) - \sum_{k=0}^{m-1} \frac{f^{(k)}(b)}{k!} (x-b)^k = \frac{1}{(m-1)!} \int_b^x (x-z)^{m-1} f^{(m)}(z)\, dz. \quad (24.2.45)$$

We make:

Remark 24.17 If $0 < \alpha \le 1$, then $m = 1$, hence

$$\left(I_{b-}^{\alpha} D_{b-}^{\alpha} f\right)(x) = f(x) - f(b) \quad (24.2.46)$$

$$= \frac{1}{\Gamma(\alpha)} \int_x^b (z-x)^{\alpha-1} D_{b-}^{\alpha} f(z)\, dz =: (\psi_1).$$

[Let $f' \in L_\infty([a,b])$, then by [4], we get that $D_{b-}^{\alpha} f \in C([a,b])$, $0 < \alpha < 1$, where

$$\left(D_{b-}^{\alpha} f\right)(x) = \frac{(-1)}{\Gamma(1-\alpha)} \int_x^b (z-x)^{-\alpha} f'(z)\, dz, \quad (24.2.47)$$

with $\left(D_{b-}^1 f\right)(x) = -f'(x)$.
Also $(z-x)^{\alpha-1} > 0$, over (x,b), and

$$\int_x^b (z-x)^{\alpha-1}\, dz = \frac{(b-x)^\alpha}{\alpha} < \infty, \quad \text{for any } 0 < \alpha \le 1, \quad (24.2.48)$$

thus $(z-x)^{\alpha-1}$ is integrable over $[x,b]$.]

By the first mean value theorem for integration, when $0 < \alpha < 1$, we get that

$$(\psi_1) = \frac{\left(D_{b-}^{\alpha} f\right)(\xi_x)}{\Gamma(\alpha)} \int_x^b (z-x)^{\alpha-1}\, dz = \frac{\left(D_{b-}^{\alpha} f\right)(\xi_x)}{\Gamma(\alpha)} \frac{(b-x)^\alpha}{\alpha} \quad (24.2.49)$$

$$= \frac{\left(D_{b-}^{\alpha} f\right)(\xi_x)}{\Gamma(\alpha+1)} (b-x)^\alpha, \quad \xi_x \in [x,b].$$

Thus, we obtain

$$f(x) - f(b) = \frac{\left(D_{b-}^{\alpha} f\right)(\xi_x)}{\Gamma(\alpha+1)} (b-x)^\alpha, \quad \xi_x \in [x,b], \quad (24.2.50)$$

where $f \in AC([a,b])$.

We have proved:

Theorem 24.18 (Right generalized mean value theorem) *Let* $f \in AC([a, b])$, $f' \in L_\infty([a, b])$, $0 < \alpha < 1$. *Then*

$$f(x) - f(b) = \frac{(D^\alpha_{b-} f)(\xi_x)}{\Gamma(\alpha + 1)}(b - x)^\alpha, \tag{24.2.51}$$

with $x \leq \xi_x \leq b$, *where* $x \in [a, b]$.

If $f \in C([a, b])$ and there exists $f'(x)$, for any $x \in (a, b)$, then

$$f(x) - f(b) = (-1) f'(\xi_x)(b - x), \tag{24.2.52}$$

equivalently,

$$f(b) - f(x) = f'(\xi_x)(b - x), \tag{24.2.53}$$

the usual mean value theorem.
We make:

Remark 24.19 In general: we notice the following

$$\left| D^\alpha_{b-} f(x) \right| \leq \frac{1}{\Gamma(m - \alpha)} \int_x^b (z - x)^{m-\alpha-1} \left| f^{(m)}(z) \right| dz$$

(assuming $f^{(m)} \in L_\infty([a, b])$)

$$\leq \frac{\left\| f^{(m)} \right\|_\infty}{\Gamma(m - \alpha)} \int_x^b (z - x)^{m-\alpha-1} dz = \frac{\left\| f^{(m)} \right\|_\infty}{\Gamma(m - \alpha)} \frac{(b - x)^{m-\alpha}}{m - \alpha} \tag{24.2.54}$$

$$= \frac{\left\| f^{(m)} \right\|_\infty}{\Gamma(m - \alpha + 1)}(b - x)^{m-\alpha} \leq \frac{\left\| f^{(m)} \right\|_\infty}{\Gamma(m - \alpha + 1)}(b - a)^{m-\alpha}.$$

So when $f^{(m)} \in L_\infty([a, b])$ we get that

$$D^\alpha_{b-} f(b) = 0, \quad \text{where } \alpha \notin \mathbb{N}, \tag{24.2.55}$$

and

$$\left\| D^\alpha_{b-} f \right\|_\infty \leq \frac{\left\| f^{(m)} \right\|_\infty}{\Gamma(m - \alpha + 1)}(b - a)^{m-\alpha}. \tag{24.2.56}$$

In particular when $f' \in L_\infty([a, b])$, $0 < \alpha < 1$, we have that

$$D^\alpha_{b-} f(b) = 0. \tag{24.2.57}$$

Notation 24.20 *Denote by*

$$D_{b-}^{n\alpha} := D_{b-}^{\alpha} D_{b-}^{\alpha} \dots D_{b-}^{\alpha} \quad (n \text{ times}), n \in \mathbb{N}. \qquad (24.2.58)$$

Also denote by

$$I_{b-}^{n\alpha} := I_{b-}^{\alpha} I_{b-}^{\alpha} \dots I_{b-}^{\alpha} \quad (n \text{ times}), n \in \mathbb{N}. \qquad (24.2.59)$$

We have:

Theorem 24.21 *Suppose that* $D_{b-}^{n\alpha} f$, $D_{b-}^{(n+1)\alpha} f \in C([a, b])$, $0 < \alpha \leq 1$. *Then*

$$\left(I_{b-}^{n\alpha} D_{b-}^{n\alpha} f\right)(x) - \left(I_{b-}^{(n+1)\alpha} D_{b-}^{(n+1)\alpha} f\right)(x) = \frac{(b-x)^{n\alpha}}{\Gamma(n\alpha+1)} \left(D_{b-}^{n\alpha} f\right)(b). \quad (24.2.60)$$

Proof By (24.2.40) we get that

$$\left(I_{b-}^{n\alpha} D_{b-}^{n\alpha} f\right)(x) - \left(I_{b-}^{(n+1)\alpha} D_{b-}^{(n+1)\alpha} f\right)(x) =$$

$$I_{b-}^{n\alpha} \left((D_{b-}^{n\alpha} f)(x) - \left(I_{b-}^{\alpha} D_{b-}^{(n+1)\alpha} f\right)(x)\right) =$$

$$I_{b-}^{n\alpha} \left((D_{b-}^{n\alpha} f)(x) - \left((I_{b-}^{\alpha} D_{b-}^{\alpha})(D_{b-}^{n\alpha} f)\right)(x)\right) \overset{(24.1.48)}{=}$$

$$I_{b-}^{n\alpha} \left((D_{b-}^{n\alpha} f)(x) - (D_{b-}^{n\alpha} f)(x) + (D_{b-}^{n\alpha} f)(b)\right) = \qquad (24.2.61)$$

$$I_{b-}^{n\alpha} \left((D_{b-}^{n\alpha} f)(b)\right) = \frac{(b-x)^{n\alpha}}{\Gamma(n\alpha+1)} \left(D_{b-}^{n\alpha} f\right)(b).$$

$$\square$$

Remark 24.22 Suppose that $D_{b-}^{k\alpha} f \in C([a, b])$, for $k = 0, 1, \dots, n+1$; $0 < \alpha \leq 1$. By (24.2.60) we get that

$$\sum_{i=0}^{n} \left((I_{b-}^{i\alpha} D_{b-}^{i\alpha} f)(x) - \left(I_{b-}^{(i+1)\alpha} D_{b-}^{(i+1)\alpha} f\right)(x)\right) =$$

$$\sum_{i=0}^{n} \frac{(b-x)^{i\alpha}}{\Gamma(i\alpha+1)} \left(D_{b-}^{i\alpha} f\right)(b). \qquad (24.2.62)$$

That is

$$f(x) - \left(I_{b-}^{(n+1)\alpha} D_{b-}^{(n+1)\alpha} f\right)(x) = \sum_{i=0}^{n} \frac{(b-x)^{i\alpha}}{\Gamma(i\alpha+1)} \left(D_{b-}^{i\alpha} f\right)(b). \qquad (24.2.63)$$

Hence it holds

$$f(x) = \sum_{i=0}^{n} \frac{(b-x)^{i\alpha}}{\Gamma(i\alpha+1)} \left(D_{b-}^{i\alpha}f\right)(b) + \left(I_{b-}^{(n+1)\alpha} D_{b-}^{(n+1)\alpha}f\right)(x) = \quad (24.2.64)$$

$$\sum_{i=0}^{n} \frac{(b-x)^{i\alpha}}{\Gamma(i\alpha+1)} \left(D_{b-}^{i\alpha}f\right)(b) + R^*(x,b),$$

where

$$R^*(x,b) := \frac{1}{\Gamma((n+1)\alpha)} \int_x^b (z-x)^{(n+1)\alpha-1} \left(D_{b-}^{(n+1)\alpha}f\right)(z)\, dz. \quad (24.2.65)$$

We see that (there exists $\xi_x \in [x,b]$:)

$$R^*(x,b) = \frac{\left(D_{b-}^{(n+1)\alpha}f\right)(\xi_x)}{\Gamma((n+1)\alpha)} \int_x^b (z-x)^{(n+1)\alpha-1}\, dz =$$

$$\frac{\left(D_{b-}^{(n+1)\alpha}f\right)(\xi_x)}{\Gamma((n+1)\alpha)} \frac{(b-x)^{(n+1)\alpha}}{(n+1)\alpha} = \frac{\left(D_{b-}^{(n+1)\alpha}f\right)(\xi_x)}{\Gamma((n+1)\alpha+1)} (b-x)^{(n+1)\alpha}. \quad (24.2.66)$$

We have proved the following right generalized fractional Taylor's formula:

Theorem 24.23 *Suppose that $D_{b-}^{k\alpha}f \in C([a,b])$, for $k = 0,1,\ldots,n+1$, where $0 < \alpha \le 1$. Then*

$$f(x) = \sum_{i=0}^{n} \frac{(b-x)^{i\alpha}}{\Gamma(i\alpha+1)} \left(D_{b-}^{i\alpha}f\right)(b) + \quad (24.2.67)$$

$$\frac{1}{\Gamma((n+1)\alpha)} \int_x^b (z-x)^{(n+1)\alpha-1} \left(D_{b-}^{(n+1)\alpha}f\right)(z)\, dz =$$

$$\sum_{i=0}^{n} \frac{(b-x)^{i\alpha}}{\Gamma(i\alpha+1)} \left(D_{b-}^{i\alpha}f\right)(b) + \frac{\left(D_{b-}^{(n+1)\alpha}f\right)(\xi_x)}{\Gamma((n+1)\alpha+1)} (b-x)^{(n+1)\alpha}, \quad (24.2.68)$$

where $\xi_x \in [x,b]$, with $x \in [a,b]$.

We make:

Remark 24.24 Let $\alpha > 0, m = \lceil \alpha \rceil$, g is strictly increasing and $g \in AC([a,b])$. Call $l = f \circ g^{-1}$, $f : [a,b] \to \mathbb{R}$. Assume that $l \in AC^m([c,d])$ (i.e. $l^{(m-1)} \in AC([c,d])$) (where $g([a,b]) = [c,d]$, $c,d \in \mathbb{R} : g(a) = c, g(b) = d$; hence here f is continuous on $[a,b]$).

Assume also that $\left(f \circ g^{-1}\right)^{(m)} \circ g \in L_\infty([a,b])$.

The right generalized g-fractional derivative of f of order α is defined as follows:

$$\left(D_{b-;g}^{\alpha}f\right)(x) := \frac{(-1)^m}{\Gamma(m-\alpha)}\int_x^b (g(t)-g(x))^{m-\alpha-1} g'(t) \left(f\circ g^{-1}\right)^{(m)}(g(t))\,dt,$$

(24.2.69)

$a \leq x \leq b$.

We saw that

$$I_{b-;g}^{m-\alpha}\left((-1)^m \left(f\circ g^{-1}\right)^{(m)}\circ g\right)(x) = \left(D_{b-;g}^{\alpha}f\right)(x), \quad a \leq x \leq b.$$

(24.2.70)

We proved earlier (24.2.35)–(24.15) that $(a \leq x \leq b)$

$$\left(I_{b-;g}^{\alpha}D_{b-;g}^{\alpha}f\right)(x) =$$

$$\frac{1}{(m-1)!}\int_b^x (g(x)-g(t))^{m-1} g'(t)\left(\left(f\circ g^{-1}\right)^{(m)}\circ g\right)(t)\,dt = \qquad (24.2.71)$$

$$\frac{1}{\Gamma(\alpha)}\int_x^b (g(t)-g(x))^{\alpha-1} g'(t)\left(D_{b-;g}^{\alpha}f\right)(t)\,dt =$$

$$f(x) - f(b) - \sum_{k=1}^{m-1} \frac{\left(f\circ g^{-1}\right)^{(k)}(g(b))}{k!}(g(x)-g(b))^k.$$

If $0 < \alpha \leq 1$, then $m = 1$, hence

$$\left(I_{b-;g}^{\alpha}D_{b-;g}^{\alpha}f\right)(x) = f(x) - f(b) \qquad (24.2.72)$$

$$= \frac{1}{\Gamma(\alpha)}\int_x^b (g(t)-g(x))^{\alpha-1} g'(t)\left(D_{b-;g}^{\alpha}f\right)(t)\,dt$$

(when $\alpha \in (0,1)$, $D_{b-;g}^{\alpha}f$ is continuous on $[a,b]$ and)

$$= \frac{\left(D_{b-;g}^{\alpha}f\right)(\xi_x)}{\Gamma(\alpha)}\int_x^b (g(t)-g(x))^{\alpha-1} g'(t)\,dt = \frac{\left(D_{b-;g}^{\alpha}f\right)(\xi_x)}{\Gamma(\alpha+1)}(g(b)-g(x))^{\alpha},$$

(24.2.73)

where $\xi_x \in [x,b]$.

We have proved:

Theorem 24.25 (right generalized g-mean value theorem) *Let* $0 < \alpha < 1$, *and* $f\circ g^{-1} \in AC([c,d])$, $\left(f\circ g^{-1}\right)'\circ g \in L_\infty([a,b])$, *where* g *strictly increasing,* $g \in AC([a,b])$, $f:[a,b] \to \mathbb{R}$. *Then*

$$f(x) - f(b) = \frac{\left(D_{b-;g}^{\alpha} f\right)(\xi_x)}{\Gamma(\alpha+1)} \left(g(b) - g(x)\right)^{\alpha}, \qquad (24.2.74)$$

where $\xi_x \in [x, b]$, for $x \in [a, b]$.

Denote by

$$D_{b-;g}^{n\alpha} := D_{b-;g}^{\alpha} D_{b-;g}^{\alpha} \ldots D_{b-;g}^{\alpha} \; (n \text{ times}), \; n \in \mathbb{N}. \qquad (24.2.75)$$

Also denote by

$$I_{b-;g}^{n\alpha} := I_{b-;g}^{\alpha} I_{b-;g}^{\alpha} \ldots I_{b-;g}^{\alpha} \; (n \text{ times}). \qquad (24.2.76)$$

Here to remind

$$\left(I_{b-;g}^{\alpha} f\right)(x) = \frac{1}{\Gamma(\alpha)} \int_x^b \left(g(t) - g(x)\right)^{\alpha-1} g'(t) f(t) \, dt, \quad x \le b. \qquad (24.2.77)$$

We need:

Theorem 24.26 *Suppose that $F_k := D_{b-;g}^{k\alpha} f$, $k = n, n+1$, fulfill $F_k \circ g^{-1} \in AC([c, d])$, and $\left(F_k \circ g^{-1}\right)' \circ g \in L_{\infty}([a, b])$, $0 < \alpha \le 1$, $n \in \mathbb{N}$. Then*

$$\left(I_{b-;g}^{n\alpha} D_{b-;g}^{n\alpha} f\right)(x) - \left(I_{b-;g}^{(n+1)\alpha} D_{b-;g}^{(n+1)\alpha} f\right)(x) = \frac{\left(g(b) - g(x)\right)^{n\alpha}}{\Gamma(n\alpha+1)} \left(D_{b-;g}^{n\alpha} f\right)(b). \qquad (24.2.78)$$

Proof By semigroup property of $I_{b-;g}^{\alpha}$, we get

$$\left(I_{b-;g}^{n\alpha} D_{b-;g}^{n\alpha} f\right)(x) - \left(I_{b-;g}^{(n+1)\alpha} D_{b-;g}^{(n+1)\alpha} f\right)(x) =$$

$$\left(I_{b-;g}^{n\alpha} \left(D_{b-;g}^{n\alpha} f - I_{b-;g}^{\alpha} D_{b-;g}^{(n+1)\alpha} f\right)\right)(x) = \qquad (24.2.79)$$

$$\left(I_{b-;g}^{n\alpha} \left(D_{b-;g}^{n\alpha} f - \left(I_{b-;g}^{\alpha} D_{b-;g}^{\alpha}\right)\left(D_{b-;g}^{n\alpha} f\right)\right)\right)(x) \overset{(24.1.74)}{=}$$

$$\left(I_{b-;g}^{n\alpha} \left(D_{b-;g}^{n\alpha} f - D_{b-;g}^{n\alpha} f + D_{b-;g}^{n\alpha} f(b)\right)\right)(x) =$$

$$\left(I_{b-;g}^{n\alpha} \left(D_{b-;g}^{n\alpha} f(b)\right)\right)(x) = \left(D_{b-;g}^{n\alpha} f(b)\right)\left(I_{b-;g}^{n\alpha}(1)\right)(x) = \qquad (24.2.80)$$

[Notice that

$$\left(I_{b-;g}^{\alpha} 1\right)(x) = \frac{1}{\Gamma(\alpha)} \int_x^b \left(g(t) - g(x)\right)^{\alpha-1} g'(t) \, dt = \qquad (24.2.81)$$

$$\frac{1}{\Gamma(\alpha)} \frac{(g(b) - g(x))^{\alpha}}{\alpha} = \frac{1}{\Gamma(\alpha + 1)} (g(b) - g(x))^{\alpha}.$$

Thus we have

$$\left(I_{b-;g}^{\alpha} 1\right)(x) = \frac{(g(b) - g(x))^{\alpha}}{\Gamma(\alpha + 1)}. \tag{24.2.82}$$

Hence it holds

$$\left(I_{b-;g}^{2\alpha} 1\right)(x) = \frac{1}{\Gamma(\alpha)} \int_{x}^{b} (g(t) - g(x))^{\alpha - 1} g'(t) \frac{(g(b) - g(t))^{\alpha}}{\Gamma(\alpha + 1)} dt =$$

$$\frac{1}{\Gamma(\alpha) \Gamma(\alpha + 1)} \int_{x}^{b} (g(b) - g(t))^{\alpha} (g(t) - g(x))^{\alpha - 1} g'(t) dt =$$

$$\frac{1}{\Gamma(\alpha) \Gamma(\alpha + 1)} \int_{g(x)}^{g(b)} (g(b) - z)^{(\alpha + 1) - 1} (z - g(x))^{\alpha - 1} dz =$$

$$\frac{1}{\Gamma(\alpha) \Gamma(\alpha + 1)} \frac{\Gamma(\alpha + 1) \Gamma(\alpha)}{\Gamma(2\alpha + 1)} (g(b) - g(x))^{2\alpha} = \frac{1}{\Gamma(2\alpha + 1)} (g(b) - g(x))^{2\alpha}, \tag{24.2.83}$$

etc.]

$$= \left(D_{b-;g}^{n\alpha} f\right)(b) \frac{(g(b) - g(x))^{n\alpha}}{\Gamma(n\alpha + 1)}, \tag{24.2.84}$$

proving the claim. $\qquad\qquad\qquad\qquad\qquad\qquad\qquad\qquad\qquad\qquad\qquad$ □

We make:

Remark 24.27 Suppose that $F_k = D_{b-;g}^{k\alpha} f$, for $k = 0, 1, \ldots, n + 1$; are as in last Theorem 24.26, $0 < \alpha \le 1$. By (24.2.78) we get

$$\sum_{i=0}^{n} \left(\left(I_{b-;g}^{i\alpha} D_{b-;g}^{i\alpha} f\right)(x) - I_{b-;g}^{(i+1)\alpha} D_{b-;g}^{(i+1)\alpha} f(x)\right) = \tag{24.2.85}$$

$$\sum_{i=0}^{n} \frac{(g(b) - g(x))^{i\alpha}}{\Gamma(i\alpha + 1)} \left(D_{b-;g}^{i\alpha} f\right)(b).$$

That is
(notice that $I_{b-;g}^{0} f = D_{b-;g}^{0} f = f$)

$$f(x) - \left(I_{b-;g}^{(n+1)\alpha} D_{b-;g}^{(n+1)\alpha} f\right)(x) = \sum_{i=0}^{n} \frac{(g(b) - g(x))^{i\alpha}}{\Gamma(i\alpha + 1)} \left(D_{b-;g}^{i\alpha} f\right)(b). \tag{24.2.86}$$

Hence

$$f(x) = \sum_{i=0}^{n} \frac{(g(b) - q(x))^{i\alpha}}{\Gamma(i\alpha + 1)} \left(D_{b-;g}^{i\alpha} f\right)(b) + \left(I_{b-,y}^{(n+1)\alpha} D_{b-;g}^{(n+1)\alpha} f\right)(x) = \tag{24.2.87}$$

$$\sum_{i=0}^{n} \frac{(g(b) - g(x))^{i\alpha}}{\Gamma(i\alpha + 1)} \left(D_{b-;g}^{i\alpha} f\right)(b) + R_g(x, b), \tag{24.2.88}$$

where

$$R_g(x, b) := \frac{1}{\Gamma((n+1)\alpha)} \int_x^b (g(t) - g(x))^{(n+1)\alpha - 1} g'(t) \left(D_{b-;g}^{(n+1)\alpha} f\right)(t) dt. \tag{24.2.89}$$

(here $D_{b-;g}^{(n+1)\alpha} f$ is continuous over $[a, b]$).

Hence it holds

$$R_g(x, b) = \frac{\left(D_{b-;g}^{(n+1)\alpha} f\right)(\psi_x)}{\Gamma((n+1)\alpha)} \int_x^b (g(t) - g(x))^{(n+1)\alpha - 1} g'(t) dt =$$

$$\frac{\left(D_{b-;g}^{(n+1)\alpha} f\right)(\psi_x)}{\Gamma((n+1)\alpha)} \frac{(g(b) - g(x))^{(n+1)\alpha}}{(n+1)\alpha} = \frac{\left(D_{b-;g}^{(n+1)\alpha} f\right)(\psi_x)}{\Gamma((n+1)\alpha + 1)} (g(b) - g(x))^{(n+1)\alpha}, \tag{24.2.90}$$

where $\psi_x \in [x, b]$.

We have proved the following g-right generalized modified Taylor's formula:

Theorem 24.28 *Suppose that $F_k := D_{b-;g}^{k\alpha} f$, for $k = 0, 1, \ldots, n + 1$, fulfill: $F_k \circ g^{-1} \in AC([c, d])$ and $\left(F_k \circ g^{-1}\right)' \circ g \in L_\infty([a, b])$, where $0 < \alpha \le 1$. Then*

$$f(x) = \sum_{i=0}^{n} \frac{(g(b) - g(x))^{i\alpha}}{\Gamma(i\alpha + 1)} \left(D_{b-;g}^{i\alpha} f\right)(b) +$$

$$\frac{1}{\Gamma((n+1)\alpha)} \int_x^b (g(t) - g(x))^{(n+1)\alpha - 1} g'(t) \left(D_{b-;g}^{(n+1)\alpha} f\right)(t) dt = \tag{24.2.91}$$

$$\sum_{i=0}^{n} \frac{(g(b) - g(x))^{i\alpha}}{\Gamma(i\alpha + 1)} \left(D_{b-;g}^{i\alpha} f\right)(b) + \frac{\left(D_{b-;g}^{(n+1)\alpha} f\right)(\psi_x)}{\Gamma((n+1)\alpha + 1)} (g(b) - g(x))^{(n+1)\alpha}, \tag{24.2.92}$$

where $\psi_x \in [x, b]$, any $x \in [a, b]$.

We make:

Remark 24.29 Let $\alpha > 0, m = \lceil \alpha \rceil, g$ is strictly increasing and $g \in AC\left([a, b]\right)$. Call $l = f \circ g^{-1}, f : [a, b] \to \mathbb{R}$. Assume $l \in AC^m\left([c, d]\right)$ (i.e. $l^{(m-1)} \in AC\left([c, d]\right)$) (where $g\left([a, b]\right) = [c, d], c, d \in \mathbb{R} : g\left(a\right) = c, g\left(b\right) = d$, hence here f is continuous on $[a, b]$).

Assume also that $\left(f \circ g^{-1}\right)^{(m)} \circ g \in L_\infty\left([a, b]\right)$.

The left generalized g-fractional derivative of f of order α is defined as follows:

$$\left(D^\alpha_{a+;g} f\right)(x) = \frac{1}{\Gamma(m - \alpha)} \int_a^x (g(x) - g(t))^{m-\alpha-1} g'(t) \left(f \circ g^{-1}\right)^{(m)}(g(t)) \, dt,$$

(24.2.93)

$x \geq a$.

If $\alpha \notin \mathbb{N}$, then $\left(D^\alpha_{a+;g} f\right) \in C\left([a, b]\right)$.

We see that

$$\left(I^{m-\alpha}_{a+;g} \left(\left(f \circ g^{-1}\right)^{(m)} \circ g\right)\right)(x) = \left(D^\alpha_{a+;g} f\right)(x), \quad x \geq a.$$

(24.2.94)

We proved earlier (24.2.22)–(24.2.25), that ($a \leq x \leq b$)

$$\left(I^\alpha_{a+;g} D^\alpha_{a+;g} f\right)(x) =$$

$$\frac{1}{(m - 1)!} \int_a^x (g(x) - g(t))^{m-1} g'(t) \left(\left(f \circ g^{-1}\right)^{(m)} \circ g\right)(t) \, dt = \quad (24.2.95)$$

$$\frac{1}{\Gamma(\alpha)} \int_a^x (g(x) - g(t))^{\alpha-1} g'(t) \left(D^\alpha_{a+;g} f\right)(t) \, dt =$$

$$f(x) - f(a) - \sum_{k=1}^{m-1} \frac{\left(f \circ g^{-1}\right)^{(k)}(g(a))}{k!} (g(x) - g(a))^k.$$

(24.2.96)

If $0 < \alpha \leq 1$, then $m = 1$, and then

$$\left(I^\alpha_{a+;g} D^\alpha_{a+;g} f\right)(x) = f(x) - f(a)$$

(24.2.97)

$$= \frac{1}{\Gamma(\alpha)} \int_a^x (g(x) - g(t))^{\alpha-1} g'(t) \left(D^\alpha_{a+;g} f\right)(t) \, dt$$

$$\overset{(\alpha \in (0,1) \text{ case})}{=} \frac{\left(D^\alpha_{a+;g} f\right)(\xi_x)}{\Gamma(\alpha + 1)} (g(x) - g(a))^\alpha,$$

(24.2.98)

where $\xi_x \in [a, x]$, any $x \in [a, b]$.

We have proved:

Theorem 24.30 (left generalized g-mean value theorem) *Let* $0 < \alpha < 1$ *and* $f \circ g^{-1} \in AC([c,d])$ *and* $\left(f \circ g^{-1}\right)' \circ y \in L_\infty([a,b])$, *where* g *strictly increasing,* $g \in AC([a,b])$, $f : [a,b] \to \mathbb{R}$. *Then*

$$f(x) - f(a) = \frac{\left(D^\alpha_{a+;g}f\right)(\xi_x)}{\Gamma(\alpha+1)}(g(x) - g(a))^\alpha, \tag{24.2.99}$$

where $\xi_x \in [a,x]$, *any* $x \in [a,b]$.

Denote by

$$D^{n\alpha}_{a+;g} := D^\alpha_{a+;g}D^\alpha_{a+;g}\cdots D^\alpha_{a+;g} \ (n \text{ times}), n \in \mathbb{N}. \tag{24.2.100}$$

Also denote by

$$I^{n\alpha}_{a+;g} := I^\alpha_{a+;g}I^\alpha_{a+;g}\cdots I^\alpha_{a+;g} \ (n \text{ times}). \tag{24.2.101}$$

Here to remind

$$\left(I^\alpha_{a+;g}f\right)(x) = \frac{1}{\Gamma(\alpha)}\int_a^x (g(x) - g(t))^{\alpha-1}g'(t)f(t)\,dt, \quad x \geq a. \tag{24.2.102}$$

By convention $I^0_{a+;g} = D^0_{a+;g} = I$ (identity operator).
We give:

Theorem 24.31 *Suppose that* $F_k := D^{k\alpha}_{a+;g}f$, $k = n, n+1$, *fulfill* $F_k \circ g^{-1} \in AC([c,d])$, *and* $\left(F_k \circ g^{-1}\right)' \circ g \in L_\infty([a,b])$, $0 < \alpha \leq 1$, $n \in \mathbb{N}$. *Then*

$$\left(I^{n\alpha}_{a+;g}D^{n\alpha}_{a+;g}f\right)(x) - \left(I^{(n+1)\alpha}_{a+;g}D^{(n+1)\alpha}_{a+;g}f\right)(x) = \frac{(g(x) - g(a))^{n\alpha}}{\Gamma(n\alpha+1)}\left(D^{n\alpha}_{a+;g}f\right)(a). \tag{24.2.103}$$

Proof By semigroup property of $I^\alpha_{a+;g}$, we get

$$\left(I^{n\alpha}_{a+;g}D^{n\alpha}_{a+;g}f\right)(x) - \left(I^{(n+1)\alpha}_{a+;g}D^{(n+1)\alpha}_{a+;g}f\right)(x) =$$

$$\left(I^{n\alpha}_{a+;g}\left(D^{n\alpha}_{a+;g}f - I^\alpha_{a+;g}D^{(n+1)\alpha}_{a+;g}f\right)\right)(x) = \tag{24.2.104}$$

$$\left(I^{n\alpha}_{a+;g}\left(D^{n\alpha}_{a+;g}f - \left(I^\alpha_{a+;g}D^\alpha_{a+;g}\right)\left(D^{n\alpha}_{a+;g}f\right)\right)\right)(x) \overset{(24.1.99)}{=}$$

$$\left(I^{n\alpha}_{a+;g}\left(D^{n\alpha}_{a+;g}f - D^{n\alpha}_{a+;g}f + D^{n\alpha}_{a+;g}f(a)\right)\right)(x) =$$

$$\left(I^{n\alpha}_{a+;g}\left(D^{n\alpha}_{a+;g}f(a)\right)\right)(x) = \left(D^{n\alpha}_{a+;g}f(a)\right)\left(I^{n\alpha}_{a+;g}(1)\right)(x) = \tag{24.2.105}$$

[notice that

$$\left(I_{a+;g}^{\alpha}1\right)(x) = \frac{1}{\Gamma(\alpha)}\int_a^x (g(x)-g(t))^{\alpha-1}\,g'(t)\,dt$$

$$= \frac{(g(x)-g(a))^{\alpha}}{\Gamma(\alpha+1)}. \tag{24.2.106}$$

Hence

$$\left(I_{a+;g}^{2\alpha}1\right)(x) = \frac{1}{\Gamma(\alpha)}\int_a^x (g(x)-g(t))^{\alpha-1}\,g'(t)\,\frac{(g(t)-g(a))^{\alpha}}{\Gamma(\alpha+1)}dt =$$
$$\tag{24.2.107}$$
$$\frac{1}{\Gamma(\alpha)\,\Gamma(\alpha+1)}\int_a^x (g(x)-g(t))^{\alpha-1}\,g'(t)\,(g(t)-g(a))^{\alpha}\,dt =$$

$$\frac{1}{\Gamma(\alpha)\,\Gamma(\alpha+1)}\int_{g(a)}^{g(x)} (g(x)-z)^{\alpha-1}\,(z-g(a))^{(\alpha+1)-1}\,dt =$$

$$\frac{1}{\Gamma(\alpha)\,\Gamma(\alpha+1)}\,\frac{\Gamma(\alpha)\,\Gamma(\alpha+1)}{\Gamma(2\alpha+1)}\,(g(x)-g(a))^{2\alpha}.$$

That is

$$\left(I_{a+;g}^{2\alpha}1\right)(x) = \frac{(g(x)-g(a))^{2\alpha}}{\Gamma(2\alpha+1)}, \tag{24.2.108}$$

etc.]

$$= \left(D_{a+;g}^{n\alpha}f(a)\right)\frac{(g(x)-g(a))^{n\alpha}}{\Gamma(n\alpha+1)}, \tag{24.2.109}$$

proving the claim.   □

Remark 24.32 Suppose that $F_k = D_{a+;g}^{k\alpha}f$, for $k = 0,1,\ldots,n+1$; are as in Theorem 24.31, $0 < \alpha \le 1$. By (24.2.103) we get

$$\sum_{i=0}^n \left(\left(I_{a+;g}^{i\alpha}D_{a+;g}^{i\alpha}f\right)(x) - I_{a+;g}^{(i+1)\alpha}D_{a+;g}^{(i+1)\alpha}f(x)\right) = \tag{24.2.110}$$

$$\sum_{i=0}^n \frac{(g(x)-g(a))^{i\alpha}}{\Gamma(i\alpha+1)}\left(D_{a+;g}^{i\alpha}f\right)(a).$$

That is

$$f(x) - \left(I_{a+;g}^{(n+1)\alpha}D_{a+;g}^{(n+1)\alpha}f\right)(x) = \sum_{i=0}^n \frac{(g(x)-g(a))^{i\alpha}}{\Gamma(i\alpha+1)}\left(D_{a+;g}^{i\alpha}f\right)(a).$$

Hence

$$f(x) = \sum_{i=0}^{n} \frac{(g(x) - g(a))^{i\alpha}}{\Gamma(i\alpha + 1)} \left(D_{a+;g}^{i\alpha} f\right)(a) + \left(I_{u+,g}^{(n+1)\alpha} D_{a\mid;g}^{(n+1)\alpha} f\right)(x) =$$

(24.2.111)

$$\sum_{i=0}^{n} \frac{(g(x) - g(a))^{i\alpha}}{\Gamma(i\alpha + 1)} \left(D_{a+;g}^{i\alpha} f\right)(a) + R_g(a, x), \qquad (24.2.112)$$

where

$$R_g(a, x) := \frac{1}{\Gamma((n+1)\alpha)} \int_a^x (g(x) - g(t))^{(n+1)\alpha - 1} g'(t) \left(D_{a+;g}^{(n+1)\alpha} f\right)(t) \, dt.$$

(24.2.113)

(there $D_{a+;g}^{(n+1)\alpha} f$ is continuous over $[a, b]$.)

Hence it holds

$$R_g(a, x) = \frac{\left(D_{a+;g}^{(n+1)\alpha} f\right)(\psi_x)}{\Gamma((n+1)\alpha)} \left(\int_a^x (g(x) - g(t))^{(n+1)\alpha - 1} g'(t) \, dt\right) =$$

$$\frac{\left(D_{a+;g}^{(n+1)\alpha} f\right)(\psi_x)}{\Gamma((n+1)\alpha + 1)} (g(x) - g(a))^{(n+1)\alpha}, \qquad (24.2.114)$$

where $\psi_x \in [a, x]$.

We have proved the following g-left generalized modified Taylor's formula:

Theorem 24.33 *Suppose that $F_k := D_{a+;g}^{k\alpha} f$, for $k = 0, 1, \ldots, n + 1$, fulfill: $F_k \circ g^{-1} \in AC([c, d])$ and $\left(F_k \circ g^{-1}\right)' \circ g \in L_\infty([a, b])$, where $0 < \alpha \le 1$. Then*

$$f(x) = \sum_{i=0}^{n} \frac{(g(x) - g(a))^{i\alpha}}{\Gamma(i\alpha + 1)} \left(D_{a+;g}^{i\alpha} f\right)(a) + \qquad (24.2.115)$$

$$\frac{1}{\Gamma((n+1)\alpha)} \int_a^x (g(x) - g(t))^{(n+1)\alpha - 1} g'(t) \left(D_{a+;g}^{(n+1)\alpha} f\right)(t) \, dt =$$

$$\sum_{i=0}^{n} \frac{(g(x) - g(a))^{i\alpha}}{\Gamma(i\alpha + 1)} \left(D_{a+;g}^{i\alpha} f\right)(a) + \frac{\left(D_{a+;g}^{(n+1)\alpha} f\right)(\psi_x)}{\Gamma((n+1)\alpha + 1)} (g(x) - g(a))^{(n+1)\alpha},$$

(24.2.116)

where $\psi_x \in [a, x]$, any $x \in [a, b]$.

References

1. G. Anastassiou, On right fractional calculus. Chaos Soliton Fract **42**, 365–376 (2009)
2. G. Anastassiou, *Fractional Differentiation Inequalities* (Springer, New York, 2009)
3. G. Anastassiou, *Basic Inequalities, Revisited*. Mathematica Balkanica, New Series, vol. 24, Fasc. 1–2 (2010), pp. 59–84
4. G. Anastassiou, Fractional representation formulae and right fractional inequalities. Math. Comput. Model. **54**(11–12), 3098–3115 (2011)
5. G. Anastassiou, The reduction method in fractional calculus and fractional Ostrowski type inequalities. Indian J. Math. **56**(3), 333–357 (2014)
6. G. Anastassiou, Advanced fractional Taylor's formulae. J. Comput. Anal. Appl. (accepted) (2015)
7. G. Anastassiou, *Univariate Left General High Order Fractional Monotone Approximation* (submitted) (2015)
8. G. Anastassiou, *Univariate Right General High Order Fractional Monotone Approximation Theory* (submitted) (2015)
9. K. Diethelm, *The Analysis of Fractional Differential Equations*. Lecture Notes in Mathematics, vol. 2004, 1st edn. (Springer, New York, 2010)
10. R.-Q. Jia, *Chapter 3. Absolutely Continuous Functions*, https://www.ualberta.ca/~jia/Math418/Notes/Chap.3.pdf
11. A.A. Kilbas, H.M. Srivastava, J.J. Tujillo, *Theory and Applications of Fractional Differential Equations*. North-Holland Mathematics Studies, vol. 204 (Elsevier, New York, 2006)
12. H.L. Royden, *Real Analysis*, 2nd edn. (Macmillan Publishing Co., Inc, New York, 1968)

Chapter 25
Generalized Canavati Type Fractional Taylor's Formulae

We present here four new generalized Canavati type fractional Taylor's formulae. It follows [3].

25.1 Results

Let $g : [a, b] \to \mathbb{R}$ be a strictly increasing function. Let $f \in C^n ([a, b])$, $n \in \mathbb{N}$. Assume that $g \in C^1 ([a, b])$, and $g^{-1} \in C^n ([a, b])$. Call $l := f \circ g^{-1} :$ $[g(a), g(b)] \to \mathbb{R}$. It is clear that $l, l', \ldots, l^{(n)}$ are continuous functions from $[g(a), g(b)]$ into $f([a, b]) \subseteq \mathbb{R}$.

Let $\nu \geq 1$ such that $[\nu] = n, n \in \mathbb{N}$ as above, where $[\cdot]$ is the integral part of the number.

Clearly when $0 < \nu < 1$, $[\nu] = 0$. Next we follow [1], pp. 7–9.

(I) Let $h \in C([g(a), g(b)])$, we define the left Riemann-Liouville fractional integral as

$$\left(J_\nu^{z_0} h \right)(z) := \frac{1}{\Gamma(\nu)} \int_{z_0}^z (z - t)^{\nu - 1} h(t)\, dt, \tag{25.1.1}$$

for $g(a) \leq z_0 \leq z \leq g(b)$, where Γ is the gamma function; $\Gamma(\nu) = \int_0^\infty e^{-t} t^{\nu - 1} dt$. We set $J_0^{z_0} h = h$.

Let $\alpha := \nu - [\nu]$ $(0 < \alpha < 1)$. We define the subspace $C_{g(x_0)}^\nu ([g(a), g(b)])$ of $C^{[\nu]} ([g(a), g(b)])$, where $x_0 \in [a, b]$:

$$C_{g(x_0)}^\nu ([g(a), g(b)]) :=$$

$$\left\{ h \in C^{[\nu]} ([g(a), g(b)]) : J_{1-\alpha}^{g(x_0)} h^{([\nu])} \in C^1 ([g(x_0), g(b)]) \right\}. \tag{25.1.2}$$

© Springer International Publishing Switzerland 2016

G.A. Anastassiou and I.K. Argyros, *Intelligent Numerical Methods:*
Applications to Fractional Calculus, Studies in Computational Intelligence 624,
DOI 10.1007/978-3-319-26721-0_25

So let $h \in C^{\nu}_{g(x_0)} ([g(a), g(b)])$; we define the left g-generalized fractional derivative of h of order ν, of Canavati type, over $[g(x_0), g(b)]$ as

$$D^{\nu}_{g(x_0)} h := \left(J^{g(x_0)}_{1-\alpha} h^{([\nu])} \right)'. \tag{25.1.3}$$

Clearly, for $h \in C^{\nu}_{g(x_0)} ([g(a), g(b)])$, there exists

$$\left(D^{\nu}_{g(x_0)} h \right)(z) = \frac{1}{\Gamma(1-\alpha)} \frac{d}{dz} \int_{g(x_0)}^{z} (z-t)^{-\alpha} h^{([\nu])}(t) \, dt, \tag{25.1.4}$$

for all $g(x_0) \le z \le g(b)$.

In particular, when $f \circ g^{-1} \in C^{\nu}_{g(x_0)} ([g(a), g(b)])$ we have that

$$\left(D^{\nu}_{g(x_0)} \left(f \circ g^{-1} \right) \right)(z) = \frac{1}{\Gamma(1-\alpha)} \frac{d}{dz} \int_{g(x_0)}^{z} (z-t)^{-\alpha} \left(f \circ g^{-1} \right)^{([\nu])}(t) \, dt, \tag{25.1.5}$$

for all $g(x_0) \le z \le g(b)$. We have $D^{n}_{g(x_0)} \left(f \circ g^{-1} \right) = \left(f \circ g^{-1} \right)^{(n)}$ and $D^{0}_{g(x_0)} \left(f \circ g^{-1} \right) = f \circ g^{-1}$.

By Theorem 2.1, p. 8 of [1], we have for $f \circ g^{-1} \in C^{\nu}_{g(x_0)} ([g(a), g(b)])$, where $x_0 \in [a, b]$ is fixed, that

(i) if $\nu \ge 1$, then

$$\left(f \circ g^{-1} \right)(z) = \sum_{k=0}^{[\nu]-1} \frac{\left(f \circ g^{-1} \right)^{(k)} (g(x_0))}{k!} (z - g(x_0))^k +$$

$$\frac{1}{\Gamma(\nu)} \int_{g(x_0)}^{z} (z-t)^{\nu-1} \left(D^{\nu}_{g(x_0)} \left(f \circ g^{-1} \right) \right)(t) \, dt, \tag{25.1.6}$$

all $z \in [g(a), g(b)] : z \ge g(x_0)$,

(ii) if $0 < \nu < 1$, we get

$$\left(f \circ g^{-1} \right)(z) = \frac{1}{\Gamma(\nu)} \int_{g(x_0)}^{z} (z-t)^{\nu-1} \left(D^{\nu}_{g(x_0)} \left(f \circ g^{-1} \right) \right)(t) \, dt, \tag{25.1.7}$$

all $z \in [g(a), g(b)] : z \ge g(x_0)$.

We have proved the following left generalized g-fractional, of Canavati type, Taylor's formula:

Theorem 25.1 *Let* $f \circ g^{-1} \in C^{\nu}_{g(x_0)} ([g(a), g(b)])$, *where* $x_0 \in [a, b]$ *is fixed.*
(i) *if* $\nu \ge 1$, *then*

$$f(x) - f(x_0) = \sum_{k=1}^{[\nu]-1} \frac{\left(f \circ g^{-1} \right)^{(k)} (g(x_0))}{k!} (g(x) - g(x_0))^k +$$

$$\frac{1}{\Gamma(\nu)} \int_{g(x_0)}^{g(x)} (g(x) - t)^{\nu-1} \left(D_{g(x_0)}^{\nu} \left(f \circ g^{-1}\right)\right)(t) \, dt, \quad all \; x \in [a, b] : x \geq x_0,$$

$$(25.1.8)$$

(ii) if $0 < \nu < 1$, *we get*

$$f(x) = \frac{1}{\Gamma(\nu)} \int_{g(x_0)}^{g(x)} (g(x) - t)^{\nu-1} \left(D_{g(x_0)}^{\nu} \left(f \circ g^{-1}\right)\right)(t) \, dt, \quad (25.1.9)$$

all $x \in [a, b] : x \geq x_0$.

By the change of variable method, see [4], we may rewrite the remainder of (25.1.8) and (25.1.9), as

$$\frac{1}{\Gamma(\nu)} \int_{g(x_0)}^{g(x)} (g(x) - t)^{\nu-1} \left(D_{g(x_0)}^{\nu} \left(f \circ g^{-1}\right)\right)(t) \, dt = \quad (25.1.10)$$

$$\frac{1}{\Gamma(\nu)} \int_{x_0}^{x} (g(x) - g(s))^{\nu-1} \left(D_{g(x_0)}^{\nu} \left(f \circ g^{-1}\right)\right) (g(s)) \, g'(s) \, ds,$$

all $x \in [a, b] : x \geq x_0$.

We may rewrite (25.1.9) as follows:
if $0 < \nu < 1$, we have

$$f(x) = \left(J_{\nu}^{g(x_0)} \left(D_{g(x_0)}^{\nu} \left(f \circ g^{-1}\right)\right)\right) (g(x)), \quad (25.1.11)$$

all $x \in [a, b] : x \geq x_0$.

(II) Next we follow [2], pp. 345–348.

Let $h \in C([g(a), g(b)])$, we define the right Riemann-Liouville fractional integral as

$$\left(J_{z_0-}^{\nu} h\right)(z) := \frac{1}{\Gamma(\nu)} \int_{z}^{z_0} (t - z)^{\nu-1} h(t) \, dt, \quad (25.1.12)$$

for $g(a) \leq z \leq z_0 \leq g(b)$. We set $J_{z_0-}^{0} h = h$.

Let $\alpha := \nu - [\nu]$ $(0 < \alpha < 1)$. We define the subspace $C_{g(x_0)-}^{\nu}([g(a), g(b)])$ of $C^{[\nu]}([g(a), g(b)])$, where $x_0 \in [a, b]$:

$$C_{g(x_0)-}^{\nu}([g(a), g(b)]) :=$$

$$\left\{ h \in C^{[\nu]}([g(a), g(b)]) : J_{g(x_0)-}^{1-\alpha} h^{([\nu])} \in C^{1}([g(x_0), g(b)]) \right\}. \quad (25.1.13)$$

So let $h \in C_{g(x_0)-}^{\nu}([g(a), g(b)])$; we define the right g-generalized fractional derivative of h of order ν, of Canavati type, over $[g(a), g(x_0)]$ as

$$D_{g(x_0)-}^{\nu} h := (-1)^{n-1} \left(J_{g(x_0)-}^{1-\alpha} h^{([\nu])}\right)'. \quad (25.1.14)$$

Clearly, for $h \in C^\nu_{g(x_0)-}([g(a), g(b)])$, there exists

$$\left(D^\nu_{g(x_0)-}h\right)(z) = \frac{(-1)^{n-1}}{\Gamma(1-\alpha)}\frac{d}{dz}\int_z^{g(x_0)}(t-z)^{-\alpha}h^{([\nu])}(t)\,dt, \qquad (25.1.15)$$

for all $g(a) \le z \le g(x_0) \le g(b)$.

In particular, when $f \circ g^{-1} \in C^\nu_{g(x_0)-}([g(a), g(b)])$ we have that

$$\left(D^\nu_{g(x_0)-}\left(f \circ g^{-1}\right)\right)(z) = \frac{(-1)^{n-1}}{\Gamma(1-\alpha)}\frac{d}{dz}\int_z^{g(x_0)}(t-z)^{-\alpha}\left(f \circ g^{-1}\right)^{([\nu])}(t)\,dt,$$
$$(25.1.16)$$

for all $g(a) \le z \le g(x_0) \le g(b)$.

We get that

$$\left(D^n_{g(x_0)-}\left(f \circ g^{-1}\right)\right)(z) = (-1)^n\left(f \circ g^{-1}\right)^{(n)}(z) \qquad (25.1.17)$$

and $\left(D^0_{g(x_0)-}\left(f \circ g^{-1}\right)\right)(z) = \left(f \circ g^{-1}\right)(z)$, all $z \in [g(a), g(x_0)]$.

By Theorem 23.19, p. 348 of [2], we have for $f \circ g^{-1} \in C^\nu_{g(x_0)-}([g(a), g(b)])$, where $x_0 \in [a, b]$ is fixed, that

(i) if $\nu \ge 1$, then

$$\left(f \circ g^{-1}\right)(z) = \sum_{k=0}^{[\nu]-1}\frac{\left(f \circ g^{-1}\right)^{(k)}(g(x_0))}{k!}(z - g(x_0))^k + \qquad (25.1.18)$$

$$\frac{1}{\Gamma(\alpha)}\int_z^{g(x_0)}(t-z)^{\nu-1}\left(D^\nu_{g(x_0)-}\left(f \circ g^{-1}\right)\right)(t)\,dt,$$

all $z \in [g(a), g(b)] : z \le g(x_0)$,

(ii) if $0 < \nu < 1$, we get

$$\left(f \circ g^{-1}\right)(z) = \frac{1}{\Gamma(\nu)}\int_z^{g(x_0)}(t-z)^{\nu-1}\left(D^\nu_{g(x_0)-}\left(f \circ g^{-1}\right)\right)(t)\,dt, \quad (25.1.19)$$

all $z \in [g(a), g(b)] : z \le g(x_0)$.

We have proved the following right generalized g-fractional, of Canavati type, Taylor's formula:

Theorem 25.2 Let $f \circ g^{-1} \in C^\nu_{g(x_0)-}([g(a), g(b)])$, where $x_0 \in [a, b]$ is fixed.

(i) if $\nu \ge 1$, then

$$f(x) - f(x_0) = \sum_{k=1}^{[\nu]-1}\frac{\left(f \circ g^{-1}\right)^{(k)}(g(x_0))}{k!}(g(x) - g(x_0))^k + $$

$$\frac{1}{\Gamma(\nu)} \int_{g(x)}^{g(x_0)} (t - g(x))^{\nu-1} \left(D_{g(x_0)-}^{\nu} \left(f \circ g^{-1} \right) \right)(t)\, dt, \quad all\ a \leq x \leq x_0,$$

(25.1.20)

(ii) if $0 < \nu < 1$, *we get*

$$f(x) = \frac{1}{\Gamma(\nu)} \int_{g(x)}^{g(x_0)} (t - g(x))^{\nu-1} \left(D_{g(x_0)-}^{\nu} \left(f \circ g^{-1} \right) \right)(t)\, dt, \quad all\ a \leq x \leq x_0.$$

(25.1.21)

By change of variable, see [4], we may rewrite the remainder of (25.1.20) and (25.1.21), as

$$\frac{1}{\Gamma(\nu)} \int_{g(x)}^{g(x_0)} (t - g(x))^{\nu-1} \left(D_{g(x_0)-}^{\nu} \left(f \circ g^{-1} \right) \right)(t)\, dt = \tag{25.1.22}$$

$$\frac{1}{\Gamma(\nu)} \int_{x}^{x_0} (g(s) - g(x))^{\nu-1} \left(D_{g(x_0)-}^{\nu} \left(f \circ g^{-1} \right) \right)(g(s))\, g'(s)\, ds,$$

all $a \leq x \leq x_0$.

We may rewrite (25.1.21) as follows:

if $0 < \nu < 1$, we have

$$f(x) = \left(J_{g(x_0)-}^{\nu} \left(D_{g(x_0)-}^{\nu} \left(f \circ g^{-1} \right) \right) \right) (g(x)), \tag{25.1.23}$$

all $a \leq x \leq x_0 \leq b$.

(III) Denote by

$$D_{g(x_0)}^{m\nu} = D_{g(x_0)}^{\nu} D_{g(x_0)}^{\nu} \cdots D_{g(x_0)}^{\nu} \quad (m\text{-times}),\ m \in \mathbb{N}. \tag{25.1.24}$$

Also denote by

$$J_{m\nu}^{g(x_0)} = J_{\nu}^{g(x_0)} J_{\nu}^{g(x_0)} \cdots J_{\nu}^{g(x_0)} \quad (m\text{-times}),\ m \in \mathbb{N}. \tag{25.1.25}$$

We need:

Theorem 25.3 *Here* $0 < \nu < 1$. *Assume that* $\left(D_{g(x_0)}^{m\nu} \left(f \circ g^{-1} \right) \right) \in C_{g(x_0)}^{\nu} \left([g(a), g(b)] \right)$, *where* $x_0 \in [a, b]$ *is fixed. Assume also that* $\left(D_{g(x_0)}^{(m+1)\nu} \left(f \circ g^{-1} \right) \right) \in C \left([g(x_0), g(b)] \right)$. *Then*

$$\left(J_{m\nu}^{g(x_0)} D_{g(x_0)}^{m\nu} \left(f \circ g^{-1} \right) \right) (g(x)) - \left(J_{(m+1)\nu}^{g(x_0)} D_{g(x_0)}^{(m+1)\nu} \left(f \circ g^{-1} \right) \right) (g(x)) = 0,$$

(25.1.26)

for all $x_0 \leq x \leq b$.

Proof We observe that $(l := f \circ g^{-1})$

$$\left(J_{m\nu}^{g(x_0)} D_{g(x_0)}^{m\nu} (l)\right)(g(x)) - \left(J_{(m+1)\nu}^{g(x_0)} D_{g(x_0)}^{(m+1)\nu} (l)\right)(g(x)) =$$

$$\left(J_{m\nu}^{g(x_0)} \left(D_{g(x_0)}^{m\nu} (l) - J_{\nu}^{g(x_0)} D_{g(x_0)}^{(m+1)\nu} (l)\right)\right)(g(x)) = \qquad (25.1.27)$$

$$\left(J_{m\nu}^{g(x_0)} \left(D_{g(x_0)}^{m\nu} (l) - \left(J_{\nu}^{g(x_0)} D_{g(x_0)}^{\nu}\right) \left(\left(D_{g(x_0)}^{m\nu} (l)\right) \circ g \circ g^{-1}\right)\right)\right)(g(x)) =$$

$$\left(J_{m\nu}^{g(x_0)} \left(D_{g(x_0)}^{m\nu} (l) - \left(D_{g(x_0)}^{m\nu} (l)\right)\right)\right)(g(x)) = \left(J_{m\nu}^{g(x_0)} (0)\right)(g(x)) = 0.$$

$\square$

We make:

Remark 25.4 Let $0 < \nu < 1$. Assume that $\left(D_{g(x_0)}^{i\nu} \left(f \circ g^{-1}\right)\right) \in C_{g(x_0)}^{\nu} ([g(a), g(b)])$, $x_0 \in [a, b]$, for all $i = 0, 1, \ldots, m$. Assume also that $\left(D_{g(x_0)}^{(m+1)\nu} \left(f \circ g^{-1}\right)\right) \in C ([g(x_0), g(b)])$. We have that

$$\sum_{i=0}^{m} \left[\left(J_{i\nu}^{g(x_0)} D_{g(x_0)}^{i\nu} \left(f \circ g^{-1}\right)\right)(g(x)) - \left(J_{(i+1)\nu}^{g(x_0)} D_{g(x_0)}^{(i+1)\nu} \left(f \circ g^{-1}\right)\right)(g(x))\right] = 0.$$
$$(25.1.28)$$

Hence it holds

$$f(x) - \left(J_{(m+1)\nu}^{g(x_0)} D_{g(x_0)}^{(m+1)\nu} \left(f \circ g^{-1}\right)\right)(g(x)) = 0, \qquad (25.1.29)$$

for all $x_0 \le x \le b$.
That is

$$f(x) = \left(J_{(m+1)\nu}^{g(x_0)} D_{g(x_0)}^{(m+1)\nu} \left(f \circ g^{-1}\right)\right)(g(x)), \qquad (25.1.30)$$

for all $x_0 \le x \le b$.

We have proved the following modified and generalized left fractional Taylor's formula of Canavati type:

Theorem 25.5 *Let* $0 < \nu < 1$. *Assume that* $\left(D_{g(x_0)}^{i\nu} \left(f \circ g^{-1}\right)\right) \in C_{g(x_0)}^{\nu} ([g(a), g(b)])$, $x_0 \in [a, b]$, *for* $i = 0, 1, \ldots, m$. *Assume also that* $\left(D_{g(x_0)}^{(m+1)\nu} \left(f \circ g^{-1}\right)\right) \in C ([g(x_0), g(b)])$. *Then*

$$f(x) = \frac{1}{\Gamma((m+1)\nu)} \int_{g(x_0)}^{g(x)} (g(x) - z)^{(m+1)\nu-1} \left(D_{g(x_0)}^{(m+1)\nu} \left(f \circ g^{-1}\right)\right)(z) \, dz$$
$$(25.1.31)$$

$$= \frac{1}{\Gamma((m+1)\nu)} \cdot$$

$$\int_{x_0}^{x} \left(g\left(x\right) - g\left(s\right)\right)^{(m+1)\nu-1} \left(D_{g(x_0)}^{(m+1)\nu} \left(f \circ g^{-1}\right)\right) \left(g\left(s\right)\right) g'\left(s\right) ds,$$

all $x_0 \le x \le b$.

(IV) Denote by

$$D_{g(x_0)-}^{m\nu} = D_{g(x_0)-}^{\nu} D_{g(x_0)-}^{\nu} \cdots D_{g(x_0)-}^{\nu} \quad (m\text{-times}),\ m \in \mathbb{N}. \tag{25.1.32}$$

Also denote by

$$J_{g(x_0)-}^{m\nu} = J_{g(x_0)-}^{\nu} J_{g(x_0)-}^{\nu} \cdots J_{g(x_0)-}^{\nu} \quad (m\text{-times}),\ m \in \mathbb{N}. \tag{25.1.33}$$

We need:

Theorem 25.6 *Here* $0 < \nu < 1$. *Assume that* $\left(D_{g(x_0)-}^{m\nu} \left(f \circ g^{-1}\right)\right) \in C_{g(x_0)-}^{\nu} \left(\left[g\left(a\right), g\left(b\right)\right]\right)$, *where* $x_0 \in [a, b]$ *is fixed. Assume also that* $\left(D_{g(x_0)-}^{(m+1)\nu} \left(f \circ g^{-1}\right)\right) \in C \left(\left[g\left(a\right), g\left(x_0\right)\right]\right)$. *Then*

$$\left(J_{g(x_0)-}^{m\nu} D_{g(x_0)-}^{m\nu} \left(f \circ g^{-1}\right)\right) \left(g\left(x\right)\right) - \left(J_{g(x_0)-}^{(m+1)\nu} D_{g(x_0)-}^{(m+1)\nu} \left(f \circ g^{-1}\right)\right) \left(g\left(x\right)\right) = 0, \tag{25.1.34}$$

for all $a \le x \le x_0$.

Proof We observe that $(l := f \circ g^{-1})$

$$\left(J_{g(x_0)-}^{m\nu} D_{g(x_0)-}^{m\nu} \left(l\right)\right) \left(g\left(x\right)\right) - \left(J_{g(x_0)-}^{(m+1)\nu} D_{g(x_0)-}^{(m+1)\nu} \left(l\right)\right) \left(g\left(x\right)\right) =$$

$$\left(J_{g(x_0)-}^{m\nu} \left(D_{g(x_0)-}^{m\nu} \left(l\right) - J_{g(x_0)-}^{\nu} D_{g(x_0)-}^{(m+1)\nu} \left(l\right)\right)\right) \left(g\left(x\right)\right) =$$

$$\left(J_{g(x_0)-}^{m\nu} \left(D_{g(x_0)-}^{m\nu} \left(l\right) - \left(J_{g(x_0)-}^{\nu} D_{g(x_0)-}^{\nu}\right) \left(\left(D_{g(x_0)-}^{m\nu} \left(l\right)\right) \circ g \circ g^{-1}\right)\right)\right) \left(g\left(x\right)\right) \tag{25.1.35}$$

$$= \left(J_{g(x_0)-}^{m\nu} \left(D_{g(x_0)-}^{m\nu} \left(l\right) - D_{g(x_0)-}^{m\nu} \left(l\right)\right)\right) \left(g\left(x\right)\right) = J_{g(x_0)-}^{m\nu} \left(0\right) \left(g\left(x\right)\right) = 0.$$

□

We make:

Remark 25.7 Let $0 < \nu < 1$. Assume that $\left(D_{g(x_0)-}^{i\nu} \left(f \circ g^{-1}\right)\right) \in C_{g(x_0)-}^{\nu} \left(\left[g\left(a\right), g\left(b\right)\right]\right)$, $x_0 \in [a, b]$, for all $i = 0, 1, \ldots, m$. Assume also that $\left(D_{g(x_0)-}^{(m+1)\nu} \left(f \circ g^{-1}\right)\right) \in C \left(\left[g\left(a\right), g\left(x_0\right)\right]\right)$. We have that (by (25.1.34))

$$\sum_{i=0}^{m} \left[\left(J_{g(x_0)-}^{i\nu} D_{g(x_0)-}^{i\nu} \left(f \circ g^{-1}\right)\right) \left(g\left(x\right)\right) - \left(J_{g(x_0)-}^{(i+1)\nu} D_{g(x_0)-}^{(i+1)\nu} \left(f \circ g^{-1}\right)\right) \left(g\left(x\right)\right)\right] = 0. \tag{25.1.36}$$

Hence it holds

$$f(x) - \left(J_{g(x_0)-}^{(m+1)\nu} D_{g(x_0)-}^{(m+1)\nu} \left(f \circ g^{-1} \right) \right) (g(x)) = 0, \tag{25.1.37}$$

for all $a \leq x \leq x_0 \leq b$.

That is

$$f(x) = \left(J_{g(x_0)-}^{(m+1)\nu} D_{g(x_0)-}^{(m+1)\nu} \left(f \circ g^{-1} \right) \right) (g(x)), \tag{25.1.38}$$

for all $a \leq x \leq x_0 \leq b$.

We have proved the following modified and generalized right fractional Taylor's formula of Canavati type:

Theorem 25.8 *Let* $0 < \nu < 1$. *Assume that* $\left(D_{g(x_0)-}^{i\nu} \left(f \circ g^{-1} \right) \right) \in C_{g(x_0)-}^{\nu} ([g(a), g(b)])$, $x_0 \in [a, b]$, *for all* $i = 0, 1, \ldots, m$. *Assume also that* $\left(D_{g(x_0)-}^{(m+1)\nu} \left(f \circ g^{-1} \right) \right)$ $\in C([g(a), g(x_0)])$. *Then*

$$f(x) = \frac{1}{\Gamma((m+1)\nu)} \int_{g(x)}^{g(x_0)} (z - g(x))^{(m+1)\nu-1} \left(D_{g(x_0)-}^{(m+1)\nu} \left(f \circ g^{-1} \right) \right) (z) \, dz \tag{25.1.39}$$

$$= \frac{1}{\Gamma((m+1)\nu)} \cdot$$

$$\int_x^{x_0} (g(s) - g(x))^{(m+1)\nu-1} \left(D_{g(x_0)-}^{(m+1)\nu} \left(f \circ g^{-1} \right) \right) (g(s)) \, g'(s) \, ds,$$

all $a \leq x \leq x_0 \leq b$.

References

1. G. Anastassiou, *Fractional Differentiation Inequalities* (Springer, New York, 2009)
2. G. Anastassiou, *Inteligent Mathematics: Computational Analysis* (Springer, Heidelberg, 2011)
3. G. Anastassiou, Generalized Canavati type fractional Taylor's formulae. J. Comput. Anal. Appl. (2015) (accepted)
4. R.-Q. Jia, *Chapter 3. Absolutely Continuous Functions*, https://www.ualberta.ca/~jia/Math418/Notes/Chap.3.pdf

Index

© Springer International Publishing Switzerland 2016
G.A. Anastassiou and I.K. Argyros, *Intelligent Numerical Methods:
Applications to Fractional Calculus*, Studies in Computational Intelligence 624,
DOI 10.1007/978-3-319-26721-0

Printed in the United States
By Bookmasters